Computer-Aided Ergonomics

Computer-Aided Ergonomics

A Researcher's Guide

Edited by

Waldemar Karwowski

Center for Industrial Ergonomics,
University of Louisville, U.S.A.

Ashraf M. Genaidy

Department of Industrial Engineering,
Western Michigan University, U.S.A

and

Shihab S. Asfour

Department of Industrial Engineering,
University of Miami, U.S.A.

Taylor & Francis

London • New York • Philadelphia

USA Taylor & Francis Inc., 1900 Frost Road, Suite 101, Bristol, PA 19007

UK Taylor & Francis Ltd. 4 John St. London WC1N 2ET

British Library Cataloguing in Publication Data
Computer-aided ergonomics.
1. Ergonomics
I. Karwowski, Waldemar II. Genaidy, Ashraf M. III. Asfour, Shihab S., *1952–*
620.8′2

ISBN 0-85066-753-4

Library of Congress Cataloging-in-Publication Data
Computer-aided ergonomics / edited by Waldemar Karwowski, Ashraf M. Genaidy, and Shihab S. Asfour.
p. cm.
ISBN 0–85066–753–4
1. Human engineering—Data processing. 2. Computer-aided engineering.
I. Karwowski, Waldemar, 1953– . II. Genaidy, Ashraf M. III. Asfour, Shihab S., 1952– .
T59.7.C66 1990
620.8′ 2′ 0285—dc20

Cover design by Jordan and Jordan, Fareham, Hampshire (based on an image generated by the SAMMIE System, with permission of SAMMIE C.A.D. Ltd.)

Phototypesetting by
Chapterhouse, The Cloisters, Formby, England.
Printed in Great Britain by Burgess Science Press, Basingstoke.

Contents

Preface

Computers in their many different forms, from mainframes and workstations to personal or laptop computers and electronic notebooks, are an indispensible part of our professional lives today. As ergonomists, we welcome the widespread use of computers that may well serve the cause of integrating many subdisciplines within the field of human factors. Although the ergonomics of working with computers is an extremely important area of study, this book is not devoted to that subject. Rather, we are focusing our attention on how computers can help us to be better specialists within our particular field, and how information technology (IT) can benefit society at large by bringing specialized ergonomics knowledge and methods to other professions and trades.

Since its birth, ergonomics (the study of work) has relied on the interaction of many disciplines such as biomechanics, work physiology, cognitive and experimental psychology, engineering and human resource development. Admittedly, a large mass of ergonomics knowledge has been generated by researchers in the subject literature, but it has not always been communicated effectively between researchers and practitioners. With the advent of mini- and microcomputers in the past two decades, ergonomists have recognized the need to develop computer-aided tools that make specialized expertise accessible to practitioners who design or redesign equipment, operations, facilities and working environments for human use. Consequently, many computer-based systems have been developed to fulfil this need. It is the aim of this book to bring together these different systems in order to advance the practical use of ergonomics knowledge in the workplace.

When we first issued the *Call for Papers* for this project, the response was overwhelming. Over sixty manuscripts were received for possible inclusion in the book, out of which thirty-nine were selected as representative, in our view, of the current and emerging research issues in ergonomics science that benefit from the application of information technology. These papers have been contributed by authors from Australia, Europe, and North America.

The book is structured as a researcher's guide, and covers a wide spectrum of topics in the broad area of applications of IT in ergonomics. It consists of nine parts. *Part I* discusses the development of human factors databases and outlines the application of information technology to the prevention of musculoskeletal injuries. *Part II* is devoted to the use of computer software in modelling the human body, while *Part III* extends these applications to workplace design. *Part IV* deals with the application of computer tools in job analysis, assessment of workload and work measurement. *Part V* discusses the design and use of software, especially knowledge-based expert systems, for analysis of posture at work, and for the design of manual material handling tasks. *Part VI* addresses the use of applied artificial intelligence techniques in safety analysis, while *Part VII* refers to computerized performance testing and decision-support issues. The last two parts of this book, *Parts VIII* and *IX*, discuss the problems of user-machine interfaces and aiding the human design process with computer tools, respectively.

The broad range of topics presented in this book reflects our view that computers should serve as the means of integrating ergonomics into the design process of other scientists or engineers who often deal with human design constraints and requirements. We hope, therefore, that this book will also reach those outside the ergonomics community, and bring our profession closer to other applied disciplines and industrial practice. Typically ergonomics attempts to optimize work with computers. This time we explore the issue of how computers can help us, ergonomists, researchers and practitioners alike, at work. To what degree we have succeeded in this quest, you will be the judge.

Finally, we would like to thank Mrs. Laura Abell, Secretary at the Center for Industrial Ergonomics, University of Louisville, for providing logistic support on this project.

March 1990

Waldemar Karwowski, Louisville, Kentucky
Ashraf Genaidy, Kalamazoo, Michigan
Shihab S. Asfour, Miami, Florida

PART I
INTRODUCTION

Chapter 1

The Computer Human Factors Database

KATHRYN E. PHILLIPS
HUSAT Research Centre
The Elms, Elms Grove
Loughborough
Leicestershire LE11 1RG, United Kingdom

Abstract. A unique bibliographic database dealing specifically with the field of computer human factors is being developed. The database brings together the diversified literature of the field while omitting areas where human factors do not apply. A user interface is being developed according to human factors principles which will cater for the needs of a wide variety of potential users. A user-centred design approach has been adopted throughout the process of development, utilizing the expertise of human factors specialists within the UK and throughout Europe.

Keywords: Computer human factors; Bibliographic databases; Information services; Thesauri; Classifications.

Introduction

Until the inception of work to be described, no single commercially available abstracting or indexing service existed dealing solely with the subject area of Computer Human Factors (CHF). Those abstracting and indexing services which do exist provide a much broader field of information than is required, yet they may fail to include certain key aspects of the CHF field. To fill this void in the information market a CHF bibliographic information service has been developed in ESPRIT project 385 – HUFIT, Human Factors in Information Technology. Potential users include human factors experts, designers, information providers and other interested specialists such as engineers and systems analysts. Always seeking to obey its own principles by being easily accessible and easy to use, the service is of an interactive nature and consequently a user-centred design approach has been adopted throughout the work. The potential users have been and will be encouraged to play an influential rôle in its future development.

The development of the Computer Human Factors Database

The field of Computer Human Factors (CHF) may be defined in the very broadest sense as ways in which people and computers interact. It is a diversified or scattered subject drawing from such fields as computer science and technology, ergonomics, psychology, physiology, sociology, education, management and information science. Material on human factors aspects of this wide range of fields is included in the CHF Database with particular reference to the use of computers and associated technology.

Sources of information

HUSAT is one of the largest research centres in Europe specializing in the field of human sciences and advanced technology. During the early 1980s work was begun on one of the contracts at HUSAT to collate and index bibliographical information in the field of CHF. The resulting set of references-with-abstracts were incorporated into an electronic journal which was itself dedicated to the field of CHF. The research was developed under the aegis of BLEND, a British Library Research and Development Department project. This work took place as a joint project between Loughborough University of Technology (HUSAT Research Centre) and Birmingham University in the UK. The 'readers' of the electronic journal included a group of some 70 UK human factors practitioners. This so-called 'LINC Community' was involved in the choice of journals which provided the majority of the references to be included on the Database. The material to be acquired retrospectively was given a start date of January 1982 which has remained the arbitrary point of commencement for items for the Database. However a large number of pre-1982 references arising from bibliographies compiled by human factors experts are included in the Database and the start date is regarded with flexibility. The HUFIT project work, which commenced in early 1985, was able to build directly upon the foundations supported by the BLEND project.

It had become clear prior to and during the initial acquisition of material that there existed no single commercially available abstracting or indexing service which dealt solely with this subject area. *Ergonomics Abstracts*, for example, covers a much wider brief within the fields of ergonomics and human factors, *Psychological Abstracts* includes psychological aspects of work with computers as a minor section and the *Institution of Electrical Engineers' Computer and Control Abstracts* indexes the whole diversified field of computer science. It may be seen then that many services include relevant aspects of the field but offer a much wider subject coverage than is required. They do not expend any additional effort in specifically acquiring CHF material, neither do they attempt to acquire all the material currently in existence.

Applied Ergonomics
Behaviour and Information Technology
Computers in Human Behaviour
Ergonomics
HCI Newsletter
Hufact Quarterly
Human-Computer Interaction
Human Factors
IEEE Transactions on Systems, Man and Cybernetics
International Journal of Man-Machine Studies
Journal of Applied Psychology
Office Systems Ergonomics Report
Office Technology and People
SIGCHI Bulletin
VDT News

Figure 1.1 The Computer Human Factors Database primary journals searched (March 1989).

ACAS Work Research Unit Information Service News and Abstracts
ACM Computing Surveys
American Federation of Information Processing Societies Conference Proceedings
Annual Review of Information Science and Technology
Annual Review of Psychology
Applied Cognitive Psychology
Aslib Proceedings
Bulletin of the British Psychological Society
Cognition
Cognitive Psychology
Communications of the ACM
Computer
Computer Bulletin
Computer Journal
Design Studies
Displays
Electrical Communication
Expert Systems. The International Journal of Knowledge Engineering
IEEE Transactions on Software Engineering
Information Design Journal
Information and Management
IPO Annual Progress Report
IT Standard News
Journal of the American Society of Information Science
Journal of Computer Assisted Learning
Journal of Documentation
Journal of Information Science
Journal of Information Technology
Journal of Occupational Psychology
Management Science
Online
Online Review
Proceedings of the IEEE
Program
Quarterly Journal of Experimental Psychology
Rutherford Appleton Laboratory Mailshot
Scientific American
Work and Stress

Figure 1.2 The Computer Human Factors Database secondary journals searched (March 1989).

Searching the literature

Searching the literature for relevant material to be included within the Database is performed on three levels. First, a series of primary journals is searched regularly and with top priority. These are journals which are highly likely to contain a large number of relevant papers because they will have been chosen as vehicles to publish research findings by CHF practitioners. The next level of searching takes place using the secondary journals. These publications are likely to contain a number of relevant articles from time to time. The complete listings of primary and secondary journals being searched as at March 1989 are to be found in Figures 1.1 and 1.2. These lists, originating from suggestions put forward by the LINC Community, continue to expand as access to further titles is obtained.

The third level of acquiring relevant information attempts to net remaining items within the literature as a whole. A variety of abstracting and indexing services, including those already mentioned, are utilized in order to widen the scope of potential source documents. It is through these information sources that references to documents in languages other than English come to be present on the Database. In addition to this formal information gathering, relevant material is attracted to the author's notice in her rôle as Librarian of the HUSAT Library, which specializes in the field of CHF and supports the work of the HUSAT Research Centre. The researchers at HUSAT are themselves a source of potential information in that through the course of their work, substantial personal files of references and documents are acquired as a matter of course. Being experts in the field, the relative merits of the items referred to may also be noted.

From these sources of information, material which is peripheral, ephemeral or irrelevant to human factors aspects of advanced technology are filtered and do not enter the Database.

Tools to aid retrieval from the Database

In the very earliest stages of the acquisition of CHF bibliographic information, consideration was given to its indexing and to the classification of the subject field. Indexing takes place whereby a string of keywords or keyphrases is assigned to a document which concisely describes its content. As each keyword is assigned, it is recorded as a main entry term in the Computer Human Factors Thesaurus.

The Computer Human Factors Thesaurus

The CHF Thesaurus is an alphabetical list of terms which also shows hierarchical links between those terms. The abbreviations BT (broader term), RT (related term) and NT (narrower term) are used to list other terms which may be linked with a main entry or preferred term. This is best illustrated by an example and Figure 1.3 shows a sample page from the Thesaurus showing examples of these links. It may be noted that there is a double entry aspect to the Thesaurus in that for every broader term there will be the converse narrower term. Many of the terms which describe concepts in the field of CHF are phrases rather than single term keywords. There is provision within the Thesaurus for rotating such phrases so that if the user enters the Thesaurus by the second or subsequent word in a phrase, a SEE reference refers to the main entry format. Where synonyms arise, the Thesaurus also seeks to aid the user by showing the synonyms (or non-preferred terms) with an accompanying USE message referring to the main entry term. The main entry term in turn lists the synonyms using the USE FOR device. Figure 1.3 shows SEE, USE and USE FOR references.

Errors		
	USE FOR	Bugs
	NT	Error correction
		Error messages
		Human errors
Errors, human		
	SEE	Human errors
Errors, keying		
	SEE	Keying errors
ESPRIT		
	USE FOR	European Strategic Programme for Research in Information Technology
	BT	Europe
		Information technology
Estimation		
	RT	Estimations
Estimations		
	RT	Estimation
Ethnic variables		
	BT	User differences
	RT	Socio cultural strata
Europe		
	NT	ESPRIT
European Strategic Programme for Research in Information Technology		
	USE	ESPRIT
Evaluation		
	NT	Evaluation tools
		Product evaluation
		User evaluation
Evaluation, personnel		
	SEE	Personnel evaluation
Evaluation, product		
	SEE	Product evaluation
Evaluation tools		
	BT	Evaluation
	NT	Field trials
Evaluation, user		
	SEE	User evaluation
Example, programming by		
	USE	Programming AND Examples
Examples		
Expectations, user		
	SEE	User expectations

Figure 1.3 Sample page from the CHF Thesaurus (1987 edition).

The Computer Human Factors Classification

When the Thesaurus had grown to such an extent that a substantial number of terms had been generated by the literature, the main entry terms were selected and re-grouped by concept rather than by alphabetization. These groupings formed the basis of the Computer Human Factors Classification. The first edition of the CHF Classification was referred for comment to human factors practitioners throughout Europe in May 1985 (Phillips, 1985). The feedback which was received was acted upon and two other schemes were investigated. One of these was a scheme which had accompanied a published collection of CHF literature

Figure 1.4 Workshop held in November 1986 to refine the January 1986 edition of the CHF Classification.

(Ergonomics Abstracts, 1984), the other was a scheme devised by Professor Brian Shackel of the Department of Human Sciences at Loughborough University of Technology, UK and a Director of HUSAT (Shackel, 1984). Neither of these schemes had been constructed with the aim of providing the detail contained within the CHF Classification but both provided groupings of broad concepts with which practitioners in the field were comfortable. For this reason both of these schemes were investigated in depth and combined with one another and then with the CHF Classification generated by the literature searching. The resulting edition of the scheme was reported to human factors practitioners throughout Europe in January 1986 (Phillips and Galer, 1986a).

After this edition of the scheme was referred to practitioners, a more focused user-centred design approach was embarked upon in which the human factors specialists consulted contributed to a series of workshops which took place in England, France and West Germany. Figure 1.4 shows Rolf Ilg of the Fraunhofer-Institut für Arbeitswirtschaft und Organisation (FhG-IAO) at the workshop help at FhG-IAO in Stuttgart during November 1986. The purpose of these workshops was to refine the CHF Classification of January 1986 in the areas of their expertise. The resulting transformations to the scheme were reported in Phillips and Galer (1986b) and Phillips (1987).

The main headings of the latest edition of the Classification are: Humans and Users; The Social and Organisational Environment; Physical Computer Human Factors; Software

Humans and Users
The user
  User characteristics, describing the user and user involvement in design

The user

User characteristics
- Users' models *(Users' internal models)*
  - User knowledge
  - User attitudes
    - Scepticism
    - Hostility
- User reactions
  - User reaction
  - User acceptance
  - User satisfaction
- User types
- User differences
  - Ethnic variables
  - Socio cultural strata
- User age
- User sex
- User performance

Describing the user

User models *(Models of the user)*
- User modelling
  - User identification
  - User assessment
  - User evaluation
  - User selection

User involvement in design

User involvement
- User control
- User requirements
- User expectations
- User experience *(with the system)*

Figure 1.5 Sample page from the CHF Classification (1987 edition).

Ergonomics and the Software Interface; Research, Evaluation and Design Methods; and General Terms. Humans and Users is the major section dealing with human physiology, psychology, communication and characteristic descriptions of the user. Society, organizations, education, work, employment and applications of advanced technology are contained within the section on The Social and Organisational Environment. Physical Computer Human Factors covers the physical environment, hardware, health hazards and characteristics of systems while all aspects of software and human–computer interaction are dealt with in the section Software Ergonomics and the Software Interface. The Research, Evaluation and Design Methods section provides for the scientific and experimental approach to the subject. General Terms is a small section containing terms which may be applied to all aspects of CHF. A sample page from the CHF Classification is to be found in Figure 1.5.

The improved hierarchies contained within the current edition of the Classification are reflected back into the Thesaurus so that the two documents are consistent with one another. Refinements to both the Thesaurus and to the Classification continue to be made as indexing generates new terms and as further consultations with subject experts develop Classification subsections.

The structure of the Database

The contents of the bibliographic database of CHF material is not restricted to references (with abstracts) to articles in journals – other information is also acquired. Books and book reviews, reports of conferences and listed abstracts of articles from other publications are noted. Other types of literature included are state of the art progress series, journal titles (those not directly available for searching), special issues of journals, standards, theses and non-book materials such as videos.

In addition to the above, details of 'grey literature' are recorded. The term 'grey literature' describes publications, for example reports and conference proceedings, which may be more difficult to obtain than journals and books. Material associated with conferences is most important due to its relative recency and both references to proceedings and to individual papers from proceedings are included on the Database.

A long term aim of the Database is to provide a full text facility concentrating on definitive items and on grey literature to offset difficulties and frustrations which subscribers may experience in attempting to acquire an original document.

The range of the Database

In addition to the many items which one would expect to find on a database specializing in human factors aspects of computer technology, it may be useful to draw attention to more specific areas which are catered for.

For information to assist with the process of product or system design, the Database provides access to quite generalized material providing there is a link with computer technology. For example, items referring to furniture design for office workstations are present. Also included are references to documents on the physical ergonomics of all levels of hardware through to the layout of workstations and office space utilization. The physical office environment is of course a major consideration but so too is the structure of organizations especially from the point of view of the introduction of new technology. The implementation of computer systems from the manager's point of view and the implications for all aspects of management in the effective use of advanced technology is included. There is also the user reaction to and user participation in the introduction and use of new technology.

Other subject areas to highlight are the ergonomics of the software interface, the training in the use of computer systems and conversely the effective use of computers during the process of education and training. There are also to be found myriad examples of human factors aspects of the applications of advanced technology, for example in the case of computer-aided design.

For general reference, the Database also contains references which may be described as background material. These are references which may prove useful to users in acquiring basic information about a new field. This aspect of the Database is especially important in a rapidly expanding subject, for example computer science.

The on-line Database

When material is initially acquired for the Database it is in paper form. In order to mount the references on a computer, funding was obtained from ALVEY (the UK Government information technology funding body) as part of a human–computer interaction consultation service at HUSAT (The HCI Service). After extensive evaluation of a range of commercial software, that chosen for the storage and retrieval of the bibliographic references was BRS/SEARCH. This was mounted on a SUN 3 computer during the winter months of 1988. The fields which are available on the Database to store information about each reference are listed in Figure 1.6. Of course not all fields will be used for each item, as, for example, a reference to a paper in a journal will require different fields to that of a book or the proceedings of a conference.

Unique Document Identifier	Series
Author(s)	ISBN
Year date	Prime author affiliation
Title	Sponsoring body/Corporate body
Citation in full	Country
Journal title in full	Language of document
Journal title, abbreviated	Location of document
Journal short code	Classification code
Abstract	Media
Contents	Comments
Introduction	Date of data entry
Notes	Conference date(s)
Keywords	Conference venue
Edition	Text
Publisher	Source

Figure 1.6 The fields contained in the Database.

In order continually to improve the quality of the Database, the user is encouraged to give an indication as to the level of satisfaction with its performance when a search session has been completed. This continues the policy of user-centred design. Users are asked to forward information about the omission of any important publications and they have the facility to addend notes to references utilizing the Comments field (this process takes place via the Systems Administrator). This feedback provides a means of tailoring the Database to suit the needs of all of its potential users and to improve its overall quality. In addition, a dialogue between users may be built up which may then be of interest to others.

Performing a search on the Database

A search of the Database for relevant references on a topic within CHF is carried out using a list of terms which describe the subject to be investigated. The user may experience varying degrees of difficulty in choosing the terminology likely to retrieve relevant items from the pool of literature. To help with this procedure, the system provides assistance with choosing the terms with which to perform the search. The initial search comprising words from the user's own terminology will be enhanced by synonyms and terms not already considered.

There is also the facility to browse the system dictionary files which are generated when references are placed on to the Database.

During searches, shortcuts may be taken by truncating search terms allowing for any combination of characters to precede, to be inserted within or to follow a chosen word stem. Searches may be made more accurate using facilities such as restricting search terms so that they are specified as being adjacent to one another or ensuring that they occur within a specified number of words of one another. Terms may be restricted to be within the same sentence, the same paragraph or to be within specified fields in each record on the Database.

The system allows the use of operators to link the chosen terms for more effective retrieval. The most commonly known of these are the Boolean AND, OR and NOT which are now widely understood. Also available is the EXCLUSIVE OR (XOR), with which for the command A XOR B, if either of the terms A or B is found in a record then that record will be retrieved but if a record contains both term A and term B then the record will not be retrieved.

When connected to the Database, the user will be able to interact with it in such a way as to refine the search strategy to provide useful references. It is possible to strike a balance such that a high degree of relevance of the retrieved references may be achieved without resorting to retrieving a vast number of references which will contain many irrelevancies. Using these facilities, it is possible to broaden a search which is not yielding a sufficient number of references and to narrow a search which is providing too many.

The process of refinement is further extended by sampling the resulting references. This may be done at any stage of the search and may be expedited by sampling say, the title only of a group of references. Adjustments may then be made to the search. The final list of retrieved references need only contain those fields of the Database record that are required and the list may be manipulated to suit the user's requirements.

Finally, when a search is working well, it may be saved in order to be re-executed on updated sections of the Database.

The development of the user interface

Users of the CHF Database have access to the assistance of an information scientist with knowledge of the Database and the search facilities. Help with search construction and refinement is readily available. There is also provided intermediary assistance from the HUSAT's own human factors experts. The service is thus operating on several different levels depending on the expertise of the user with respect to knowledge of the subject field and to experience with on-line information retrieval systems.

It is envisaged that, after preliminary training, users will wish to work alone. To this end, in order that they may fully avail themselves of the facilities offered by the system with the minimum of difficulty, a user interface has been developed and evaluated by HUSAT according to human factors principles. The design of the interface is such as to provide online help and self-evident search process guidance.

The interface uses the facilities provided by a windows, icons, mouse and pull down menu system and provides help at all stages of the interaction. On entering search mode, the user is presented with on-screen access to the word-list files (the Classification, the Thesaurus and the Dictionary file associated with the references as input on to the Database) and to the Boolean and positional operators. It is possible to construct searches off-line using the interface and when a degree of satisfaction has been reached, the interface performs the connection with the Database including passing through the various levels of security associated with access to the Database. When the Database host system has responded to the search statements they may be amended as the user desires. The interface allows the user to choose the format of the results of the search and the references are then received by the interface either for testing the search performance or for final recording and downloading of useful references. Off-line prints of remaining references are sent through the post. The interface takes different forms according to the expertise of the user. For a new user there is a great deal of help at every stage of the search and limited access to the facilities of the software in order to reduce any confusion. The intermediate level of interface accounts for the user who is gaining in confidence and the expert interface provides access to all the sophisticated facilities that the system has to offer.

In addition to the personalized search facilities offered by the CHF Database, it is envisaged that a series of generalized bibliographies will be produced and a news bulletin about information within the field will be issued. Access to hard copy format of a range of the references on the Database will be made available to the users via the HUSAT Library which is being developed towards this goal. It is also envisaged that full text versions of a range of the references will be made available on-line (subject to copyright restrictions).

A pilot search on the Computer Human Factors Database. Topic: Usability

A request from a subject expert was made for a search to be carried out for bibliographic information in the field of usability. This field is fairly broad and was considered to be wide ranging enough to be one of a number of pilot searches to be executed on the Database. At the time of the pilot search the Database was in both paper form and machine-readable form. An interview with the subject expert requesting the search was conducted using the CHF Thesaurus as a guide to the indexing terminology to be applied. The following list of keywords to be searched upon was thus constructed. Such a keyword list tailored to an individual information requirement is called a search profile.

Usability		
Cognitive psychology		
Cognitive factors		
Cognitive processes	AND	Software
Cognitive skills		
Cognitive strategies		
Cognitive style		
Human performance		
User performance		
User friendly interfaces		
User acceptance		
User satisfaction		
User characteristics		
User expectations		
User identification		
User knowledge	NOT	Expert systems

User learning (*in the context of ease of learning*)
User modelling
User requirements
User reactions
Opinions
Attitudes
Preferences
Annoyance
Evaluation
Methodologies
Criteria
End users

Experts		Casual users		
Experienced users	AND	Naive users	AND	Comparison
		Novices		

Human behaviour
Graceful interaction
Assessment

An AND between two terms means that each term must be present in the reference record for the reference to be retrieved while an AND between two sets of terms means that at least one term from each set must be present for the reference to be retrieved. NOT indicates that if the term or concept following the NOT is present then the article is not to be retrieved. By far the most commonly used logical device is OR. All the keywords listed in the profile above are subject to the OR device so that if any of those keywords are present in the reference record (assuming that all other conditions imposed are adhered to) then the article is to be retrieved.

Over 600 CHF machine-readable references dating from January 1982 were accessible at the time of this pilot search; 89 references were retrieved from this part of the database and the system provided printed details of the articles, including the associated abstracts. A further 475 references in paper form were scanned manually and 81 references were obtained in this way.

Details of the final set of 170 references retrieved were sent to the subject expert together with a request for information about the degree of relevance of the references. The following list shows a sample of 20 items from the 110 relevant references selected. The references are presented in the order and format specified by the expert.

Towards an Ergonomic Design of Software
Behaviour and Information Technology
5(1), Jan–Mar 1986, 63–70
Cakir A

Expert–Novice Differences for Software: Implications for Problem-Solving and Knowledge Acquisition
Behaviour and Information Technology
5(1), Jan–Mar 1986, 15–29
Barfield W

Human Factors Design Considerations for Public Videotex Input Devices
Behaviour and Information Technology
4(3), Jul–Sep 1985, 189–200
Goodman D; Dickinson J and Francas MJ

The Effect of Computer Response Time on User Performance and Satisfaction: A Preliminary Investigation
Behaviour Research Methods and Instrumentation
15(2), Feb 1983, 213–216
Dannenbring GL

Making Life Easier for Professional and Novice Programmers
Byte
8(12), Dec 1983, 155–160
Pope A; Kates G and Fineberg D

Designing for Usability: Key Principles and What Designers Think
Communications of the ACM
28(3), Mar 1985, 300–311
Gould JD and Lewis C

Operators' Activities at CRT Terminals: A Behavioural Approach
Ergonomics
26(4), Apr 1983, 329–340
Delvolve N and Queinnec Y

Assessing the Usability of Human–Computer Interfaces
IEEE Software
2(1), Jan 1985, 74–88
Lindquist TE

The Influence of Troubleshooting, Education and Documentation on Computer User Satisfaction
IEEE Transactions on Systems, Man and Cybernetics
16(1), Jan–Feb 1986, 165–168
Rushinek A and Rushinek SF

Goodness-of-Fit in the User–Computer Interface: A Hierarchical Control Framework Related to "Friendliness"
IEEE Transactions on Systems, Man and Cybernetics
16(1), Jan–Feb 1986, 158–162
Thesen A and Beringer D

An Approach to the Informal Analysis of User Complexity
International Journal of Man–Machine Studies
22(4), Apr 1985, 365–394
Kieras D

A Structured Approach to Designing Human–Computer Dialogues
International Journal of Man–Machine Studies
21(2), Aug 1984, 105–126
Benbasat I and Wand Y

A Methodology for Interactive Evaluation of User Reactions to Software Packages: An Empirical Analysis of System Performance Interaction and Run Time
International Journal of Man–Machine Studies
20(2), Feb 1984, 169–188
Rushinek A; Rushinek S F and Stutz J

Towards an Understanding of the Role of Experience in the Evolution from Novice to Expert
International Journal of Man–Machine Studies
19(5), Nov 1983, 497–518
Kolodner J L

Programming Problem Representation in Novice and Expert Programmers
International Journal of Man–Machine Studies
19(4), Oct 1983, 391–398
Weiser M and Shertz J

Steps Towards Graceful Interaction in Spoken and Written Man–Machine Communication
International Journal of Man–Machine Studies
19(3), Sep 1983, 231–284
Hayes P J and Reddy D Raj

Detour Routes to Usability: A Comparison of Alternative Approaches to Multi Purpose Software Design
International Journal of Man–Machine Studies
18(4), 1982, 391–400
Wright P and Bason G

Managing User Expectations
Journal of Systems Management
34(6), Jun 1983, 6–12
Doll W J and Ahmed W V

Development of a Tool for Measuring and Analysing Computer User Satisfaction
Management Science
29(5), May 1983, 530–545
Bailey J E and Pearson S W

The User Friendly Index: A New Tool
Online
8(3), May 1984, 31–34
Matthews J R and Williams J F

A service for ergonomists

Computer science and technology plays a major rôle in the world today. Computers are to be found within many aspects of modern life and they have long since ceased to be the domain of the computer expert. People need to interact effectively with them and this means that human factors expertise is rapidly gaining the respect that it deserves. Given this increase of interest in the field, it is envisaged that ergonomists will welcome the service as a means of saving the time taken up by accessing information from the diversified literature of the subject. Added to this, most potential information sources contain material to which human factors do not apply and it is additionally time-consuming to filter out this material. Users will have the benefit of access to many more information sources than would be possible for an individual to acquire or to search on a regular basis without sacrificing time spent on the prime aspects of their work. The Database makes available to its users the more obscure items which are present as a result of the specialist information acquisition procedures which have been developed.

Conclusion

The CHF Database brings together relevant material on computer human factors from a very wide range of sources in a large number of scientific areas. In addition, the CHF Database will maintain currency in an area of rapidly advancing technological development, namely information technology. These two principal facets of the CHF Database will enable ergonomists to search an up to date specialist database which will provide them with easily accessed relevant material.

With the increasing interest in and demand for human factors input to information technology developments, the ergonomist will benefit from a tailored information service, readily available to aid and complement their skills and techniques and to enhance their contribution in this rapidly advancing field.

References

Ergonomics Abstracts, 1984, Special issue on 'Human-Computer Interaction' **16**(1), iv–v.

Phillips, K. E., 1985, Classification of domains of human factors in information technology. *ESPRIT Project 385 HUFIT Working Paper*, A3.1a. HUFIT/1–HUS–5/85.

Phillips, K. E., 1987, The 1987 Computer Human Factors Classification and Thesaurus and the collation of computer human factors literature. *ESPRIT Project 385 HUFIT Deliverable*, A3.1d/A3.2c. HUFIT/12–HUS–11/87.

Phillips, K. E. and Galer, M. D., 1986a, The development of a computer human factors classification and collation of human factors knowledge. *ESPRIT Project 385 HUFIT Working Paper*, A3.1b/A3.2a. HUFIT/2–HUS–1/86.

Phillips, K. E. and Galer, M. D., 1986b, The continuing development of a computer human factors classification and collation of computer human factors literature. *ESPRIT Project 385 HUFIT Working Paper*, A3.1c/A3.2b. HUFIT/6–HUS–11/86.

Shackel, B., 1984, Ergonomics information technology in Europe – a review. A report for the Commission of the European Communities. *HUSAT Memo* 309, A5.1–A5.4.

Chapter 2
Computer-aided ergonomics: A tool for control of musculoskeletal injuries

ASHRAF M. GENAIDY, WALDEMAR KARWOWSKI[1], and S. HOSSEIN MOUSAVINEZHAD[2]

Department of Industrial Engineering
College of Engineering and Applied Sciences
Western Michigan University
Kalamazoo, MI 49008
U.S.A.

[1]*Center for Industrial Ergonomics*
Speed Scientific School
University of Louisville
Louisville, KY 40292
U.S.A.

[2]*Department of Electrical Engineering*
College of Engineering and Applied Sciences
Western Michigan University
Kalamazoo, MI 49008
U.S.A.

Abstract. There is an urgent need to document the mass of knowledge generated in the ergonomics literature in order to assist occupational health and safety practitioners in the control and prevention of musculoskeletal injuries. This paper presents a conceptual framework for a computer-aided ergonomics system in order to fulfill the above need. Such a system is based on the integration of principles of ergonomic job design, worker selection, and worker training/education.

Keywords: Computer-aided ergonomics; Job design and redesign; Worker selection; Worker training and education.

Introduction

According to *Injury in America* (1985), injury is the most costly major health problem in the United States (about $75 to $100 billion per year). Injury has been the leading cause of physician contact. In 1980, there were 99 million contacts due to injury, compared with 72 million for heart disease and 64 million for respiratory disease, the second and third leading causes, respectively. Injury is also a leading cause of short- and long-term disability. In 1981, people spent 144 million days in bed because of injuries.

NIOSH (the National Institute for Occupational Safety and Health) (1986) reported that musculoskeletal injuries are one of the ten leading work-related diseases and injuries in the United States. Musculoskeletal injuries include disorders of the back, trunk, upper extremity, neck, lower extremity, and traumatically induced Raynaud's phenomenon.

Musculoskeletal injuries account for 18% of occupational injuries in the USA (Morbidity and Mortality Weekly Report, 1983). They are the leading cause of disability of people in their working years, afflicting 19 million persons (NIOSH, 1986). Musculoskeletal injuries rank first among the health problems in the frequency with which they affect the quality of life as indicated by the extent of activity limitation. Nearly one-half of the US workforce is affected at some time during their working life. High risk industries include manufacturing, construction, and food processing.

Musculoskeletal injuries are overexertion in nature where the worker usually exceeds his or her physical capability. Job demands of many industrial tasks requiring manual materials handling, repetitive motions, and vibration impose a high degree of stress on the musculoskeletal system making it highly vulnerable to a variety of occupational injuries and illnesses.

Manual materials handling

Manual materials handling (MMH) tasks, which involve the act of unaided lifting, lowering, carrying, pushing, pulling, and holding are the principal source of compensable work injuries in the USA (NIOSH, 1981). MMH injuries affect primarily the low back (Federal Register, 1986). According to the National Safety Council (1983), approximately 35% of all worker compensation claims are related to back injuries. Snook (1988) reported that, in a study conducted for the American Academy of Orthopaedic Surgeons (Holbrook *et al.*, 1984), the annual direct and indirect costs of back pain was almost $16 billion. Bases on estimates derived from the Bureau of Labor Statistics (1982a), about one million workers in the USA suffered back injuries in 1980.

From an epidemiological perspective, the NIOSH (1981) cited studies showing that frequency rates (number of injuries per man-hour on job) and severity rates (number of hours lost due to injury per man-hour on job) of back injuries increase significantly when: (i) heavy objects are lifted, (ii) the object is bulky, (iii) the object is lifted from the floor, (iv) objects are frequently lifted, and (v) loads are lifted asymmetrically (by one hand or at the side with the torso twisted).

Repetitive motions

Repetitive motions can cause cumulative trauma disorders (CTD) including carpal tunnel syndrome, tendinitis, ganglionitis, tenosynovitis, bursitis, and epicondylitis (Putz-Anderson, 1988). Persons employed in construction, food preparation, clerical work, product fabrication, and mining are at a high risk of CTD. NIOSH (1977) reported that 15–20% of workers in these jobs are at a potential risk of CTD. The Bureau of Labor Statistics (1982b) estimated that approximately 23 900 occupational injuries are associated with CTD.

In a review of the literature, Armstrong (1986) and Armstrong *et al.* (1986) reported that the frequently cited risk factors of CTD are: (i) repetitive exertions; (ii) posture – shoulder (elbow above mid-torso reaching down and behind), forearm (inward or outward rotation with a bent wrist), wrist (palmar flexion or full extension), and hand (pinching); (iii) mechanical stress concentrations over the base of palm, on the palmar surface of the fingers, and on the sides of the fingers; (iv) vibration; (v) cold; and (vi) gloves.

Vibration

Haber (1971) estimated that seven million workers in occupations such as vehicle operation are intermittently exposed to whole-body vibration which significantly stresses the musculoskeletal system. The vibration syndrome, which is characterized by intermittent numbness and blanching of the fingers with reduced sensitivity to heat, cold, and pain, affects up to 90% of workers in occupations such as shipping, grinding, and chain sawing (Wasserman *et al.*, 1974; Taylor and Pelmear, 1976).

The vibration syndrome is primarily caused by vibration of a part or parts of the body of which the main sources are hand-held power tools such as chain saws and jack hammers. Gruber (1974) estimated that 1.2 million workers are exposed to the risk of vibration syndrome.

Techniques used to control musculoskeletal injuries

Traditionally, the control of musculoskeletal injuries has been achieved through ergonomic job design and redesign, worker selection, and worker training and education. A brief description of each technique is given below.

Ergonomic job design and redesign

The essence of ergonomic job design/redesign is to maintain job requirements within the capabilities of the work-force. Industrial tasks are evaluated using the concept of ergonomic job design and redesign through the development of a job stress index (JSI) which is defined as the ratio of job demands and worker capacity. This index has been shown to correlate positively with the statistics of back injuries (Chaffin *et al.*, 1977; Ayoub *et al.*, 1978; Ayoub *et al.*, 1983; Liles *et al.*, 1984; Liles, 1986). Thus, a reduction in JSI leads to a drop in injury statistics.

Job demands are often expressed in terms of a task parameter that is a function of other task variables (e.g. frequency of handling), human variables (e.g. gender), and environmental variables (e.g. temperature). The worker capacity can be determined on the basis of his or her physical capability based on physiological, biomechanical, or psychophysical criteria (Ayoub *et al.*, 1980a, b; Garg and Ayoub, 1980; NIOSH, 1981; Chaffin and Andersson, 1984; Genaidy and Asfour, 1987; Genaidy *et al.*, 1988a, b; Asfour *et al.*, 1988; Putz-Anderson, 1988).

Worker selection

The main goal of worker selection is to screen the potential employee on the basis of his or her physical capability and match it with job demands. In order to evaluate an employee's capability, the following criteria should be applied when selecting between alternative screening methods (NIOSH, 1981):

1. Safety in administering,
2. Capability of giving reliable, quantitative values,
3. Relation to specific job requirements,
4. Practicality,
5. Ability to predict the risk of future injury or illness.

Isometric strength testing has been advocated as a means to predict the risk of future injuries resulting from jobs that require a high amount of force. Chaffin *et al.* (1977) reported that both frequency and severity rates of musculoskeletal problems were about three times greater for workers placed in jobs requiring physical exertions above that demonstrated by them in isometric strength tests, when compared with workers placed in jobs having exertion requirements well below their demonstrated capabilities. The literature on worker selection has been reviewed by NIOSH (1981), Ayoub (1983), and Chaffin and Andersson (1984).

Worker training and education

Training and education are the oldest approaches in industry to control MMH injuries (Snook, 1988). Traditionally, training and education are designed to instruct workers in proper methods and work procedures such as training in safe lifting. The basic concept of training in safe lifting has centred on instructing the worker to lift an object from the floor level using the straight back-bent knees technique. Later, additional principles of safe lifting have evolved based on the concept of kinetic method of manual handling (Anderson, 1970; Himbury, 1967; National Safety Council, 1971).

In recent years, the concept of back schools has been advocated as a means to educate the workers in all aspects of back care. The back school is a more comprehensive approach than the traditional training in safe lifting. The components of a back school may include anatomy and physiology of the lower back, body mechanics, posture, physical fitness, nutrition, advice on stress management, coping with pain, relaxation, drug use and abuse, first aid and acute care, epidemiology, activities of daily living, vocational guidance, and job simulation (Snook, 1988).

The control of musculoskeletal injuries: the need for computer-aided ergonomics

A comprehensive approach is needed for the total control and prevention of musculoskeletal injuries. The use of one technique may not be a totally satisfactory solution to the problem. An insurance company found that ergonomic job design can be used to control up to one-third of the MMH injuries (Snook, 1978). Thus, a comprehensive approach should be based on a combination of ergonomic job design, worker selection, and worker training and education. This means that a great deal of ergonomic knowledge should be put together in order to arrive at guidelines for use by practitioners of health and safety in the control of musculoskeletal injuries in industrial settings.

Since ergonomics knowledge relies on the interaction of many disciplines (e.g. psychology, biomechanics, work physiology, human resource development), it is evident

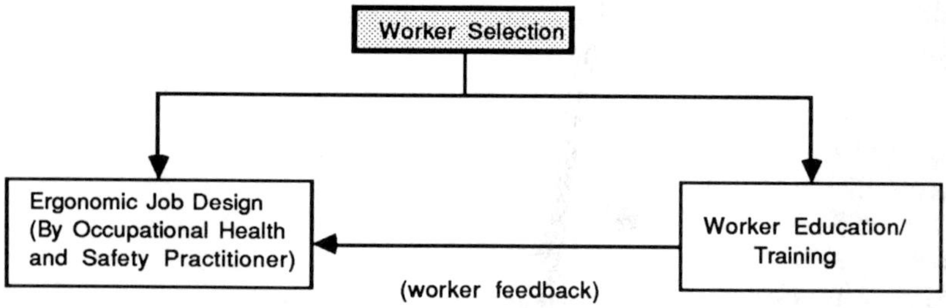

Figure 2.1 Conceptual framework for a comprehensive approach to control of musculoskeletal injuries.

that there is a need to document the wide range of knowledge generated in the ergonomics literature. This may be achieved through the development of a computer-aided ergonomics system. Examples of early work on computer-aided ergonomic systems in the United States are those that have been developed for the control of back injuries at the University of Michigan (Chaffin, 1987; Evans *et al.*, 1988), University of Louisville (Karwowski *et al.*, 1986, 1987), University of Miami (Kabouka *et al.*, 1988), and Western Michigan University (Genaidy *et al.*, 1989).

An attempt is made in this paper to present a conceptual framework for a comprehensive computer-aided ergonomics system. There are different ways to design such a system. A possible framework is depicted in Figure 2.1. A worker may be first screened to determine if he or she can be selected for the job (e.g. medical examination, strength testing). If the worker is fit for the job, he or she may be given an educational training programme in industrial ergonomics. Simultaneously, the occupational health and safety practitioner will evaluate specific jobs demands with respect to worker capability in order to design and redesign the job. Once the worker completes the educational training programme, he or she can give feedback to the practitioner with respect to the design and redesign of the job.

Ergonomic job design

As shown in Figure 2.2, the application of ergonomic job design and redesign in an industrial setting includes several steps in the following sequence:

1. Quantify job demands in terms of a measurable output. The amount of weight handled on the job by a female worker given a certain frequency over a specific range with containers of certain size, is an example of job demands. For an existing operation, task demands are determined on the job. For a new operation that does not exist, demands may be planned so that they match the capabilities of the potential employees.
2. Compute employee capability based on physiological, biomechanical, or psychophysical techniques. The outcome of employee capability should be a measurable output identical to that used in the quantification of job demands. Employee capability may be determined using tables, regression models, or it may be measured directly on the job.
3. Calculate the job stress index (JSI_{comp}) as a ratio of job demands and employee capability. The job stress index is a unitless measure.

4. Set the desired job stress index range in the workplace (JSI_{range}). This will be based on how much risk a company is willing to accept. Initially, the occupational health and safety practitioner may decide on this range. Later, it may be revised as needed (see step 8).
5. If JSI_{comp} exceeds JSI_{range}, apply engineering and administrative controls. An example of engineering controls is the reduction of the value of a task variable such as frequency of handling. The application of a job rotation program is an example of administrative controls.
6. If JSI_{comp} is less than JSI_{range}, increase job demands to match employee capability.
7. If JSI_{comp} is within JSI_{range}, no correction is needed.
8. Test if JSI_{range} results in a minimum risk of musculoskeletal injury. This can be accomplished by conducting a prospective epidemiological study. If different, go back to step 4.

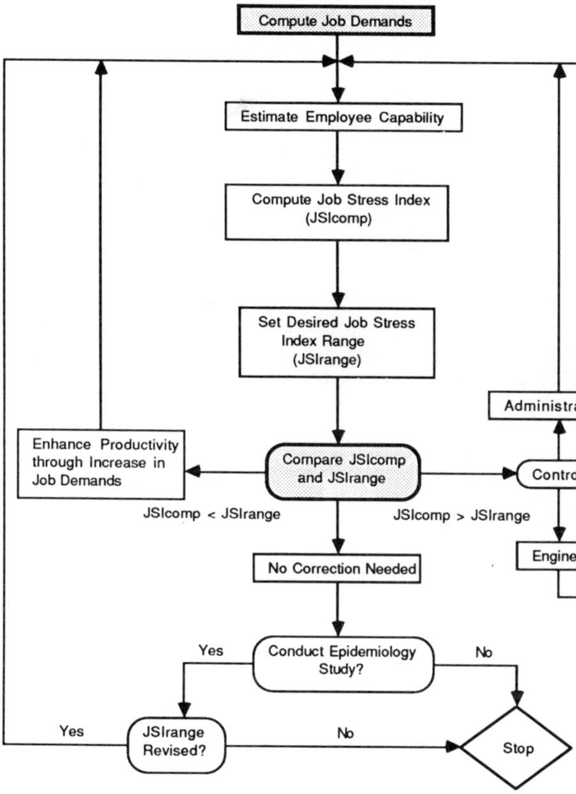

Figure 2.2 Ergonomic job design flowchart.

Educational programme

A beneficial educational programme should be implemented in order to train the worker in the identification and control of musculoskeletal injuries. Such a program should consist of two parts. The components of the first part, which is designed to be taught in a classroom environment, are as follows (Figure 2.3):

1. (a) Musculoskeletal injury statistics.
   (b) Anatomy of the musculoskeletal system.
2. Physiology of the musculoskeletal system.
3. Description of musculoskeletal injuries and case studies.
4. Review and evaluation of session #1.
5. The biomechanical basis of ergonomics in the work place (use of proper body mechanics).
6. The psychophysical basis of ergonomics in the work place (use of one's own perception to adjust existing workload).
7. The physiological basis of ergonomics in the workplace (use of physiological measures to adjust existing workload).
8. Review and evaluation of session #2.
9. Application of ergonomic guidelines at work.
10. Application of ergonomic guidelines at leisure time.
11. Nutrition and physical fitness at work and at home.
12. Review session and evaluation of #3.

Following the first part, a follow-up on-the-job training may be given to transfer the knowledge gained by the worker in the classroom environment to the job environment.

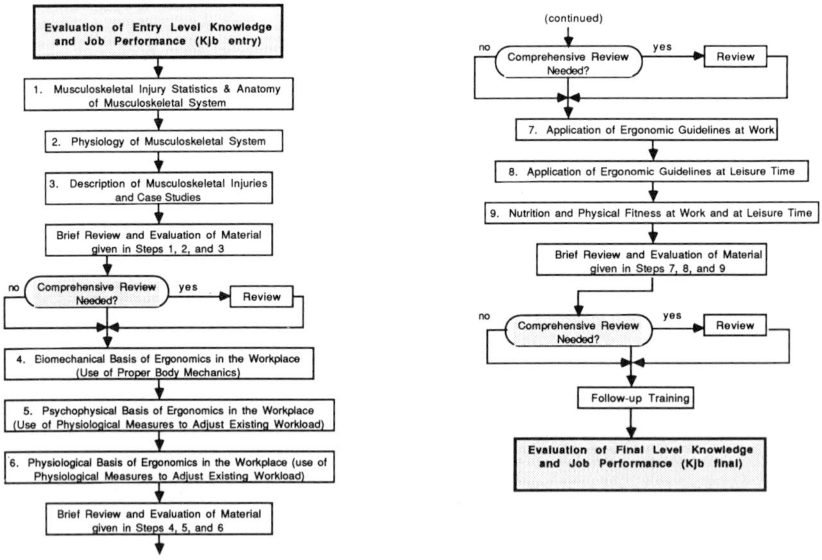

Figure 2.3 Computer-based tutorial system for worker training and education.

Worker selection

A worker selection flowchart is given in Figure 2.4. Some of the basic concepts used in the ergonomic job design and redesign approach applies to worker selection. The only difference is that, in the worker selection technique, general screening tests are given to the worker such as back X-ray and strength testing. On the other hand, ergonomic job design is very specific to the operation under consideration.

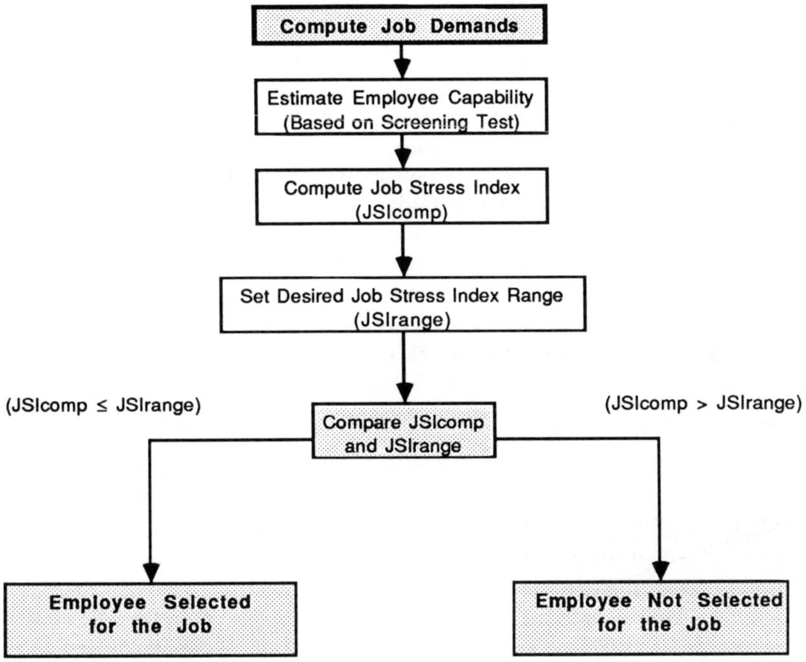

Figure 2.4 Worker selection flowchart.

Conclusions

A conceptual framework for a comprehensive computer-aided ergonomics system has been presented in this paper for the control of musculoskeletal injuries in industrial settings. This system is based on the integration of principles of ergonomic job design and redesign, worker selection and worker training and education.

References

Anderson, T. M., 1970, Human kinetics in strain prevention. *British Journal of Occupational Safety*, **8**, 248–250.

Armstrong, T. J., 1986, Ergonomics and cumulative trauma disorders. *Hand Clinics*, **2**(3), 553–565.

Armstrong, T. J., Radwin, R. G., Hansen, D. J. and Kennedy, K. W., 1986, Repetitive trauma disorders: job evaluation and design. *Human Factors*, **28**(3), 325–336.

Asfour, S. S., Genaidy, A. M. and Mital, A., 1988, Physiological guidelines for the design of manual lifting and lowering tasks: the state of the art. *American Industrial Hygiene Association Journal*, **49**, 150–160.

Ayoub, M. A., 1983, Design of a pre-employment screening program. In *Ergonomics of Workstation Design*, edited by T. O. Kvalseth (London: Butterworth and Company).

Ayoub, M. M., Bethea, N. J., Deivanayagam, S., Asfour, S. S., Bakken, G. M., Liles, D., Mital, A. and Sherif, M., 1978, *Determination and modeling of lifting capacity*. DHEW (NIOSH), Grant No. 5R010H00545-02.

Ayoub, M. M., Mital, A., Asfour, S. S. and Bethea, N. J., 1980a, Review, evaluation, and comparison of models for predicting lifting capacity. *Human Factors*, **22**(3), 257–269.

Ayoub, M. M., Mital, A., Bakken, G. M., Asfour, S. S. and Bethea, N. J., 1980b, Development of strength and capacity norms for manual materials handling activities: the state of the art. *Human Factors*, **22**(3), 271–290.

Ayoub, M M., Liles, D. H., Asfour, S. S., Bakken, G. M., Selan J., Mahajan, P. and Bethea, N. J., 1983, *Effects of task variables on lifting capacity*. DHHS (NIOSH), Grant No. 5R010H00798-04.

Benson, J. D., 1984, Ergonomics and safe lifting training. *Professional Safety*, July, 33–36.

Bureau of Labor Statistics, 1982a, *Back Injuries Associated with Lifting*. Bulletin 2144, (Washington, DC: US Department of Labor).

Bureau of Labor Statistics, 1982b, *Occupational Injuries and Illnesses in the United States by Industry, 1980*. Bulletin 2130 (Washington, DC: US Department of Labor).

Chaffin, D. B., 1987, Biomechanical strength models in industry. In *Ergonomic Interventions to Prevent Musculoskeletal Injuries in Industry* (Chelsea, MI: Lewis Publishers), pp 27–45.

Chaffin, D. B. and Andersson, G., 1984, *Occupational Biomechanics* (New York: John Wiley).

Chaffin, D. B., Herrin, G. D., Keyserling, W. M. and Foulke, J. A., 1977, *Pre-employment Strength Testing*. DHEW (NIOSH), Publication No. 77-163.

Evans, S. M., Palmiter, S. L. and Elkerton, J., 1988, The EDGE system: ergonomic design using graphic evaluation. In *Proceedings of the Human Factors Society 32nd Annual Meeting*, Santa Monica, CA, pp 612–616.

Federal Register, 1986, vol. 51, No. 191, October 2, pp 35241.

Garg, A. and Ayoub, M. M., 1980, What criteria exist for determining how much load can be lifted safely? *Human Factors*, **22**(4), 475–486.

Genaidy, A. M. and Asfour, S. S., 1987, Review and evaluation of physiological models for manual materials handling. *Human Factors*, **29**, 465–467.

Genaidy, A. M., Asfour, S. S., Mital, A. and Tritar, M., 1988a, Psychophysical modeling in manual materials handling. *Human Factors*, **30**, 319–337.

Genaidy, A. M., Bafna, K. M. and Asfour, S. S., 1988b, Ergonomic job design in manual materials handling. *Proceedings of the Annual IIE Conference*, Orlando, Florida, May, pp 115–120.

Genaidy, A. M., Ayoub, M. M. and Duggal, J. S., 1989, A microcomputer-based ergonomic software for the design and evaluation of frequent manual lifting tasks. *Computers and Industrial Engineering*, **16**(3), 455–468.

Gruber, G. J., 1974, *Relationship Between Whole-Body Vibration and Morbidity Patterns Among Interstate Truck Drivers*. Publication No. 75–104, September (Cincinnati, OH: NIOSH).

Haber, L. D., 1971, Disabling effects of chronic disease and impairment. *Journal of Chronic Disease*, **24**, 269–487.

Himbury, S., 1967, *Kinetic Methods of Manual Handling in Industry* (Geneva: International Labour Office, Occupational Safety and Health), Series No. 10.

Holbrook, T. L., Grazier, K., Kelsey, J. L. and Stauffer, R. N., 1984, *The Frequency of Occurrence, Impact, and Cost of Selected Musculoskeletal Conditions in the United States*. The American Academy of Orthopaedic Surgeons, 154–156.

Injury in America. A Continuing Public Health Problem, 1985, (Washington, DC: National Academy Press).

Kabouka, M., Genaidy, A. M. and Asfour, S. S., 1988, A knowledge-based system for the design of manual materials handling tasks. *Applied Ergonomics*, **19**(2), 147–155.

Karwowski, W., Mulholland, N. O., Ward, T. L., Jogannanthan, V. and Kirchner, R. L., Jr., 1986, LIFTAN. An Experimental Expert System for Analysis of Manual Lifting Tasks. *Ergonomics*, **29**, 1213–1234.

Karwowski, W., Mital, A., Palenque, L. E. and Ward, T. L., 1987, Development of a microcomputer-based expert system for the analysis of manual materials handling tasks in industrial settings. *International Journal of Industrial Ergonomics*, **2**, 49–59.

Liles, D H., 1986, The application of the job severity index to job design for the control of manual materials handling injury. *Ergonomics*, **29**, 65–76.

Liles, D. H., Deivanayagam, S., Ayoub, M. M. and Mahajan, P., 1984, A job severity index for the evaluation and control of lifting injury. *Human Factors*, **26**, 683–694.

Morbidity and Mortality Weekly Report, 1983, Musculoskeletal Injuries. 15 April, 32 (14).

National Institute for Occupational Safety and Health, 1977, *National Occupational Hazard Survey, 1972–1974*. DHEW (NIOSH) Publication No. 78–114 (Cincinnati, OH: NIOSH).

National Institute for Occupational Safety and Health, 1981, *Work Practices Guide for Manual Lifting*. DHHS (NIOSH) Publication No. 81–122, March (Cincinnati, OH: NIOSH).

National Institute for Occupational Safety and Health, 1986, *Proposed National Strategies for the Prevention of Leading Work-related Diseases and Injuries. Part 1*. DHHS (NIOSH), PB87–114740 (Cincinnati, OH: NIOSH).

National Safety Council, 1971, Human kinetics and lifting. *National Safety News*, June, 44–47.

National Safety Council, 1983, *Accident Facts*. Chicago, IL.

Putz-Anderson, V. (ed), 1988, *Cumulative Trauma Disorders. A Manual for Musculoskeletal Diseases of the Upper Limbs* (London: Taylor & Francis).

Snook, S. H. 1978, The design of manual handling tasks. *Ergonomics*, 21, 963–985.

Snook, S. H., 1988, The costs of back pain in industry. *Occupational Medicine: State of the Art Reviews*, **3**(1), 1–5.

Snook, S. H., Campanelli, R. A. and Hart, J. W. 1978, A study of three preventive approaches to low back injury. *Journal of Occupational Medicine*, **20**, 478–481.

Taylor, W. and Pelmear, P. L., 1976, Raynaud's phenomenon of occupational origin: an epidemiological survey. *Acta Chiropractica Scandinavica* (Supplement), **465**, 27–32.

Wasserman, D. E., Badger, D. W., Doyle, T. E. and Margolies, L., 1974, Industrial vibration – an overview. *ASSE Journal*, **19**, 38–43.

PART II
ANTHROPOMETRY AND COMPUTER MODELS OF MAN

Chapter 3
SAMMIE: a man and workplace modelling system

KEITH CASE, J. MARK PORTER[1] and MAURICE C. BONNEY[2]

Department of Manufacturing Engineering,
Loughborough University of Technology,
Loughborough, Leics, LE11 3TU, UK

[1]Department of Human Sciences
Loughborough University of Technology,
Loughborough, Leics, LE11 3TU, UK

[2]Department of Production Engineering and Production Management,
University of Nottingham,
Nottingham, UK

Abstract. SAMMIE (System for Aiding Man–Machine Interaction Evaluation) is a long established computer-aided design (CAD) system which assists in the design and evaluation of human workplaces. This paper describes the currently available commercial system, illustrated by examples taken from the many successfully completed design studies. The future of such systems in relation to developments in computer hardware, software, and the general use of CAD techniques is considered. The active lines of research being pursued at both Loughborough and Nottingham Universities and which will potentially contribute to future versions of SAMMIE are described.

Keywords: Man modelling; Workplace design; Anthropometry; Biomechanics.

Introduction

SAMMIE is possibly the longest established and most widely available man modelling system, and is also notable for its effective method of modelling workplaces. The man model is based upon what are now established methods using a pin-joint and rigid link representation combined with joint articulation limited by constraints. For visualization and fit evaluations, the model is fleshed by reference to the somatotyping technique of body shape classification. All aspects of anthropometry are controlled by data-driven techniques, and there are several methods of specifying population data and the selection of appropriate subsets for particular design studies. Several different methods are available for placing the man model in a working posture and predicting new postures on the basis of reach requirements. Important methods of representing the results of a series of reach tests include reach paths and reach volumes (which have been used as the basis of a British Standard for agricultural tractors).

SAMMIE takes standard viewing methods which are available in most three-dimensional CAD systems and applies them to the needs of ergonomics evaluation. Thus the model can be viewed in a wide variety of ways to give the scene as seen by an outside observer, the man model itself, and reflected views. These techniques are typically used in visibility studies for the production of two- and three-dimensional vision charts and mirror views.

SAMMIE recognizes that a man modelling system can only be truly useful if it is closely integrated with a workplace modelling system which is capable of representing the equipment being designed or evaluated. Techniques which are well established in general CAD systems are used together with special purpose facilities which assist the objective of ergonomic design.

SAMMIE was designed as a general purpose system and is not limited to any particular application area. To illustrate this and to demonstrate the various facilities available within the system some examples of completed design studies are given. These include the design and layout of a power station control consoles, interior package design for a new automobile, visualization studies for the Channel Tunnel, design of financial trading facilities, concept design studies for helicopters and visibility studies for underground trains.

The future of SAMMIE is discussed in terms of both the needs of commercial exploitation, and current and future university-based research. The impact of emerging and existing standards on these kinds of systems are discussed particularly in relation to the use of engineering workstations with the Unix operating system, X-windows based user interfaces, and PHIGS graphics. University research is principally concerned with fundamental enhancements to the capabilities of the man model such as improvements in the anthropometric representation and strength modelling, the addition of intelligence through the use of expert systems and geometric reasoning, and improvements in the user interface. Advances in these areas are reported together with an assessment of their importance for the future of man modelling systems in general.

Man modelling

SAMMIE uses what is now the normal method of modelling the human body by an approximation consisting of a collection of pin-joints connected together by rigid links. This method is ideally suited to mechanisms of the open chain link variety, and can be traced back to the work of Dempster (1955) in designing a physical mannikin. Many of today's man models as reviewed in Dooley (1982) and Rothwell and Hickey (1986) use the technique.

In these models the way in which volumetric aspects of the body are modelled varies according to the application area and the type of computer output required. For kinematic modelling with a graphical representation on the screen (as with SAMMIE and most other modellers concerned with workplace design and evaluation) it is normal to place a three-dimensional geometric model about the link model to represent spatial relationships. Modellers which are aimed at non-geometric criteria (for example those solely concerned with strength evaluations) might dispense with the flesh representation entirely.

Similarly, the method by which the body is driven to assume a particular posture is a function of the area of application. Kinematic modelling, as is to be expected, is normally associated with spatial evaluations of workplaces where either the human operator or parts of his physical environment assume different positions over time. This requires a modelling technique which is able to evaluate spatial considerations such as the interference between two solids in three-dimensional space. However, dynamic modelling is usually associated with assessing the body's response to large external forces (such as those experienced by the rapid decelerations in car crash simulations). Here a different form of modelling is likely to be used where more consideration may be given to the body segment parameters such as mass, centre of gravity, and moments of inertia.

Pin-joints and rigid links

SAMMIE uses a pin-joint representation as shown in Figure 3.1. A right-handed axis set with its origin located at the centre of articulation of the link is used to describe the link's extension away from a neutral position aligned with the z-axis. The modeller uses the angles phi (ϕ), theta (θ) and epsilon (ψ) together with the link length to determine the location of the distal end of the link and the orientation of the next set of joint axes. The angle set (ϕ,θ,ψ) are an Eulerian set (i.e. they are independent angles) and thus are mathematically consistent. However these angles are not those which are normally collected in surveys and are not generally familiar to ergonomists and designers. An 'orthogonal' set (α,β,γ) would be more normal where for example α would be an angle of joint abduction/adduction and β an angle of flexion/extension. These are provided as the normal user interface and internally mapped onto the Euler set whenever required. This provides the necessary familiar interface but at the expense of mathematical consistency.

The joint can be constrained to operate within feasible limits by the provision of joint constraint angles describing the maximum articulation in both directions away from the neutral position (Figure 3.1b). These constraints are contained in user-accessible data files and thus can be modified to represent varying conditions (e.g. constraining clothing, physical disabilities, or population variations). In a similar way a pair of 'normal' constraints can be defined for each degree of freedom at a joint through a user-accessible data file. The purpose of this facility is to provide a method by which the system can flag a condition where a joint goes outside some preferred range of movement. This preference could be related to postural comfort in a general sense so that, for example, the upper arms were constrained not normally to go above shoulder height, or the facility could be used to define explicit design criteria. Thus for example the 'normal' range of flexion at the knee could be limited to 20° for car pedal operation.

SAMMIE currently approximates to the complex skeletal structure by modelling the ankles, knees, hips, lumbar and thoracic spine, shoulders, elbows, wrists, neck, head and eyes (Figure 3.2). This choice was based on the importance of these joints in assessing human workplaces such as cabs and cockpits whilst keeping the modelling sufficiently simple to ensure a fast response at the screen. Greater realism would be provided by a more intricate spine model such as that developed by Freidlich (1987) and the study of manual assembly tasks would be possible if the fingers were modelled. The interactive display and manipulation of the increased complexity is a realistic proposition, but unfortunately we do not yet have an adequate understanding of how the spine and hand contribute to tasks to be able to build the necessary predictive algorithms.

a

b

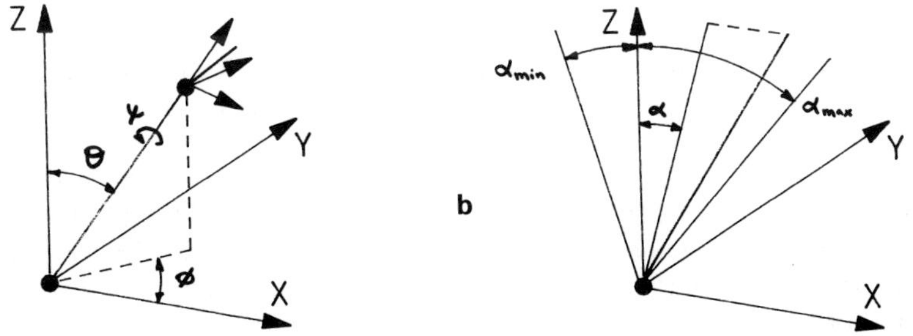

Figure 3.1 The pin-joint model. (a) Euler; (b) orthogonal.

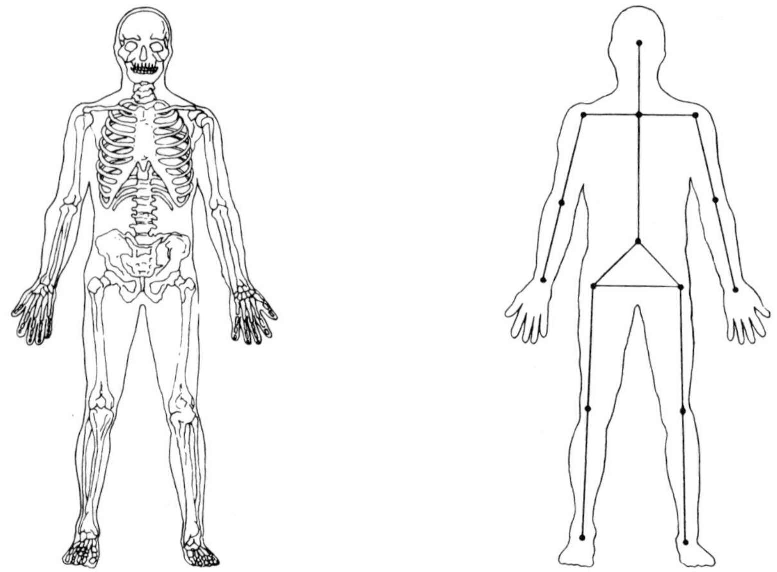

Figure 3.2 'Stick man' model in relation to the skeletal structure.

The joints are logically and spatially related to each other using a hierarchical data structure which was originally developed for the workplace model. Two facilities are required of a joint and link model: first, it must be possible to predict the location of an end-point given a knowledge of all the joint extensions and the link lengths. This is the 'forward' evaluation of a link structure and is used for example to predict the location of the hand with a given set of arm joint extension angles. Second, it should be possible to perform the 'inverse' evaluation where the joint extensions can be predicted given the required location for an end-point (hand) and a knowledge of the link lengths. This permits the prediction of a posture that needs to be adopted to perform some task. In SAMMIE the 'forward' prediction is accommodated by representing a joint's extension in up to three degrees of freedom (e.g. abduction/adduction, flexion/extension, and medial/lateral rotation) as a single transformation matrix (see Korein, 1984 or Craig, 1986 for mathematical details). Concatenation of all such matrices contributing to a posture and application to a vector representing the link will then result in the required end-point coordinates. The 'inverse' problem is considerably more difficult and more useful. The human body is highly flexible in the way in which it can adopt different postures to perform some task. In mathematical terms we would say that there are redundant degrees of freedom which need to be constrained before a posture can be predicted. Thus there may be many ways of reaching to the same point with a fingertip, and in the real situation the choice of a particular way might be made by reference to the context of the reach. Hence it may be that the elbow is kept on a worksurface to stabilize the posture. In modelling we have the choice of either including large numbers of such contextual rules or using more general algorithmic methods. The rules approach is clearly suited to artificial intelligence techniques, but at the current time an adequate description of the geometric context is beyond us. (Research work on a geometric reasoning for CAD is progressing, see Bonney *et al.*, 1986; Howe and Fothergill, 1988). SAMMIE therefore adopts the approach of general algorithms which attempt to choose a suitable posture related to a minimization of joint extension. This is described more fully later as part of the 'Reach' facilities.

Flesh description

The joint and link model described above forms the basic structure on which the model operates, but for visualization and assessment of fit this is encased in a solid model representation of the flesh shape. Once again responsiveness of the system to the designer's commands is maintained by using relatively simple flesh shape representations based on plane faceted polyhedra. However the size of these flesh modules can be controlled using the somatotype classification devised by Sheldon (1940). This technique seeks to relate body shapes to three seven-point scales which define the extent of endomorphy, mesomorphy, and ectomorphy. Of the 343 theoretically possible shapes, Sheldon's work discovered a total of 76 and determined their frequency of occurrence in the sample population. Sheldon studied body shape by taking 19 body measurements mainly concerned with volumetric dimensions (circumference of the chest, waist, etc.) but also including height and weight. The somatotype number is then determined from the relationship between the volumetric dimensions and height and weight. For flesh modelling purposes SAMMIE reverses this procedure so that the designer specifies a somatotype (in a variety of ways, but most usefully from a set of 16 generalized shapes such as 'extreme endomorph') which is used to determine the volumetric dimensions with a knowledge of the current stature and weight of the man model. These dimensions are then used together with a knowledge of the link lengths to generate the detailed dimensional information required to locate every vertex on the man model's flesh shape. Figure 3.3 illustrates some of the body shapes obtained in this fashion.

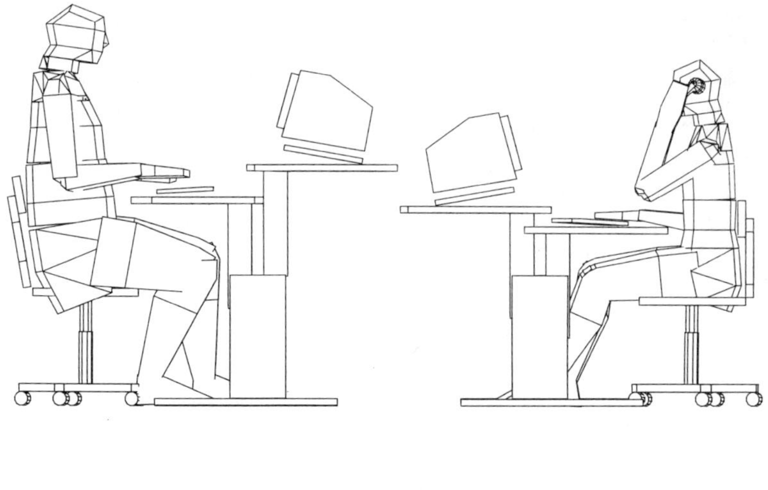

Figure 3.3 A range of body shapes defined by somatotype.

Anthropometry

The somatotyping technique previously described creates some of the anthropometric data used by the model. However this is based upon the individual link dimensions and the stature of the model before 'applying' a somatotype. Control of anthropometry through the link (joint-to-joint) dimensions is possible in several different ways. The important point to understand is that the model is data-driven and has no dependency whatsoever on any particular set of anthropometry collected during surveys. Indeed a major objective of anthropometric modelling within SAMMIE is to allow the input of data from a variety of sources and to this end a simple data base facility is accessible from within the system. Files describe each link in terms of a mean length and standard deviation, and these files can be created and modified by the designer. One of these files could be used to create the 'current' anthropometry within the man model. As there can be as many as eight man models, this implies that it is feasible simultaneously to display and control man models which might be based on widely different anthropometry. This could be useful if, for example, an automobile was being assessed for the world market, when models representing different national groups could be simultaneously displayed.

Once the anthropometric data has been introduced into the man model it can be manipulated in a variety of ways. The most immediately available is by the specification of an overall percentile. This will apply standard techniques to generate a dimension for each link based on the mean and standard deviation information and using a normal distribution. Figure 3.4 illustrates some of the possibilities.

The use of overall percentiles can be useful in some of the gross evaluation stages early in

Figure 3.4 Varying anthropometry by percentile control.

design, but can be dangerously misleading. It has been well known for a long time (Hertzberg, 1960) that there is poor correlation between link dimensions and this implies that a simplistic percentile approach will invariably produce an unrealistic model. Hence other more detailed facilities are provided for more precise control of anthropometry. Thus it is possible to control each dimension individually by reference to percentiles rather than simply as a total body percentile. This can be very useful as a way of encapsulating design criteria within the model. (An example in our own experience involved models with extremes of dimensions in the arms, legs and trunk, which was used to evaluate reach to controls and head clearance in an automobile under various conditions of seat adjustment.) It is also possible to move entirely away from the percentile approach and explicitly to define link dimensions so that a model is defined which relates to an individual.

One of the difficulties faced in specifying anthropometry is that the model is based on the internal joint-to-joint dimensions whereas most surveys quite naturally determine external measurements. Facilities are therefore provided to relate external measurements such as anterior arm reach to the appropriate internal link dimensions (upper arm, forearm, and hand, in this instance). These use regression equations based on the work of Dempster *et al.* (1964). It would be a simple matter to use a different set of regression equations provided that they were capable of generating the required information. For example Pheasant (1986) claims that all dimensions can be related to stature and sex. If this was deemed acceptable then it could be accommodated by the model.

The anthropometry definition within SAMMIE is currently under review with the potential of a fundamental change which would produce a more detailed and precise model which was still capable of fast response on modern computer equipment. This is described more fully under 'Research and development'.

Postures

One of the principal requirements of a man modelling system is that it be capable of representing working postures. This implies the control of the joint extension angles, i.e. a

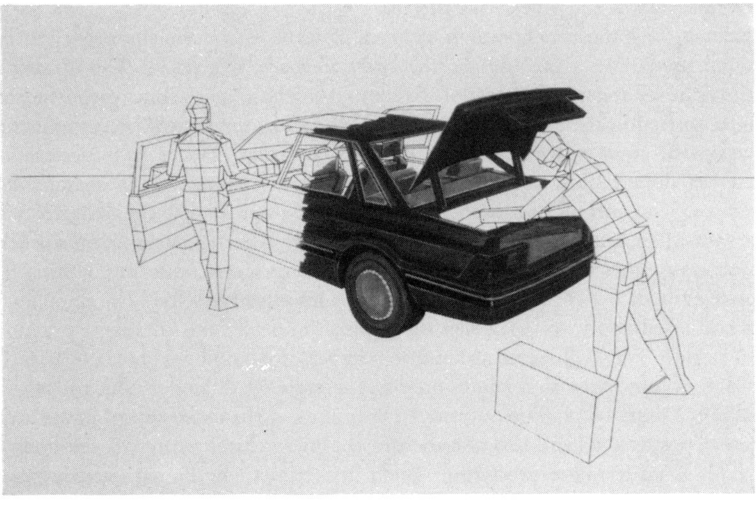

Figure 3.5 Illustration of various postures.

posture is considered to be a set of all joint angles across the man model, and is thus independent of dimensional aspects of anthropometry. Hence it is possible to describe certain useful postures as sets of joint angles which are stored within a data base for future use. Initial creation of these postures occurs by several methods. Those which are commonly used as starting points for modification into working postures are always available, and postures such as 'sit', 'stand', 'prone' and 'supine' are included in this category. A particular posture can be generated by interactive manipulation of the joint angles within the joint constraints described above. Each joint potentially has three degrees of freedom, each of which can be directly controlled from the 'joint' menu of commands. At any time a posture can be entered into the off-line data base for future use by any SAMMIE user. Some typical postures are shown in Figure 3.5.

Reach assessment

A major evaluation tool is provided by various assessments of the reach capabilities of the man model. At the very simplest level the attainment of a posture by the methods described above implies that the point at, say, the fingertip is within reach (as the posture can be achieved within the current link lengths and without exceeding joint constraints). However it is normally more convenient to test reach by defining a 'reach point' in three-dimensional space and then attempting automatically to assume a posture which achieves this, i.e. the 'inverse' evaluation method described above is used as a predictive tool.

The reach point itself can be defined by a number of methods including a definition of three-dimensional coordinates with the model space, incremental moves from the current position for the hands or feet, or most usefully by the specification of a 'touchpoint'. A touchpoint is some point on the model of the item being evaluated. Each geometric entity within the workplace model has one such point represented by its origin or 'anchor' point. Further named points can be defined as a subpart of a geometric entity, so that for example two points representing the grip points on a steering wheel could be specified and assigned identifying names. As they are subparts of the steering wheel they would maintain the correct relationship as the wheel was rotated.

In whatever way the touchpoint is defined, it is necessary for the model to have the capability of predicting a posture for the body to reach this point. The human body is provided with redundant degrees of freedom which give us our great flexibility in comparison with most mechanical devices, and this means that in general any touchpoint can be reached with an infinite number of different postures. Hence it is necessary to use algorithms to determine a single posture which is both feasible and reasonable in the circumstances. The current approach adopted by SAMMIE is that the designer is charged with the task of making the strategic decisions within which the algorithms work. Thus gross considerations such as whether the model should be seated, standing, leaning forwards and so on are under the control of the designer. This effectively fixes the shoulder and hip locations and orientations so that reach algorithms can determine suitable arm and leg joint extension angles. This still leaves an infinite number of possibilities, but a relatively simple optimization of joint extension angles produces a single set of angles which is both feasible and realistic for most tasks. The optimization minimizes the extension of joints away from their neutral position whilst also considering the hierarchical nature of the body linkage system. This is an iterative procedure which first tries to find a satisfactory posture for proximal joints (i.e. within the 'normal' joint constraints) before allowing the extension of more distal joints. If necessary joints are progressively permitted to move into the 'maximum' extension range and a position is only deemed to be out of reach when all contributing joints are at their maximum appropriate extension.

This algorithm is useful for 'single-shot' reach evaluations of existing designs, but when creating new workplace layouts it is desirable to be able to predict areas or volumes which are suitable for the location of controls or whatever. For these purposes it is possible to define a 'reach path' which consists of an ordered sequence of coordinate locations in three-dimensional space. The man model can then be given a single instruction to reach to each of these points in turn and determine a suitable reach posture. This can result in a re-display of the model at each point or alternatively the storage of information on the 'reachability' of each point. These reach paths are most useful if they are organized to relate directly to the design proposal. For example, if the design task were to produce a layout for instruments and controls on a console then a regular pattern of reach points would be laid across the operational surfaces. As this reach path is a modelling entity which is associated with the particular surface, it will keep the same relationship during design changes. Thus if console surfaces are moved or angles of inclination changed there is no need to regenerate the reach path. After reach evaluations have been carried out across the reach path, an envelope of reach can be generated and displayed. The extent of the envelope can represent the maximum reach or the maximum normal reach (where joint extension is constrained to remain within the normal joint constraints).

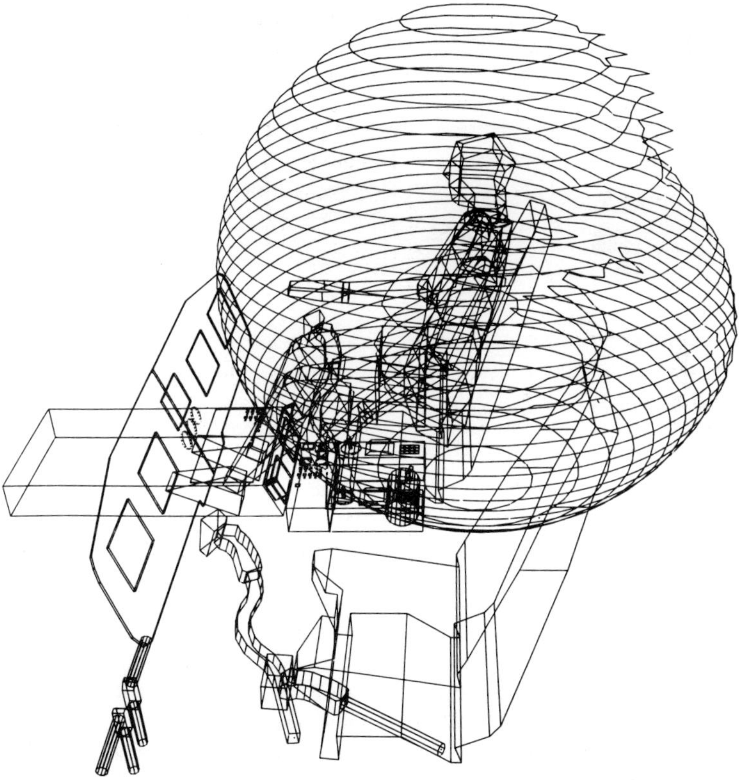

Figure 3.6 A reach envelope.

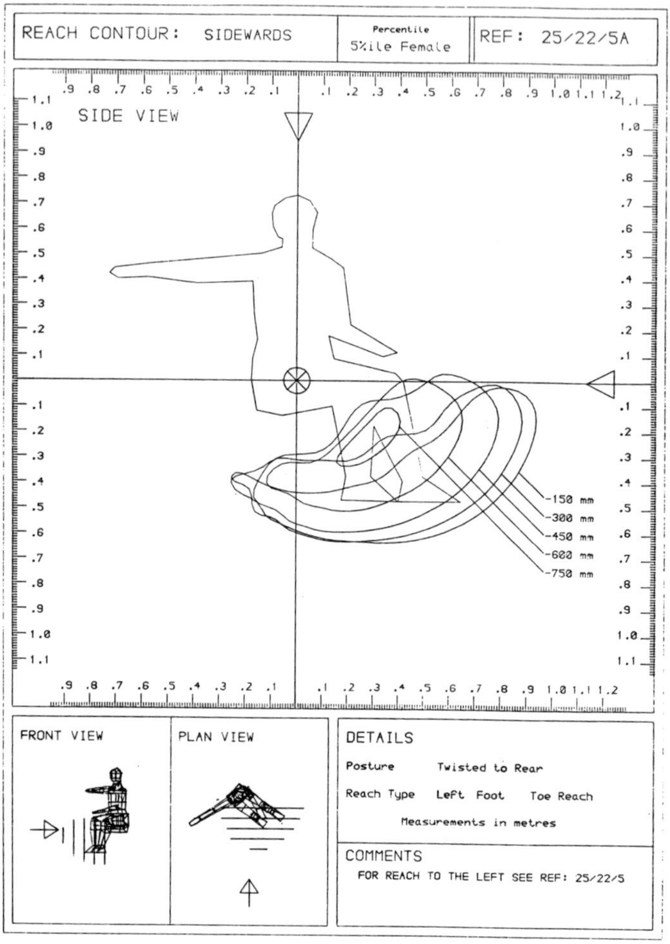

Figure 3.7 Reach volumes for tractor design.

In a similar fashion it is possible to predict volumes within the workspace which can be reached (Figure 3.6). These have been useful in many areas including the location of controls within a tractor cab (Figure 3.7) and have formed the basis of a British Standard (BSI, 1987).

Fit assessment

The man model and the workplace model use the same internal representation and hence it is possible to determine spatial relationships between them. In particular the geometric modeller has an interference checking capability which can be used to evaluate fit in confined circumstances such as those frequently found in cockpit type situations. Control of the anthropometry and somatotype permits a variable man model to be checked against the workplace model so that any areas of interference can be discovered and highlighted. Such techniques need not only be applied to this static situation but could, for example, form part of a kinematic evaluation as the man model moves about the workplace performing a task. However, it should be realized that this is a computational intensive activity and is not normally suited to interactive use in this kinematic sense.

Viewing

A major strength of SAMMIE is the extensive viewing facilities which are available. As the system is based on a solid CAD modeller it contains all the viewing capabilities that would be expected of such a system. However, these have been modified and extended to provide a range of facilities which offer direct ergonomics evaluation capabilities. The basic capabilities include plane parallel projections, perspective, multiple views, zooming, scaling and panning and are more fully described later under the 'Workplace modelling' section.

It is normal in CAD work to view the model as a detached observer, but the SAMMIE viewing facilities make it possible to display views as would be seen by the man model within the workplace model. The views obtained will obviously depend upon the location of the man model and anthropometry which is used to determine an eye position. Perspective views from the left, right, or an imaginary central eye position can be displayed and would be typically used for example in the evaluation of the visibility characteristics of vehicles.

Extensions to the control of perspective views (man's views or otherwise) provide several useful evaluative methods. Perspective is controlled by the 'acceptance angle' which determines the field of view. For normal viewing purposes this is calculated to provide a true (if limited) perspective view on the screen, but if a large angle of 180° is specified then a highly distorted view is generated (180° of view are compressed into the small size of the graphics screen). If this is made to occur whilst in a plan view then a vision plot is produced such as that shown as Figure 3.8. This has proved to be extremely useful in evaluating the visibility characteristics of vehicles where the traditional approach involves a physical analogy which places a light at the driver's eyepoint and records the positions of shadows generated

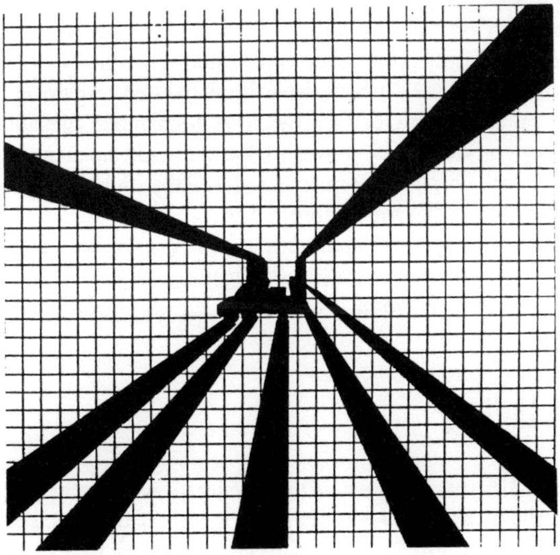

Figure 3.8 Two-dimensional vision plot.

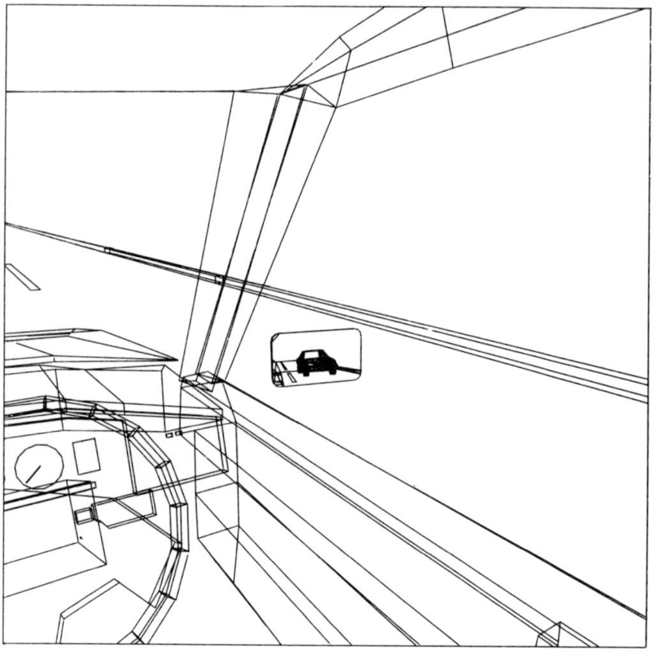

Figure 3.9 Mirror view.

by the bodywork of the cab. Using the computer simulation it is possible to ensure that vehicles removed from the prototype development programme to undergo this approval test have a high probability of being successful.

The technique has been extended to include three-dimensional vision charts (Porter *et al.*, 1982), and more recently Aitoff projections as favoured for military evaluations of visibility against MIL standards (MIL, 1970) have been implemented. Again principally related to vehicle design and evaluation, mirror facilities have been made available. The face of any object in the workplace model can be nominated as a mirror and assigned a focal length to make it concave, convex, or plane (Bonney *et al.*, 1980). Subsequent displays will show appropriate reflected views as shown in Figure 3.9.

The technique has also been used to investigate the occurrence of unwanted reflections such as those on the windscreen of a vehicle, the display surfaces of control consoles, or the screen of a computer graphics terminal.

Workplace modelling

From the outset SAMMIE was designed with close integration between the man modelling and the modelling of the workplace being designed or evaluated. Several of the other available man models do not have this close link, and thus although they may be sophisticated in man modelling terms their usefulness can be severely limited. Only a brief description of the workplace (or geometric) modelling is given here.

Solid modelling

Several varieties of solid modelling are available including Boundary Representation, Constructive Solid Geometry, and Occupancy Enumeration (Requicha, 1980). The choice of technique to use is largely determined by the application and SAMMIE uses a Boundary Representation (BRep) to take advantage of its inherent speed in drawing on the graphics screen. A BRep model is topologically explicit in that the data structure holds information on the location of vertices together with the structure of the solid. The structural information defines how pairs of vertices form edges, how ordered sets of edges form faces, and how sets of faces form a complete closed solid. Other varieties of solid modelling do not hold this information explicitly and thus require an 'edge evaluation' each time an object is drawn. Thus although they may have benefits in other areas, Constructive Solid Geometry and Occupancy Enumeration tend to be slow in operation. SAMMIE uses a relatively simple form of BRep modelling which limits the shapes to plane faceted convex solids. This is adequate for the objectives of ergonomics evaluation, and allows solid evaluations such as hidden line removal and interference checking without producing unduly complex models which would affect response times.

Primitives

It is possible to define any solid shape by defining the location of vertices and stating the edge and face relationships. However it is usually very much more convenient to use primitive shapes which can be parametrically defined. SAMMIE provides simple input or interactive creation techniques for cuboids, and a range of prisms including regular- and polygon-based, cylindrical, and conical. In most cases these are sufficient to generate a model of adequate accuracy for an ergonomics evaluation. If greater precision or realism is required then it is necessary to resort to the base method.

Hierarchical data structure

The geometric data is entered into a hierarchical data structure, implying that relationships can be established between different entities within the model. Each entity will be owned by an entity at the next highest level, and may itself own any number of subordinate entities. This can be clearly seen in the man model itself where the natural hierarchy of say the arm is reflected in a data structure where the upper arm owns the forearm which in turn owns the hand. The importance of this is that, just as in the human body, when the upper arm is rotated the forearm and hand move appropriately without any intervention by the user. It is of course still possible to move the forearm without affecting upper arm location. This definition of the structure of mechanisms can be usefully extended into the workplace model. For example, see Figure 3.5, a car may be modelled such that the doors function in the expected way.

Geometry editing

Recognizable objects such as desks and chairs in an office layout design consist of a number of primitive shapes arranged in an appropriate spatial relationship. Design and evaluation objectives more normally call for control over the gross dimensions of composite items so that actions such as changing the height, width and depth of a desk can be initiated. In SAMMIE this is known as geometry editing and is always available, requiring no special

Figure 3.10 Geometry editing.

modelling activity on the part of the user. Entities can be scaled (uniformly or differentially along one or more axes), dimensioned, sheared, or reflected. Figure 3.10 shows an example.

Structure editing

It has already been stated that SAMMIE is particularly strong in the representation of the logical or functional relationships between items by the use of a hierarchical data structure. Design and evaluation objectives often require that these relationships be changed to reflect differing requirements, for example during the execution of some task within the workplace. This is known as structure editing and permits re-ordering of the data structure under the interactive control of the user, or in some cases by program control. Any entity (including any subordinate entities) can be assigned a new owner whilst either maintaining its original location and orientation or alternatively assuming some new spatial relationship determined by the new owner. The latter facility allows, for example, a model of a book on a table to be modified so that the book changes ownership to some other table and also moves to a position on the new table that is equivalent to its old position. In general the facility is useful for modelling relationships that change over time. Thus, for example, if an assembly line were being evaluated, incoming components would be 'owned' by a conveyor system and thus move with it, until such time as they were picked up by the man model when they would be owned by the hand. Placing the component in some assembly jig would then result in a further change of ownership.

Viewing

Viewing has already been described as a set of special purpose applications which directly relate to ergonomics evaluations. To provide a usable design tool it is also necessary to give the designer some ability to gain an insight into the three-dimensional nature of the model with which he is working. One of the most effective ways of achieving this is by the provision of extensive and effective viewing capabilities.

SAMMIE stores a full scale representation of the three-dimensional scene and the task of the viewing system is to provide a means of displaying the desired view of this on a two-dimensional screen. The principle of having a centre of interest which is viewed along a line of sight defined by a viewpoint is used. Either or both of these points can be re-positioned in a variety of ways (specification of new coordinates related to the global axis system or to the axis system of an object, by screen interaction, under program control for certain special effects, or by reference to the eyepoint for man's view). The type of projection used would normally be either plane parallel or perspective with mapping onto the limited size of the screen achieved by scaling control in the first case and acceptance angle in the latter.

View scaling for parallel projections is implemented via zooming techniques which allow any area of the screen to be nominated for enlargement to the full screen area. Similar techniques involving uniform movement of both the centre of interest and viewport provide a panning facility.

Acceptance angle control for perspective views relates to the field of view presented on the screen. In normal circumstances the user sees an image of his three-dimensional model in much the same constricted way as a passenger looking out of a ship's porthole, i.e. a true perspective is seen but only for the limited part of the scene that can be seen through the porthole. In the case of the CAD system the screen is the 'porthole' onto the three-dimensional model world and the angle subtended at the user's eye by the screen is the acceptance angle. Some assumptions have to be made about how far from the screen the user sits, but otherwise a mathematically accurate perspective view is presented. On occasion the acceptance angle can be changed to obtain unrealistic but useful perspective projections. Thus increasing the angle has the effect of a wide angle lens and consequent exaggeration of perspective whilst decreasing the angle gives a telephoto lens effect.

Whatever the projection, it is always possible to move the viewpoint and centre of interest relative to each other. Typically the centre of interest would remain fixed and the viewpoint moved around the surface of a sphere defined by the viewing distance. Any oblique view can be obtained in this way as can the especially useful plan, front and side views. The radius of the viewing sphere can also be changed to modify the viewing distance. This obviously has no visible effects on the parallel projections, but in perspective it produces views closer to or further away from the model.

Applications

SAMMIE has been used by the authors in over fifty case study applications during the last 10 years (SAMMIE, 1987). A selection from these are briefly described here to illustrate some of the points made and to demonstrate the flexibility of the system.

Power station control consoles

Figure 3.11 illustrates a project involving the design of a power generation control room layout. Part of the work concerned the design of an optimal control panel profile, given that the control and display surface dimensions had already been fixed. There was a requirement for the 5th to 95th percentile male operators to be able to reach across the panels over a 1-m wide area to push-button controls. An additional requirement was that the task had to be carried out without the need to lean fowards or to rest the hands on the front panel (to avoid inadvertent control operation). Figure 3.11 shows the 95th and 5th percentile overlaid, and the use of SAMMIE's reach contour facility. Contours at $Y=0$ mm (directly in front of the operator) were swept out (a and b) for the 95th and 5th percentiles respectively, giving the extent of normal forward reach. Contours at $Y= \pm 500$ mm (c and d) were swept out for the

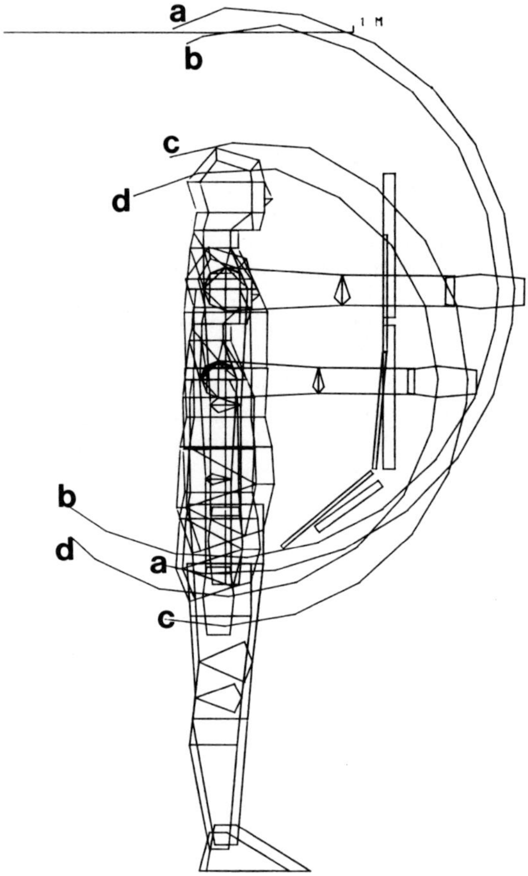

Figure 3.11 Reach contours in control console design.

95th and 5th percentiles respectively, and the fixed size panels were manipulated such that the top edge fell within normal comfortable 5th percentile reach, and the bottom edge fell within 95th percentile reach. Finally the desk front edge was limited to 20–30 mm in front of the operator (again to ensure that inadvertent operation does not occur by clothing etc. making contact with the panel surface).

Car package design

SAMMIE was used to evaluate fit, reach to controls, postural comfort, visibility and mirror design for a concept automobile design. In the particular view shown in Figure 3.12, a small female driver is reversing the vehicle. Of concern was rearwards visibility and how the head restraint and passengers might affect this, together with constraints placed on movement by the design of the seat and reach to controls. It was important in terms of safety and operational feasibility to ensure that smaller users were able to depress the brake and clutch pedals fully so as to remain in control of the vehicle. Thus sufficient forward seat travel is

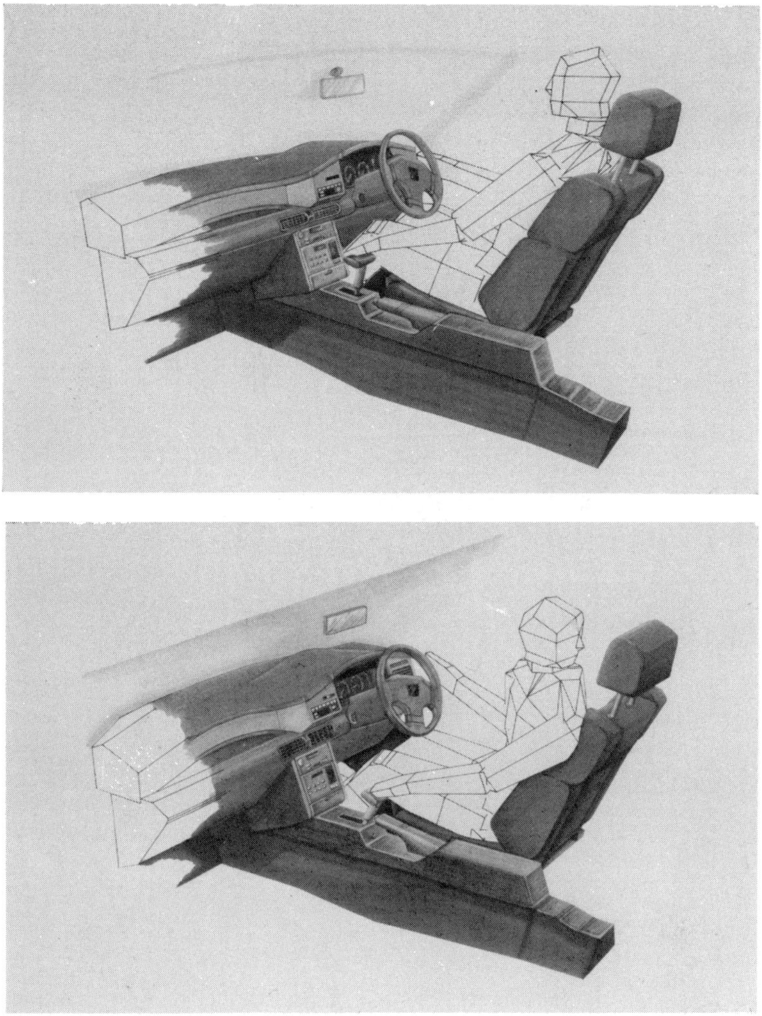

a

b

Figure 3.12 Posture assessment in car design.

essential. Note also that the reversing posture requires the left shoulder to push well into the seat back. This area of the seat should therefore contain soft, compressible padding to allow the driver to turn easily.

Channel Tunnel visualization

Figure 3.13 is taken from part of a visualization study concerning the design of 'RO-RO' double-deck trains for passenger car transport in the cross-channel tunnel (a 25 mile tunnel under the English Channel between Britain and France which is currently under construction). The customer wished to be able to visualize typical views that car drivers would see when approaching and loading their cars on to the trains. Their requirement was

a

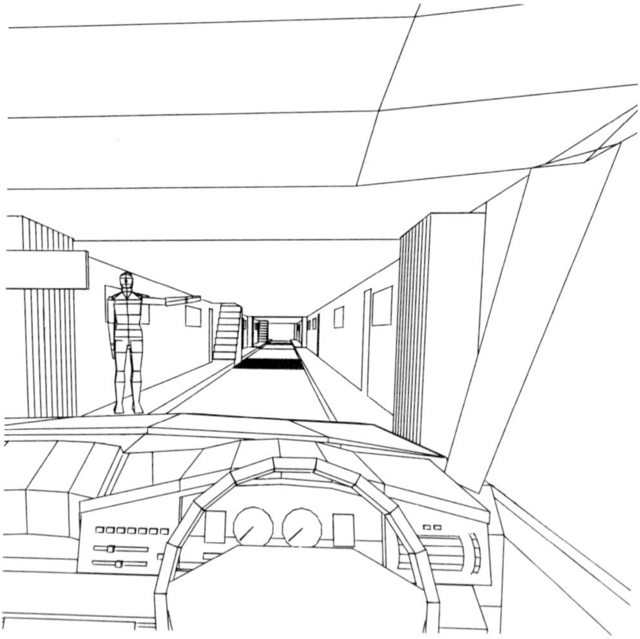

b

Figure 3.13 Visualization studies for the channel tunnel. Driver's view with (a) a preceding car; (b) as the first car to load.

to be able to see how a 1.5 km long train, itself essentially a mobile tunnel, would appear to a range of different sized male and female drivers, and to get some idea of how it might feel to drive into such a vehicle. Figure 3.13a shows a driver's view along the approach to the train's loading bay with a preceding car on the upper deck's loading ramp. Figure 3.13b is a driver's view from the first car to load as he or she is directed through the length of the train on the lower deck.

Financial trading facilities

Two views of a financial dealing room are shown in Figure 3.14. Figure 3.14a shows an *ad hoc* collection of equipment, where standard desks and chairs form a confused and unsuitable

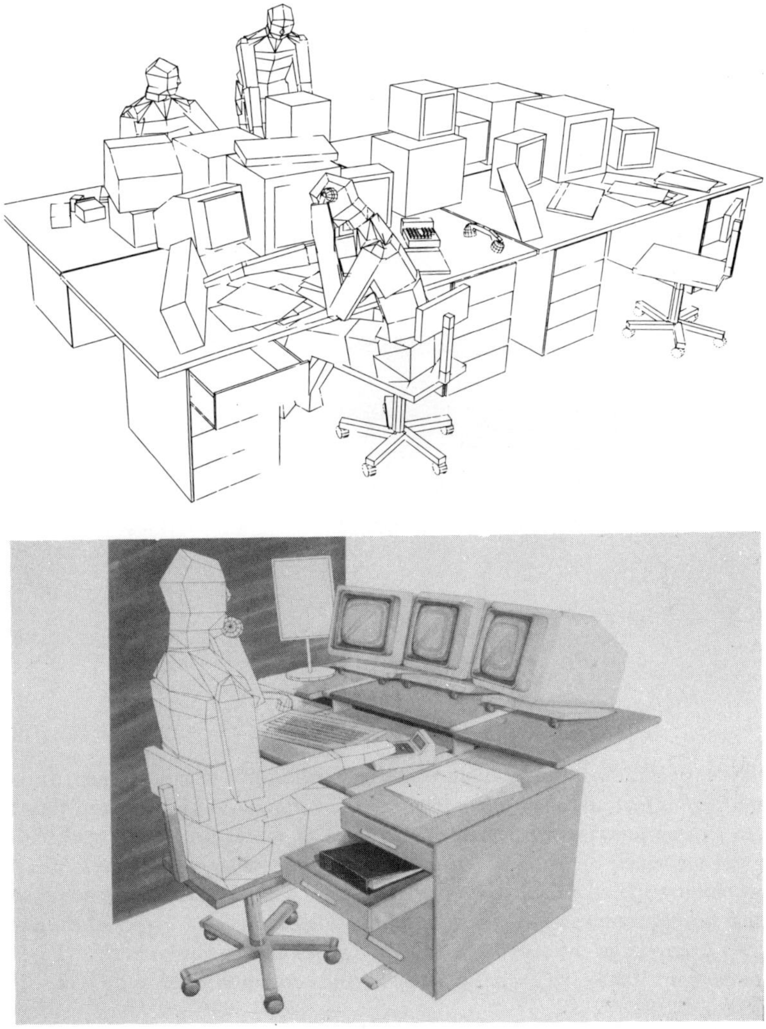

a

b

Figure 3.14 Financial dealing room furniture. (a) Randomly chosen furniture; (b) adaptable furniture.

workplace. A study was conducted to define furniture and equipment incorporating various ranges of adjustability that would suit the full range of male and female users. The main evaluation criteria used were reach, fit and postural comfort. The result, Figure 3.14b, incorporates height adjustable desks, and visual display units, seats and seat backrests all with height and tilt adjustment. Users can adapt the workstation to suit their particular requirements and can also adopt a variety of postures to maintain comfort over long periods. Rendering of the SAMMIE-produced line drawings by an industrial designer adds some styling to aid effective communication and prove that the ergonomics contribution to the design does not have to detract from the aesthetic appearance of the product.

Concept design for helicopters

An ergonomics evaluation was conducted on a concept design for a helicopter. Figure 3.6 demonstrates SAMMIE's volumetric reach capability (available for both hands and feet), where hand reach for a 50th percentile pilot is being assessed. From such assessments control locations can be evaluated and optimum locations assigned on the basis of frequency of use,

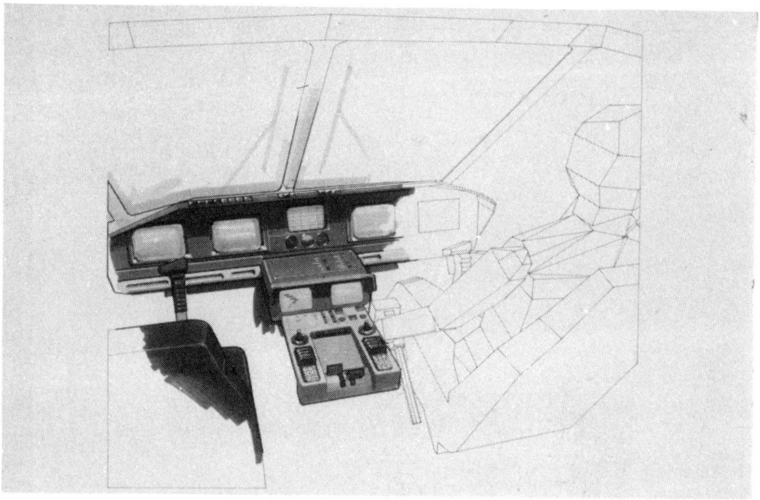

Figure 3.15 Helicopter concept design.

importance, operational sequences etc. These are critical in a situation where the operator is harnessed to his seat and may suffer from extreme gravitational forces and thus cannot be expected to lean forwards or overreach. Figure 3.15 shows the pilot's workstation where assessment was made of visibility, especially fields of view to ground level during a typical range of landing and take-off scenarios. Ground level vision is critical in these situations since the visual field is dramatically limited due to the angled 'nose-up' aspect of the aircraft. The combined strengths of the ergonomist and stylist are clearly shown in this view where the ergonomics contribution to the design is powerfully communicated using three-dimensional graphics.

Figure 3.16 *(a) An underground train with a selection of users. (b) The driver's workstation.*

Underground trains

Figure 3.16a shows a London Underground train with a selection of user types. The principal evaluations carried out were concerned with the driver's workstation which was evaluated for ease of access and for reach to controls. Vision through the windscreen to mirrors or display monitors mounted on the platform was also important to ensure that all drivers could easily view passengers entering and leaving the train without requiring the driver to leave his seat. This criterion was met by allowing the driver to lean forwards by up to 40° and leftwards by up to 20°, and these postures were considered satisfactory when the train is stationary (Figure 3.16b).

Research and development

In addition to being a long-established system which is being commercially exploited, SAMMIE is also actively being researched and developed. This activity falls into two major categories of improving the functionality of the system as a CAD system and fundamental enhancements to the man modelling and ergonomics evaluation aspects.

In the first category there are a large number of developments which simply reflect the general direction in which CAD is progressing. Thus as more capable hardware and wider-ranging software becomes available it is possible, for example, to contemplate the inclusion of colour rendering of images as a practical tool in helping the designer to evaluate proposals. Such developments form a natural part of the evolution of any CAD system and are not discussed here. However, there is one direction being taken by CAD which has the potential of being a major and fundamental influence on the way in which design systems like SAMMIE are used. This is the growing interest in the implementation of standards across the whole range of activity in CAD from low-level graphics to the exchange of product information. The use of standards has many beneficial effects, some of which are felt directly by the end-user of software and some which only have an indirect effect. The objective with SAMMIE is to write the system using standardized computer languages (Fortran77 and C), to take advantage of standard graphics, and to implement on computer systems using the *de facto* standard operating system, Unix.

This has been achieved with the exception of the use of standard graphics. Suitable implementations of the various international standards are only now becoming available and the intention is to convert SAMMIE to use PHIGS (Programmers Hierarchical Interactive Graphics System, Shuey, 1987) for screen graphics and the Computer Graphics Metafile (Mumford and Skall, 1988) for transfer of information to other graphical systems for desktop publishing, colour enhancement, or plotting. The principal purpose behind this is to provide a version of SAMMIE which can operate on a wide variety of computer equipment, and to reduce the support and maintenance overheads. The indirect benefit to the user of SAMMIE is that this will free resources to extend the ergonomic functionality of the system. Of more direct importance to the user is a proposal to use X-Windows as the basis on which the user interaction is built. X-Windows is a standard which provides methods of handling multiple windows particularly on the screen of engineering workstations such as the Sun or Apollo. Its use would ensure that a standard method of interaction would be available whatever computer SAMMIE was running on. The standard of course only provides a mechanism for describing interactions and there is much work to be done in defining suitable interfaces such as menus, messages, and direct graphical interaction.

Perhaps the single most important aspect of international standards however relates to the exchange of data between different CAD systems. Standards such as IGES (Initial Graphics Exchange Specification, Smith, 1983) have existed for some time and provide a means by which systems may communicate geometric information, in the form of two-dimensional, three-dimensional, solid, or surface models together with limited product information. PDES/STEP (Owen and Bloor, 1987) is a more recent standard which extends the scope of IGES particularly in aspects of product and manufacturing information. The importance of these developments is that they provide a means of establishing SAMMIE in the correct place in the design process. SAMMIE is ideally suited to the early pre-concept stages of design where few, if any, of the eventual dimensional and functional constraints have been established. Accuracy to the levels required by CAD systems aimed at producing manufacturing information is not necessary, and a high degree of flexibility and speed of response assume considerable importance. At the end of this early design stage it is necessary to move onwards to other more rigid CAD systems for activities such as engineering analysis and draughting. The intention is to provide an IGES/PDES communication facility so that geometric information produced by SAMMIE can be used as a basis for work with subsequent CAD systems. There is a difficulty in this area in that SAMMIE is frequently used as an evaluative tool rather than a conceptual one. This implies that it will be required to receive (geometric) information from some other system, evaluate and possibly modify it, and return it to the originating system. A mismatch between the two systems will now almost certainly be evident. For example, SAMMIE cannot deal with curved surface information whereas the originating system might well do so. On the other hand it is unusual for other CAD systems to have the capability of describing logical and functional relationships as SAMMIE does, and obviously no general purpose system provides the same kind of man model and anthropometric data. This kind of conflict is by no means limited to this rather special purpose situation, and the hope must be that continuing research work across the world will progressively remove these problems.

Research on man modelling and ergonomics evaluation has centred on two areas with a third due to commence shortly. Research aimed at investigating certain specific aspects of strength modelling experimentally (Tracey *et al.*, 1987), has been complemented by Kennedy (1989) with the new implementation of a range of strength modelling techniques within prototype new versions of SAMMIE. The objective of the work is initially to provide task evaluation methods ranging from the simple recommendations of NIOSH (1981) to the results of more sophisticated two-dimensional and three-dimensional models. This would provide straightforward and useful assessments of the suitability or safety of the task for manual execution. A longer term, and as yet unattempted, objective would be to include strength as one of the factors to be used in the posture prediction algorithm. Thus in appropriate circumstances strength requirements for a posture would be optimized together with the current joint extension minimization to determine a set of joint angles representing a posture.

It has been mentioned that some change in the methods of defining and representing anthropometry are planned for the future. The current method of controlling anthropometry through the linear dimensions of the link model and the volumetric measures from the somatotype classification has provided a simple and effective tool. However it is now possible to contemplate a more precise method of modelling which could also lead to a more detailed geometric representation on the screen. It is hoped that this would permit a far more detailed analysis of problems of fit into the workspace and would perhaps even allow detailed examination of such aspects as the interaction between seats and the man model. At the presentation level it would be possible to produce much more realistic images and thus improve the designer's perception of the model. The precise details of the modelling method are yet to be determined, but it is likely to be based on a large number of external body measurements such as those normally measured during anthropometric surveys. Regression equations would be used to determine the internal link lengths which would remain the main 'driving' mechanism of the model. The capabilities of the anthropometric data base would be extended to cope with the considerably increased amount of data, and methods would be provided for creation and modification of the data base. A potential problem is seen in that the objectives of the change require that the man model be defined by hundreds rather than the present tens of data items. This may overwhelm the designer so it will be essential to provide interpolation and extrapolation techniques so that a full set of data can be derived from a smaller number of supplied items.

An area of current research which spans the general development of CAD systems and the needs of ergonomics evaluation is the provision of intelligence within SAMMIE. To date SAMMIE has been regarded as a design tool which provides assistance to the designer, and although predictive algorithms have been provided in some areas, no genuine intelligence resides within the system. Clearly there are areas which would benefit considerably from techniques originating from artificial intelligence. The simplest approach is to formalize specific design problems into a set of design rules which can then be implemented using expert systems. Hence some work (Azarkadeh, 1988) has been done on devising a 'design shell' for control and instrument layout within automobiles. This is a particularly appropriate area of application as there is a mix of legislative requirement (which varies from country to country), standard or recommended practice, good ergonomics, and subtleties introduced by the needs of marketing and image. KES (1987), a proprietary expert system shell, has been interfaced to SAMMIE and provides a monitoring and advising facility which removes from the designer the need to keep a detailed control over the dimensions which are important to the design rules.

Two major difficulties exist with such an approach. First, it is application specific in that the rule base only knows about control and instrument layout within automobiles, and another application effectively requires reprogramming. The second problem relates to the expert system's source of geometric knowledge. To embed geometric capability within the expert system will merely duplicate the geometric modelling within SAMMIE and will almost certainly be inferior. However, in common with most CAD systems, SAMMIE has difficulty in responding to inquiries about the geometry it represents. Simple questions about dimensions of objects and distances between objects cause no problems, but those requiring a geometric reasoning capability are beyond SAMMIE's current ability. Thus geometric questions such as 'does the steering wheel visually obscure the third quadrant of the speedometer' cannot be answered in any truly general sense. Some work has been done in the general field (Howe and Fothergill, 1988) and some work specific to SAMMIE (Bonney *et al.*, 1986) has been started, but a comprehensive and general implementation of such techniques must be regarded as a long-term aim.

Conclusions

This paper has attempted to demonstrate that SAMMIE and the general field of conceptual design and ergonomics evaluation by man modelling techniques is 'alive and well' after some 20 years of research and development. It seems that acceptability of the approach has arrived in the last 5 years, probably because of the wider acceptance of general CAD systems. Evidence of usage is provided by the considerable number of design consultancies carried out, and the relatively small but growing number of companies which have purchased SAMMIE or similar systems for in-house use. These systems are providing real and valuable solutions to today's problems but it is also possible to see vast areas of application which await appropriate research and development.

Acknowledgements

The authors acknowledge the original funding of research by the Science and Engineering Research Council, the support of the University of Nottingham and Loughborough University of Technology, and exploitation support from the British Technology Group. The SAMMIE system is currently marketed on a range of computer equipment by SAMMIE CAD Limited, Quorn, Loughborough, Leicestershire, UK.

References

Azarkadeh, S., 1988, Integration of Expert Systems with a Geometric Computer Aided Design Environment. MSc Dissertation, Loughborough University of Technology.

Bonney, M. C., Case, K., Porter, J. M. and Levis, J., 1980, Design of mirror systems for commercial vehicles. *Applied Ergonomics*, **11** (4), 199–206.

Bonney, M. C., Taylor, N. K. and Case, K., 1986, Using CAD and expert systems for human workplace design. Proceedings of the UK Conference on Geometric Reasoning. In *Geometric Reasoning*, IBM, Winchester, UK.

BSI (British Standards Institution), 1987, Reach volumes for location of controls on agricultural tractors and machinery. BS 6375.

Craig, J. J., 1986, *Introduction to Robotics Mechanics and Control*. (Reading, MA: Addison–Wesley).

Dempster, W. T., 1955, *Space Requirements of the Seated Operator*. WADC Technical Report 55–159 (Ohio: Wright Patterson Air Force Base).

Dempster, W. T., Sherr, L. A. and Priest, J. G., 1964, Conversion scales for estimating humeral and femeral lengths and the lengths of functional segments in the limbs of American caucasoid males. *Human Biology*, **36** (3), 246–262.

Dooley, M., 1982, Anthropometric modeling programs – A survey. *IEEE Computer Graphics and Applications*, **2** (9), 17–25.

Freidlich, T., 1987, Unpublished Research Report. Department of Production Engineering and Production Management, University of Nottingham.

Hertzberg, H. T. E., 1960, Dynamic anthropometry of working postures. *Human Factors*, **2** (3), 147–155.

Howe, J. A. M. and Fothergill, A. P., 1988, Solid modelling, sensors and the effects of uncertainty on the programming of robots. *Proceedings of the Application of Computers in Manufacturing Engineering Conference*, University of Nottingham.

Kennedy, S. J., 1989, Computer Models of Human Strength. Unpublished PhD Thesis, Loughborough University of Technology.

Knowledge Engineering System (KES), 1987, *Knowledge Base Author's Manual* (A&E Limited).

Korein, J. U., 1984, A Geometric Investigation of Reach. An ACM Distinguished dissertation, MIT Press.

MIL, 1970, Aircrew Station Vision Requirements for Military Aircraft. Standard 850B.

Mumford, A. and Skall, M. (eds), 1988, *CGM in the Real World*. (Berlin: Springer).

NIOSH (National Institute for Occupational Safety and Health), 1981, *A Work Practices Guide for Manual Lifting*. Technical Report 81–122 (Cincinnati, OH: US Dept. of Health and Human Services).

Owen, J. and Bloor, M. S., 1987, Neutral formats for product data exchange: the current situation. *Computer Aided Design*, **19** (8), 436–443.

Pheasant, S., 1986, *Bodyspace: Anthropometry, Ergonomics, and Design*. (London: Taylor & Francis).

Porter, J. M., Case, K. and Bonney, M. C., 1982, A computer generated three dimensional visibility chart. In *Human Factors in Transport Research*, volume 1, edited by D. J. Oborne and J. A. Levis (New York: Academic Press) pp. 365–374.

Requicha, A. A. G., 1980, Representations for rigid solids: theory, methods and systems. *Computing Surveys*, **12** (4), 437–464.

Rothwell, P. L. and Hickey, D. T., 1986, Three-dimensional Computer Models of Man, *Proceedings of the Human Factors Society 30th Annual Meeting*, Dayton, Ohio, pp. 216–220.

Sheldon, W. H., 1940, *The Varieties of Human Physique*. (New York: Harper & Bros).

Shuey, D., 1987, PHIGS: a graphics platform for CAD application and development. *Computer Aided Design*, **19** (8), 410–417.

Smith, B. M., 1983, IGES: a key to CAD/CAM systems integration. *IEEE Computer Graphics Applications*, **3** (8), 78–83.

Tracey, M., Haslegrave, C. M. and Corlett, E. N., 1987, Automating the measurement and biomechanical analysis of posture. In *New Methods in Applied Ergonomics* edited by J. R. Wilson, E. N. Corlett and I. Manenica (London: Taylor & Francis).

Chapter 4

WERNER: a personal computer implementation of an extensible anthropometric workplace design tool

WILHELM BERNHARD KLOKE

Institut für Arbeitsphysiologie an der Universität Dortmund
Abteilung Ergonomie
Ardeystraße 67
D–4600 Dortmund 1, Federal Republic of Germany

Abstract. This paper presents a computer-aided design tool which uses personal computer technology to support graphic and computational assessment of ergonomic problems.

Keywords: Interactive graphics; Man model; Workplace design.

Introduction

Almost from the beginning of computer supported graphics, the idea of using the computer to design the layout of human workplaces has been very attractive. In the military and high-tech domains, as well as occasionally in the academic domain, there have been several important attempts to give graphic supported simulation of man in a man–machine system. A survey of some of them is found in Dooley (1982). In some existing systems the question of their inner ergonomics seems to have been secondary to the considerations of technical usability of the produced output.

Some special problems found in this application of computer-aided design (CAD) technology are: (i) the graphic modelling of man; (ii) the fuzziness in the size and shape of the man model; (iii) the interactive use of a man model in the design process; (iv) the non-definitive, game-like character of the work done; (v) the influence of the time-dependency of the work process; (vi) the openness to future ergonomic assessment methods; (vii) the need for an external representation language.

The not yet satisfactory solution of these problems, as well as limitations of computer hardware on account of power and price in the past, precluded the actual introduction of existing workplace modelling programs into the toolbox of workplace design. However, sufficiently powerful computer workstations are now in widespread use, and the use of workplace simulation tools at the planning stage of everyday workplaces is going to be possible. At the Institut für Arbeitsphysiologie an der Universität Dortmund (Institute of Occupational Health at the University of Dortmund) a program named 'Werner' is being developed as a tool for workplace design on small computers. The solutions it provides for some of the problems will be discussed in this article.

Considerations regarding graphic representation

The decision about which graphic representation of the man model to choose depends on the solution or non-solution of some of the other problems; the choice will heavily influence the performance of the program and thus the limitations in interactive use. The sense of space provided by the model is also of great importance, since there is little point in trying to perform workplace design two-dimensionally. The possible methods can be classified according to whether they use a volume- or surface-oriented approach, whether they visualize by showing a transparent cage of straight lines, surface modelling splines, surface shadowing or colouring, or showing a linear outline of the body. The latter is done in conventional template methods such as that found in the German national standard DIN 43116 (1985), and in WERNER as well (Figure 4.1). Figure 4.2 shows the man model used in a context of ergonomic assessment of an assembly workplace.

Advantages of the chosen method are (i) equally optimal resolution on all devices, either screen or hardcopy; (ii) no overloading of the pictures; (iii) limited calculation effort; (iv) minimal aliasing effect caused by irrelevant details; (v) relatively clear function oriented body image; (vi) utilization of a simple and concise external representation language; (vii) straightforward extension of a conventional ergonomic method. Some of these are

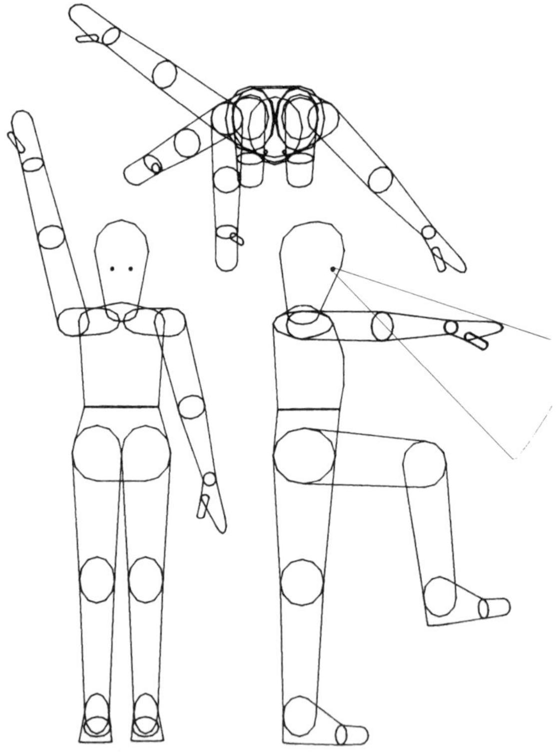

Figure 4.1 The man model WERNER in positions taken from a widely used drawing template conforming to the German National Standard DIN 43116.

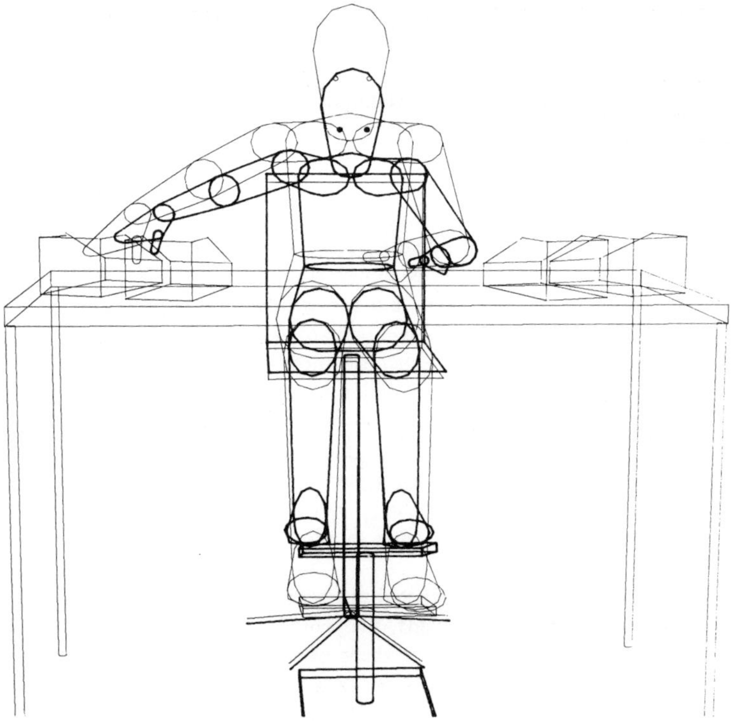

Figure 4.2 Ergonomic assessment of anthropometrics. Note the different height of the footrest, and the different position of the right shoulder in the two models.

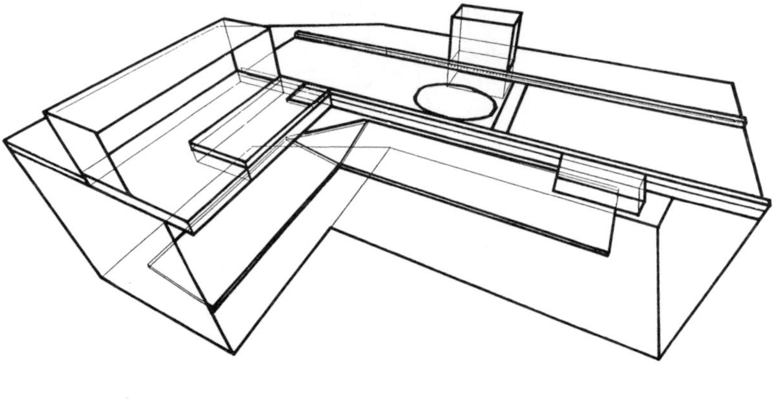

Figure 4.3 Drawing of a cash-desk workplace, designed to show how certain ergonomic proposals concerning the position of the scanning device, the cash, footrest and others could be realized together, with manual enhancements.

considered essential to the eventual usefulness of computer-aided workplace design tools and will be explained further.

In WERNER, spatial feeling is given to the user by three different, combinable features. First, all graphics is done by central projection in order to support discrimination of far and near parts of most geometrically regular objects, especially of tables and boxes (Figures 4.3 and 4.4). Second, the recalculations of not too heavily loaded scenarios can be done fast enough to give a feeling of a directly manipulatable object if the model is turned using a trackball or similar device. Third, stereoscopic presentation is possible if red and green coloured eye-glasses are used to separate the pictures for the left and the right eye. The loss of the colour distinction of the objects is almost compensated for by the more transparent configuration of lines to spatial objects, but the calculation effort and time is doubled.

A common practice in all known human body models is to take the model of a branching chain of rigid body segments connected by more or less constrained pin-joints. This approximation can be accepted for almost all joints except the shoulder where the joint has no fixed location in the trunk segment. One possible solution, at the cost of a varying shape trunk, is to use an additional clavicular segment. However, there are more reasons for providing varying shapes for the body segments, such as deformations from outside, e.g. the deformation of the thigh on a seat, or the range of different shapes caused by body type or pregnancy or physical handicap. Such cases should not be excluded in the consideration of workplace design.

Since WERNER's shape is constructed differently from most other models, especially those based on standard CAD programs, the principle of its construction requires further explanation. Using a rigid skeleton, the shape of each segment is defined by some simple solids, mostly ellipsoids, located in fixed or changeable positions in this or its neighbouring

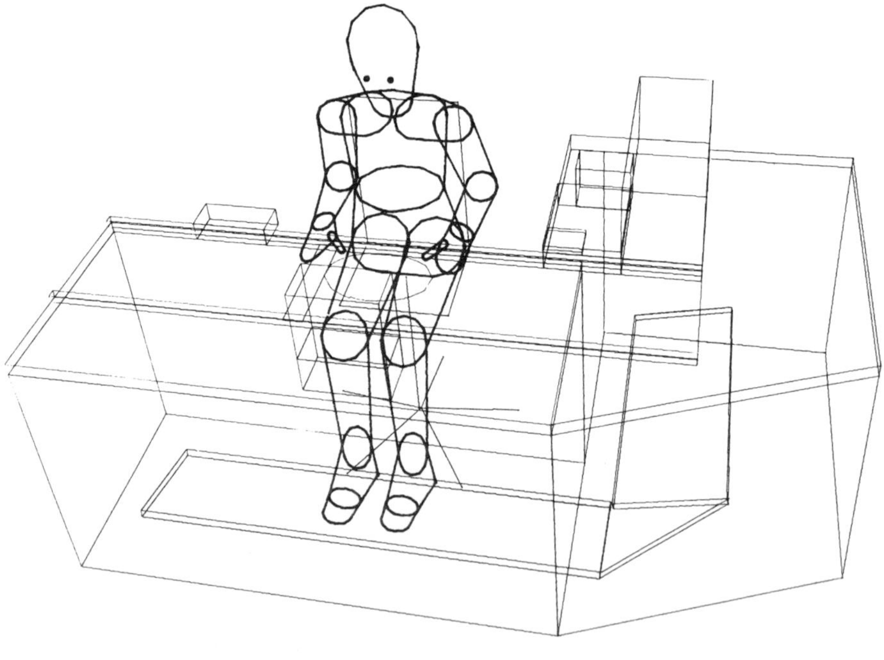

Figure 4.4 A 5th percentile female manikin positioned in the proposed workplace from Figure 4.3.

segments. A convex hull, whose outline is shown in the final projections, is constructed over these solids. This construction avoids defining irrelevant edges resulting from a fixed multi-faceted approximation of the body surface, and also the gaps needed in some volume-oriented models to allow for bending of the body and movement of the arms and legs. Deformations, either for the body shape or as a result of external forces, can be defined in a quite natural way, as long as the restriction of convexity is not violated, just by adding or changing some ellipsoids. Allowing the use of solids in segments other than their own will also adjust the shape of a segment smoothly when needed, as in the case of rotation of the forearm.

User interaction in WERNER

The main possibilities built into the program are (Figure 4.5): (i) selection of parts for further work; (ii) selection of data files for insertion into a scene at a selected position; (iii) rotation of a joint; (iv) definition of a range of allowed rotation; (v) movement of parts, and possibly changes to their shape; (vi) determination of a position from a file of joint positions; (vii) saving the scene, a part of it, or the joint positions, in a data file; (viii) saving the graphics for two-dimensional work with other programs.

They can be best used in an interactive environment by means of devices such as a trackball for two combined angles, and position wheels for separate angles, in combination with a mouse and a keyboard for the menu and dialogues. Additionally, a pure window-oriented interface with simulated controls in a window and an interactive editing window, which can

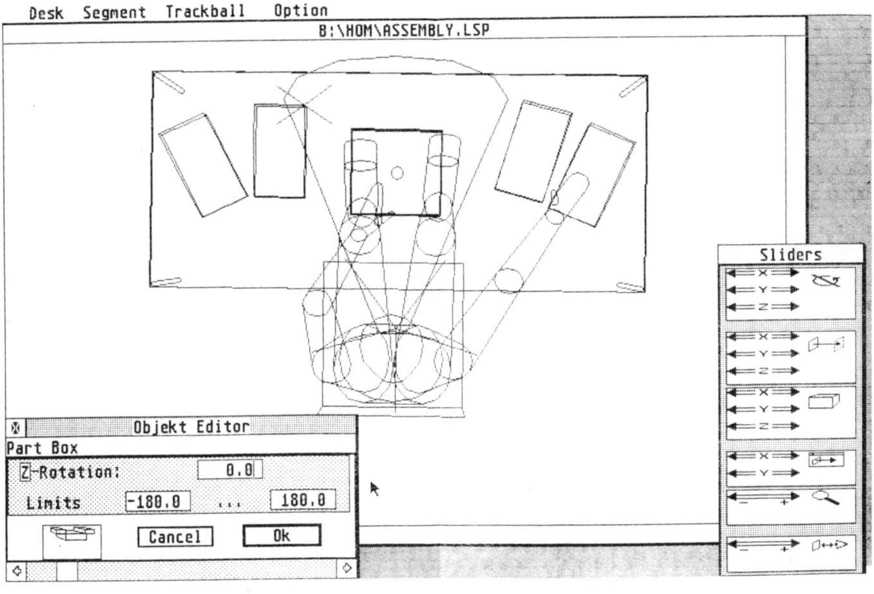

Figure 4.5 The window-oriented interaction instruments in WERNER: interactive knobs for changing angles and positions of segments, viewing point and scaling of the scene. Numerical treatment of selected picture elements, here of the angle of vertical rotation of the second part box from the left, can be done using an interactive editing window.

be used either alternatively or intermixed with the direct controls, is implemented. Purely data-oriented working on the external language data files using a standard text editor is possible and sometimes recommendable.

A very simple approach is used to define the approximate joint positions which are needed in most contexts. From a catalogue of partially determined postures, such as grasp locations for one arm (Figure 4.6), positions can be selected by name, adjusted for the actual context

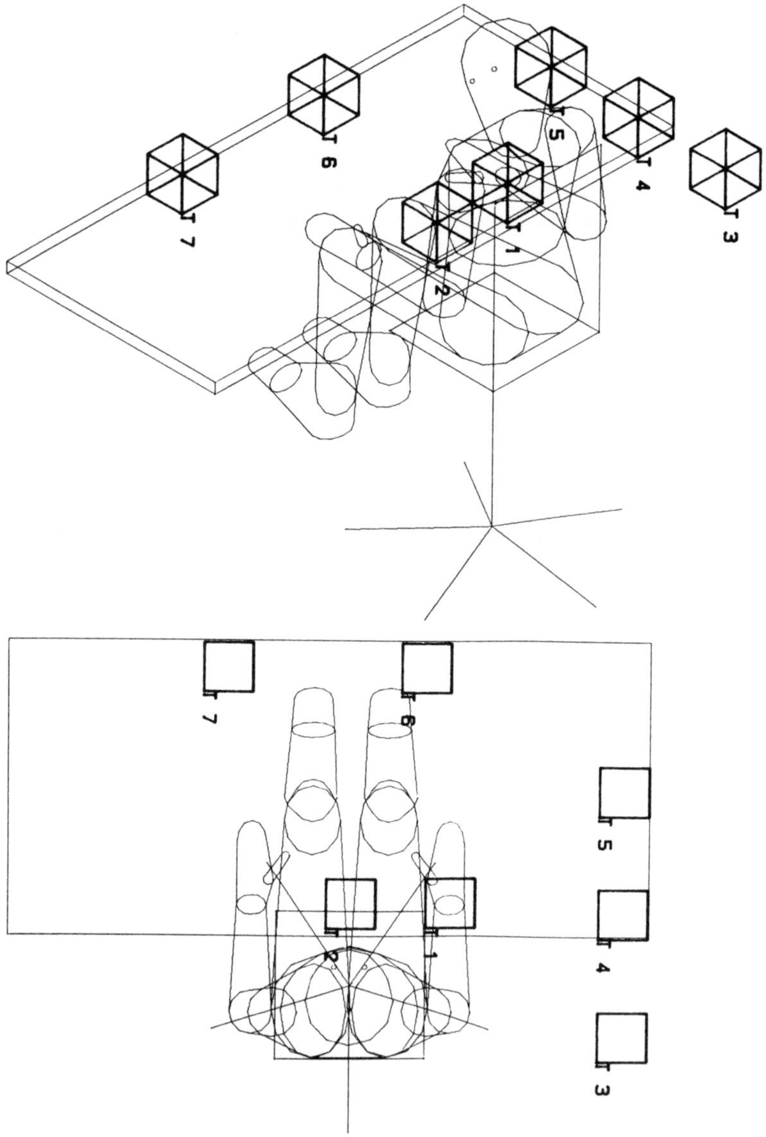

Figure 4.6 Distribution of grasp positions T1–T7 for the right arm on the work table.

Figure 4.7 Four phases of a jump, marking the turning points for interpolation to animation.

and further combined to simulate the working man. WERNER contains an option for showing these actions animated in almost real time by doing recalculations in advance. The four positions in Figure 4.7 have been used in the animation of jumping simply by means of linear interpolation of the angles from one subposition to another.

Using knowledge-base technology, even more abstract transformations will be feasible in the future. Using the elements of a predetermined time system such as Measurement Time Methods (MTM) for classification and process description, and the more precisely specified locations of body positions and movements in WERNER, both analyses can be carried out in one step. For example, the MTM-element 'Reach' uses a parameter 'length' and a classification of motion precision. The former can be easily calculated using a program. The latter, however, should be explicitly settable to account for modelling errors, while still allowing support of determination by some characteristics of the geometric context as well.

The external representation language

WERNER's external representation language uses a Lisp-like format and can, in fact, be considered as an extension to the Lisp computer language like AutoCAD's AutoLisp. As a result, the model user can easily create or adapt a wide range of model variations which are in compliance with the internal data handling, out of a toolbox of simple parts and simple constructions. For example, the structure of the biomechanical model by Jäger (1987) for the calculation of mechanical stress on the lumbar spine has been successfully translated to WERNER, and the calculations of this model will be optionally included. This can be done in spite of the fact that Jäger employs a mixed coordinate approach, using floor relative coordinates for most segments and a segment relative coordinate system for the five segments of the lumbar spine whereas WERNER's standard approach is always to use segment relative coordinates. Therefore, WERNER is a tool suitable for even providing a visual

```
( segment
  ( name "leg" )
  ( translation 0 -26 -372 )
  ( c-hull ( ellipsoid 39 43 61 ) )
  ( mass 2.25 )
  ( translation 0 26 0 )
  ( x-rotation -89 -135 -45 )
  ( y-rotation 0 -12 30 )
  ( z-rotation 0 -30 60 )
  ( segment
    ( name "foot" )
    ( mass 0.7 )
    ( translation 0 -65 -109 )
    ( c-hull ( ellipsoid 44 30 16 ) )
    ( segment
      ( name "toes" )
      ( translation 0 0 -65 )
      ( c-hull ( ellipsoid 39 26 22 ) )
    )
  )
  ( segment
    ( name "heel" )
    ( translation 4 -93 34 )
    ( c-hull ( ellipsoid 53 0 35 ) )
  )
)
```

Figure 4.8 Part of an external definition file for a right leg. As appropriate, data are given in millimetres, kilograms or degrees.

representation of approaches in the field of work physiology, which did not consider the special demands of body drawing in their design.

Figure 4.8 shows a part of the man model description which defines the right leg. The description is built up in the form of lists starting with descriptive keywords from the elements 'translation', 'x-rotation' etc. which define coordinate system transformations from one body segment to the next, thus giving the data structure for a rigid body skeleton. Each segment has geometrical and mechanical properties defined by elements such as 'mass' and 'ellipsoid'. The keyword 'c-hull', as abbreviation of convex hull, defines the elements over which the outline is to be drawn. By means of the attribute 'name' both the selection of certain segments in interactive work and the cooperation of shape and position descriptions in separate data files are possible. The flexibility of the construction method allows easy modification of any data structure, for instance to give full details of fingers. Of course, there is a price for this in computing time and storage needs. However, there are no arbitrary limits on the number of workplace elements, number of segments of the man model and even on the number of man models used. In all cases the data files will remain small to fit on diskette storage.

Another function of the external representation language is cooperation and data exchange with other programs. Practically, the most important example is communication with AutoCAD. Although WERNER itself is a sort of CAD program there is a natural division in the work done. The demands on interactive usability imply that objects in WERNER have to be quite simple and sketch-like, whereas the main strength of the construction

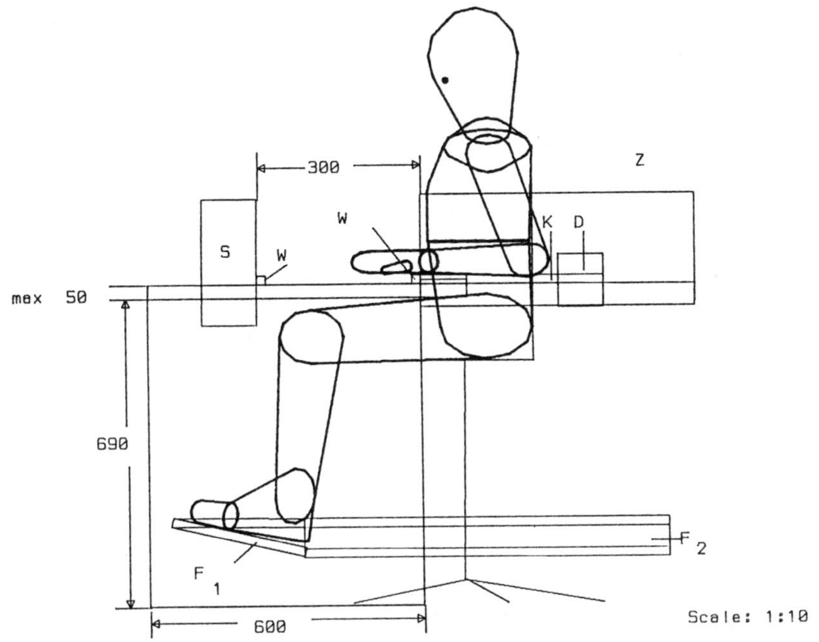

Figure 4.9 The workplace from Figure 4.3 shown in parallel projection, originally scaled 1:10, comments added using a standard drawing program. Characters have been used to reference details in an accompanying text.

program is in adding the details and providing the technical drawings. An example of work done with a cooperating program is given in Figure 4.9. Here the standard GEMDraw program has been used to add illustrating comments to the raw picture from WERNER. The workplace shown was added as an example to a report on ergonomic demands for scanner equipped cash-desks for West Germany to prove their realizability. The demands include vertical positioning of the scanner, minimal manual load handling, a uniform direction of work flow independent of the need to use the keyboard, adaptability to a wide range of body sizes as well as compact size and good visibility range.

As a first step towards the integration of anthropometric and biomechanical approaches, the most obvious mechanical calculations are done in the current version of WERNER. Each segment may have a mass located at a fixed point. External forces may also be applied. The system is brought to equilibrium by adding a resulting force at the root of the internal segment tree. The geometric constraints and forces propagated in the links of the segment chain need to be compensated for by torques in the joints in order to achieve equilibrium. These torques, which have to be built up by muscle forces specific to the joint, are an indicator of muscular workload resulting from posture and static external forces. WERNER aids this kind of assessment by calculating and visualizing these data. Figure 4.10a shows the torques calculated as vectors starting in the respective joint and normal to the plane of force and lever, the length indicating the magnitude, in Figure 4.10b the respective forces, all vectors are in vertical direction, because only weight forces have been considered.

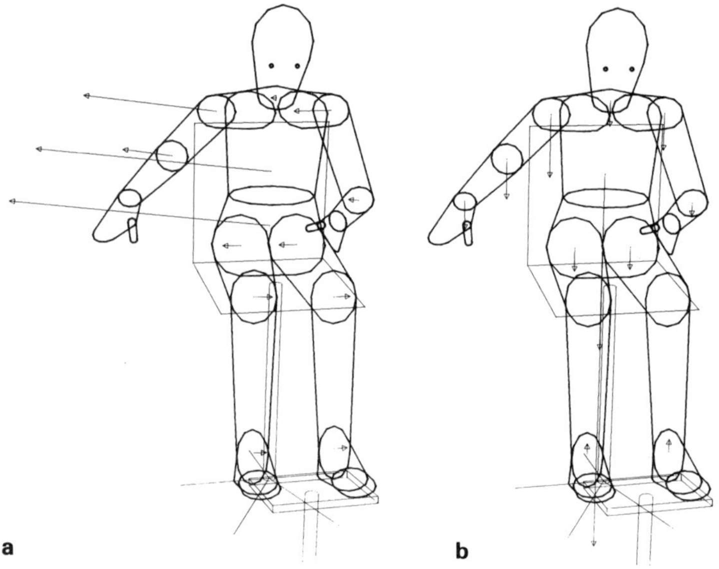

Figure 4.10 Forces and torques calculated for a sitting position with external forces acting from the footrest against the feet, shown as vectors in the respective joints.

Remarks on implementation details

WERNER is implemented on an Atari ST personal computer and uses its native window manager GEM for interaction handling, whereas the three-dimensional data processing is done by internal routines. Demonstration versions with limited usability exist for the IBM PC/AT and for the microVAX/GPX under Ultrix with XWindows V11. On the ST there is no need for a hard disk or memory above the most basic version of 512 kByte. All GEM-supported graphics devices can be used, and there is no restriction to Atari supplied resolutions. In fact, at the author's system, a simple modification of the hardware is used, which gives 50% better resolution (768 × 536 pixels in black and white, 768 × 268 or 384 × 268 pixels in 4 and 16 colours, respectively). On larger screens, however, support by graphic processors such as the GPX is strongly advisable.

The implementation language of WERNER is C, a widely accepted standard for high-performance portable programs. It should also be noted that the program sources will be available to the public as soon as the main data structures have been settled. The work done with this program can therefore be used to support further research, especially in biomechanics, where extension from the static to the dynamic analysis should be a desirable goal.

Discussion

In a workshop on ergonomic models, Kroemer *et al.* (1988) discussed requirements for an integrated model of the human body and gave recommendations for further research. They emphasized the need for standardization. Successful standardization, however, is not only a matter of convention, but is also a result of broad discussion and public availability, as is known from examples from the history of computing (Tanenbaum, 1987). It is therefore thought that approaches like WERNER which employ personal computers will contribute most to the efforts to integrate a human body model. Also, some of the ideas involved in WERNER itself, mainly the external representation language and the strategy of breaking down the model into simple parts and constructions, may be a step towards a unified and standardizable treatment of human body models.

References

DIN 43116, 1985, Zeichnerische Darstellung der menschlichen Gestalt in typischen Arbeitshaltungen (Berlin: Beuth Verlag).

Dooley, M., 1982, Anthropometric modeling programs – a survey. *IEEE Computer Graphics and Applications*, **2** (9), 17–25.

Jäger, M., 1987, *Biomechanisches Modell des Menschen zur Analyse und Beurteilung der Belastung der Wirbelsäule bei der Handhabung von Lasten* (Düsseldorf: VDI Verlag).

Kloke, W. B., 1987, Entwicklung eines rechnergestützten Werkzeugs für ergonomische Gestaltung. *Zeitschrift für Arbeitswissenschaft*, **41**, 153–156.

Kroemer, K. H. E., Snook, S. H., Meadows, S. K. and Deutsch, S., 1988, *Ergonomic Models of Anthropometry, Human Biomechanics, and Operator–equipment Interfaces* (Washington, DC: National Academy Press).

Tanenbaum, A. S., 1987, *Operating Systems: Design and Implementation* (Englewood Cliffs, NJ: Prentice-Hall).

Chapter 5
Man models in the ergonomic design of workplaces with the microcomputer

M. LAUNIS and J. LEHTELÄ

Department of Physiology,
Institute of Occupational Health,
Haartmaninkatul,
SF-00290 Helsinki 29,
Finland

Abstract. Two man model systems were developed for the computer-aided design of workplaces, i.e. two-dimensional ergoSHAPE to supplement a widely used micro-CAD program, and three-dimensional ergoSPACE as an independent program. The systems are based on the anthropometry and biomechanics of the human body, and also include schematic recommendation charts. The study was carried out in three enterprises to investigate how the systems could be integrated to the present design practice of the workplaces, how the man models suited the different design tasks, and how the different groups were able to use the man models. The methods consisted of interviews, the subjects' assessments of the technical drawings of workplaces, design tasks, and determinations of workplace dimensions. The study revealed that workplaces were poorly documented, being restricted to the layout sketches. The arrangements of the work and workplace were left mainly to the supervisors. In the design of highly mechanized workplaces the documentation was complete, but the physical human work was not experienced as an important or critical design factor. The man models were of great help to the occupational health and safety personnel in assessing the drawings. The planners expected simple, schematic recommendations but, on the other hand, were interested in the most natural and detailed man models. The simplified three-dimensional system is not yet useful for designing workplaces, due to its overreduced unnatural and confusing visual appearance. In general, the man model is a strong reference for workplace measurements, leading to a similar result within different occupational groups.

Keywords: Man models; Anthropometry; Biomechanics; Workplace design; Computer-aided design.

Introduction

The difficulty of conveying ergonomic knowledge, even when it is trivial, to the planning of workplaces has been widely experienced. Constant haste in the design process, constantly changing specifications and demands for the design, and the technical complexity of modern workplaces create a situation in which ergonomic considerations are simply disregarded by industrial engineers. In such a situation, the available ergonomic literature usually only creates confusion by its heterogeneity. The engineers evidently lack coherent and appropriate ergonomic knowledge, or it is in the wrong form to be used in routine design work. The following criteria can be recommended for better ergonomic design tools (Launis *et al.*, 1988):

1. They should be easy and fast to use, at exactly the right phase of designing.
2. They should be universally applicable to different problems and new situations.
3. They should be capable of tackling many design goals and tasks simultaneously, e.g. analysis, synthesis, documentation and evaluation.
4. They should be illustrative in order to facilitate cooperation between different experts and occupational groups.
5. They should have a clear and rational basis in order to be applicable to difficult or critical situations.
6. Special training should not be required of the user but the tools should be utilizable for expert consultations, too.

As a universal ergonomic design tool, two-dimensional anthropometric manikins and drawing templates have been introduced. But this equipment has never been taken into wide design usage, probably because of practical problems caused by different body sizes and various drawing scales. On the other hand, the sophisticated man model programs on the market so far have been too expensive to be widely used in design practice.

Man model systems

To solve the dilemma mentioned previously, two easy-to-use microcomputer-based man model systems have been developed at the Institute of Occupational Health in Finland. The

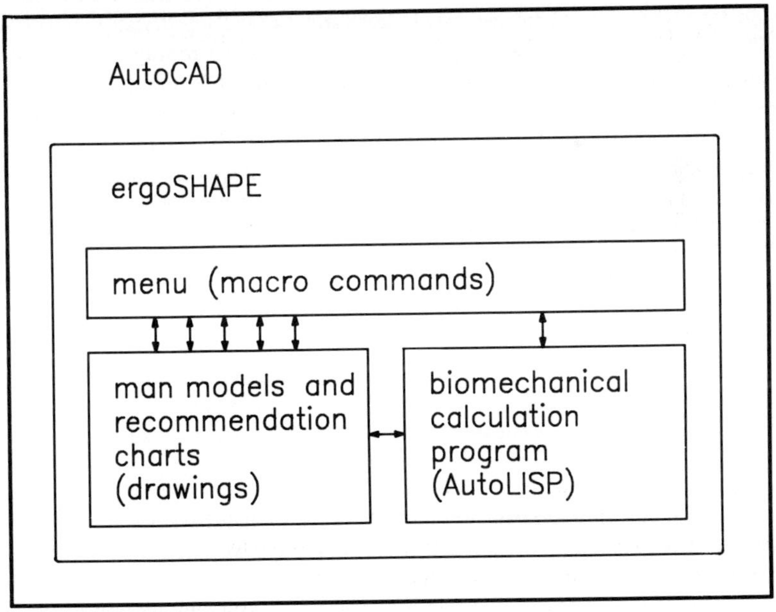

Figure 5.1 Block diagram of the ergoSHAPE system.

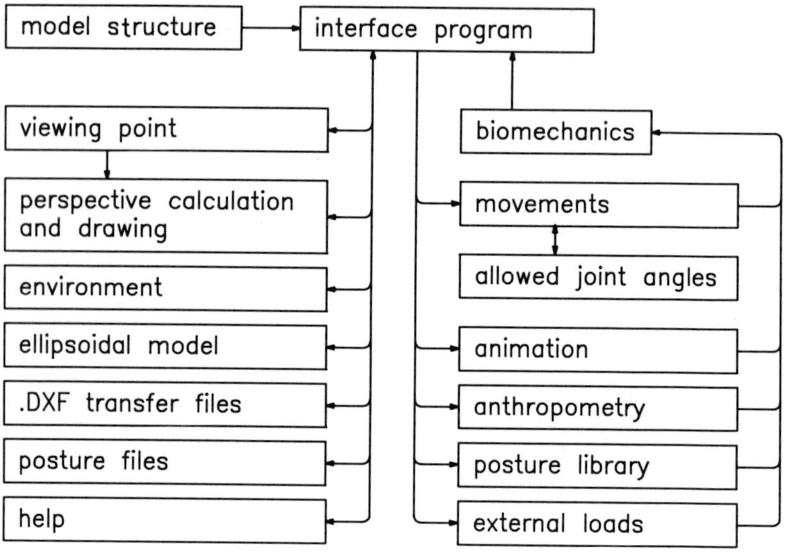

Figure 5.2 Block diagram of the ergoSPACE system.

first one is the two-dimensional ergoSHAPE, which is constructed inside the AutoCAD* design program, as drawing files, menu files and AutoLISP program files. The development of a two-dimensional man model system is supported by practical arguments. Two-dimensional design is easy to perceive, easy to learn and use, and appropriate in normal design work. The ergoSHAPE system is intended to be the engineer's routine working tool.

The other man model system is three-dimensional ergoSPACE, which is an independent microcomputer program programmed by Turbo–Pascal. The system was developed to comply with the restrictions of microcomputers and therefore the graphic presentation of the man models and the environment is greatly simplified. The main aim was to study the possibilities of a three-dimensional design program in designing workplaces.

Functional block diagrams of the systems are presented in Figures 5.1 and 5.2. Both systems have the same anthropometrical and biomechanical basis.

ErgoSHAPE system

The ergoSHAPE system consists of three parts: (i) anthropometric man models, with the help of which working space can be fitted to human dimensions; (ii) biomechanical calculations, which enable the evaluation of postural stress in manual materials handling or in static postures; (iii) recommendation charts, which give direct design guidelines to particular work situations and workplaces.

One hundred and eighty different versions of human models have been constructed. The reason for the great number of models is to assure that the most suitable model can be found for all design purposes. The models combine the following characteristics (Figure 5.3):

*AutoCAD is a registered trademark of Autodesk, Inc.

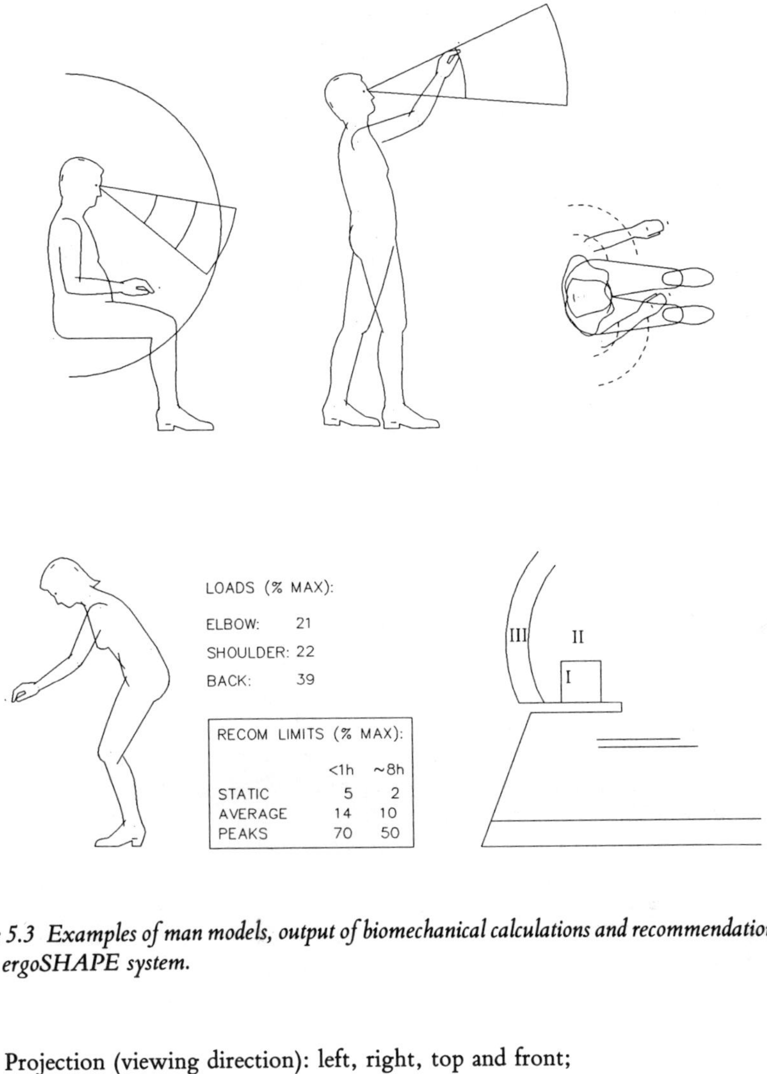

Figure 5.3 Examples of man models, output of biomechanical calculations and recommendation charts of the ergoSHAPE system.

1. Projection (viewing direction): left, right, top and front;
2. Complexity (movability): fixed one-part models in typical working postures, models with six separate parts (trunk, head, two arm and leg segments), or nine parts (lower and upper trunk, head, three arm and leg segments);
3. Basic posture: sitting and standing including some minor anthropometric differences;
4. Sex: male and female;
5. Anthropometric size: small, average and large (5th, 50th and 95th percentiles), and a user-specified individual size.

The models can be supplied with curves indicating the viewing angles and distances and the reach zones. The models and other data can be manipulated by built-in macro commands which are displayed on the screen or on the digitizing tablet. Thus the technical use of the ergoSHAPE system does not require any special training.

The anthropometric data are based on Finnish, North European and North American populations (e.g. NASA, 1978; Heliövaara and Aromaa, 1980; DIN 33402, 1986). The man models can, however, be easily linearly scaled to other populations.

The evaluation of postural stress is based on mechanical modelling of the human body (e.g. Chaffin and Andersson, 1984; Jonsson, 1984). In the actual posture with the determined external load, the biomechanical program calculates the stress as a percentage of the maximal static muscle strength. The percentage values can be referred to the displayed recommendations for various situations or time ranges. Biomechanical calculation at present takes into account only vertical loads.

The ergoSHAPE system is created with the CAD program and therefore it can be easily expanded by institutions and single users of the system. Special recommendations, such as layout and dimensioning charts for different purposes can be added to the system. In this respect the system differs considerably from earlier computerized man models, which are rather closed programs for the user's own applications.

ErgoSPACE system

The ergoSPACE system is a completely three-dimensional program, in which the environment and human models can be viewed from any direction in real perspective (Figure 5.4) or, simultaneously in three standard projections and in a dimetric projection each in one quarter of the screen (Figure 5.5). To get a fast enough response, a stick model is used for

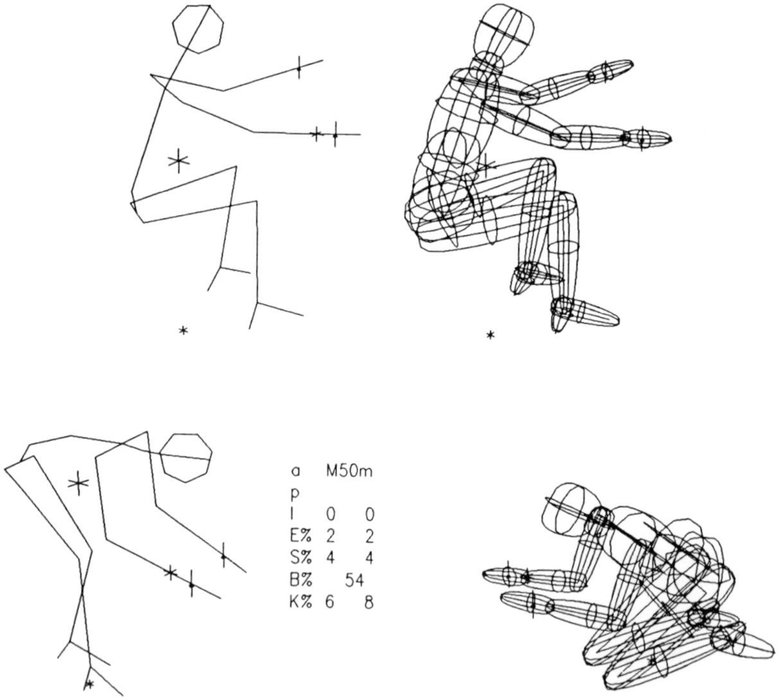

Figure 5.4 Examples of man models and output of biomechanical calculations of the ergoSPACE system.

moving the model. The model can be 'filled' to form an ellipsoidal wire model for evaluating space requirements.

ErgoSPACE models have 17 movable joints. A joint can be moved by selecting the joint from the menu, and turning the joint in small steps with cursor keys. Different postures can be chosen from the posture library, too. The postures of the model can be saved in a file and shown as an animation series.

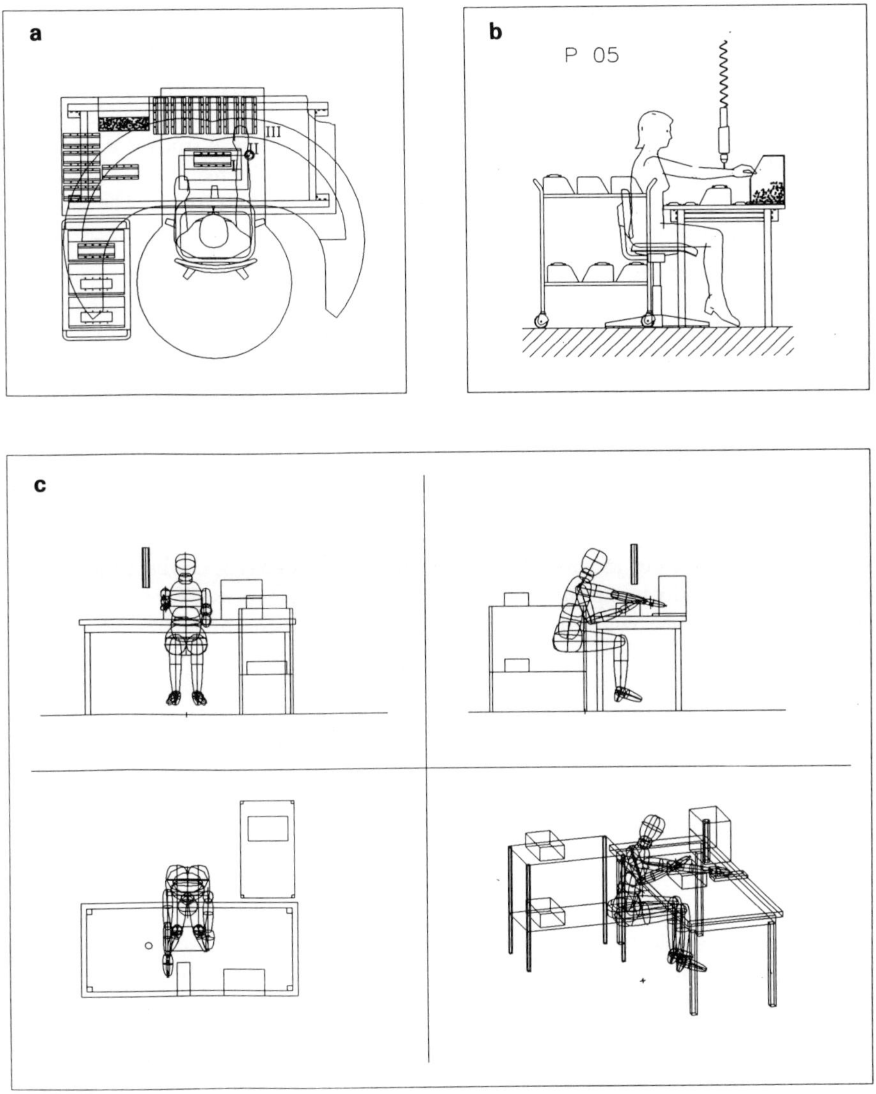

Figure 5.5 Examples of the assessed workplace plans: (a) recommendation chart in top projection; (b) 9-part model in side projection; and (c) quadrant screen of the three-dimensional system. These were also the most preferred models.

The environment is structured as simplified wire models, i.e. as lines, squares and boxes. Hidden lines cannot be removed. Environmental data can be taken into the system also as a DXF-transfer file, and drawings can be saved in the same format. By using these features, the independent ergoSPACE program can be used with a CAD program, too.

Study of utilization

Aim of the study

The development project was intended not only to ensure the ergonomic considerations among the industrial engineers and other planners, but also to promote cooperation between different experts, and to create preconditions for participative design. A study was carried out in three enterprises to evaluate the preconditions for utilizing the man model systems for these purposes, and to gather detailed information for developing the systems. The aim of the study was to investigate:

1. The present design practice of the workplaces (e.g. how the plans are documented and how ergonomics is taken into consideration) and the role of different occupational groups in the design process;
2. How the different types of man models suit the different design tasks;
3. How the different occupational groups are able to use the man models and to assess the plans of the workplaces with them; and
4. How strong a reference the man model is for the dimensioning of a workplace.

Material and methods

The study was carried out in three different metal-product industries, where the so-called just-in-time (JIT) production principle had been adopted. CAD was used for the designing of workplaces only in one enterprise. All those occupational groups which influence the designing of the workplace in one way or another, i.e. industrial engineers, supervisors, occupational health and safety personnel and workers, were represented in the study.

The study was divided into four phases in which the following methods were applied:

Phase I

Interviews of all subjects about the present design practice.

Phase II

Assessment of the technical drawings with all subjects. The technical drawings were supplied with different kinds of man models and recommendation charts. The subjects were asked to generate improvements to the plans and to evaluate the usefulness of the models for this purpose. Some examples of the assessed drawings are shown in Figure 5.5.

Phase III

Design tasks with the computer. The industrial engineers applied the different man models and recommendation charts to their own specific design problem. The subjects were asked to evaluate the models, i.e. the (i) ease of technical use; (ii) reliability in assessing the measurements; (iii) functionality in the intended purpose; and (iv) experienced applicability to their own work.

Phase IV

Experiments to determine the optimal and acceptable workplace dimensions for a particular task with a man model. All subjects participated. One of the experiments concerned working height for a lifting task. The subjects chose the optimal and acceptable alternatives from four comparative series of drawings and photographs, where the man model or an actual man lifted an object from either near or far (Figure 5.6). The series of varying working heights

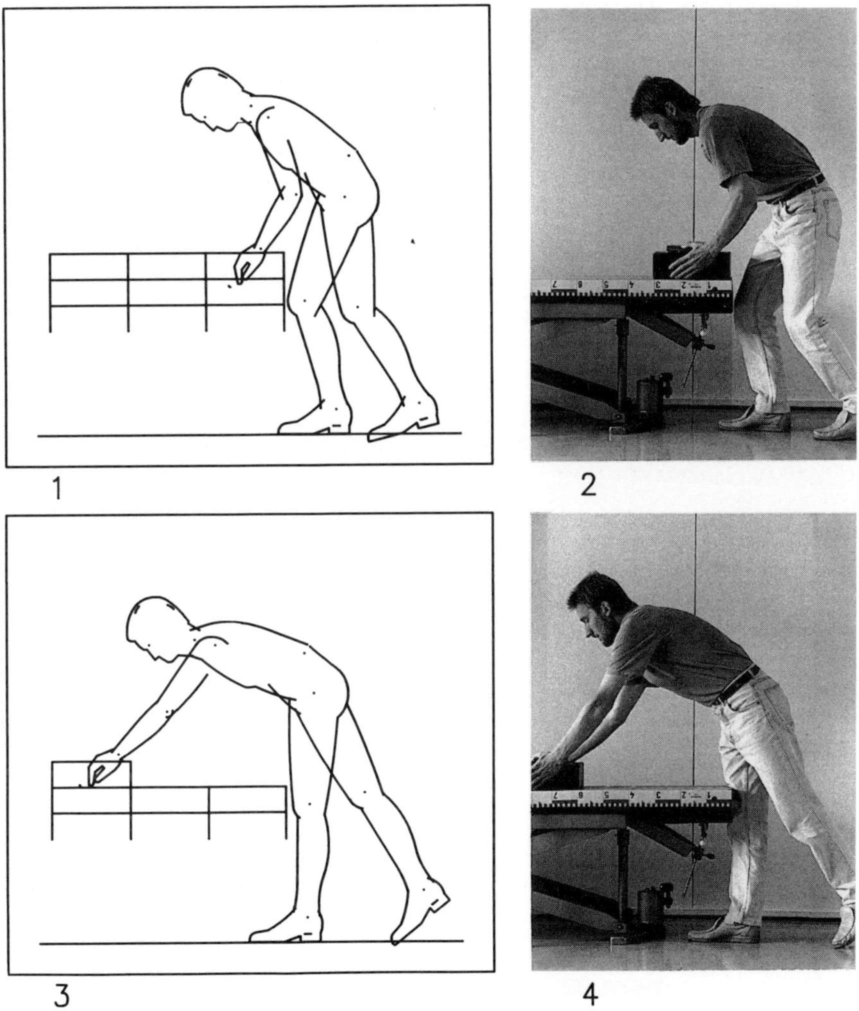

Figure 5.6 Examples of the picture series of the lifting task used in the determination of workplace dimensioning: ergoSHAPE CAD models and photographs of an actual man, lifting from near and far.

(from 10 to 120 cm, each step 10 cm), were presented to the subjects with a slide projector. The task was defined in such a way that it was not practical to adjust the hydraulic table to the optimal height at all times. The subjects applied their own comfort criteria to the determination of the minimal and maximal acceptable lifting heights.

As a reference, a larger group of engineering students participated in the experiments.

Results

Phase I

The interviews revealed that the work task and workplace were not generally well perceived in the design phase. The workplace plans were poorly documented, being restricted to occasional layout sketches. Detailed plans of a workplace were rare, normally the only documentation was some space reserved in a general layout plan. The workplace was a result of separate solutions and detailed plans concerning production technology and material handling. The arrangements of the work and workplace were left mainly to the supervisors.

In the case of designing a highly mechanized workplace, the documentation was complete (including side view projection), but the physical human work was not experienced as an important or critical design factor. The occupational health and safety personnel did not generally participate in the design phase, except in cases known to be problematic.

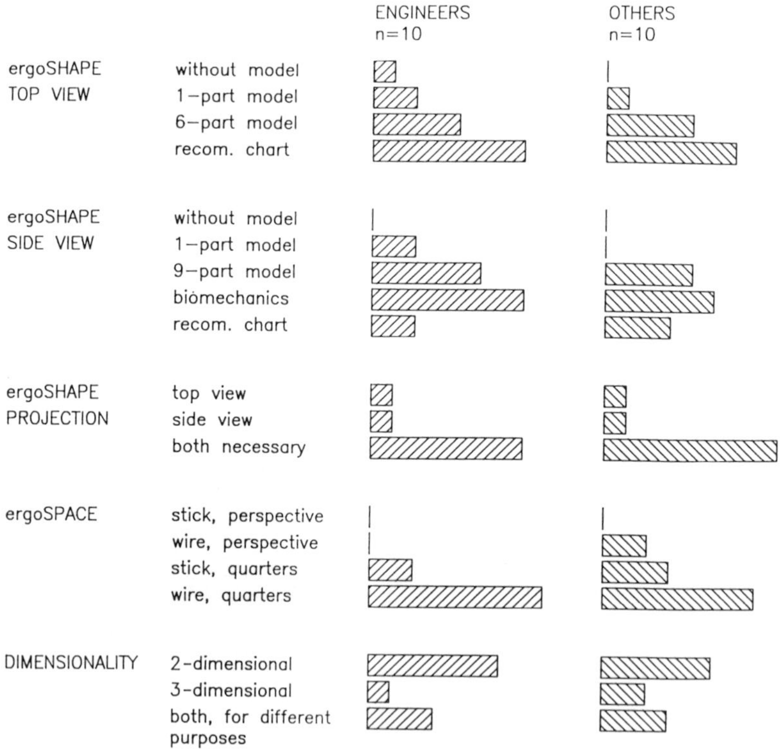

Figure 5.7 Preferences of models according to assessments of the workplace plans.

Phase II

The man models were of great help to the occupational health and safety personnel in the assessment of the technical drawings. The engineers and supervisors were able to find rough ergonomic deficiencies in the drawings even without the man models, which were mainly useful for the fine adjustment of the workplace dimensions.

The preferences concerning the different kinds of man models and different projections were quite similar among the industrial engineers and other occupational groups (Figure 5.7). In the two-dimensional top-view projection of the ergoSHAPE system, the recommendation charts were preferred to the man models, whereas in side-view projection of the nine-part models (including the model for biomechanical calculations) were preferred. Most of the subjects refused to give a priority to top and side views, although in practice they used only top-view plans. They considered both views necessary in ergonomic assessments.

In the three-dimensional ergoSPACE system the simultaneous four projections were found necessary, and ellipsoidal presentation was preferred to stick model presentation. The majority of the subjects found the three-dimensional design confusing, whereas some of them considered it the best way to get an overall picture of the work situation. Generally, the detailed two-dimensional plans were preferred for accurate ergonomic assessments.

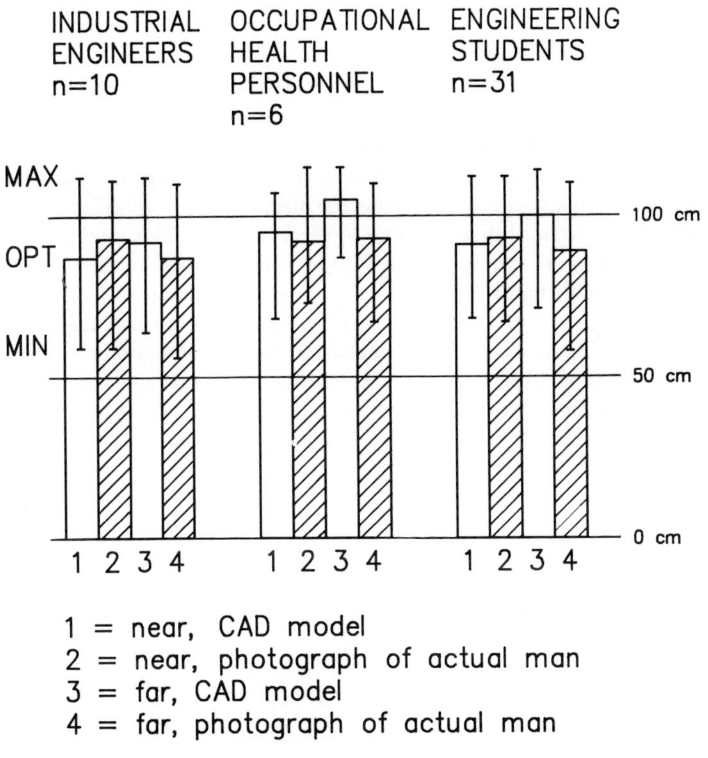

Figure 5.8 Optimal, minimal and maximal heights in lifting work in the determination of workplace dimensioning: means of determinations of three occupational groups in the four experiment series (see Figure 5.6).

Phase III

The results of the design tasks supported the above observations. The dichotomy of preferences was strengthened; the schematic recommendation charts were well accepted but, on the other hand, the most complicated and thus most natural nine-part model and biomechanical calculation aroused great interest. The three-dimensional system was not considered suitable for design purposes in its present form.

Phase IV

In the determination of the optimal and acceptable working height with the man model, only slight differences were found between the different occupational groups (Figure 5.8). The differences between the four different series of pictures within the occupational groups (see also Figure 5.6) were more obvious. The results of the industrial engineers were the most even in all respects, whereas the results of the occupational health personnel were the most uneven (their difference between experiments 1 and 3 was statistically significant, $p = 0.012$, Student's t-test). Ideally, according to the task description, the results should be the same in the different experiments.

Conclusions

The utilization of man model systems is based on graphic design documentation. Although the study comprised only a few cases, it can be concluded that in the enterprises there is considerable discrepancy between the actual design documentation of the workplaces and the ergonomic man model presentation. The design practice in enterprises concentrates mainly on horizontal action (material flow), whereas the main interest from the ergonomic point of view lies often in the vertical dimensioning of the workplace. This implies that the design documentation must be at a more advanced stage before the man models can be utilized.

Because the industrial engineers are generally not very interested in the detailed designing of workplaces, it is obvious that the main area of application of the man model systems is still limited to the designing of workplace products, e.g. big machines, cabins and furniture. On the other hand, there is a trend to use ready-made components and construction entities. Workplace design in enterprises is already to a great extent, and will be more so in future, the buying of machines and equipment and installing them without detailed technical drawings.

The designing of work and workplaces is holistic by nature, and therefore any separate ergonomic system does not motivate the planners. The integration of ergonomic knowledge to the whole of design data, which will be inside the CAD system in future, is therefore essential. The fitting of technical elements to each other and to the worker is the most complicated phase in designing, and it takes place in 'one single head', that of an industrial engineer. Expert cooperation with, for example, occupational health personnel, and also user participation, is natural before and after this phase. The man models in workplace documents are experienced as very useful in facilitating the feedback to the planner.

The man model is a very strong reference for workplace measurements, leading to a quite similar result within different occupational groups. Basic education and general experience have little influence on the ability to assess plans ergonomically and to determine dimensions with the man model. Familiarity with the actual working situation and the work process is far more important. The natural appearance of the man models tends to be more important than the small differences in the ease of their technical use. To cover all design situations, both the most versatile man models and the schematic recommendation charts are needed in the ergonomic design system, most preferably so that they can easily be combined.

The simplified three-dimensional system is obviously not useful for designing workplaces, partly because it is a separate system with no connections to existing ways of documenting the workplaces, and partly because of its overreduced, unnatural and confusing visual appearance. A more detailed modelling of the environment would probably require too much work input, especially if the design problem is not expected to be critical, and the three-dimensional designing is not felt to be necessary.

Although the use of human models releases designers from following rigid recommendations, it presumes a greater understanding of human functions. The biomechanical calculations can be considered as a means to teach the planner a more ergonomic way of thinking, and to activate him to take the human being into better consideration in the design stage.

References

Chaffin, D. B. and Andersson, G., 1984, *Occupational Biomechanics* (New York: John Wiley).

DIN 33402, 1986, Teil 2: Körpermasse des Menschen. Werte.

Heliövaara, M. and Aromaa, A., 1980, Suomalaisten aikuisten pituus, paino ja lihavuus. *Kansaneläkelaitoksen julkaisuja* ML:19, Helsinki.

Jonsson, B., 1984, Rörelseorganens funktionella anatomi och biomekanik. Arbetarskyddsstyrelsen, *Utbildning* **12**, Solna.

Launis, M., Lehtelä, J. and Kuusisto, A., 1988, Kaksiulotteinen antropometrinen ihmismallijärjestelmä työpaikkojen tietokoneavusteiseen suunnitteluun (CAD). *Työ ja ihminen* **1**, 68–83.

NASA, 1978, *Anthropometric Source Book* vol. I, edited by Webb Associates. Reference Publication 1024. (Washington DC: National Aeronautics and Space Administration, Scientific and Technical Information Office).

Chapter 6
Anthropometric data for a driver's workplace design in the AutoCAD system

JERZY GROBELNY
Institute for Production Engineering and Management,
Technical University of Wroclaw,
50–370 Wroclaw, Poland

Abstract. This chapter presents a simple method of implementation for anthropometric data using microcomputer-aided design system AutoCAD. The anthropometric data, including proper differentiations in the population of adult males, are represented in the form of flat manikin elements in the AutoCAD graphical database. The elaborate 'menu-based' control program makes it possible to build easily defined manikins in the required positions during the design process. The present proposals are illustrated by excerpts from a truck driver's workplace design study.

Keywords: Anthropometry; Computer-aided design.

Introduction

The idea of computer-aided ergonomics is fairly recent (Gigure *et al.*, 1986) although computer models assisting design from the viewpoint of ergonomics were already being built in the 1960s (Roebuck *et al.*, 1975). The rapid development of microelectronics has allowed widespread dissemination of earlier concepts and methods as well as the preparation and spread of the new instruments of computer-aided ergonomics. The usefulness and even the necessity of taking advantage of these developments may be confirmed by two observations. First, it is a fact (given by P. Kern and W. Bauer, personal communication), that currently only about 1% of workplace designers make use of computer models; second, there is a widespread opinion that the use of such models increases the quality of designs. This opinion was documented by Evans and Chaffin (1986).

In this chapter a simple design-assisting program of a driver's workplace is proposed. The program was built into the AutoCAD system of the Autodesk firm (as the 'overlay'). The following questions are discussed: (i) the location of the presented problems in the process of workplace design; (ii) the method of representation of the population anthropometric data; (iii) the installation of graphical models in the AutoCAD system; and (iv) the construction of a supervisory program. A simple example using this program in a real design problem is shown.

A workplace design process

Figure 6.1 shows a simplified diagram of logical time interdependencies of the basic components of the workplace design process. The arrows determine a logical sequence of particular stages. Possibilities of recurrences were marked in the case of unsatisfactory joint estimation of the performed design operations. Basic problems being solved in each stage are pointed out in the form of blocks. The location of two problems discussed in this chapter are marked by asterisks; first, models for representation of the population anthropometric data, and second, a technique of presenting these data in the AutoCAD system.

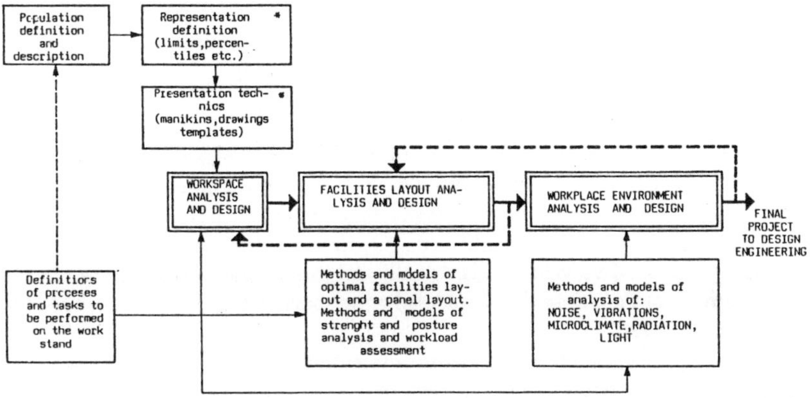

Figure 6.1 Simplified diagram of basic activities (double lines) and methods of the work-stand design process. Locations of discussed methods are marked by asterisks.

A method of population anthropometric data representation

A basic task of the presented program is to assist the design and analysis of a workspace for a truck driver. From the viewpoint of engineering anthropometry this kind of a workplace is a complex operator stand. The complexity results from the simultaneous influence of multidimensional factors on the quality and correctness of the spatial relations. In this case a traditional 'percentile' approach for anthropometric data representation turns out to be unsatisfactory. Bittner *et al.* (1987) proposed the method of representing the population by means of 17 manikins with parameters estimated from population data by factor analysis. These manikins represent a dimensional differentiation taking account of, for example, different length proportions for 'tall' individuals, different weights of individuals having similar height, and so on. In the approach proposed here a similar idea is applied which was described in detail by Grobelny and Cysewski (1989).

Since a man is the 'normal' user of the driver workplace, the features of the adult male Polish population are used as the starting point for finding a suitable representation of the anthropometric data. The following stages facilitate the determination of representatives of the adult male population (in Poland):

1. On the basis of the correlation coefficients table of anthropometric quantities for males (Roebuck *et al.*, 1975), sets of three anthropometric data are fixed, using heuristic methods. These data are markedly correlated with other quantities but not with one another.
2. The sets of data chosen in the previous stage are validated by a multi-dimensional model of linear regression for each set. In each of these models the set of three variables is a system of independent variables. The basis for the estimation of quality of each model is a mean multiple correlation coefficient. The highest one was obtained for the following set of independent variables: weight, height, and a sitting height (about 0·8).
3. Assuming a normal distribution of the variables: weight, height, sitting height and standardizing these variables (0–1 covariance matrix) one may find a radius of the sphere inside which, in the system of independent variables, a given percentage of population measurements will be included (95% of the population is assumed here). Appropriate data are taken from the study by Zenkteller *et al.*, 1983).
4. On the surface of this sphere evenly distributed points are determined which represent the differentiation of the measurements included inside.
5. In the process of variable destandardization, the points situated in a real system on the ellipsoid are obtained which 'correspond' to the following differentiations: big height, big weight, big sitting height; big height, big weight, small sitting height; big height, small weight, big sitting height, etc. Obviously, there are exactly eight possible combinations for two value levels: 'small' and 'big'. Additionally, as the ninth case, a central point of the ellipsoid is considered which corresponds to the mean size of each of the three features.
6. While substituting each of nine (three-dimensional) points to the model obtained in stage 2 one may receive nine different sets of anthropometric data representing the population (of adult males in Poland).

Algorithms for calculations and proceedings in every stage are programmed and presented in detail in Grobelny and Cysewski (1989). Having the data concerning correlation coefficients of the anthropometric quantities, means and standard deviations of particular quantities in a given population, one may find more or less complex representations of this population based on the number of chosen independent variables.

Anthropometric data in the AutoCAD system

The basic anthropometric data obtained by the previously described method are entered into the AutoCAD system in the form of respective elements of the flat manikins.

AutoCAD belongs to small, professional CAD systems and is predestined for personal computers (Raker and Rice, 1985; Voisinet, 1986). Since it also belongs to the most widespread systems of the CAD type, its supplement with ergonomic data and procedures makes possible the propagation of these data among wide groups of designers.

The process of formation of the anthropometric data subsystem may be divided into two stages (Grobelny, 1988): Stage 1 – formation of a graphical data base, i.e. a set of respective drawings of all human body elements which may be of interest to the designer in the AutoCAD system; Stage 2 – elaborating the way of using this base.

In Stage 1 the drawings of three projections of body elements are made for the nine population representatives obtained by the previously described method. Each element is drawn by means of a digitizer and is stored as a block. The axes of each element are marked by

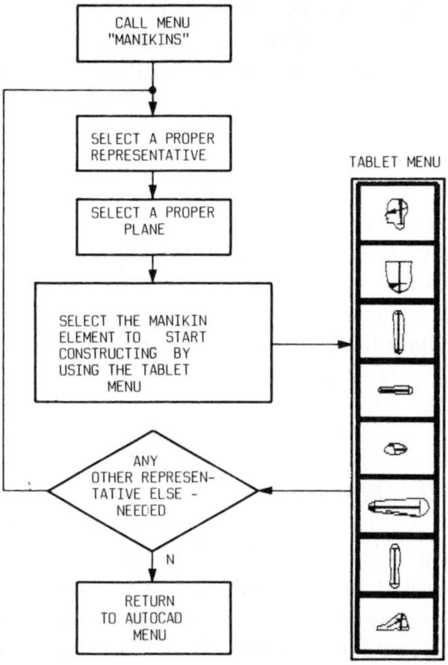

Figure 6.2 Simplified flowchart of the anthropometric data base control program and the tablet-menu scheme.

thickened lines to facilitate the use of the drawn elements. The areas of optimal comfort angles for drivers are also specified as well as angles of vision according to Rebiffe, using the data presented by McCormick (1976).

In the whole concept of the presented program the possibilities of the AutoCAD system are utilized which consist in calling the blocks stored on a disk and inserting them into the drawing. The insertion consists in dragging an element onto a particular place on the screen (or determining this place from the keyboard by means of coordinates) and rotating it, in relation to the stored insertion point, to the required position.

Having the drawings of all elements of the human body the manikins (or their parts) can be built on the screen in any place and mutual spatial relations. However, 'manual' calling of each element (by means of the INSERT instruction) is ineffective. Moreover, it requires the designer to remember the names of inserted elements. In Stage 2 a program based on the AutoCAD menu is constructed which considerably improves the use of the anthropometric data. Construction using this program is based on an idea presented in detail in Grobelny (1988), and a simplified flowchart of this program is shown in Figure 6.2.

In the first step (by means of navigational key buttons on the keyboard) a representative of the population, characterized by three quantities (height, weight and a sitting height), is chosen. Each of the above variables may assume the two values: B – big and S – small as described earlier. While the ninth representative is characterized by the 'mean' value (M) of the quantities.

In the second step the respective plane of the drawing is chosen by means of the navigational key buttons (sagittal, frontal or transversal).

Virtual work begins in the third step using only the digitizer. In any part of the tablet a graphical diagram (tablet-menu), presented in Figure 6.2, should be pasted. By calling (once at the beginning of the work) the TABLET command from the AutoCAD it is possible to position the diagram by means of the digitizer pointer. Activating any box of the diagram (through setting the indicator and pushing the key button 'enter'), initiates the respective instruction of the program. Each box starts the generation of a sequence of manikin's elements from the part drawn in it.

Each element box corresponds to a different type or design problem or different 'starting' point in the design of workplaces. The activation of the first element box allows the building of manikins from a determined eye position. The last element box corresponds to a situation where a foot position is determined; the second element box corresponds to a situation where the seat position is determined. The tablet-menu diagram itself is also a drawing of the AutoCAD and may be generated in any scale according to the size of the tablet and the required active area on the tablet.

Design problem: a case study

The first design problem undertaken with the described system was a search for the essential spatial relations in the cabin of newly-designed truck (of the Polish Star firm). Restrictive assumptions concerned the preservation of the shape and size of the floor and the location of the clutch and accelerator pedals. The manufacturer anxious to propose regulation ranges of particular elements (driver's seat, control panels) which would make it possible for the majority of potential drivers (95% of the population) to maintain a comfortable position. Figures 6.3–6.8 present fragments of tests on the regulation ranges of the driver's seat and control panels.

The stability assumption for the shape and size of the existing floor and the control pedals determined the use of the last box of the tablet-menu (construction of manikins from the feet to the head) in the presented fragments of tests.

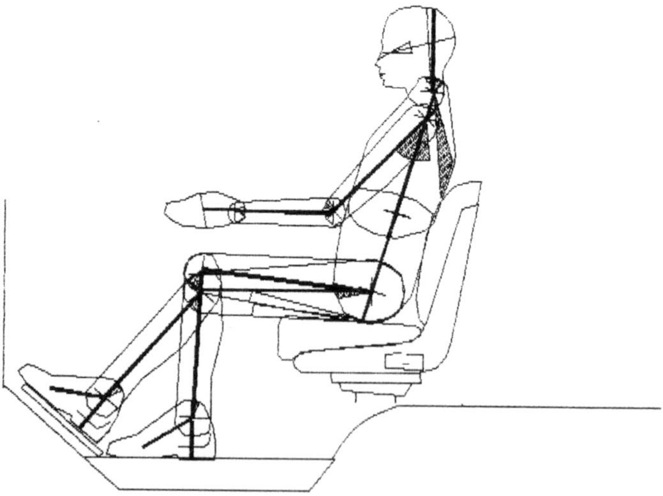

Figure 6.3 Manikin 2 in the basic position for the seat accommodation test.

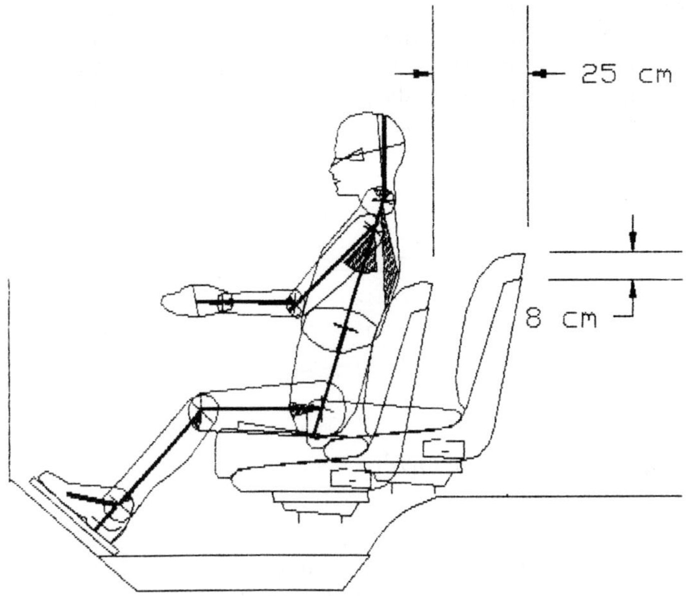

Figure 6.4 Manikin 7 in the basic sitting position and accommodation ranges between individuals 7 and 2.

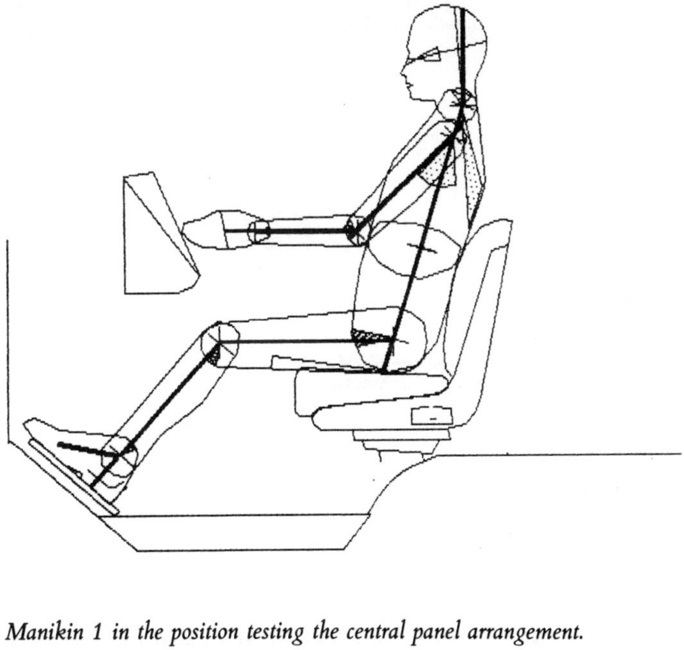

Figure 6.5 Manikin 1 in the position testing the central panel arrangement.

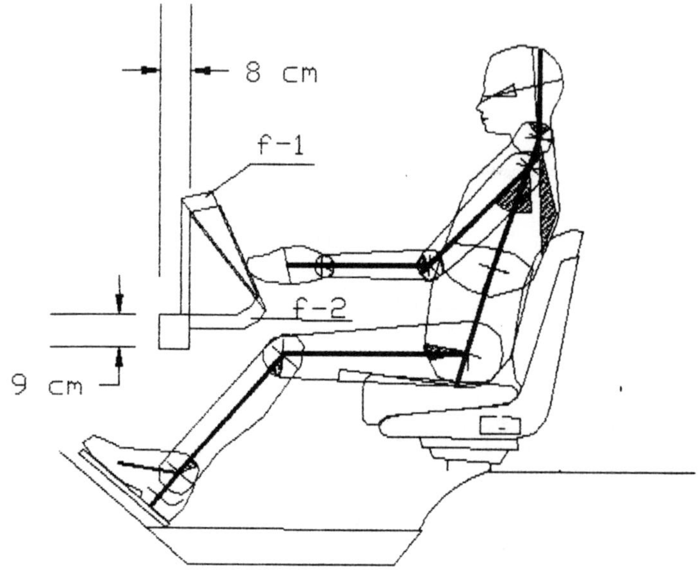

Figure 6.6 Manikin 2 in the position testing a central panel location and the two locations determined by manikins 1 and 2 (f-1, f-2).

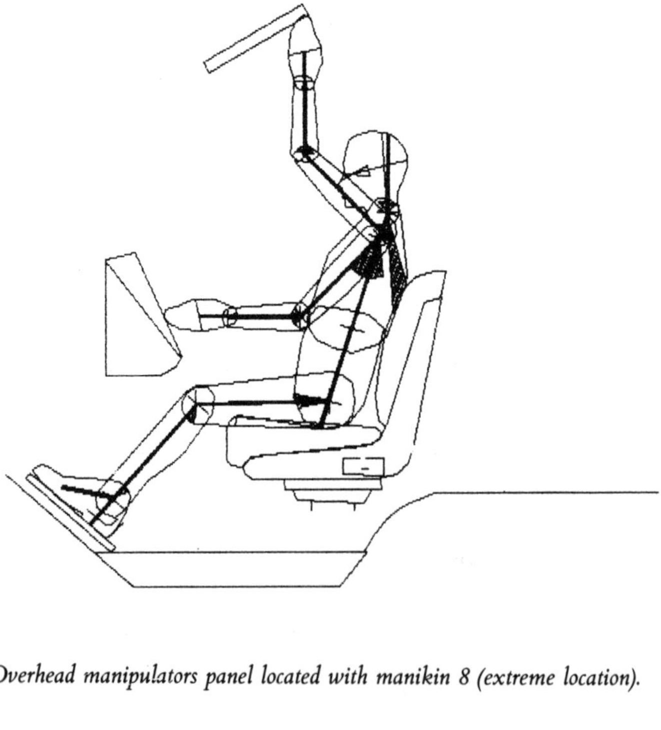

Figure 6.7 Overhead manipulators panel located with manikin 8 (extreme location).

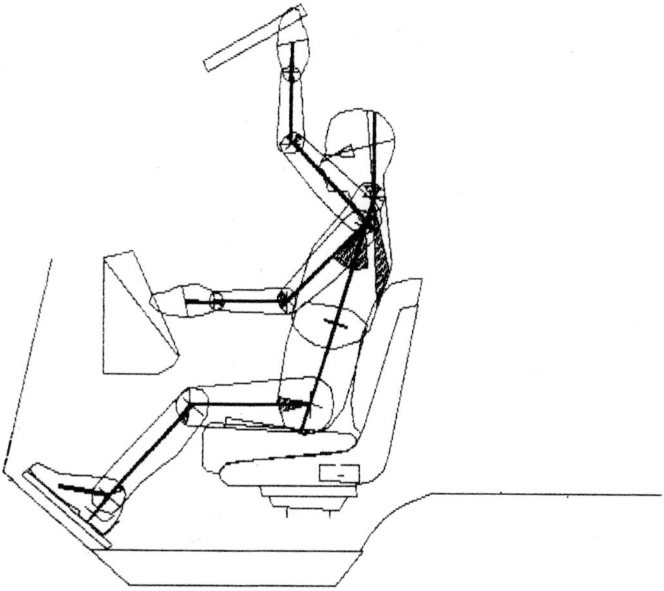

Figure 6.8 Manikin 7 at the arrangement of panel made by manikin 8.

According to the manufacturer's suggestion for these investigations, it is assumed that ensuring 'comfortable' spatial conditions for 95% of drivers should mean the facility to take up a comfortable position during normal driving. The comfort position in the presented tests is ensured by building manikins with the mutual positions of the body within marked optimal comfort ranges. The basic test position is shown in Figure 6.3. Manikins 1–8 built the same way result in getting the required seat adjustment range (Figure 6.4), which, moreover, is determined by dimensions of individuals 7 and 2, i.e. adequate representatives with big height, big weight and small sitting height as well with small height, small weight and big sitting height. The ranges obtained are much wider than those used so far by the manufacturer and which were obtained using the traditional 'percentile' approach.

Tests for basic control panel location ranges illustrated in Figures 6.5 and 6.6 are carried out for the manikin position shown in Figure 6.5. Since some control elements installed within the panel are quite seldom used, a hand location is assumed which gives the maximum range (according to optimal comfort angles). As shown in Figure 6.6, manikins 1 and 2 determine the upper right limit for adjustment ranges of the central panel for the assumptions taken. These are the representatives with all 'big' features (manikin 1) and big height, big weight and small sitting height (manikin 2).

The lower left limit of the central panel adjustment range is determined by manikin 7 with small weight, small height and big sitting height, and manikin 8 with all features at the level 'big'.

The extreme overhead panel location (Figures 6.7 and 6.8) is determined by manikin 8 whose features are all small. In this case the hand position is assumed arbitrarily in respect of very rarely-used elements in the overhead panel.

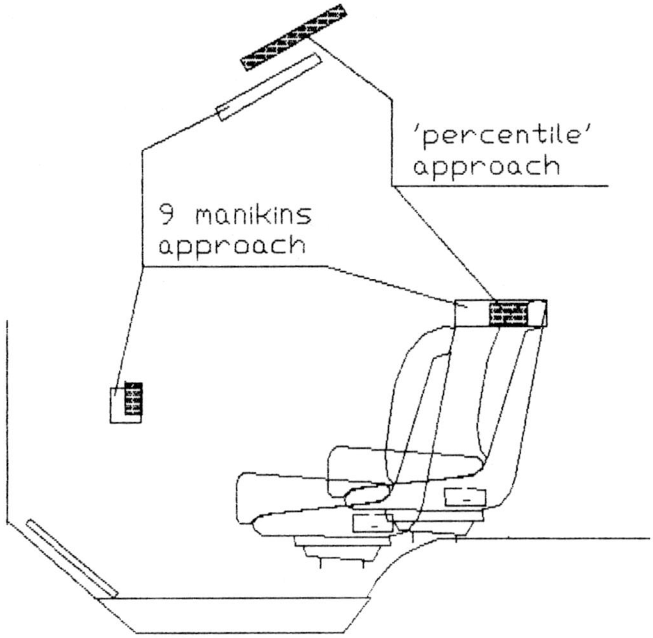

Figure 6.9 Approximated adjustment ranges for the proposed and 'percentile' approaches.

The ranges attained this way are based on differentiations in anthropometric features of the population of adult men in Poland. Figure 6.9 shows a comparison of the qualitative results of the same sets from the 'percentile' approach and from the method proposed in the present chapter. Obtained operating ranges and the overhead panel extreme location using the percentile approach are marked with dots. As may be seen the proposed approach considerably changes traditional workspace design standards.

The above simple examples of real design problems show the usefulness of the concept for representing population differentiations. In the problems considered above, four different representatives are used while in the 'percentile' approach only two representatives determine parameters of that type. The information received on the basis of the proposed representation is essentially different from that which results from the traditional percentile approach.

Analysis of the problem presented here and further more complex problems, has also shown the efficiency of the proposed system for anthropometric data representation in AutoCAD. With a little practice, the construction of any manikin in any position takes only a few tens of seconds. Together with the capabilities of the AutoCAD system, the proposed program is an effective instrument for assisting a driver's workplace design.

Summary and conclusions

The proposed approach allows enrichment, in a relatively simple way, of the widespread CAD system (AutoCAD) with an effective instrument for ergonomic driver's workplace analysis and design.

Although this system will not replace investigations of mockups, prototypes and models (with real people), it may cause a considerable reduction in research costs by producing the correct design in terms of dimensional relations and leaving only those problems in later design stages which cannot be solved without testing on a real scale.

The efficiency of the drawings presentation system and the described population representation of the method allow the formation of a set of 'representatives' from any population according to real needs. However, each element needs drawing and its respective supervisory program requires modelling. There is also a possibility of using the AutoLISP language (with which the latest version of AutoCAD is equipped) to build more flexible supervisory programs. However it involves some more expenditure of labour and the indispensable development of facilities in existing versions of this language (e.g. a possibility for interruptions in programs and access to all variables of the AutoCAD).

References

Bittner, A., Glenn, F., Iaveccia, H., Harris, R. and Wherry, R., 1987, CADRE – a family of manikins for workstation design. In: *Trends in Ergonomics/Human Factors IV*, edited by S. S. Asfour (Amsterdam: North-Holland), pp. 733–741.

Evans, S. M. and Chaffin, D. B., 1986, Using interactive visual displays to present ergonomic information in workspace design. In: *Trends in Ergonomics/Human Factors III*, edited by W. Karwowski (Amsterdam: North-Holland), pp. 17–25.

Gigure, D., Jequier, J. C. and Durand, P., 1986, Toward computer-assisted ergonomics: some realizations and exploration of possible avenues. In: *Trends in Ergonomics/Human Factors III*, edited by W. Karwowski (Amsterdam: North-Holland), pp. 9–16.

Grobelny, J., 1988, Including anthropometry into the AutoCAD-microcomputer system for aiding engineering drafting. In: *Trends in Ergonomics/Human Factors V*, edited by F. Aghazadeh (Amsterdam: North-Holland), pp. 77–82.

Grobelny, J. and Cysewski, P., 1989, ANTHROPOLIN – A method for constructing anthropometric data for complex workstation analysis. In: *Advances in Industrial Ergonomics and Safety I*, edited by A. Mital (London: Taylor & Francis), pp. 169–179.

McCormick, E. J., 1976, *Human Factors in Engineering and Design* (New York: McGraw-Hill).

Raker, D. and Rice, H., 1985, *Inside AutoCAD* (Thousand Oaks, Ca: New Riders Publishing).

Roebuck, J. A., Kroemer, H. E. and Thomas, W. G., 1975, *Engineering Anthropometry Methods* (New York: John Wiley).

Voisinet, D., 1986, Introduction to Computer Aided Drafting (New York: McGraw-Hill).

Zenkteller, B., Buczek, T., Switluk, L., Szczotka, H. and Welon, Z., 1983, Wymiary ciała dorosłej ludności Polski. Monografie Zakładu Antropologii PAN, Wrocław.

Chapter 7
TADAPS: a three-dimensional CAD man model

J. WESTERINK, H. TRAGTER, A. VAN DER STAR and D. P. ROOKMAAKER

Ergonomics Group/CAD Group,*
University of Twente,
University for Technical and Social Sciences,
The Netherlands

Abstract. This chapter describes the structure, application and future developments of a three-dimensional computer man model called TADAPS (Twente Anthropometric Design Assessment Program System), in development at the University of Twente (UT). The program is based upon ADAPS, developed at Delft University of Technology (DUT), and in cooperation rewritten, restructured and extended at the UT. ADAPS is a CAD tool for anthropometrical workspace evaluations and comparative biostatic evaluations of postures, implemented on a PDP-11 computer. The aim of the development of TADAPS was to make a VAX-computer implementation of ADAPS with extended possibilities regarding interactive workspace design, model definition, biomechanical assessment and program structure.

Within the anthropometrical assessment a man model and a workplace model are used. The man model is based upon a simplified model of the human skeleton. It consists of a number of logically linked segments, around which a wire frame man model is created. The data structure allows variation in the number of segments and surface points. Thus simpler man models or more sophisticated man models can be created. The workplace model can be constructed by generating and composing simple structures, called primitives: node, wire, block, prism, pyramid, extrusion, rotation.

Various man models (males, females, children, etc.) are available or can be created if data are available. Man model percentile or length can be chosen. The anthropometric assessment of the workspace involves the transformation of workplace objects and user analyses of 'body clearance and interference', 'reach', 'visual field', for various percentiles, postures, positions and points of view.

Within the biomechanical assessment the anthropometric man model with a whole body mass distribution (ADAPS) is extended with a muscle model for the lower back at the transversal intersection Lumbar 5 – Sacral 1 (L5–S1) (TADAPS). This extension, together with mechanical and mathematical calculating techniques, results in values for the compression and shear forces on the intervertebral disc L5–S1. This calculation can be performed for various (non)symmetrical postures and external loads.

Anthropometrically the computer model can be applied for workspace evaluations of new or existing workplaces. The biomechanical model can be applied in the evaluation of new or existing tasks with physical components, in which physical complaints (might) occur. Further validation, sophistication and extension of the anthropometrical and biomechanical man models as well as assessment options are planned for the near future.

Keywords: Anthropometry; Biomechanics; CAD; Man modelling; Workspace evaluation.

*The Ergonomics Group is an inter-departmental cooperation between the departments of Mechanical Engineering and Social Sciences.

Introduction

The program can be divided in two main parts: the man model and the workplace model. Within the man model, an anthropometrical model and a biomechanical model can be distinguished. In this Chapter the structure of both man models and workplace models will be described. The application of the program is dealt with and the application value of the models and developments for the near future are also discussed. First, however, tools for workspace evaluation, of which TADAPS is one, will be introduced.

The need for tools for workspace evaluations

Currently there are various tools available for anthropometrical workspace evaluations. These tools can be applied on newly-designed workplaces as well as for analyses of existing workplaces. In an overview (White and Van der Star, 1987) we have the following, most common, man (and workplace) modelling possibilities:

1. Tables with anthropometric data (Figure 7.1), which have a great flexibility in their application but are time-consuming and require sketching skills.
2. French Curves (Figure 7.2) which are easy to use but have no flexibility (only two-dimensional (2-D), fixed posture and limited sizes and scales).
3. Two-dimensional man models with posturing possibilities (Figure 7.3), which are easy to use but have limited flexibility (only two-dimensional, limited sizes and scales).

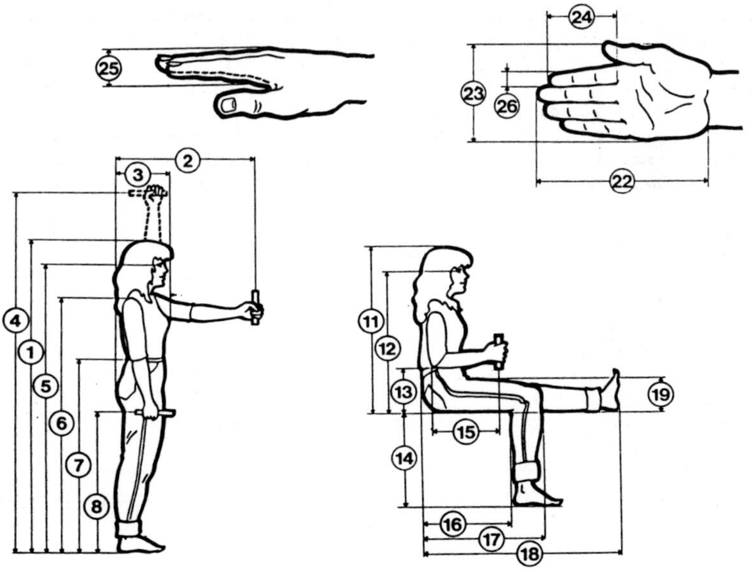

Figure 7.1 The DINED-table (Molenbroek and Dirken, 1986).

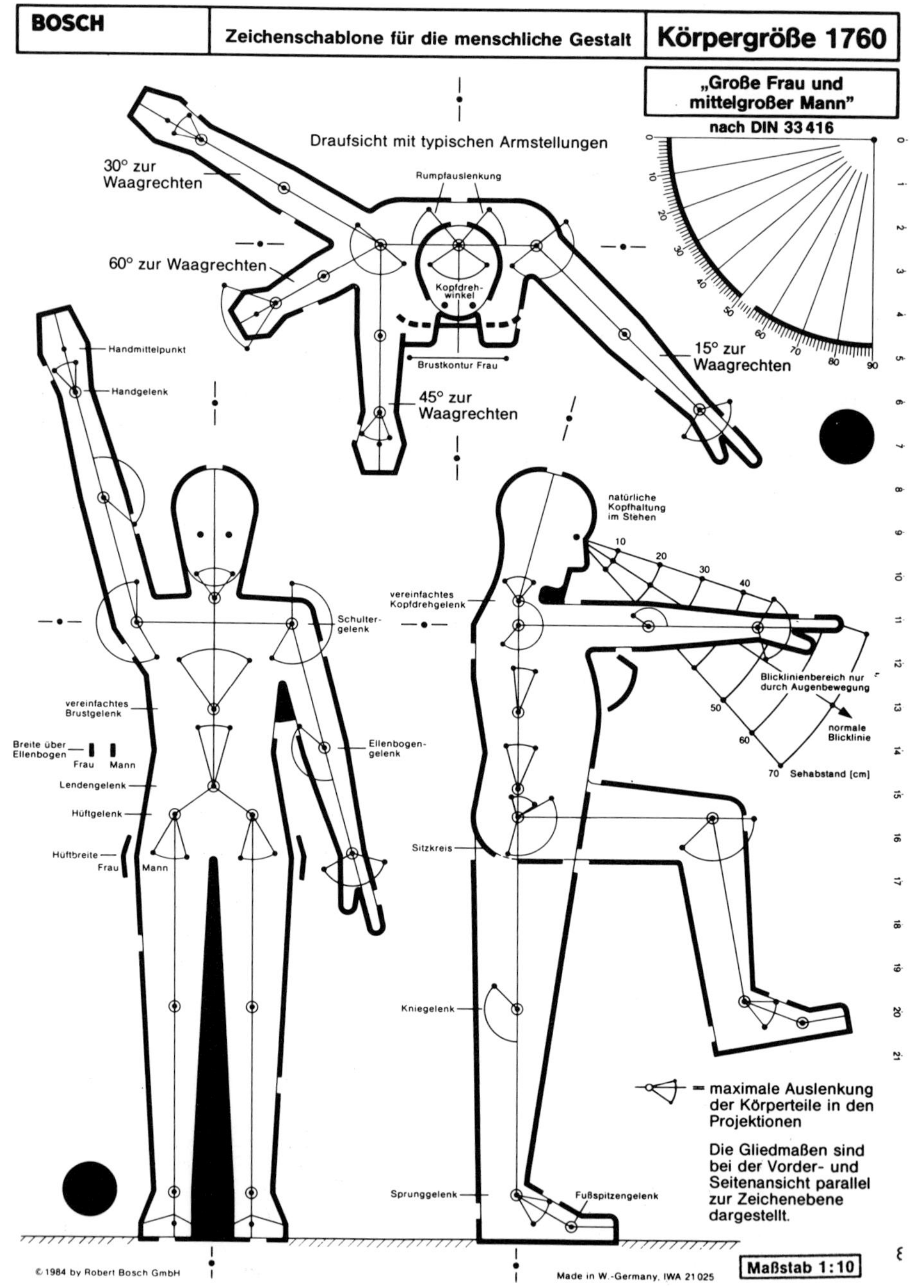

Figure 7.2 Example of French Curve (BOSCH).

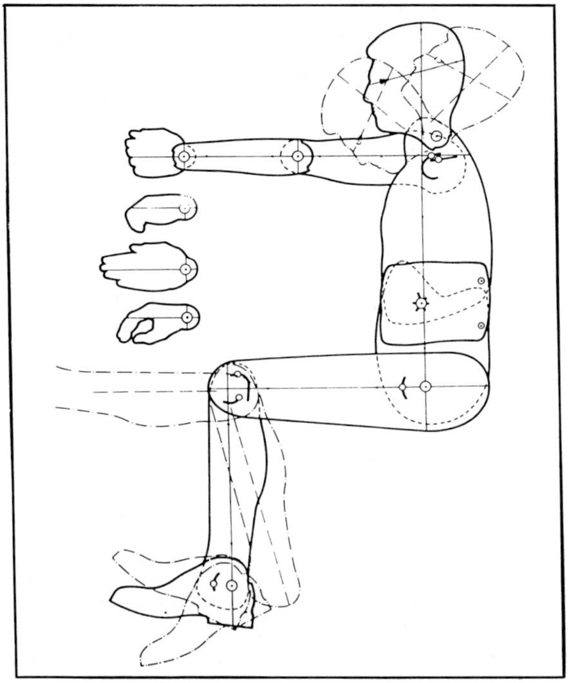

Figure 7.3 Example of 2-D man model (Kieler Puppe).

4. Two-dimensional computer models with posturing and scaling possibilities. These models are very flexible but demand experience with computer graphics of the user; both workplace and men are simulated. Two-dimensional computer models are regarded as restricted compared with the remaining types.
5. Three-dimensional (3-D) computer models with posturing and scaling possibilities. Here even more experience with computer graphics is required of the user. These models are very flexible and offer a good opportunity to assess anthropometrical problem areas in a relatively short time and at low costs (if the computer and the man model are available).
6. Mock-up models. These are real-scale models of the workplace. They allow evaluations of the workspace on anthropometrical and other features with real test-persons. The disadvantages are: mock-up experiences are time consuming and expensive. The results depend upon the selection of test persons. Sometimes maquettes (scale workplace models) are also used to evaluate the workspace.

Combinations of the tools listed above, for anthropometric assessment, are possible. For example a 3-D computer model can be applied in the early phases of the development process and a mock-up can be built later on. Because of the problem areas discovered with the computer model the design can be changed and the mock-up and further development costs can be diminished.

All the previously mentioned man–workplace models have in common that they deal strictly with human body sizes in relation to the workspace. Anthropometry includes data about human sizes, forces, postures and motions, both statically and dynamically. In addition, anthropometry can be linked to biomechanics by application of mechanical principles on the man model. For biomechanical assessment in TADAPS anatomical data of the body, the whole body mass distribution (e.g. see Post and Smeets, 1981) and muscle model for the lower back primarily according to Andersson *et al.* (1980), had to be implemented in the man model. In our practical experience with the evaluation of workspaces, including the physical demands to the workers, there is a great need for an integrated tool combining anthropometrical data in a wider sense with biomechanics. The TADAPS model described has been developed originating from this need (Kroonenberg, 1989). For that reason TADAPS offers extended assessment possibilities in relation to well-known existing man modelling CAD programs like SAMMIE and COMBIMAN (Hickey *et al.*, 1985; Lombaers *et al.*, 1986).

Introduction to TADAPS

The Delft University of Technology (DUT), Department of Industrial Design Engineering, developed ADAPS in the late 1970s (Post and Smeets, 1981). The ADAPS man model consists of an anthropometric body-size model and a biostatic model for calculation of the forces and moments on the model junctions. In 1987 the University of Twente (UT), Departments of Mechanical Engineering and Social Sciences, started to create a new version of ADAPS in co-operation with DUT, called TADAPS.

ADAPS is a CAD tool for anthropometrical workspace evaluations and comparative biostatic evaluations of postures, implemented on a PDP–11 computer. The aim of the development of TADAPS was to make a VAX-computer implementation of ADAPS with extended possibilities regarding interactive workspace design, model definition, biomechanical assessment and program structure. The main reason for this was that the hardware and software, which ADAPS required, would not be supported in the future (hardware and software requirements are dealt with towards the end of the chapter).

The TADAPS man–workplace model was realized in 1988. In that year the development of the biomechanical model was started. The first results of that research and development are discussed in this section. For the anthropometrical part the aim was to develop a tool for anthropometrical workspace evaluations. For the biomechanical part the first aim was to develop a model with a predictive value for the shear and compression forces on the intervertebral disc L5–S1 for various postures and external loads.

TADAPS structure

Man model

The anthropometric man model is constructed around a simplified model of the human skeleton. It consists of a linked-segment system with (24) segments as shown in Figure 7.4. The segments are logically linked originating from the root segment (L1) to others in junctions with limited rotation possibilities. Each segment has a local coordinate system in which a number of surface points are defined. These surface points are mutually connected by line elements: the surface traces. These surface traces create a wire frame man model as presented in Figure 7.5. The segment structure and the wire frame model are based on ADAPS (e.g. see Hoekstra, 1985a).

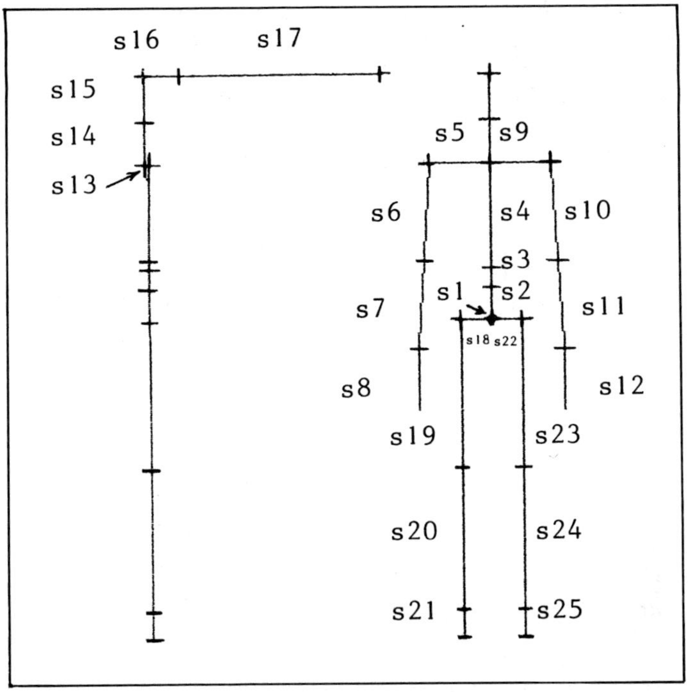

Figure 7.4 (T)ADAPS segment system (see e.g. Hoekstra, 1985a).

The extension of the TADAPS data structure now allows easy variation in the number of skeleton segments and surface points. This enables the user to create a more simplified model, a more sophisticated model and models of disabled people.

Postures can be stored and retrieved independently of the man model type. This enables the user to copy a posture from one man model to another and to use default postures (sitting, standing, etc.). Another aid in the posturing process is the availability of 'anchor points' in the man model. These anchor points can easily be connected to points in the working environment, for example, an anchor point on the middle of a hand can be connected to a point on a steering-wheel. This method prevents complicated posturing procedures. A man model can have any number of anchor points.

The biomechanical TADAPS man model consists of the anthropometric man model with a mass distribution over the segments extended with a muscle configuration model for the transversal intersection at lumbar 5-sacral 1. Mass-distribution is based on the literature data of Dempster (1955) implemented in the ADAPS man model (e.g. see Post and Smeets, 1981). The muscle model is based on Andersson *et al.* (1980) and Schultz *et al.* (1982) updated with data found by computed tomography (CT) scanning (Kumar, 1988; McGill *et al.*, 1988). A load distribution optimalization model for the lower back muscles (Bean *et al.*, 1988) is implemented in the program. The implementation of the muscle model and the load distribution model are new extensions of the TADAPS model (Kroonenberg, 1989).

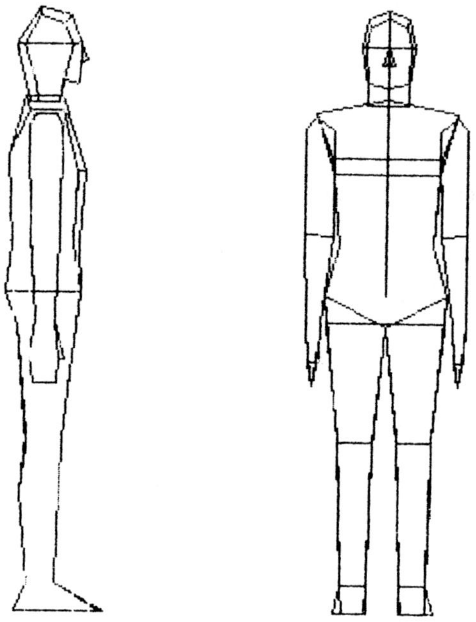

Figure 7.5 (T)ADAPS wireframe man model (see e.g. Hoekstra, 1985a).

Workplace model

The workplace model can be constructed by generating and composing simple structures, called primitives. The following primitives, are available: node, wire, block, prism, pyramid, extrusion, rotation. Examples of these primitives are present in Figure 7.6. They are all composed of straight line entities, making fast computer-image generation possible.

By dimensioning, positioning and composing these primitives the working environment can be simulated and designed. Several primitives can be grouped to an object. These objects can interactively be inserted, translated, rotated and scaled. The primitives and objects are structured as surface models to make hidden line removal on the workplace model possible.

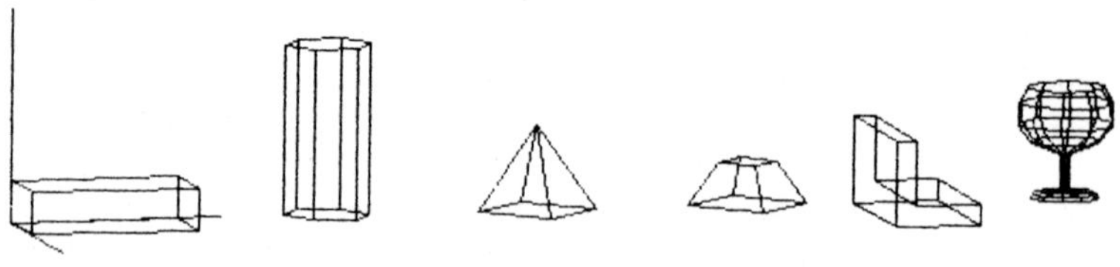

Figure 7.6 Examples of workplace primitives.

TADAPS application

Program functioning

The program can be operated by activating the menu commands by mouse or by keyboard. For skilled users the keyboard operation is preferred, since it works faster.

The menu ccnsists of a main menu and various submenus. The submenus are activated by selecting a main menu item. The submenus contain commands related to the selected main menu item. The restricted size of this chapter does not allow a discussion of all the submenu commands. The description will be confined to the main menu items with a brief mention of the submenu command possibilities. The following main menu items are available.

Manikin

With the commands of this submenu it is possible to load, position, posture, save or delete a man model. Any number of man models can be placed in one workplace but only one man model can be manipulated at a time. Parts of the man model, such as the skeleton, the surface wireframe and also segments can be set visible/invisible. To simplify the positioning and posturing of the man model the local axis and the global axis can be visualized on the screen for the segments and the whole man model, respectively.

Workplace

With this submenu a workplace can be composed out of defined objects. It is possible to insert, translate, rotate, scale, colour and delete objects. Additional, previously created workplaces can be loaded and new workplaces can be saved.

View

With the commands in this item it is possible to choose or define views in three dimensions. Standard views of normal X, Y and Z and a 3-D user defined view are available. The user-defined view can be changed by giving a viewing eyepoint in the workspace, a view direction, a view-up vector and choosing between parallel presentation or perspective presentation. To create a clear and well-positioned view there are also commands for zoom and pan operations available.

Utility

Below this main menu item various commands are available. The command input can be changed from mouse menu to keyboard. A workplace, once created, including man model (s) can be completely saved or retrieved. It is possible to ask for the status information of the loaded workplace model and man model, in addition all data can be cleared.

There is a command for executing macrofiles (command sequences) and it is possible to delay the executions which is useful for demonstrations. A print command can also be given to list all the data of the man model and the workplace. Hard copies of the displayed image are made by invoking the hard copy command. The program can be terminated by the 'end' command in this submenu.

Mechanics

In this main menu item the commands for the biomechanical model are listed. The body mass and its distribution over the segments can be loaded from here. The external load is imposed by giving values for seizing point, magnitude and direction. The calculation of forces and moments on junctions, the shear and compression force of the intervertebral disc L5–S1 is invoked here. The results can be shown graphically or in notes on the display.

Help

Help information is available about all the main menu and submenu items.

Application of the program

The application of the program will be discussed briefly here. The two main application areas, anthropometry and biomechanics, will be dealt with separately.

Anthropometry

The user of this tool needs to have either (i) a dimensioned new workplace design, at least a sketch with indicative dimensions, or (ii) a drawing of an existing workplace, in three dimensions. The workplace model will be introduced to the computer by a user-generated data file. The computer is now able to generate an image of the workplace on the graphics screen. At this point it is useful to load the man model. There are man model data files available for a Dutch male, Dutch female, Dutch boy, aged 4 years, and an American pilot.

The Dutch male, female and boy are based on the ADAPS reports of Hoekstra (1985b, 1986). The American pilot was the first model created in ADAPS, and served as a starting point for other models. The anthropometric data used for this model originate from Churchill *et al.* (1977) and Dreyfuss (1967).

If the right anthropometric data are available it is relatively easy to create a man model for any other population. At this moment a connection to the European Anthropometric Database is being researched.

When loading a man model the system asks either for a percentile or a length of the man model. All percentiles between the 0·5th and 99·5th percentile can be chosen, the man models are linearily scaled out of the 50th percentile model. When the man model and its percentile or length are given the system asks for the position (x,y,z) and the default posture of the man model in the workplace. Once the position and posture are given the man model appears in the workplace on the screen. It might be necessary to edit interactively the position and posture before the assessment can start. To prevent unnatural postures, limits for the junction angles, based on ADAPS (Hoekstra, 1985a) are included in the program.

When workplace and the right man model are loaded and well positioned the following features can be analyzed by the user (TADAPS is still no expert system). Various percentiles and postures should be examined from different viewpoints:

1. Body clearance: examination of the body clearance and interference.
2. Reach analyses: examination of whether controls are within the reach range.
3. Visual field: examination of the visual field (e.g. dead angles) of the man model.

During and after the analysis, the workplace design can be interactively changed and optimized in an iterative process. An example of the application of the anthropometric man model is reviewed in the next section.

Biomechanics

The biomechanical man model is an extension of the anthropometrical model as previously described. As input data the program needs a posture, the person's mass, and the external load according to the structure given by Chaffin (1988). Symmetrical and asymmetrical postures and external loads can be calculated. At first the program calculates the distribution of the mass over the segment model. Together with the external load the forces and moments on the segment junctions are calculated. With the value for the junction representing L5–S1 the program calculates the load distribution over the muscles in the transversal intersection L5–S1. These values lead, by a minimalization method (Bean *et al.*, 1988), to a calculated value for the compression and shear force of the intervertebral disc L5–S1. External validation is to be expected next year.

Application examples

The following is an example of using the anthropometric man model for the analysis of an existing workplace. The most important, present operational features will be reviewed and the future options will be mentioned.

An interesting common object for anthropometrical analyses is the automobile. A whole list of items can be analyzed with the TADAPS anthropometric man model, e.g.:

1. Body clearance: any interference of parts with the steering-wheel, dashboard, roof, other passengers or opened fifth door. The fitting of the chair, head support and seat-belt (example in Figure 7.7).
2. Reach: are the needed controls within reach, e.g. dashboard, gear lever, fastening of seat-belts, mirrors and pedals (example in Figure 7.8).
3. Visual field: for both the inside and the outside is all the information needed within the direct field of view: dashboard displays, mirrors, crosspoint of a viewline from the eyepoint with the road should be within a defined distance, the visual limitation of the A-, B- and C-styles should be within defined limits (examples in Figures 7.7 and 7.9).

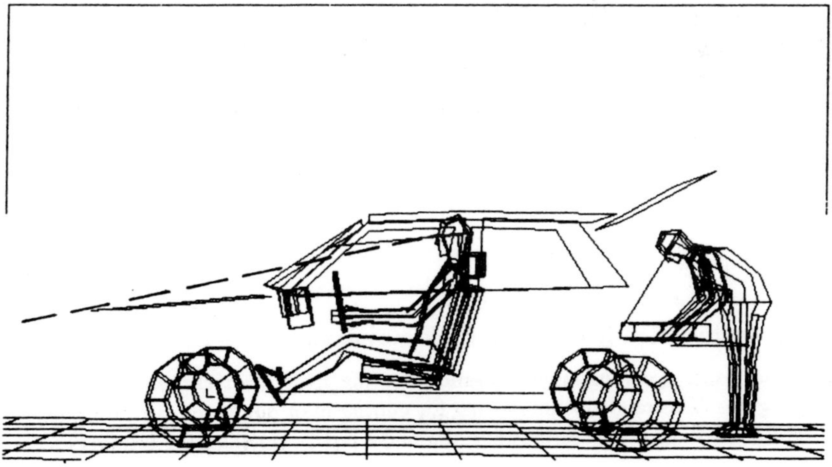

Figure 7.7 Example of body clearance analysis. Sitting male, 95th percentile; standing female, 5th percentile.

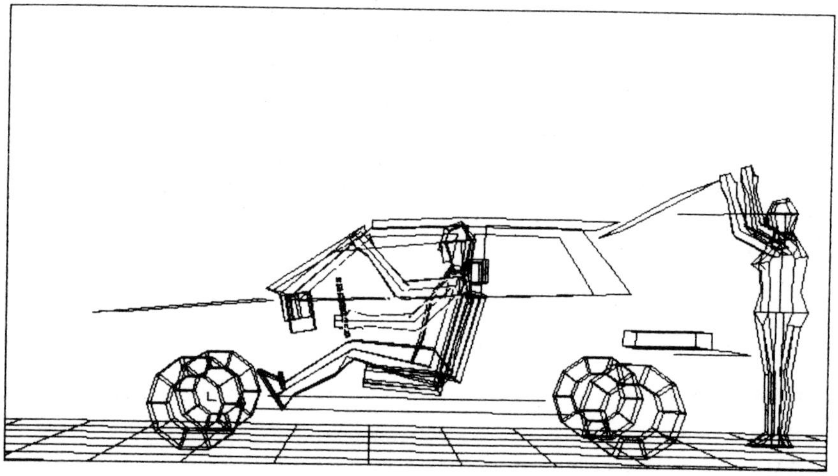

Figure 7.8 Example of reach analysis.

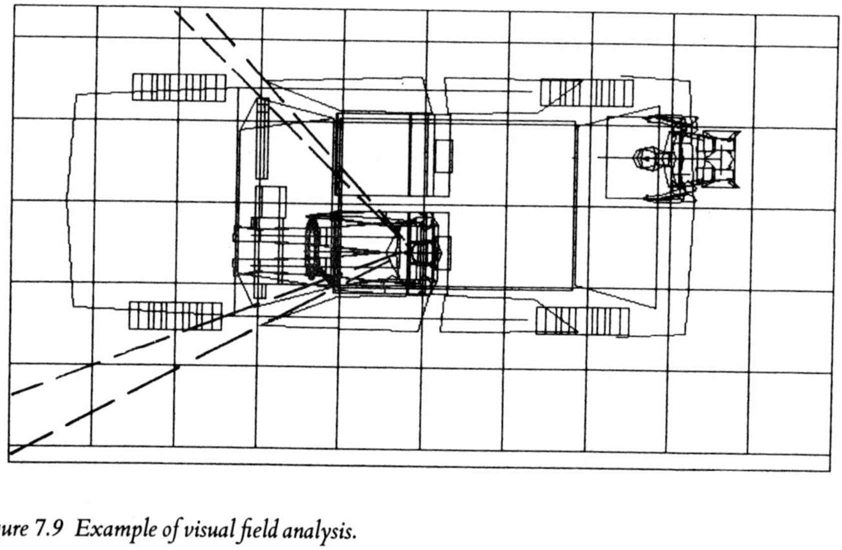

Figure 7.9 Example of visual field analysis.

For biomechanical analyses this object offers some aspects that can be analyzed already.

1. Joint loads: a comparative indication for the joint loads can be calculated by the program. This might be interesting for getting in or out of the car, lifting loads in or out of the luggage space, pushing the brake pedal or pulling the handbrake.

The above mentioned analyses should be performed for various percentiles of males and females and eventually for various postures. In the near future the analysing possibilities are planned to be extended (i) for the field of anthropometry: individual man models;

somatotypes; and increased accuracy; and (ii) for the field of biomechanics with: lower back load values (intervertebral disc L5–S1); NIOSH lifting calculations; and OWAS posture targeting.

The example discussed is only one in a wide range of application areas where the computer man models can be used. Any industry, institute or governmental agency that is involved in designing or analyzing workspace for men (especially with restricted room), e.g. transport, off-shore, industrial production, architecture, etc., is involved in the field of anthropometry. In the field of biomechanics, applications can be found wherever designing or analyzing tasks with physical load aspects is involved, e.g. the building industry, rail and road construction, maintenance workplaces for all transportation means, dust-loading, etc.

Hardware and software requirements

The hard- and software requirements for operating the TADAPS program are listed below:

Hardware:
Computer : any VAX (DEC).
Terminal : Tektronix colour graphics.
Printer : compatible with VAX computer.
Hardcopier : compatible with Tektronix graphics terminal.

Software:
Operating system : VMS.
Dedicated software : TADAPS modules.

For further development, the source codes and the following compilers are needed:

Compiler : Fortran 77
Graphics compiler : Tektronix – Plot 10 – IGL (Interact. Graph. Libr.).

Discussion

The predictive value of the application of these computer models should be considered prior to and whilst using them. As with every model, the accuracy is restricted! Here the anthropometrical accuracy of the man model is restricted to an extent which cannot be ignored (see e.g. Hoekstra, 1985a), because of several factors.

The model is an approximation of a non-existent percentile-man with defined proportions. Linear scaling of percentiles from the 50th percentile model does not resemble reality. The link-system differs substantially from the real skeleton: less links and junctions instead of joints. Besides, the space needed for clothing, shoes, hair, and so on is not included in the model.

The anthropometrical man model can be applied in an early phase of the development process of the working environment, when there is still no mock-up or prototype of the workplace available (White and Van der Star, 1987). Anthropometrical problem areas will be detected easily and design changes can be realized. The mock-up and/or prototype costs can be reduced substantially through the computer model analyses. Another option is to analyze existing workspaces.

The biomechanical accuracy of the calculated values for the compression and shear forces of the intervertebral disc L5–S1 is still being compared with practical medical tests. Results of this validation study will be published in the near future. Internal validity has been proved by comparing the model output with EMG of the m.erector spinae by sagittal force-exertion according to Mouton *et al.* (1988).

A great effort will be put in making the biomechanical models applicable in the design of new tasks with physical components, for the analyses of existing tasks in which physical complaints occur and to determine the influence of workplace design, from an anthropometric point of view, on working posture and lower back load.

The framework of TADAPS has been realized now and by using the present models it will occur that validation, extension and sophistication of these models is necessary. A research project-group was formed at the University of Twente for the research and development of TADAPS. The activities of this project-group will mainly be research-oriented and situated in the field of ergonomical design assessment. The main areas of interest in this project are anthropometry, biomechanics and CAD applications both for workplace design and analysis as well as physical task design and analysis.

The following developments are planned for realization in the early 1990s. In the field of anthropometry, the possibility of creating individual models (especially disabled people) is aimed at, as well as further sophistication of the present man model and improvement of its validity. The biomechanical part needs external validation. Already tests are carried out in this field and the results of these will be published in the near future. Further research will be aimed at creating judgement tools (expert systems) in the field of biomechanics: predictive L5–S1 load model, implementation of NIOSH-lifting calculations and OWAS posture targeting.

Within the workplace model an aim is to provide further sophistication for creating and presenting the workplace model as well as the connection to other CAD systems, using an IGES interface. That would make the exchange of workplace models with other CAD systems possible.

Acknowledgements

We thank the project group members: N. T. van Harpen, Ch. Talens, T. N. White, and the student A. J. van den Kroonenberg for their contributions and cooperation. Special thanks are due to the ADAPS project group of the University of Delft for their data and co-operation.

References

Andersson, G. B. J., Oertengren, R. and Schultz, A., 1980, Analysis and measurement of the loads on the lumbar spine during work at a table. *Journal of Biomechanics*, **13**, 513–520.

Bean, J. C., Chaffin, D. B. and Schultz, B. A., 1988, Biomechanical model calculation of muscle contraction forces: a double linear programming method. *Journal of Biomechanics*, **21**, 59–66.

Chaffin, D. B., 1988, Biomechanical modelling of the low back during load lifting. *Ergonomics*, **31** (5), 685–697.

Churchill, E., Kikta, P. E. and Churchill, T., 1977, *The AMRL Anthropometric Data Bank Library*, Vols. I–V (AMRL-TR-77-1) (Yellow Springs, OH: Webb Associates) (AD A047 314).

Dempster, W. T., 1955, *Space Requirements of the Seated Operator*. Technical report WADC-TR-55-159 (OH: Wright–Patterson Airforce Base). (AD 87 892).

Dreyfuss, H., 1967, *The Measure of Man: Human Factors in Design*, 2nd edition (New York: Whitney Library of Design).

Hickey, D. T., Pierrynowski, M. R. and Rothwell, P. L., 1985, *Man-modelling CAD programs for Workspace-evaluations* (University of Toronto, Ontario M5S 1A1 and Defence and Civil Institute of Environmental Medicine, Downsview, Ontario M3M 3B9).

Hoekstra, P. N., 1985a, *ADAPS–Data sources of the anthropometric model* (Delft University of Technology, Department of Industrial Design Engineering) (in Dutch).

Hoekstra, P. N., 1985b, *ADAPS–MAN84: Provisional Anthropometric Model of the Dutch Male* (Delft University of Technology, Department of Industrial Design Engineering) (in Dutch).

Hoekstra, P. N., 1985c, *ADAPS–WOM84: Provisional Anthropometric Model of the Dutch Female* (Delft University of Technology, Department of Industrial Design Engineering) (in Dutch).

Hoekstra, P. N., 1986, *ADAPS–BOY4: Provisional Anthropometrical Model of a Four Years Old Dutch Boy* (Delft University of Technology, Department of Industrial Design Engineering) (in Dutch).

Kroonenberg, A. J. van den, 1989, *The Development of a Biostatic TADAPS-model* (Enschede: University of Twente, Department of Mechanical Engineering (internal thesis report)) (in Dutch).

Kumar, S., 1988, Moment arms of spinal musculature determined from CT scans. *Clinical Biomechanics*, **3**, 137–144.

Lombaers, J. H. M., Molenbroek, J. F. M. and Osinga, D. S. C., 1986, *Anthropometric Models, a Comparison of Their Usability* (Delft University of Technology, Department of Industrial Design Engineering) (in Dutch).

McGill, S. M., Patt, N. and Norman, R. W., 1988, Measurement of the trunk musculature of active males using CT scan radiography: implications for force and moment generating capacity about the L4–L5 joint. *Journal of Biomechanics*, **21**, (4), 329–341.

Molenbroek, J. F. M. and Dirken, J. M., 1986, *Dutch Body-sizes for Designing, DINED-table* (Delft University of Technology, Department of Industrial Design Engineering) (in Dutch).

Mouton, L. J., Besseling, S., Hidding, A. and Van der Bij, J., 1988, Different EMG-force relationships of the low-back muscles during standing and sitting. In *Elektrophysiological Kinesiology* (Amsterdam: Elsevier) pp. 459–462.

Post, F. H. and Smeets, J. W., 1981, ADAPS: Computer aided anthropometrical design. *Tijdschrift voor Ergonomie*, **6**, (4), 11–18 (in Dutch).

Schultz, A. B., Andersson, G. B. J., Haderspeck, K., Ortengren, R., Nordin, M. and Bjork, R., 1982, Analysis and measurement of lumbar trunk loads in tasks involving bends and twists. *Journal of Biomechanics*, **15**, 669–675.

White, T. N. and van der Star, A., 1987, Anthropometric computer-models for ergonomical design. In *Integration of the Computer in the Product Development Process* (Enschede: Stichting Industrieel Ontwerpen-oost), pp. III.1–III.19 (in Dutch).

Chapter 8
Anthropometric and biomechanical man models in computer-aided ergonomic design – structure and experiences of some programs

ARTO KUUSISTO and MARKKU MATTILA
Tampere University of Technology,
Occupational Safety Engineering,
P.O. Box 527, 33101 Tampere, Finland

Abstract. This paper illustrates why computer-aided design should take human factors into consideration and describes the current trends in the development of computerized ergonomic design systems. The programs developed in Finland are described and experiences of their use are given.

Keywords: Ergonomics; Man model; Biomechanical man model; Work environment design.

Introduction

Ergonomic analyses are carried out mostly for already existing workplaces. Several analytical methods have been developed for this purpose. Today, however, ergonomic features need to be taken into consideration during the design stage.

Standards (e.g. ISO 6385) and some ergonomic handbooks (e.g. McCormick and Sanders, 1982) help designers to draft the ergonomic design. Some national and international acts obligate designers to use the existing ergonomic standards and directives. In Finland, according to the current Occupational Safety Act, the designer is responsible for the safety and ergonomics of the workplace he has designed. This principle is now common in many countries.

Current trends

Ergonomic tools are available for the conventional design and drawing, the plastic 'Humanscale' plates being one example (Diffrient *et al.*, 1981). These tools help the designer to draw human models by hand, in various postures and and anthropometric sizes. They also recommend how some typical workplaces should be dimensioned and indicate the forces a worker can produce.

The traditional design of working environments is intended to ensure that *work* can be performed effectively. *Human factors* have been considered to be less important. Computer-aided design can put these two requirements into the same category by making it easy to check the human demands of work.

Human parameters can be integrated into computer-aided design in several ways. Checklists are easy and quick to use at any design stage. Currently developed expert systems for ergonomic workplace design give more detailed information than pure checklists (DeGreve and Ayoub, 1987). Graphic reach area and comfort area curves (e.g. SAE J898 JUL82, 1983) are examples of graphic checklists.

Man models are used for more sophisticated evaluation of man–machine systems. Most of these models simulate human anthropometry and postures (Dooley, 1982). Evaluation of biomechanical strength is also a common part of many man model programs. Data for the calculations of human strength are often derived from the studies of Chaffin and Andersson (1984).

Most man models are three-dimensional, and can be used only in minicomputers. OSCAR is one of the few three-dimensional models which can be run in an IBM-compatible microcomputer (Lippman, 1986). ErgoSHAPE is one of the two-dimensional model programs which runs on an IBM microcomputer (Launis *et al.*, 1988).

Lack of user-friendliness is a problem common to many of the three-dimensional man model programs, and which prevents their use in industrial design applications. Difficulty is caused, e.g. by the various joint angle determinations when the working posture is modified. Models involving more than 20 joints are difficult to handle, especially if the program has no automatic functions for moving (e.g. reach). Because of these factors the three-dimensional man models are quite rough, and workspace dimensioning can be accomplished only at a rough level.

Ergonomic programs for use of microcomputers are relatively cheap; they are also easy to learn and run. For this reason they are especially suited for use in education. Compatibility between microcomputers is also better than between minicomputers. This makes the inexpensive two-dimensional programs easy to deliver.

Ergonomic programs developed in Finland

This section contains detailed descriptions of some Finnish ergonomic computer programs and experiences of their use.

MINTAC (Man Machine Interaction) program

MINTAC was developed in 1984–5 jointly by Kuopio Regional Institute of Occupational Health and the University of Oulu. The initiative for developing this program arose from the need to evaluate the difficult working postures occurring in Finnish agriculture and forestry. At that time, there were some commercially available ergonomic programs, but they were not usable because of their poor compatibility with the available hardware.

MINTAC was developed for the Computervision CAD/CAM system. The system consisted of Computervision Designer V hardware and CADDS4 modelling program. Posture modifications for the man model and the mathematical man–machine interaction operations were carried out using a Newvar programming language (Kuusisto *et al.*, 1987).

Structure of the program

The program creates a three-dimensional geometric man model which can be modified and placed in a three-dimensional model of the working environment. The anthropometric data are derived from the studies of Dreyfuss (1966). The man model was assumed to be dressed in winter clothes.

The structure of the model is simple. It contains six links: lower limbs (one rigid block, which can be selected), back, upper arms and forearms. Thirteen blocks, all in different postures, represent the lower limbs. In each case the user defines which block is best suited to the situation under study.

Compatibility with the OWAS (Karhu *et al.*, 1977) working posture analysis system was one of the criteria when the model was created.

The building process of the developed model is interactive. The Newvar program asks for the parameters needed in the process. First, the user selects the proper block of the lower limbs and places it into the modelled environment. The next task is to give posture parameters (three angles) for the back. The body can now be seen on the screen. The program then gives the distance value between the elbows and a user-defined working object. If the hands reach the indicated object, they will be added to the body automatically. Otherwise the program asks the user to change the position of the legs or back.

When the posture is finished and no longer needs to be modified, the program lists the following parameters:

1. The man model's standing height percentile (user-defined);
2. Back posture (the angles of the back);
3. OWAS-classified working posture;
4. OWAS action category;
5. Height of the working level (from ground level);
6. Height of the working object (from working level);
7. Distance between the man model's chest and the working object.

In addition to these parameters, visual contact with the working object can be evaluated using a line drawn from the eyepoint to the working object (Kuusisto *et al.*, 1987).

Experiment

In an experiment, a forest tractor was used as the studied working environment. Such tractors call for several maintenance tasks which usually take place outdoors and most frequently in the winter. This kind of work today presents many ergonomic and safety problems (Hansson *et al.*, 1978; Väyrynen, 1986).

In the experiment, nine different maintenance situations were studied. The situations involved daily checking and greasing tasks. The working postures were evaluated using the OWAS method. The postures were also compared with actual working postures observed from a video tape. All studied tasks were simulated using the man model's percentiles of 5, 50 and 95 (Figure 8.1).

The OWAS-classified working postures created by the MINTAC program were identical to those observed in the real situation. Some difficult postures could be corrected by finding a better place for the model to stand. This was possible because some new methods for performing the task were revealed by the CAD simulation. In some cases, the need for platforms or the need to adjust them were observed. Poor working methods were also found; the man model was able to do a task without twisting the back, which occurred in the actual situation (Kuusisto *et al.*, 1987).

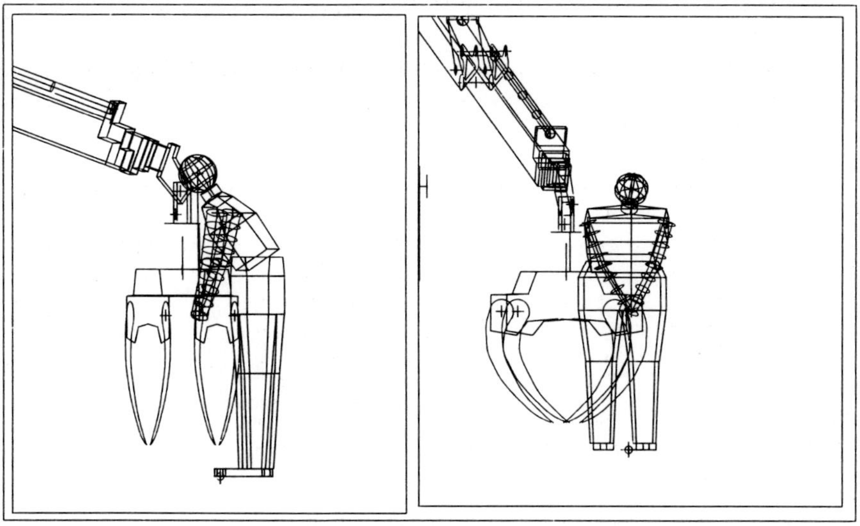

Figure 8.1 MINTAC model lubricating grease nipples of a forest tractor's grab (Kuusisto, unpublished).

Usability of the MINTAC program

The results of the experiment were encouraging. The designer seems to get reliable information about the ergonomics of the studied working situation, at least as far as body postures and some anthropometric criteria are concerned.

The features now preventing widespread use of the MINTAC program are: (i) suitability only for the analysis of heavy work; (ii) the limited degree of the model's posture modifications; and (iii) great time consumption due to the hardware.

It is, however, possible to compile the MINTAC program so that it can use the CADDS4X drawing program and SUN computer workstations. For greater accuracy, the OWAS method should be replaced by a more precise posture analysis method, and, finally, the model should contain more links in order to be usable and more precise in various working tasks.

The ergonomic program developed at Tampere University of Technology

A research project has been carried out at Tampere University of Technology, the aims of which were:

1. To study the possibilities of including ergonomic and safety aspects in CAD,
2. The development of CAD programs to evaluate and improve the ergonomics and safety of work systems, and
3. To test the programs in different design situations.

The program was developed for the Computervision CAD/CAM system, which consists of Designer V workstations and the CADDS4 modelling program. The programming languages used in the system are Newvar and Varpro2 (Mattila and Leppänen, 1987).

Structure of the program

The program is divided into two parts, which can also be used separately. The first part graphically evaluates the recommended location areas for the controls included in a work system. The other part uses a three-dimensional man model for the visualization of a human subject at work (Leppänen and Mattila, 1987).

Evaluation of control locations

The program has a data bank of recommendations for the location of controls (SAE J898 JUL82, 1983). The location areas are shown on the screen; thus the designer can immediately see the accepted locations of controls from the ergonomic point of view. On the other hand, this program analyzes the ergonomic acceptability of control locations. The evaluation criteria are the user's reach, the location of controls and their frequency of use. The evaluation is based on Schmidtke's (1976) method for the ergonomic evaluation of a work system.

The primary input includes: (i) proportions of the product model; (ii) the user's height; (iii) the user's location; (iv) working posture; and (v) working direction. The number of controls to be used should be entered next. The location of the controls is determined (digitized) and their frequency of use is then entered. With this information the program calculates a numerical value (from 1 to 6, representing from excellent to poor) for each control separately, and a total value for the whole system.

On the basis of these numerical values, it is easy to recognize which controls have a poor location ergonomically. By modifying the product model, the designer can make the work system more user-friendly (Mattila and Leppänen, 1987).

Man model

The graphic three-dimensional man model has 16 joints. Its segments are constructed of ellipses and B-spline curves. The measurements of the model are based on human anthropometric data (Roebuck *et al.*, 1975; Panero and Zelnick, 1979; McCormick and Sanders, 1982).

The model can be placed into the work system's model in the desired size. Its posture can be changed interactively in three dimensions, according to the work task being evaluated. Changes in posture are made with the CADDS-level 'rotate' command. In rotation, angles for the x-, y- and z-axis are given first, then the segments to be rotated are shown, and the base point of the rotation is shown last. It usually takes one to ten rotations to compose the desired working posture. If necessary, the postures can be stored, and they can be recalled later for other work system models (Figure 8.2).

Experiments

The program for evaluating control locations has been used in the design of a locomotive cabin. The results were satisfactory. The numerical values given by the program make it easy to recognize controls placed in an unsuitable location. Often only simple changes are needed to make the work system much better for the worker. The program also makes it easy to compare several alternative product models, and to choose the most acceptable one (Mattila and Leppänen, 1987).

The man model has been used to visualize the worker's reach to controls in a supervising room. In another experiment, the fields of vision from eight different tower crane cabins

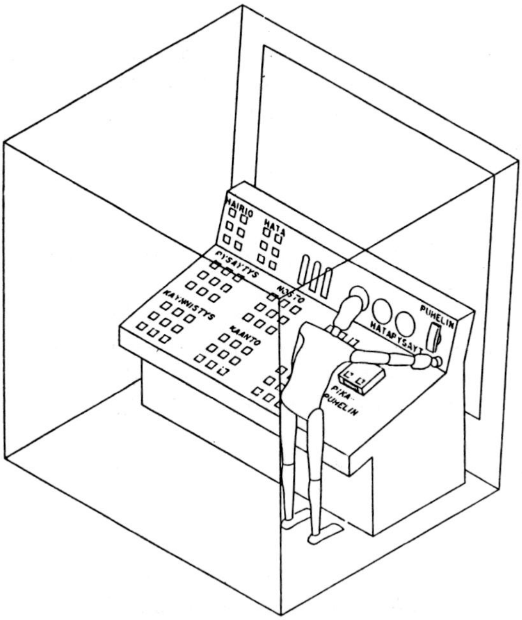

Figure 8.2 Evaluating worker's reach to controls in a supervising room (Mattila and Leppänen, 1987).

were compared to standard recommendations using the man model. The fields of vision to ground level were determined when the cabin was at heights of 20 and 40 m. According to the study, the standard recommendation corresponds closely to actual practices (Leppänen and Mattila, 1987).

Usability of the programs

The programs can presently be used for determinations of control locations, workplace dimensioning, working posture evaluation, and fields of vision. The man model is a good aid for the designer in evaluating work systems, but it is still rather complex and time-consuming to use (Mattila and Leppänen, 1987).

Ergonomics Checklist program for computer-aided workplace design

The Ergonomics Checklist program was created at Tampere University of Technology using the Computervision Designer V system and the CADDS4 modelling program. A graphic model of the workplace created using a computer-aided design system, can be evaluated from the ergonomical aspect by means of the program. The graphic model can be examined from different views, and specific details of the model can be focused on (Järvinen, 1987).

Structure of the Ergonomics Checklist program

The Ergonomics Checklist program consists of a checklist, files containing information and instructions associated with the list, subroutines evaluating specific areas of ergonomics, and the main program processing this information. The checklist can be changed according to the type of designed application. The checklist can be, e.g. a general inspection instruction or a detailed list made for a special object of design.

The program displays one point of the checklist at a time on the terminal. The designer is able to consider the designed workplace model with respect to each point of the checklist. Should any defects be found, corrections can be made immediately. It is also possible to retrieve information from all existing files during any stage of the program, and to have this information displayed.

The files associated with the program may be used alone as a data source without running through the checklist. At the end of the session, a summary of the terminated session, or of all previous sessions, can either be displayed on the terminal, or a computer printout can be made. The date, designer, object of design, modifications made, and uncorrected defects are included in the summary (Järvinen, 1987).

Usability of the Ergonomics Checklist program

The checklist was tested for the design of an electronics assembly line. The checklist used was a set of general inspection instructions based on an ergonomics design manual (Clark and Corlett, 1984).

Testing of the program proved that it is an operative tool for workplace design. The modifications made during the testing phase also proved that use of the program improves design quality. The computer and a checklist together enable systematic examination, the effect of this being that the ergonomic properties are considered thoroughly. When the files are well-organized, it is easy and quick to retrieve the ergonomic information needed.

Exact modelling of the workplace using CAD is time-consuming and laborious. The complicated graphic model is slow to handle. Therefore, it is not reasonable to create a graphic model of the workplace just for the program to evaluate its ergonomics properties.

Although the designer makes all the decisions, the program may restrict planning in some cases. Thus the checklist should be as versatile as possible (Järvinen, 1987).

Computer-aided evaluation of the ergonomic properties of a tractor cabin

The aim of this project was to create a CAD program for the evaluation of the ergonomic properties of a tractor cabin. The main principles when designing the use of this program were that it should be easy to learn, easy to run and easy to remember (Leppänen, 1988).

The program runs on an AutoCAD drawing program which operates in a microcomputer. AutoCAD provides two-dimensional wire-frame modelling and a three-dimensional visualization of a product model. It also enables the designer to build screen menus and to compile programs using an interactive AutoLISP programming language (Autodesk, 1986).

Structure of the program

The evaluation program consists of four parts, which are:

1. Text files containing information about: (i) work space, (ii) visibility, (iii) controls, (iv) displays and (v) use of this program.
2. Graphic files containing: (i) a two-dimensional man model, and (ii) recommendations for the location areas of hand-operated and foot-operated controls.
3. Analyzing programs, which enable analysis of the ergonomic acceptability of hand-operated control locations.
4. A screen menu, which makes it easy to use this program.

Text files

There are 15 text files, which can be displayed interactively on the screen. The text files contain information about the ergonomic evaluation and design of work space, access, visibility, controls, and displays.

Two-dimensional man model

The man model is presented from the side view. It is a wire-frame model consisting of eight segments which can be interactively rotated around joints until the desired posture is found. The man model can be used to evaluate, e.g. work space, working postures, reach, and the fields of vision available to the driver (Figure 8.3).

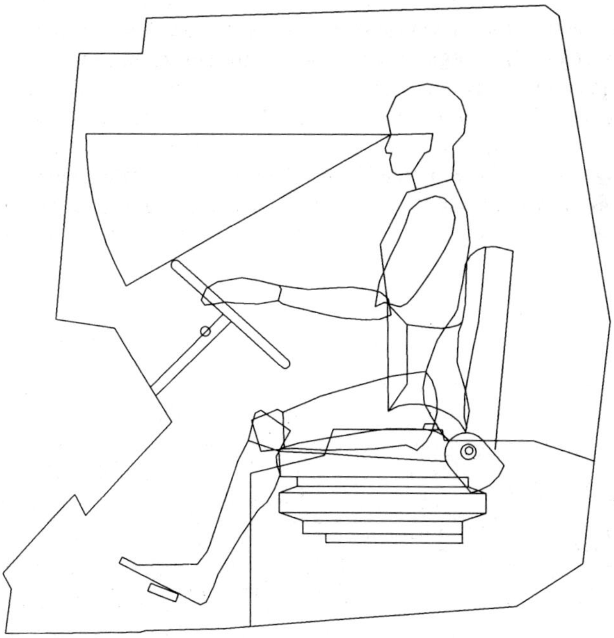

Figure 8.3 Two-dimensional man model in the Valmet 505 tractor cabin (Leppänen, 1988).

Recommended location areas of controls

The graphic presentation of the recommended location areas for hand-operated and foot-operated controls is based on the SAE J898 recommendation (SAE J898 JUL 82, 1983) which separates into five views: one from the front, two from the side (left/right) and two from the top (left/right).

Analyzing programs

The analyzing programs are compiled in an interactive AutoLISP programming language. There are three separate programs: (i) one for analyzing the side view, (ii) one for the top view, and (iii) one for the front view. The evaluation criteria are the driver's reach, the location of controls and the frequency of use of controls. The evaluation is based on Schmidtke's (1976) method for the ergonomic evaluation of a work system.

The results of the analysis can be seen on the display, it can be printed out, or it can be stored on disk for later use.

Experiment: evaluation of the Valmet 505 tractor cabin

Tractor drivers often face problems of difficult work postures, noise, vibration, and dust (Matthews, 1977). Access to the cabin and exit from it are problematic. Although these usually take only a short time, most accidents occur while entering and leaving the cabin (Häkkinen *et al.*, 1988).

The object of this evaluation was a Valmet 505 tractor. This tractor model belongs to the 'middle-series' (3190–4180 kg, 50–77 kW), which is the most popular tractor group in Finland.

For this evaluation, two-dimensional models of the tractor cabin from side, top and front views were created using technical drawings. For the evaluation, the dimensions of the cabin were measured from the models. The evaluation of controls was made for 11 hand-operated controls in two different work postures, the seat angle having a swivel of 0° and 30° clockwise.

According to the evaluation, the Valmet 505 tractor cabin is ergonomically acceptable. Minor changes in the location of some controls would make it ergonomically well-designed (Leppänen, 1988).

ErgoSHAPE program

The Institute of Occupational Health in Helsinki developed a program, called ErgoSHAPE, in 1987–8. ErgoSHAPE runs in a widespread micro-CAD program called AutoCAD. ErgoSHAPE consists of geometric two-dimensional human models which can be manipulated using commands selected from a menu file. Both male and female models can be called up. The hight of the models can be selected from three anthropometric sizes, the 5th, 50th and 95th percentiles (Launis *et al.*, 1988). A biomechanical static strength calculation is also provided. This facility tells the designer the load which affects the elbow, shoulder and the back at each working posture. More information about ErgoSHAPE is presented elsewhere in this book.

Conclusions

All these described programs are developed to help the designer take into consideration the 'human' parameters needed during the design process. The programs are designed to be:

1. Quick to use at every design stage,
2. Equipped with simple and effective interaction,
3. Usable without any previous experience of ergonomics,
4. Able to give reliable and unambiguous design parameters, and
5. Equipped only with the features the designer really needs.

In Finland, very few industrial designers now use computer-aided ergonomic design systems. The situation is probably the same worldwide.

To make these programs suitable for everyday design tasks, they should interact better with the existing design systems and methods. The ergonomic parameters should also be more task-oriented than at present. This would help designers select the essential information required for their plans.

References

Autodesk, 1986, *AUTOCAD Reference manual*, Autodesk Inc., USA.

Chaffin, D. and Andersson, G. B. J., 1984, *Occupational Biomechanics* (New York: Wiley).

Clark, T. S. and Corlett, E. N., 1984, *The Ergonomics of Workspaces and Machines* (London: Taylor & Francis).

DeGreve, T. B. and Ayoub, M. M., 1987, A workplace design expert system. *International Journal of Industrial Ergonomics*, **2**, 37–48.

Diffrient, N., Tilley, A. R. and Bardagjy, L. C., 1981, *Humanscale 1/2/3* (Cambridge, MA: Massachusetts Institute of Technology).

Dooley, M., 1982, Anthropometric modeling programs – A survey. *IEEE Computer Graphics and Applications*, **2**, 17–25.

Dreyfuss, H., 1966, *The Measure of Man. Human Factors in Design* (New York: Whitney Library of Design).

Hansson, J.-E., Klusell, L., Löfroth, C., Pettersson, B. and Zylberstein, M., 1978, Skogsmaskinen som arbetsplats 1977. (Ergonomic evaluation of logging machines). *Redogörelse*, No. 6 (Skogsarbeten, Stockholm).

Häkkinen, K., Väyrynen, S. and Pesonen, J., 1988, *Traktoreiden kulkuteiden arviointi*. SF, Työsuojeluhallitus, Selvityksiä 5.88.

ISO 6385, 1981, *Ergonomic principles of the design of work systems*. 4 pp.

Järvinen, J., 1987, Ergonomics checklist program for computer aided workplace design. In *New Methods in Applied Ergonomics*, edited by J. R. Wilson and E. N. Corlett (London: Taylor & Francis), pp. 77–80.

Karhu, O., Kansi, P. and Kuorinka, I., 1977, Correcting working postures in industry: A practical method for analysis. *Applied Ergonomics*, **8**, 199–201.

Kuusisto, A., Väyrynen, S. and Ojanen, K., 1987, The CAD system as a tool for analysing ergonomic features of maintaining a forest tractor. In *IXth Joint International Ergonomics Symposium "Working postures in agriculture and forestry"* (Kuopio, Finland: Regional Institute of Occupational Health), pp. 61–62.

Launis, M., Lehtelä, J. and Kuusisto, A., 1988, Kaksiulotteinen antropometrinen ihmismallijärjestelmä työpaikkojen suunnitteluun (CAD). *Työ ja ihminen*, **2**, 68–83. (With English summary.)

Leppänen, M., 1988, *Traktorin ohjaamon ergonomian tietokoneavusteinen suunnittelu ja arviointi*. Report 51. (Tampere: Tampere University of Technology, Occupational Safety Engineering).

Leppänen, M. and Mattila, M., 1987, Computer aided ergonomic design. In *New Methods in Applied Ergonomics*, edited by J. R. Wilson and E. N. Corlett (London: Taylor & Francis), pp. 81–87.

Lippman, R., 1986, Arbeitsplatzgestaltung mit Hilfe von CAD. *REFA-Nachrichten*, **3**, 13–16.

Matthews, J., 1977, The ergonomics of tractors. *ARC Research Review*, **3**, 59–65.

Mattila, M. and Leppänen, M., 1987, Computer aided ergonomics design – a program for suitable control locations. In *Human–Computer Interaction – INTERACT '87*, edited by H.-J. Bullinger and B. Shackel (Amsterdam: North–Holland), pp. 997–999.

McCormick, E. J. and Sanders, M. S., 1982, *Human Factors in Engineering and Design* (New York: McGraw-Hill).

Panero, J. and Zelnick, M., 1979, *Human Dimensions and Interior Space* (New York: Whitney Library of Design).

Roebuck, J. A. Jr., Kroemer, K. H. E. and Thompson, W. G., 1975, *Engineering Anthropometry Methods* (New York: John Wiley).

SAE J898 JUL82, 1983, *Control Locations for Off-Road Work Machines* (Society of Automotive Engineers Inc.).

Schmidtke, H., 1976, *Ergonomische Bewertung von Arbeitssystemen* (München: Carl Hanser Verlag).

Väyrynen, S., 1986, Mechanical safety and ergonomic factors in the maintenance of forest machinery with special reference to the accessibility of maintenance objects. Työterveyslaitoksen tutkimuksia 4, lisänumero 1, Helsinki, 198 pp. Doctoral thesis. (In Finnish with an abstract and summary).

PART III
WORKSPACE DESIGN

Chapter 9
ERGOSPEC: a prototype expert system for workstation design

LOUIS BRENNAN, LORRAINE FARRELL, and DERVILLE MCGLENNON

Department of Industrial Engineering and Information Systems,
Northeastern University,
Boston, MA 02115, U.S.A.

Abstract. This paper is concerned with the development of ERGOSPEC, which is a prototype expert system to aid the ergonomist in designing a workstation, addressing particularly the needs of the operator and the aspects of the task to be performed. The nature of ergonomics is examined and the abundance of facts, rules and heuristics as the basis of ergonomics is highlighted. It is argued that ergonomics provides an appropriate application area for expert systems.

Keywords: Ergonomics; ERGOSPEC; Expert systems; Work design.

Introduction

Expert systems have been designed to serve a number of roles including those of designers, problem-solvers and tutors. Many of the problems that have been felt to require human management are now amenable to expert systems. Modelling the way the expert performs the task rather than modelling the task itself is the primary characteristic of an expert system.

This paper describes a prototype expert system for ergonomic based workstation design. A prototype implies a system which is capable of providing samples of the capability that can be expected from the ultimately constructed system. The role of the prototype is to illustrate what has been specified for the system and help to identify shortcomings in the system before the major design work is commenced.

Rule-based systems are introduced and the match between production systems and ergonomics described. OPS5, the development tool used in ERGOSPEC, is outlined (Brownson *et al.*, 1985) and its appropriateness for use in ERGOSPEC is explained. The nature of the prototype expert system is outlined, its usefulness is evaluated and the possibilities for extending the design are explored.

Expert systems

Expert- or knowledge-based systems are designed to represent and apply factual knowledge of specific areas of expertise to solve problems. The term expert system is used to refer to an intelligent computer program that is able to perform within a specific and limited task domain (a field or application area) at the level of a human expert in that domain.

Expert systems derive their specific identity by performing difficult tasks at expert levels of performance, by emphasizing domain-specific problem solving strategies, by employing self-knowledge to reason about their own inference processes and providing explanations or justifications for the conclusions reached. Finally, the problems they solve generally fall into one of the following categories: interpretation, prediction, diagnosis, debugging, design, planning, monitoring, repair, instruction and control (Hayes-Roth *et al.*, 1983). Rule-based systems use collections of rules to solve problems. Rule-based systems, also known as production systems, have been used extensively to model human problem solving. A major reason for developing rule-based expert systems is that human experts find it intuitively appealing to express their domain knowledge in terms of situation action pairs. Rules are a relatively uniform way to represent knowledge without complex programming constructs. Highly motivated domain experts can sometimes extend and debug knowledge bases without being trained experts. Because expert systems are built incrementally as the expert's knowledge accrues in piecemeal fashion, the knowledge must be encoded in a readily extendable and modifiable form.

Expert systems can provide an audit trail of the data and references on which their recommendations are based. The prototype system ERGOSPEC is designed to provide not only a static recording of expertise, but also the capability to develop and extend that expertise to local situations.

Building expert systems

To build an expert system successfully, two types of resources are required – a methodology and a set of tools.

The transfer and transformation of problem-solving expertise from a knowledge source to a program is the heart of the expert system development process. The knowledge engineer and expert must define the problem, discover the basic concepts involved and express the relationship that exists between concepts. Thus the major stages in building an expert system are identification, conceptualization, formalization, implementation and testing. Finally the developed prototype should solve the initial sub-problem and provide the desired interface features. Ideally following the development of the prototype the expert system should evolve from simple to increasingly hard tasks, by improving the organization and representation of knowledge.

The previous section has dealt briefly with the methodology of the problem. The other element of building expert systems relates to the tools available. In selecting from the tools available the predominant consideration involves matching the problem characteristics to necessary tool features. The problem characteristics suggest certain solution features, and these together with the desired expert system features, suggest tool features that would facilitate development. The desired tool features then form the basis for choosing a particular tool.

Production systems

A production system is an appropriate term when the knowledge to be programmed occurs naturally in rule form. Ergonomic expertise contains factual knowledge (i.e. knowledge learned from books and/or gained from experience) and heuristic knowledge (i.e. rules of thumb, intuitions and association). These two facets of ergonomic expertise match the requirements of an expert system or, more specifically, a production system.

The program in a production system consists of an ordered collection of basic units called productions, production rules or simply rules. The architecture of a production system includes three major components.

The first component is the working memory which is a data base representing facts and assertions about the problem. The second component is the rule memory containing production rules which are really condition–action pairs. Each rule has a condition part usually indicated by the keyword 'IF' in an English description, which describes the data configuration for which the rule is appropriate. The second part of the rule is the action part, denoted by the keyword 'THEN' in an English description which gives instructions for changing the data configuration. The third component of the production system is the inference engine whose function is to execute or 'fire' the rules.

Application areas of expert systems

Expert systems lend themselves to applications where the nature of the problem has the following characteristics:

1. A solution to the problem has so high a payoff that it warrants the development of a system. Solutions are needed in the area and other methods of obtaining them have not worked.
2. The problem can be solved only by an expert's knowledge rather than by using a particular algorithm.
3. There is access available to an expert who can help formalize the knowledge needed to solve the problem.

The problem addressed by expert systems does not necessarily have a unique solution. In fact, expert systems can work best for problems that have a number of acceptable solutions. Expert systems can be useful in solving routine problems freeing the expert for more challenging tasks. Expert systems can also bring expertise to locations where a human expert is unavailable (e.g. ergonomic expertise where there is no in-house ergonomist). Expert systems have explanation facilities that tell how they reached their answers and also the reasons for choosing one alternative over another. Expert systems are also useful in dealing with problems where the information is not always entirely consistent or complete and still produce satisfactory answers.

Expert system technology applied to ergonomics

Traditionally ergonomics has not been a well understood and applied methodology. A variety of reasons explains this phenomenon. For example, ergonomics has not been viewed as an area for cost–benefit payback and so areas like ergonomic workstation design have received relatively little attention. The lack of ergonomic expertise has also been a difficulty.

Ergonomic expertise is gained from many years of experience of working in the field coupled with knowledge acquired from studying and researching in the application area. Thus, ergonomic expertise consists largely of two varieties – factual and heuristic. As such, there is a clear match with the contents of an expert system. In fact, ergonomics is an ideal application area for an expert system due primarily to it being a production system, i.e. ergonomics is based virtually entirely on facts, rules and heuristics.

Examples of expert systems in ergonomics

There are some examples of successful application areas in the field of ergonomics or areas similar in nature to ergonomics. For example, Kengskool *et al.* (1987) have developed an expert system for human operators' workplace design activities by reducing the time and effort required of human experts in the industrial engineering field. The system is microcomputer based and is designed to assist a human operator with the design of work systems. The problem domain of this system is one of manual jobs performed by one operator at one workplace where the operator totally controls the cycle time of the operation.

Another example is a prototype expert system for physical work stress analysis (Chen *et al.*, 1987). The prototype expert system was built by applying models and guidelines from experts to build a knowledge based system to diagnose and solve physical work stress problems.

A final example is LIFTAN (Karwowski *et al.*, 1986) which is an experimental expert system for analysis of manual lifting tasks. The system allows a non-expert in the field of manual lifting to utilize relevant knowledge and apply it to analyse specific work situations.

These examples illustrate the type of opportunities that expert systems bring to the task of widespread application of ergonomic principles.

Selection of a development tool for prototype system

In considering our choice of languages/tools, LISP, PROLOG and OPS5 were reviewed. With LISP and PROLOG the expert system must be built from scratch using the raw programming language. By contrast, OPS5 is a higher level knowledge representation language.

LISP is a highly flexible, user written/controlled inference procedure which makes use of recursive procedures. It is difficult to interface and requires a long lead time to proficiency.

PROLOG is a backward chaining language which is especially suitable for diagnosis or query systems. From the perspective of expert systems applied to ergonomics, the disadvantage lies in its query orientation as opposed to a data driven orientation.

OPS5 is a production system programming language. It is primarily used as a production rule, forward chaining system and therefore it can be classified as a narrowly focused tool which can aid a developer in building rule-based, forward chaining systems – thus the match with our particular problem of workstation design. Forward chaining is a problem solving method which starts with an initial knowledge and applies inference rules to generate new knowledge until either one of the inferences satisfies a goal or no further inferences can be made.

OPS5 was chosen as the development tool for the following reasons:

1. OPS5 allows for easy access to other software tools. It is a very general programming environment.
2. It involves English-like knowledge representation making it user-friendly.
3. It is ideally suited for data driven modelling – user specified data is easily accepted – a generic model is applicable.

4. OPS5 is a production rule, forward chaining language which is ideal for our problem of workstation design. The optimum solution, namely the final workstation layout, can be obtained from user specified data.
5. OPS5 is one of the most widely used application languages for commercially successful, large expert systems.

OPS5 has been referred to above as a programming environment. These are special packages of pre-written code that lie above the high level languages such as LISP and PROLOG. They are usually associated with a particular high level language and contain a chunk of the code written in that language that is useful for particular programming tasks. OPS5 is closely associated with LISP.

Knowledge engineering tools such as OPS5 offer two advantages over knowledge system developers: (i) they provide for rapid system development by providing a substantial amount of computer code that would otherwise need to be written, tested, debugged and maintained; and (ii) tools provide specific techniques for handling knowledge representation, inference and control that help knowledge engineers to model the salient characteristics of a particular class of problem.

When a tool is combined with knowledge in a specific domain, the result is a knowledge-based expert system. Such a knowledge system delivers advice similar to that which a human expert in a specific problem domain might give.

OPS5, which consists of working memory and rule memory, has been chosen as the appropriate development mechanism for this particular expert system as it allows simple representation of the problem domain in rule format.

ERGOSPEC

ERGOSPEC is a prototype expert system which illustrates the basic process by which an expert system is constructed and applied in the area of ergonomic workstation design. Specifically facts and alternatives are first accumulated from appropriate sources. Secondly, relationships between these facts are identified and used in the construction of rules for solving specific problems. Finally, an appropriate expert system tool (OPS5) is selected to implement a process of forward chaining inference to apply the knowledge base to individual problems. Since the problem of workstation design involves an unstructured and complex decision making process typically involving the use of domain dependent knowledge, implementations of appropriately designed expert systems provide the potential for extensive and economical distribution of expertise for this problem.

In ERGOSPEC, OPS5 is employed to interpret user inputs. User inputs involve descriptions of the characteristics of the job to be performed at the workstation, operator characteristics and a number of other factors. Screen Management (SMG) and BASIC are used to keep the user up to date while interacting with the system. The system was developed and implemented on a VAX II/780 computer.

The main features of the ERGOSPEC system are a system specification ability/menu driver, an external data base which contains a variety of ergonomic information, BASIC code which controls all the input/output operations, the production/rule base which acts on the user specification and the generated report which provides workstation design specifications to match the user specified requirements (see Figure 9.1).

The motive behind the development of ERGOSPEC was to investigate whether computer methods, in particular, an expert system could assist with such problems in workstation design as: (i) consideration of population stereotypes; (ii) comfort, safety, aesthetics and

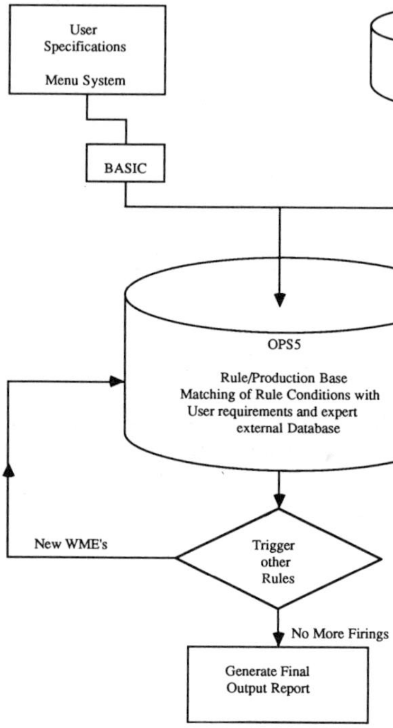

Figure 9.1 ERGOSPEC execution mechanism.

styling; (iii) location of controls for ease of use; (iv) balance of work between limbs; (v) avoidance of operator overload conditions; (vi) the need to satisfy the wide range of operator sizes; and many other factors, possibly not quantifiable, which an experienced designer (an expert) would bring to bear in the design of a new workstation arrangement.

In designing a protocol or navigation mechanism for ERGOSPEC the following characteristics of the problem domain area were considered to be pertinent to the task and therefore were incorporated as part of ERGOSPEC's design methodologies.

1. Job characteristics (broken down by task),
2. Operator characteristics,
3. Operator sitting/standing,
4. Time constraints,
5. Population percentile statistical factors.

From observations and questioning on the above aspects it can be deduced that ERGOSPEC's production system is necessarily data-driven and forward chaining with applicability across a spectrum of different workstation designs. In developing this prototype expert system it is intuitive that the system must be extremely user-friendly as well as allowing for easy update and modifications as the system is further developed into a working system.

Development and use of ERGOSPEC

Essentially the development of ERGOSPEC required two major activities. The first was the construction of a prototype knowledge base to support the design of a workstation. The second was the construction of a prototype expert system to utilize this knowledge base.

The expert system is used to proceed from a simple description of a workstation to a specific ideal workstation using a forward chaining process. The system developed is considered a prototype because the knowledge base supporting the expert system considers only a limited range of aspects. The prototype system is merely meant to be representative of the type of much larger scale systems that can potentially be developed to deal with all types of aspects of workstation design.

The external working memory data base is loaded into the system at run-time. It contains information concerning human capabilities and anthropometry, the visual and auditory environments and controls and displays. Specifically the working memory contains anthropometric data, optimum chair and workbench design dimensions, characteristics of dials, etc. and corresponding optimum viewing distances, the maximum muscular force capabilities of males and females and the maximum power capabilities in various working positions and some standard recommendations.

The production memory contains 18 rules relating to workspace dimensions, task design, setting of environmental variables and rest periods and design of controls and displays. The first two rules relate to the design of the workspace. These rules relate operator's height and sex to a variety of population dimensions and workspace and seat design. The second group of rules pertain to the positioning of the operator when performing operations requiring the exertion of force, hand operations and visual tasks. This group also deduces reach capabilities for a particular operator profile and acceptable shelf heights.

Five rules deal with the auditory environment and auditory displays. These rules specify maximum noise intensity and duration and suitable auditory displays for different types of auditory signals. For example, the rule which addresses maximum noise intensity and duration suggests a maximum noise duration of 5 h for a noise intensity of 85 dB with shorter durations suggested for more intense noise levels.

The visual environment is covered by a single rule which specifies illuminance values and visual displays for different types of tasks. Thus a value for illumination of 400 lux is suggested for work involving recognition, while tasks involving scanning and precise reading are recommended illumination values of 600 and 800 lux respectively. The rule associates, for example, circular dials with comparative work, digital displays with work where accuracy and quantitative data are emphasized and linear dials with work requiring the reading of ranges.

Two rules deal with the selection and location of controls for both hand and foot operations. The remaining rules are concerned with communication between operators and the determination of rest periods related to the type of work being undertaken by the operator. The user interacts with the system to provide the specification for the workstation. After the user responds to each inquiry from the system, the rules are examined and, if appropriate, executed to build up a design for the required workstation.

Interaction with the expert system occurs at a number of levels (see Figure 9.2). At level one, the user is offered the main menu. This consists of the following options:

1.1 Introduction
1.2 Specify operator
1.3 Specify task description
1.4 Specify percentiles

1.5 Specify positions
1.6 Specify rest periods
1.7 Specify communications
1.8 Specify time
1.9 Generate workstation design
1.10 Exit

Selection of option one at level one provides the user with a general overview of ERGOSPEC. Option nine is selected to produce the design for the workstation from the expert system. However options two to eight generate level two options which allow the user to provide more information to the system.

At level two the following options follow from the earlier selections at level one.

1.2 Operator: M/F
1.3 Task description: visual, hand, food, auditory
1.4 Percentile: 5th, 50th, 95th
1.5 Position: sit, stand, sit/stand
1.6 Rest periods: every hour, two hours, three hours
1.7 Communication: with operator/machine
1.8 Time: in seconds/minutes/hours

In some cases further options follow at level three. These relate to the operator and the task description. For example, for each of the four task descriptions the possible roles of various operators can be identified. Thus in the case of a visual operation, the various possibilities which the operators may encounter are displayed for selection by the user. These include:

Read precisely
Read a range
Go–No Go Gauge
Quantitative information
Colour patterns
Sequences, symbols
Light pattern
Warning indicators

Depending on the factors specified by the user and using the relationships and rules pertaining to this set of factors, the system provides recommendations as to the design of an optimal workstation. For the prototype system these recommendations are in text format. The recommendations specified include suitable workspace dimensions, control and display designs and environmental requirements. An example of the type of output which the system generates is illustrated by the extract given in Table 1.

The output is generated by the user merely specifying the operator sex and height. As the user inputs more information to the system, other aspects of the workstation design are addressed by the system and output.

The prototype expert system described here is by no means thorough enough for most real world applications. However, it does illustrate the basic processes by which such a system can be constructed. Furthermore as the system has been developed in OPS5 and in the language BASIC, the software can be easily modified and added to. Coupled with the report which the system generates on the workstation, it also creates another file which essentially traces the firing of the rules within the system execution, thus providing a simple explanation facility.

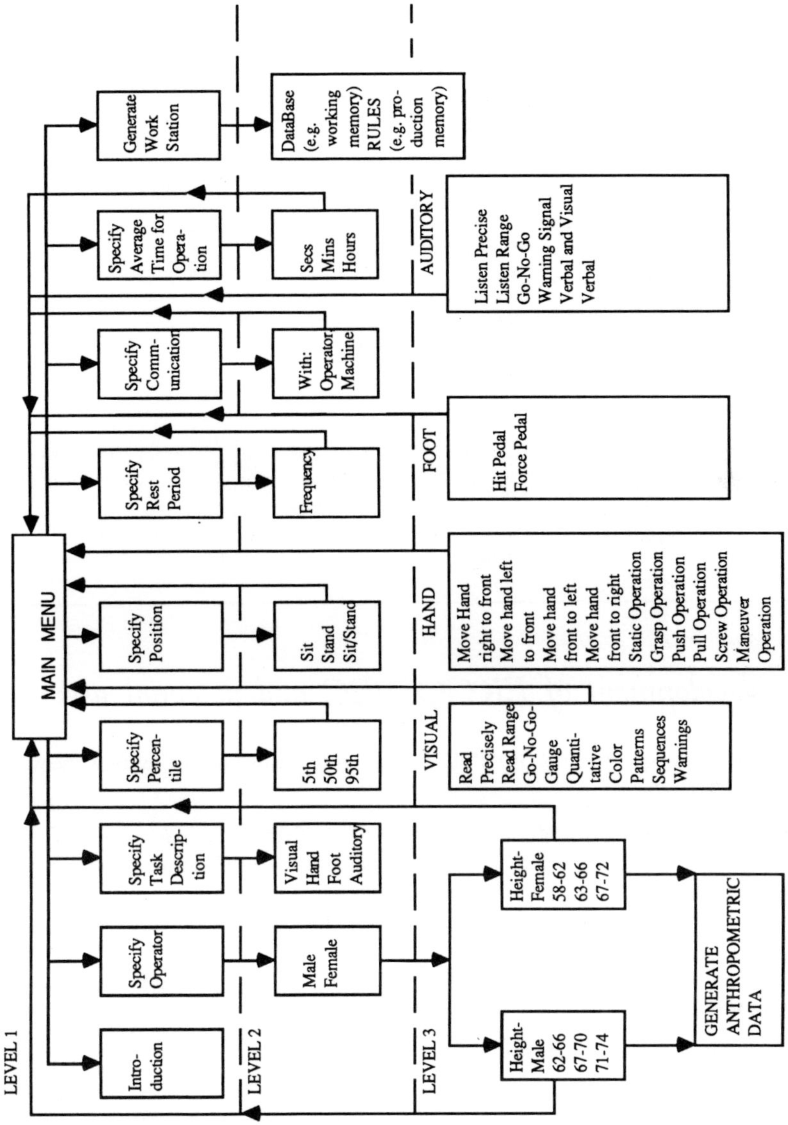

Figure 9.2 *User interaction with the prototype system.*

Table 9.1 Extract of sample output from a prototype system. Given the fact that one of the operators who will use this workstation is 63 inches tall and is female, the following characteristics are derived from population percentile factors gained from in-depth studies in the field of ergonomics by experts.

Overhead reach is	79·0 ins
Body width is	17·3 ins
Sitting height is	33·4 ins
Eye height when sitting is	29·3 ins
Eye height when standing is	60·3 ins
Buttock popliteal length is	18·9 ins
Knee height is	20·5 ins
Thigh clearance is	5·4 ins
Forward reach is	31·0 ins
Elbow to fingertip distance is	17·1 ins
Shoulder breadth is	15·7 ins
Chair height for 95% of population is	17·0 ins
Popliteal sitting height is	19·6 ins
Workbench height for standing is	41·0 ins
Workbench height for sitting is	36·0 ins
Maximum height of overhead shelf is	65·0 ins
Minimum height of overhead shelf is	50·0 ins
Knee clearance is	6·0 ins
Length of foot control is	4·0 ins
Width of foot control is	4·0 ins
Hand clasp distance is	28·0 ins
Appropriate pedal angle is	10·0 deg
Push force	8·0 kps
Pull force	20·0 kps
Kicking with knee (force)	37·0 kps
Stretching the back (force)	74·0 kps

Future development of ERGOSPEC and potential uses

Developments envisaged for ERGOSPEC which would enhance the system include the following:

1. Add more 'expertise' by adding more rules to the OPS5 rule base.
2. Provide an enhanced interface for more detailed user specifications by adding more menus and submenus to the user interface.
3. Generate a graphical output from the system supplemented by an extensive textual report on the workstation design.
4. Provide more help/user-friendliness for the non-expert user, by explaining some of the ergonomic terminology contained within the system.

It is possible to incorporate the above suggestions into the current software with relative ease and little or no changes.

With the enhancement of the prototype expert system to incorporate the above suggestions, the system would be highly valuable in the following roles: (i) as a confirmation of the ergonomist's beliefs; (ii) as an aid to the industrial engineer; (iii) as a decision support tool for the non-expert/lay user; (iv) as a quick reference guide for workplace design; and (v) as a training aid for student ergonomists/industrial engineers.

References

Brownson, L., Farrell, R., Kant, E. and Martin, N., 1985, *Programming Expert Systems in OPS5* (Reading, MA: Addison-Wesley).

Chen, J., Schlegel, R. E. and Peacock, J. B., 1987, Prototype expert system for physical work stress analysis. In *Trends in Ergonomics/Human Factors IV*, edited by S. S. Asfour (Amsterdam: Elsevier).

Hayes-Roth, F., Waterman, D. A. and Lenat, D. B., 1983, *Building Expert Systems* (Reading, MA: Addison-Wesley).

Karwowski, W., Mulholland, N. O., Ward, T. L., Jagannathan, V. and Krichner, R. L., 1986, LIFTAN - an experimental expert system for analysis of manual lifting tasks. *Ergonomics*, **29**(10), 1213–1234.

Kengskool, K., Goldman, J. and Leonard, M. S., 1987, An expert system for human operators workplace design. In *Trends in Ergonomics/Human Factors IV*, edited by S. S. Asfour (Amsterdam: Elsevier).

Chapter 10
CAD-video-somatography: a method for the anthropometric design of workplaces

H-J. BULLINGER and D. LORENZ[1]
University of Stuttgart and Fraunhofer Institute for Industrial Engineering, Stuttgart, Federal Republic of Germany

[1]*Fraunhofer Institute for Industrial Engineering, Stuttgart, Federal Republic of Germany*

Abstract. CAD systems are used more and more in designing manual and partly automated work systems. Ergonomic aspects are often not taken into consideration because of a lack of usable ergonomic design instruments. Therefore a wide variety of computer-assisted models of man for the anthropometric design of workplaces have been developed and demonstrated worldwide. As different from these methods, which are based on the modelling of man, CAD-video-somatography calls upon a real human being. By using video systems a trick picture is established, that allows a human to 'move' in a drawing of a workplace produced by a CAD system. CAD-video-somatography is largely independent of the type of the CAD-systems used.

Keywords: Anthropometric design; Workplace design; Body sizes; CAD-video interface.

Introduction

Workplace design is an essential task of ergonomics in industral practice. It pursues the goal to adapt the working conditions to the anatomical and psychophysiological conditions of man and to enhance the effectiveness and productivity of the work system. In so doing, the use of the conceptive methods of work design, in which design rules and principles are being considered as early as in the stages of planning and designing of workplaces, should be given absolute priority over corrective measures.

Planning and design aids making allowance for the requirements of ergonomics will be able to be used successfully in the future, if they are compatible with the methods and aids used in engineering and work design. This is where CAD systems are increasingly gaining in importance. The aids for anthropometric work design such as body size tables and templates, still frequently used today, will thus lose importance. Because of this predictable trend, a wide variety of computer-assisted models of man for the anthropometric design of workplaces are being developed and demonstrated worldwide (e.g. Kroemer *et al.*, 1988). As different from these methods, which are based on the modelling of man, video-somatography

(Martin, 1981; Lorenz, 1983) has been successfully used for the anthropometric and movement-oriented design of workplaces. Using video systems, video-somatography calls upon a real human being for establishing the dimensional design of workplaces in sketches and design drawings. CAD-video-somatography now also permits the advantages of this method to be used in the development and design of workplaces on computer-assisted graphics systems (generally called CAD systems).

Functional description of CAD-video-somatography

CAD-video-somatography (CADVS) is a method of analysis and design for dimensioning the most varied types of workplaces with respect to measures and movements. It can be used in the very early stages of the development and design process on CAD systems. The functional principle of CADVS can be seen from Figure 10.1.

CADVS is divided into three functional areas: (i) subject shooting area (SSA); (ii) control area (COA); and (iii) CAD area (CAA). In the SSA, orthogonal shots of a subject are taken for all three spatial planes by video cameras (camera 1, 2, 3 in Figure 10.1) and made available to the COA. In the CAA, the view of the workplace to be analysed and designed will be prepared as a sketch or design drawing or selected from an existing set of drawings. This drawing will also be at the disposal of the COA through a CAD-video interface.

In the COA, the associated view of the subject will now be admixed to the selected drawing or faded over and displayed on several video monitors. Zoom lenses used on the video cameras permit the size of the image of the subject to be adapted to the scale conditions of the drawing. This geometrical adjustment is under control of COA. On suitable selection of the subject, the zoom lenses also permit the image of the subject to be scaled up or down in such a way that the adjustment range from the 5th to the 95th percentile which is commonly used in ergonomics can be steplessly covered. The subject can watch his or her image as a trick picture on one or several video monitors installed in the SSA in conjunction with the drawing of the workplace produced by the CAD system. This visual feedback now enables the subject

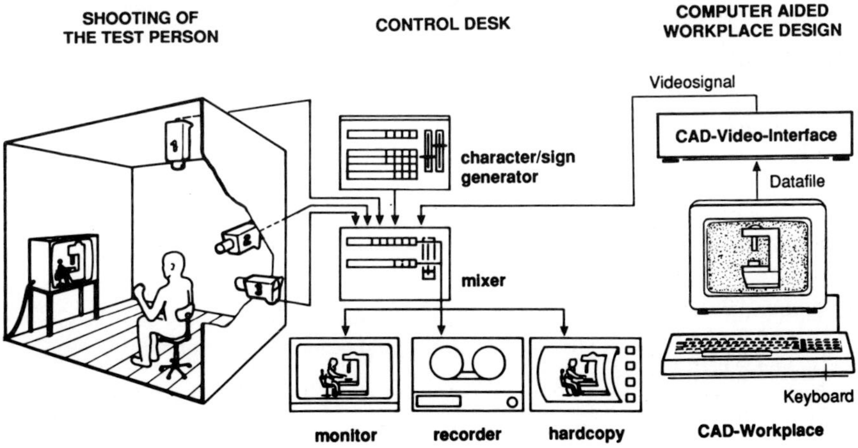

Figure 10.1 Configuration of a CAD-video-somatography system.

to assume the desired postures at the hypothetical workplace and to go through the movements required. This permits, for example, the simulation of a complete work cycle, in which the required workpieces and tools can be made available to the subject. Video recorders and hard copiers are used for documentation of the analysis and design steps. Individual steps of the analysis can be identified and explained via a text generator. While all movements of the subject can be dynamically documented on the recorder, the hard copier permits the instant reproduction of the individual work situations. As a further explanation of the CADVS, the equipment required in the CADVS laboratory and the approaches to performing anthropometric design tasks will be briefly described below.

Set-up of a CAD-video-somatography laboratory

Analogous to the functional areas of CADVS represented in Figure 10.1, the laboratory required and the necessary hardware and software can also be sub-divided for these three areas. The hardware configuration described below essentially reflects the laboratory installed at the Fraunhofer-Institut for Industrial Engineering (IAO), Stuttgart, Federal Republic of Germany. Figure 10.2 shows a condensed representation of the system configuration of the CADVS; in addition, Figure 10.2 also shows the mode of functioning of the specifically developed CAD-video interface in more detail.

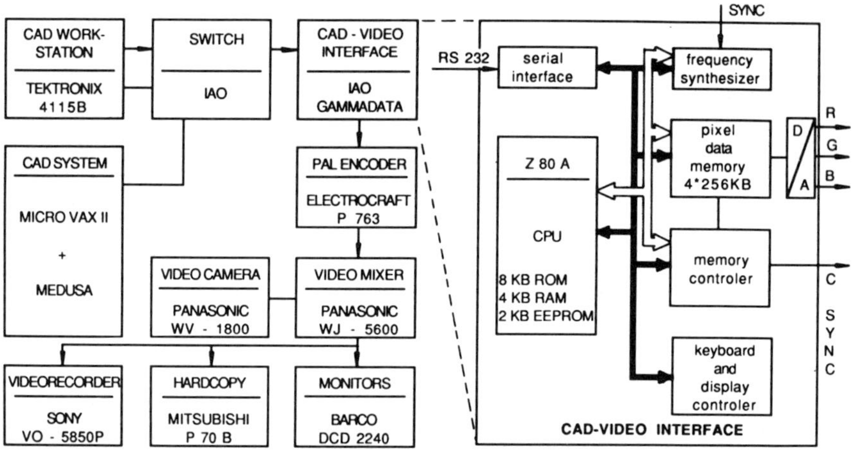

Figure 10.2 System configuration CAD-video-somatography.

In setting up a CADVS laboratory, it must be ensured that the SSA and the COA are accommodated in one room; ideally, the CAA should be located in neighbouring rooms, but this is not absolutely necessary.

CAD area

The primary task of the CAD system is to assist a design engineer or work designer in his development and design work. For this purpose, a CAD system must be selected and installed which meets the specific requirements and expectations of the user. There are no special requirements for the use of the CAD system in the CADVS laboratory. However, it is of advantage to be able to generate perspective representations of workplaces under freely selectable directions of observation and perspectives. The sketches or design drawings produced on the CAD display screen must be translated into the television standard used in each case via suitable converters. In the CAD system used as a basis for the development of the CADVS and the CADVS laboratory installed at the IAO, a MicroVAX II (Digital Equipment Corporation) is used as a host computer, using the CAD software CIS-MEDUSA (Computervision). The main memory capacity of the MicroVAX II is 13 Mbytes. Besides the hard disk storage (with three disks) with a capacity of 1·1 Gbytes, a magnetic streamer tape device with a capacity of 95 Mbytes is connected. At the CAD workplace, there is a graphics raster display screen (with a resolution of 1280 × 1024 pixels, colour in four picture levels = 16 colours), an alphanumeric display for checking the entered texts and instructions, a joy stick for movement of the cross hairs and a menu tablet handled by a joy stick.

The transition to the video area (see Figure 10.2 on the right 'CAD-video interface') is the serial standard interface RS 232 C (V 24) with a downstream video storage/compressor, specifically developed for this purpose by the IAO together with Gammadata. The task of this CAD-video interface with four video storages is the conversion of the video signals coming from the CAD graphics display screen into the German TV standard (PAL system with a maximum resolution of 800 × 625 pixels). Some of the newer CAD workstations already have a video interface. In these cases a special CAD-video interface (video storage/compressor) is not needed.

Subject pick-up area

The subject pick-up area covers the video cameras equipped with zoom lenses, the lighting fixtures and the actual shooting area, the so-called Blue Box. Dependent on the room conditions, three video cameras (one each for top view, front and side view of the subject), but at least two cameras (one for top view and another for the serial shooting of the front or side view) are required. The minimum distance to be observed between camera and subject is of great importance for a distortion-free image and for ensuring adequate depth of motion space for the subject (influence of perspectivity on allowable mistakes). For the high resolution three-tube colour camera with a focal length *f* 9·5–143 mm used at the IAO, this minimum distance is 4·5 m for top view and 6·5 m for front and side views. The cameras are equipped with a motorized focal range adjuster and mounted on a camera dolly which can also be adjusted in height by an electric motor. These functions can be activated from the control area. Lighting of the subject should preferably be by means of discharge lamps, whose colour temperature is closely similar to daylight, so that combined lighting – artificial light and daylight – will prevail in the laboratory. During shooting, the subject is in the so-called Blue Box. This is a section of the laboratory room which is delimited in such a way that the subject will appear against a uniform background in the primary colour blue in each of the selected views of all cameras. For this purpose, the floor and the walls concerned are covered with suitable textile material (e.g. floor covering) or done in colour. Furthermore, all aids which should not appear within the analysis picture (e.g. surfaces for laying down workpieces or rests for the subject) must also be painted in this colour. Several video monitors which can be readily watched by the subject have to be arranged on the periphery of the Blue Box in such a way that they are free of glare.

Control area

The required analysis and design steps are controlled and documented on the equipment installed in the control area. For this purpose, a preview monitor is available for the pictures coming from the CAA and the SSA. These pictures are mixed or punched on a colour trick mixer and passed on to a control monitor in the COA and the video monitors in the SSA. The use of colour cameras in connection with the Blue Box permits the picture of the subject to be 'separated' from his or her background and to be punched into the picture coming from the CAA. This permits a considerably better picture quality to be obtained than in simple mixing of the pictures; and the image of the subject will not be crossed by any lines of the sketch or design drawing (nearer-to-reality picture). Comments can be made on the video pictures by means of a character and text generator. For documentation of the video pictures, a video recorder (dynamic documentation) and a hard copier (static documentation) are available.

Steps to be taken

At least two persons are needed for using the CADVS:

1. The subject who is representative of the potential user group (e.g. male, female, ethnological origin) and who has been reasonably informed on the activities to be performed at the workplace to be designed.
2. The design engineer or work designer who has been trained in the operation and application of the CADVS and at least has a basic knowledge of ergonomic work design.

The steps necessary to make an anthropometric analysis of an existing design sketch prepared on CAD will be briefly described below.

The design engineer or work designer selects the view of the workplace to be investigated at the CAD workstation and enters these views into the video storage/compressor. Dependent on the existing scale of the design sketch or drawing, the associated percentile scale will be called in and also entered into the video storage/compressor. Different percentile scales are available for the geometrical size adjustment of the subject (e.g. arm's reach forward, thigh length while sitting, body height while sitting or standing) stored at the CAD workstation. The body sizes covered originate from the collections of anthropometric data of the population group to be observed. The percentile scales are selected in dependence on the work task and the view chosen. Now, the picture of the percentile scale required will then first be called in the COA from the video storage/compressor and superimposed with the image of the subject with the help of the video mixer.

The subject is adjusted to the percentile required for the analysis through the motorized adjustment of the focal length of the video camera's lens and the motorized height adjustment of the camera. Examples of this are shown in Figure 10.3 for the arm's reaches forward (top view) and in Figure 10.4 for the body height while standing (side view) for a 95th percentile male.

The position thus assumed by the subject in the SSA must not be changed any more. In the COA, the required view of the workplace is transferred from the video storage/compressor to the video mixer and superimposed with the image of the subject. In this connection, it is essential that the base line of the workplace coincides with that of the percentile scale. An example of a trick picture thus obtained is represented in Figure 10.5.

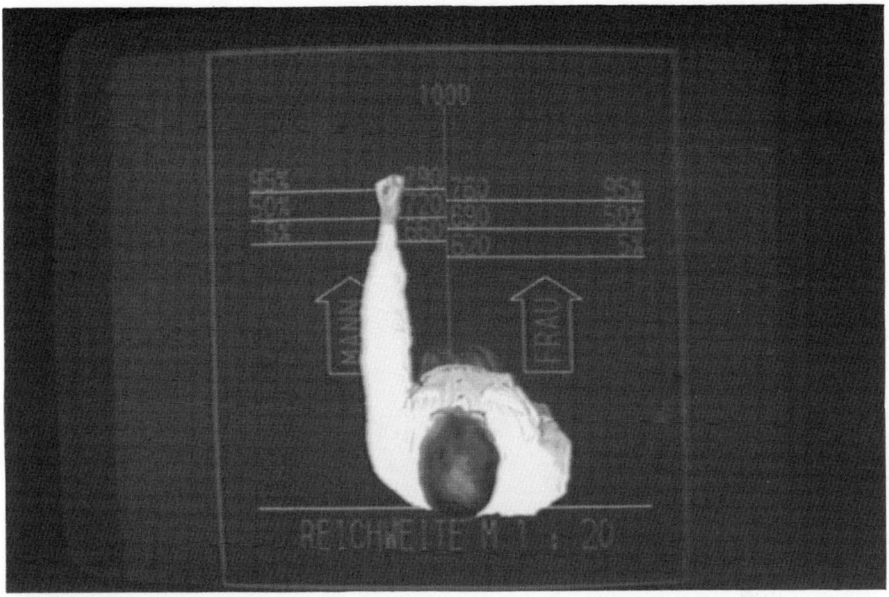

Figure 10.3 *Adjustment of body sizes on the percentile scale: arm's reach forward (photo from screen).*

Figure 10.4 *Adjustment of body sizes on the percentile scale: body height while standing (photo from screen).*

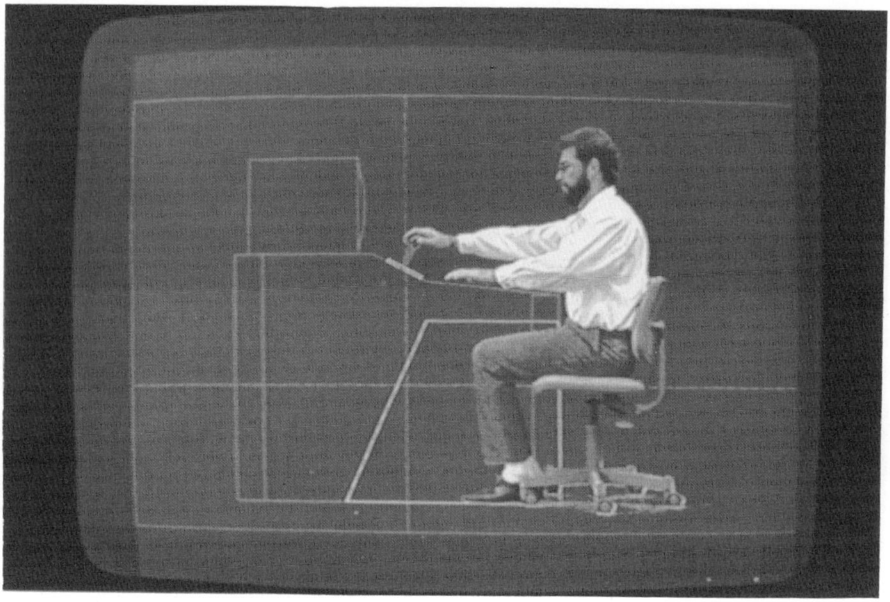

Figure 10.5 Checking the dimensional design of a control desk with the help of CADVS (photo from screen).

Figure 10.6 Front view of a press workplace 95th percentile, male, 'workpiece grasping' (photo from screen).

To simplify the adjustment of body sizes to the dimension of the workplace, the percentile scale can be part of the CAD drawing. An example is given in Figure 10.6. The percentile scale for the body height while standing is represented by two dotted lines (95%, 5%) on the upper right frame of the press machine. The subject will coordinate the movements of his and her hands, arms, legs or of the entire body by observation of the trick picture. This permits the movements and activities necessary while working at the workplace to be completely performed within the shortest time. The coordination of the movements via the exclusively visual feedback from the video monitor in the SSA will largely correspond to the movements to be made at a real workplace after a period of practising for a few minutes.

Natural movements and postures will be found during the analysis, since tools and workpieces can also be used in their original sizes and weights during analysis. The total analysis will be recorded by a video recorder. Insufficient design can be recognized at the workplace under investigation as early as during recording.

Valuable information on the design state of the workplace can also be supplied by the subjects who, for example, are exposed to stress and strain due to unfavourable posture and movements. The videotape now permits the design state of the workplace to be analysed in depth. Insufficient design can be marked with the text and character generator and hard copies of individual pictures can be made. Based on the knowledge thus gained, the workplace can be changed accordingly in the CAB.

The changes necessary in the design sketch or drawing can also be made interactively, i.e. as soon as design shortcomings are recognized during analysis, the analysis will be interrupted. The necessary changes will be made at the CAD workstation and re-tested in the CADVS laboratory. However, this approach can only be recommended, if the design changes can be made quickly. After completion of the analysis for a selected percentile (e.g. 95th percentile, male), the analysis will be repeated for any other percentiles (e.g. 5th percentile, male), so as to be able to investigate the total adjustment range. As described previously, the image of the subject will again be adjusted to this purpose based on the percentile scales.

Under certain marginal conditions, it is permissible to scale the image of a subject up or down for representation of the 5th to 95th percentile of a user group. In this connection, consideration has been given to numerous error sources. When selecting a subject in the mean percentile range of normally proportioned body structure and with a carefully made percentile adjustment of the body parts essential to the analysis, stepless scaling up and down of the image of the subject will be permissible. For certain analyses, tight-fitting clothes may be of importance.

Perspective views (three-dimensional representation) of workplaces lend themselves particularly well to obtaining an overall impression of the man–workplace interface. In this connection, it must be taken into account that the workplace dimensions, which may change with the perspective, do so in the same proportion as those of the image of the subject which has the same scale at any point of the perspective representation of the workplace. An example of a perspective representation is given in Figure 10.7.

As shown in Figure 10.6, the subject is able to simulate the work process with the tools (blacksmith's tongs) and workpieces (blank) used at the workplace. This has the advantage that natural postures are assumed at the workplace due to the influences of the tool to be handled and the weights involved.

Expanded polystyrene blocks of different sizes are suitable for the simulation of the workpiece staging and lay-down areas. These aids must be painted in the same colour as the Blue Box so that they can no longer be discerned in the trick picture. This leads to a true representation of the workplace situation and the movements gone through by the subject.

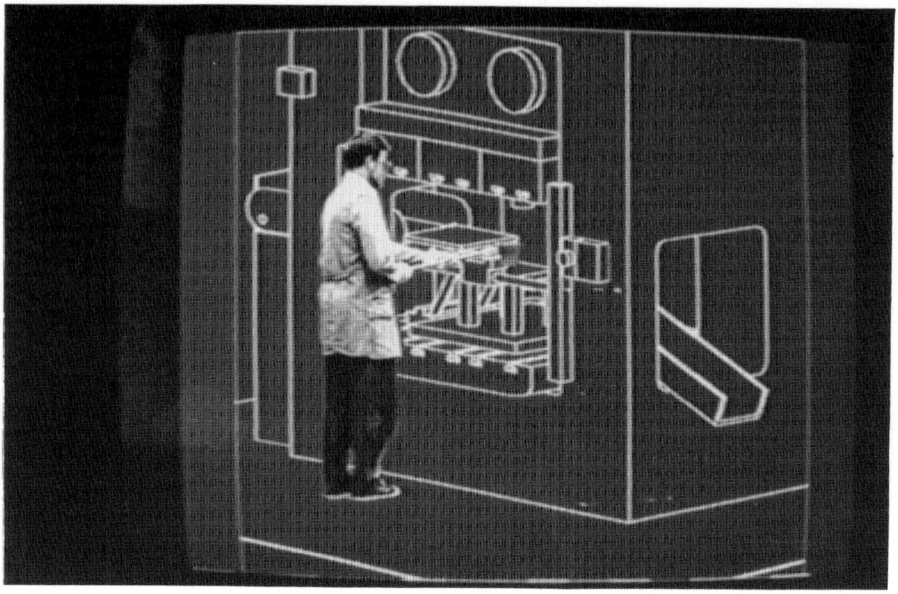

Figure 10.7 Example of a representation of a perspective view of a trimming press, using CADVS.

The subject shown in Figure 10.7 can thus grab the workpiece to be worked in the machine from the staging area (bottom left) with the tongs at the correct height and position. The workpiece with a weight of 72 N will then gradually be accelerated at the simulated workplace and loaded into the machine in the same way as at the real workplace. This supplies valuable information on the movements and the motion space required at the workplace. For assessment of the movements, the picture recorded on videotape can be called in picture by picture. This permits statements to be made on the stresses and strains to which the worker is subjected more reliably.

Applications of CAD-video-somatography

CADVS has primarily been developed for the anthropometric design of machine and assembly workplaces in industrial practice. On principle, the method is suitable, however, for many tasks of: (i) anthropometric product design; and (ii) layout planning.

In the product design range of application, CADVS is used for designing various man–product interfaces. Besides the design of machine and assembly workplaces, CADVS is equally suitable for the design of cabs or vehicles and for the design of office and kitchen furniture. Special advantages are offered by CADVS particularly in the design of products for specific population groups, such as pregnant women, the handicapped, wheelchair users or children. In this connection, it offers the advantages of the evaluation of design solutions for individuals, without the need to make complex mock-ups. On the other hand, product design can, with few restrictions, also be made for those groups of people for whom no anthropometric data is available.

In the field of layout planning, CADVS can only be used for rooms or room sections not larger than the floor area of the Blue Box. Through the maximum possible pick-up area, the Blue Box defines the maximum area of movement of the subject or subjects. This permits the layout planning for rooms or parts of a room with a floor area of up to 5 × 5 m to be performed in the laboratory at IAO. Examples of such layout planning are office rooms, kitchens and laboratory rooms. The arrangement of individual pieces of furniture, workplaces or the like inside the room can be optimized by allowing for necessary changes of location and movements and also for necessary minimum sizes of passages. Both the floor plan and a perspective view of the room as an elevation can be used for this purpose.

Summary

CAD systems are being increasingly used for the design of products, workplaces and room layouts. However, aspects of ergonomics are often neglected, as there are still too few universally applicable problem solution aids available for these computer systems. The linking of video-somatography, which has been used successfully for years, with CAD systems will contribute to bridge this gap. CAD-video-somatography offers the advantage of using a real human during the analysis and design process and is largely independent of the type of CAD system used. In this method, the data and information required for the anthropometric design of workplaces, products, etc. are supplied by a subject. This obviates the need for the complicated modelling of movements and posture (e.g. under weight load), but these can be implied as natural. As this method, apart from a CAD system, requires a special room with complex video hardware configurations, its use will only be economic, if utilized more frequently. Therefore, it is particularly suitable for big companies, specialized business consultants and research institutes which, in their business and research work, have committed themselves to work design fit for humans.

References

Kroemer, K. H. E., Snook, S. H., Meadows, S. K. and Deutsch, S. (Eds), 1988, Ergonomic models of anthropometry, human biomechanics and operator–equipment interfaces. Proceedings of a Workshop (Washington, DC: National Academy Press).

Lorenz, D., 1983, Gestaltung von Maschinenarbeitsplätzen mit Hilfe von Video-Somatographie. *FhG-Berichte*, **3/4**, S. 13–16.

Martin, K., 1981, Video-Somatographie: Ein neues Hilfsmittel bei der Arbeitsplatzgestaltung. *Fortschrittliche Betriebsführung und Industrial Engineering*, **30**(1), 21–26.

Chapter 11
Models for ergonomic analysis and design: COMBIMAN and CREW CHIEF

JOE W. MCDANIEL
AAMRL/HEG, Wright-Patterson AFB, OH 45433, U.S.A.

Abstract. Early identification of potential design-induced maintainability problems is essential to correct a problem before mock-up, fabrication, or production. To facilitate early identification of design problems, the Air Force Human Systems Division has developed two computer models: COMBIMAN (COMputerized BIomechanical MAN model, a model of an aircraft pilot) and CREW CHIEF (a computer-aided design (CAD) model of an aircraft maintenance technician) which allows the designer to perform the functions of an expert ergonomist. The designer may simulate operations or maintenance activities on the computer-aided design image to determine if the activity is feasible.

Expert system software automatically creates the correct body size and proportions for males and females, the encumbrance of clothing, personal protective equipment, and mobility. Physical access for reaching into confined areas (with hands, tools, and objects), visual access, and strength.

This chapter describes these models and demonstrates their utility through a series of case studies. The expert system software allows evaluation of ergonomic accommodation without having to interact with the complex and interrelated data bases describing operator and maintainer characteristics. Analyses can be done in minutes with these models which would require weeks or months by experimental methods.

Keywords: Computer-aided; Design; Models; Expert systems; Human models; Ergonomics.

Introduction

The availability of modern CAD systems provides a new environment for implementing and transferring ergonomics knowledge and its applications to the designers and evaluators of systems and equipment. Modern CAD systems create conceptual designs that can be used as 'electronic mock-ups.' New ergonomic design tools are being developed which provide a quick and accurate means of evaluating the interaction between an operator or maintenance technician and a system design. Without such tools, these types of evaluations were rarely cost-effective, and often of insufficient quality.

This chapter discusses two computer-graphics tools for ergonomic analysis: COMBIMAN (COMputerized BIomechanical MAN model, a model of a seated operator) and CREW CHIEF (a model of a maintenance technician). These two models are actually expert systems which allow the user to perform the functions of an expert ergonomist by simply defining the task to be performed. Both models perform four types of accommodation analyses: fit, reach, strength, and vision. Extensive empirical data bases were used to develop these models. The models are, in effect, a way to access and use the complicated and interactive data describing the physical capabilities and characteristics of the human operator and maintainer.

Traditional anthropometric and biomechanical accommodation analyses require a system prototype or mock-up which has a mature design as a prerequisite and which is difficult to change. Even a hardware mock-up has a significant cost. Designs on a CAD system, on the other hand, have little or no additional cost because modern designs are already created on CAD systems.

Customarily, mock-up evaluations have tested only four or five subjects, too few to represent the user population. Because of variation in proportionality among body dimensions, there is a significant possibility of erroneous conclusions. Considering the world community, adult human stature varies from 2 to 8 ft and adult body weight ranges from approximately 20 to 1000 lb. Fortunately for the designer, these extreme individuals are so rare, they need not be considered in general purpose designs. The great range of body sizes is not found in the military populations because the military has admission standards that screen out the rare individuals. Typically, military populations have statures ranging from 4·5 to 6·5 ft, and weights from just below 100 lb to slightly over 250 lb. But even these ranges are still too large for some systems to accommodate. More stringent body size requirements are necessary for aircraft pilots, and many of the physically demanding jobs now have minimum strength requirements as well.

Because the body size and strength characteristics of military populations has been studied more thoroughly than that of the civilian population, designers sometimes erroneously apply these data to civilian designs. Military populations do not include the older and obese individuals found in the civilian sector. Also, it is rare to find a military worker over age 35 performing strenuous physical work.

Accommodation analysis with human subjects placed in an actual mock-up is risky because of the lack of proportionality among people. Two people with exactly the same stature will have different length arms, legs, etc. Typical design specifications may require accommodation of a size range from a 5th percentile (small) female to a 95th percentile (large) male. There are, however, no 5th percentile women and there are no 95th percentile men. The term 'percentile' refers to a statistical property of a single measure. Fifth percentile refers to that measure which 5% of the people are at or below and 95% are above. People near the extremes of a distribution in one dimension are usually nearer the average in other dimensions. So, if one selects a person with a 5th percentile stature (short), that person's arms (and therefore arm-reach characteristics) are likely to be much longer than a 5th percentile arm reach, and can even be much longer than average. If one evaluates reach to controls with such a subject, one will likely over-estimate the arm reach and arrive at an erroneous conclusion.

A major obstacle to anthropometric accommodation has been finding a sample of subjects which represents the exact measures on one or more dimensions which are pertinent to the design. The major advantage of using computer models of the human body to evaluate accommodation is that all dimensions can be adjusted to the required values. Thus, the computer-generated subjects chosen to evaluate the design are as good as one's ability to define the needed values.

Another reason anthropometric evaluation has been difficult is because the body-size measures are uni-dimensional segment lengths and circumferences, not the three-dimensional (3-D) motion envelopes required by workplace designers. Each is the measure of an important body segment or characteristic, but the measures are not attached to each other nor to any common coordinate system. For example, the measure 'shoulder–elbow length' is measured to the bottom surface of the elbow while 'elbow-wrist length' is measured from the back of the elbow. Neither is measured to the centre of rotation of the elbow joint. Even if a design engineer were given these measures, he or she would not be able to assess arm reach capability.

Anthropometric measures are made on the surface of the body, while the engineer usually needs internal dimensions associated with the skeletal link system, that is, from one joint centre to another. The science of relating external and internal body dimensions is too complicated for a designer to apply routinely, but it is not too difficult to incorporate into a computer model and calibrate against empirical task analyses. Development of task-related data is usually very difficult because it is also task-specific. While computer models are based on more general data, the designer must beware not to apply them to unusual designs which fall outside the region of the model's data bases. For this reason alone, there will continue to be a need for some accommodation analyses with human subjects.

Although considerable data are found in published studies, the data were gathered for specific applications and can rarely be generalized to other applications. COMBIMAN and CREW CHIEF model developers have discovered a technique for using benchmark tests performed on large samples of the population to adjust the data gathered on small samples, but with large numbers of interacting variables. However, all of the task-related data bases in the 3-D CAD models COMBIMAN and CREW CHIEF were gathered especially for these models. The modellers have compiled and developed the most accurate representations of the link system and task representations available. The engineer can now use the model without having to develop the model.

Because of the powerful and efficient capabilities of modern CAD systems and the ready availability of body size data, many people have started developing human models. It is not difficult to develop a two-dimensional (2-D) human model, and not much more difficult to develop a 3-D human model. Most developers run aground when attempting to install a skeletal link system with high-fidelity relation to the enfleshment. Almost all developers are stopped by strength and task analytic capabilities. The best advice for would-be modellers is 'if you do not have the data, do not assume that they are available, for in most cases, they are not.'

COMBIMAN and CREW CHIEF have realistic size, strength, mobility, and automated task analysis capabilities for simulating problems with operability and maintainability early in the design. These models function as an electronic mock-up, simulating the human operator or maintainer throughout the entire range of 1st through 99th percentile body dimensions.

The military aircraft cockpit is a particularly difficult design problem because the need to save weight and volume requires the cockpit to have the minimum acceptable size and adjustment. When designing a complex workplace such as an aircraft cockpit, several dimensions must be accommodated simultaneously. To be specific, pilots must be able to see out of the window and to reach the hand controls and the foot controls simultaneously. This is referred to as *multivariate accommodation*. While the concept of multivariate accommodation is not new (see Moroney and Smith, 1972), the procedures for accomplishing it have been costly and laborious. Computer models can facilitate multivariate accommodation by allowing a designer to represent several specific individuals lying at the limits of a multi-dimensional distribution of critical body dimensions. Each individual in this set represents the worst case of several critical dimensions. For example, one model may have the shortest arms and legs appropriate for a person with a tall sitting height. Another model in the set may have short sitting height as well as short arms and legs, etc. By defining a population with combinations of realistic extremes, a multivariate accommodation can be verified if all the members of the set are accommodated. CAD models are the only practical way to represent the unusual dimensions needed for multivariate analyses. To achieve a similar quality of evaluation with human subjects, a large number would have to be sampled to find the critical ones. Computer models can do this by searching existing surveys for the required values, then dimensioning the model with those values.

Demands for improved quality, reliability, and performance have placed additional requirements upon the design process itself. Early in the design of a new system, decisions are made which commit the project to an approach with implicit limitations in quality, reliability and performance. These CAD models are actually expert systems which allow the system designer to perform the functions of an expert ergonomist. These models also make it easy to evaluate the accommodation of different population subgroups. Other 3-D models provide complex animation with multiple human figures to aid in task visualization.

In the sections that follow, the COMBIMAN and CREW CHIEF CAD tools for ergonomic evaluation will be briefly described. Then, through a series of actual case studies, the major features of each tool will be discussed in the context of the complex issues of anthropometric accommodation.

COMBIMAN and CREW CHIEF are not simply 3-D anthropometric human models, although they are that, too. The unique capabilities are their ability automatically to simulate the performance of a task while accounting for the complex and interacting effects of body size, joint mobility, clothing restraints, harnessing restraints, grip type, tool type, type of object, obstacles, posture and the operation performed. Because of the large number of interactions which are considered in these models, the research effort to develop the empirical data involved more than 100 person-years of effort.

Both of these models use a new enfleshment technique to represent the irregular surface necessary to depict clothing and person protective equipment. A wire-frame mesh surface is created by an array of small, interconnected triangles, similar to the technique used in finite element analysis. An algorithm determines which lines comprise the profile view (from any view direction) and also which lines are essential (such as facial features, etc.) and eliminates all other lines before generating the displayed image. The result is a profile view of a high-fidelity depiction of the irregular human figure. Although presented on a two-dimensional display, the crew station and human model are three-dimensional and can be rotated for viewing at any angle.

Description of COMBIMAN

COMBIMAN was developed by the Armstrong Aerospace Medical Research Laboratory (AAMRL) as a computer-aided design (CAD) model of an aircraft pilot (Korna and McDaniel, 1985). COMBIMAN is a three-dimensional interactive computer graphics model of an aircraft pilot used to evaluate the physical accommodation of the pilot in existing or conceptual crewstation designs. COMBIMAN is an expert system which performs four types of analyses: fit analysis, visibility analysis, reach analysis, and strength for operating controls with the arms and legs. The user of COMBIMAN does not have to be an expert in ergonomics or anthropometric accommodation because that expertise is automated in the software. The user defines and directs a task to be performed, just as one would a human subject in a mock-up review. The program generates accurate body sizes, proportions, strength, clothing restraints, harness restraints and vision obscuration.

COMBIMAN has been distributed to the major aerospace industries since 1978. It has been used by both the US Army and the US Air Force to evaluate design changes, saving the costs associated with hardware mock-ups and prototypes. COMBIMAN is the predecessor of CREW CHIEF (a model of a maintenance technician), and much of the COMBIMAN technology was the basis of the CREW CHIEF program. The link system, joint system, reach algorithms, vision analysis, strength algorithms and other features originated in COMBIMAN.

COMBIMAN provides an alternative to the traditional crewstation evaluation. Because it uses a computer graphics generated crewstation, it is possible to analyse many conceptual designs and variations of designs early in the program. COMBIMAN has two methods for dimensioning the human model. First, one can specify one or two critical dimensions, and the model will generate the other dimensions using regression equations. Second, where multivariate analysis is desired, one may read-in dimensions for a series of human models. Postures can be manipulated automatically by task-driven commands, or manually by positioning individual joints. Every link dimension and every joint angle of the human model may be set to any desired value.

COMBIMAN gives the user complete control over sizing the human model and provides a number of alternatives for dimensioning and changing the dimensions of the model. These dimensions can be defined as percentiles (in the range of 1st through 99th percentiles) of a known population or actual measures. The typical evaluation session begins with dimensioning the human model by one of the following techniques:

1. Direct measure. Twelve measurements of an individual are entered from the keyboard. This method is used to dimension COMBIMAN to represent a specific individual. In addition to its use in validation of the model, this method can be used to enter individuals with extreme combinations of dimensions for multi-dimensional accommodation.
2. Computer-aided dimensioning. Statistically correct human models can be generated from anthropometric survey data. A critical body dimension which is relevant to the task being simulated is used to construct a proportioned human model from regression equations. For example, arm length is used in evaluating reach capabilities. Sitting height is used for ceiling clearance. Leg length is used to reach to pedals. The model's size and proportions are computed from the selected size measure plus body weight, providing a statistically accurate human model for task analyses.
3. Changing individual dimensions. Beginning with a predefined model, the user can change any dimension or joint angle individually.
4. Data base summary statistics. Percentiles computed from large samples are stored in the model and used to define the human model. Individual segments may be set to the desired percentile of the selected population. Currently, there are six anthropometric surveys included in the COMBIMAN model: 1964 Survey of Navy Pilots, 1967 Survey of USAF Pilots, 1968 Survey of USAF Women, USAF Female Pilots, 1970 Survey of Army Pilots and 1977 Survey of Army Women. These surveys have been edited to reflect recent changes in body size and weight standards for the military. As newer surveys become available, they are added or substituted, as appropriate. The user is provided with a utility program which allows the addition of new surveys.

Description of CREW CHIEF

CREW CHIEF, a computer graphics model of a maintenance technician, was developed by the Armstrong Aerospace Medical Research Laboratory (AAMRL) and the Human Resources Laboratory (HRL) (McDaniel and Askren, 1985; Korna *et al.*, 1988). This model is interfaced to several existing commercial CAD systems employed by aerospace manufacturers and may be used to evaluate the maintainability of aircraft and other complex systems. CREW CHIEF is an expert system which lets the designer perform the functions of an expert ergonomist. The CREW CHIEF model allows the designer to simulate a maintenance activity on the computer-generated image of the design and to determine if the required maintenance activities are feasible. As with COMBIMAN, the user of CREW

CHIEF need only define the task to be performed, just as one would define a task for a human subject in a mock-up review.

Approximately 35% of the lifetime cost of a military system is spent for maintenance. Excessive repair time is caused by failure of the system design to consider maintenance adequately. The maintenance technician will spend hours making a repair which could have been completed in minutes if accessibility had been adequate. Ultimately, development engineering costs and acquisition time as well as life cycle costs and maintenance time will be reduced.

CREW CHIEF generates 10 sizes of human models (five male and five female) with the encumbrance of four types of clothing and personal protective equipment, mobility limitations for 12 postures, physical access for reaching into confined areas (with hands, tools, and objects), automated obstacle avoidance, visual accessibility, and strength analysis (for using hand tools and manual materials handling of objects).

Designers, human factors engineers, and maintainability engineers may use CREW CHIEF as an electronic mock-up. Analyses, which would have required weeks or months to perform by conventional methods, can be performed in minutes. Version 1 of the CREW CHIEF system of programs became available for use in May 1988.

Sample applications

Rather than provide a detailed description of these tools, actual case studies of real design problems and the difficult issues of anthropometric accommodation will be presented to illustrate how the CAD tools assist the designer.

COMBIMAN applications

Case study 1

PROBLEM

Can raising the entire pilot crew station overcome limitations in over-the-nose vision? When testing a prototype helicopter, it was discovered that when landing, the helicopter maintained a significant nose-up attitude. To compensate for this, pilots would approach the landing pad at an angle to see out the side window, since the raised nose obscured forward visibility. The helicopter has a tandem seating arrangement, with the pilot sitting behind the copilot-gunner. The copilot-gunner had to be placed in front to use a direct view optical sight. The pilot's vision is limited by the copilot-gunner's helmet which is approximately 6·5 ft forward of the pilot's eye position and the nose of the helicopter which is about 12 ft forward. There were two proposed solutions to the pilot's vision problems. The less expensive of the two involved raising the entire pilot crew station about 6 in to improve forward vision. The more expensive option involved redesigning the tail of the aircraft to permit a level landing attitude.

SOLUTION

The COMBIMAN program has a function which plots the crewstation from the pilot's viewpoint. MIL-STD-850 requires such visibility plots to ensure adequate vision over the nose and around the window frames. While MIL-STD-850 requires only plots from the arbitrary 'design eye position,' COMBIMAN vision analysis additionally considers realistic

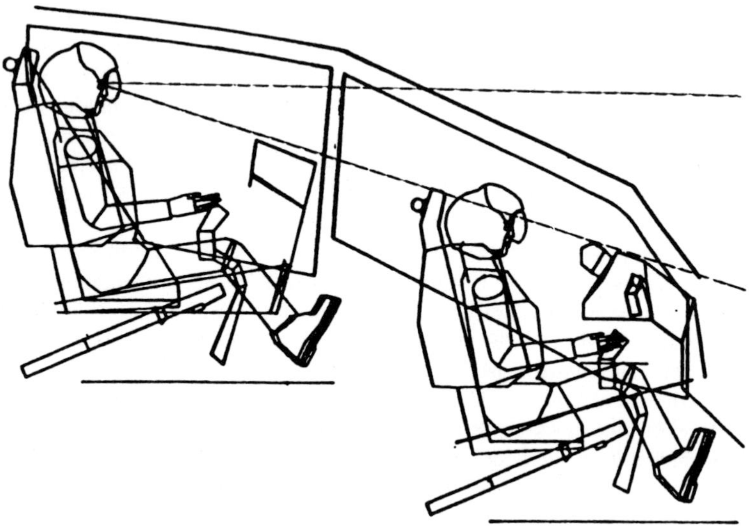

Figure 11.1 *Electronic mock-up of helicopter crewstation represented in the COMBIMAN model. Although projected on a 2-D display, the model itself is 3-D and can be viewed from any angle.*

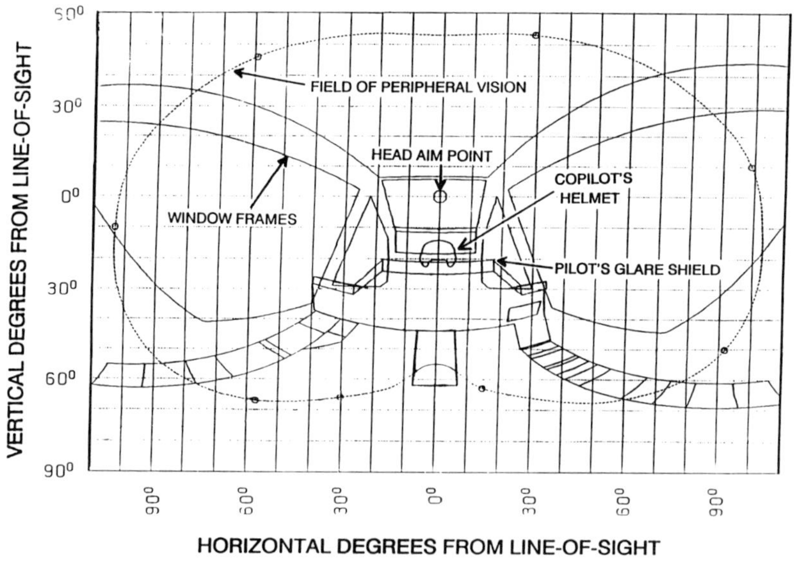

Figure 11.2 *Visibility plot from COMBIMAN showing the landing pad is not visible in the over-the-nose view of the pilot even in a raised cockpit. The overlay shows the features inside the pilot's peripheral vision envelope. COMBIMAN allows the user to install additional obscuration envelopes for head gear such as helmets and night vision goggles. If problems are evident in a design, changes may be quickly and inexpensively made to the electronic mock-up, and the evaluation repeated.*

combinations of body size and seat adjustment, as well as realistic offset of the eye position due to head motion and obscuration overlays which represent the masking of helmets, respirators, night vision goggles, and other head gear. COMBIMAN's vision plots can also show objects outside the aircraft, such as other aircraft, runways, landing pads, tanker aircraft, and so on.

To evaluate the proposed alternative, a graphics model of the pilot's and copilot-gunner's crewstations were entered into the COMBIMAN model (Figure 11.1). To simulate the landing pad, several pads at different altitudes and attitudes relative to the crew stations were also entered. Notice that for the purpose of this type of analysis, the graphics representation is quite simple, containing only the relevant features, such as instrument panels, controls, and windows. Once the crewstation model is created and entered into the program, the analysis effort typically requires less than 2 h. In performing the analysis, visibility plots similar to Figure 11.2 were made using different sized crew members and different seat adjustments. After the standard configuration was evaluated, the pilot's crew station was raised 6 in and a second set of visibility plots was generated. The resulting plots clearly showed that the landing pad was still obscured by the nose, and that raising the pilot's crew station was not the answer. The cost savings of using COMBIMAN to evaluate the problem, rather than modifying the prototype helicopter, as the contractor had proposed, was quite significant.

Case study 2

PROBLEM

How does lack of seat adjustment affect reach to controls? Aircraft ejection seats adjust vertically, but usually not horizontally, so the pilot's knees or feet will not strike the instrument panel or canopy frame when ejecting. This would happen if a long-legged pilot adjusted the seat too far forward. Since the result of a knee strike is catastrophic, ejection seats adjust only vertically. The 'ejection clearance line' is based on the large pilot's buttock-knee length. However, when flying with the shoulder harness locked, as often happens during manoeuvres, the small pilot's arms may not be long enough to reach controls on the forward instrument panel.

SOLUTION

The seat and human model in COMBIMAN can be adjusted without limit in three dimensions. The Reach Analysis Model is based on data gathered with three types of controls, six types of clothing, three types of harnessing, and seven reaching planes. The data collection required over 2 years of full-time effort, providing the most comprehensive description of arm reach available anywhere.

The COMBIMAN program evaluates arm reach with three different harness configurations: (i) lap belt only, representative of some non-pilot crew members, (ii) shoulder harness on but not locked, representing the non-critical situation, and (iii) the shoulder harness on and locked, representing the situation where manoeuvring or inverted flight is anticipated. The harness is usually locked when practising for combat. When the shoulder harness is locked, arm reach envelopes are limited to arm and shoulder movement.

COMBIMAN has three different hand postures for the types of controls found in aircraft and other vehicles: (i) grip centre for reaching to hand grips, (ii) functional grip for turning knobs, and (iii) finger tip pushing buttons. Each type of hand posture has a different reach envelope.

One of the most important features of the COMBIMAN model is its ability to generate accurate body sizes and proportions representing the characteristics of a population. First, there are six anthropometric populations built into COMBIMAN: Male Air Force Pilots, Female Air Force Pilots, Air Force Women, Male Army Pilots, Army Women and Male Navy Pilots. The user can also add additional populations. The program user need only select the population and dimensions critical to the task and COMBIMAN will create realistic body size, proportion, and clothing characteristics. For example, the user may select arm length as the critical variable for evaluating reach to hand controls, or select sitting height for analysis of head clearance.

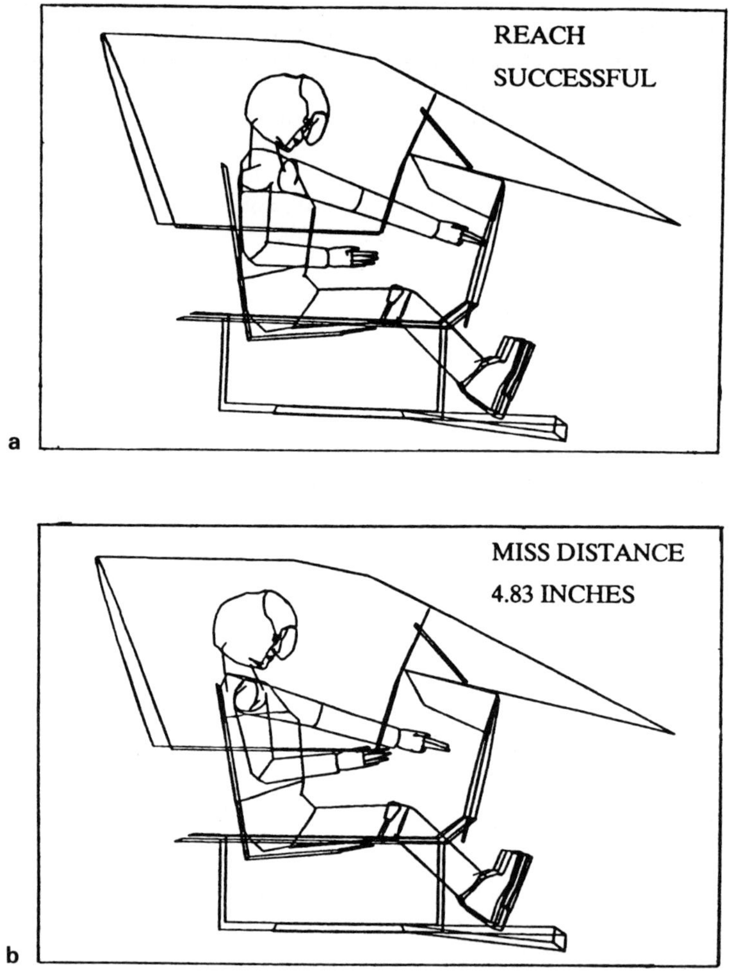

Figure 11.3 (a) Reach analysis with 50th percentile arm length and shoulder harness unlocked. In this condition, the pilot can lean forward with the upper body. Note the message 'REACH SUCCESSFUL'. (b) Reach analysis with shoulder harness locked, allowing only arm and shoulder movement, but not upper body leaning. Note the message 'MISS DISTANCE 4·83 INCHES.' The picture and the data provide details of the nature and magnitude of the problem.

The encumbrance of clothing also affects the reach envelope. To allow the designer to evaluate the effects of clothing accurately, COMBIMAN has six different types: summer-weight flying suit; summer suit with survival vest; winter-weight suit; winter suit with survival vest; winter suit with survival vest, jacket, and flotation device; and the chemical defence protective clothing.

In this example, the summer-weight flight suit and 50th percentile arm length are selected. Figure 11.3a shows a reach to a forward instrument panel with the shoulder harness unlocked, allowing upper body movement. Figure 11.3b shows the same pilot failing to reach the forward instrument panel with the shoulder harness locked. For aircraft with ejection seats, this illustrates why critical controls which must be operated when the shoulder harness is locked cannot be placed on the forward instrument panel. The preferred location for such controls is on the stick and throttle (or cyclic and collective for a helicopter). The result is that these handgrips are being overloaded with controls, some having as many as 18 functions on a single grip.

Case study 3

PROBLEM

Can the pilot reach controls located on the side consoles if the seat is adjusted upward to achieve adequate outside vision? As in the previous example, body size, seat position, and vision requirements provide conflicting design requirements. A large pilot may have to adjust the seat down to achieve head clearance beneath the canopy. For the large pilot, reach and vision are seldom problems; however, the fit may be cramped in extreme cases. For the small pilot, everything is a problem. If the small pilot adjusts the seat up to get adequate vision through the windows, he or she is moving farther from the controls and side consoles.

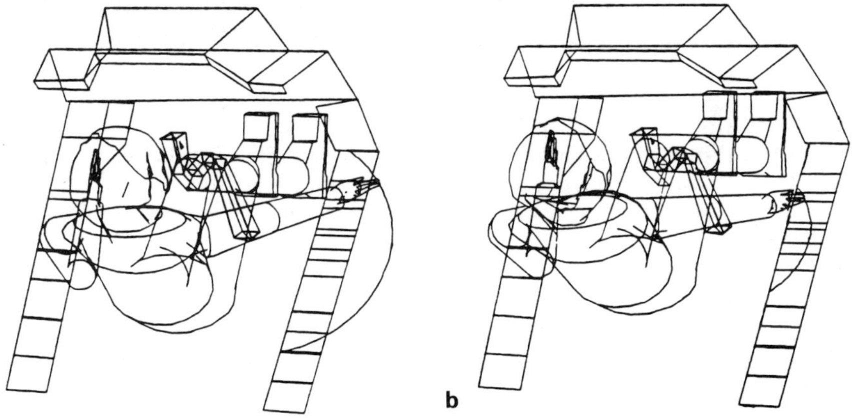

Figure 11.4 (a) The result of a COMBIMAN reach envelope analysis to the right side console of a helicopter for a male pilot with a 50th percentile arm length and the seat adjusted to the full down position. The reach curve superimposed on the side console defines the area that can be reached with a functional grip. (b) The same conditions for the seat adjusted to the full up position. Comparing the two reach envelope plots, the difference in the part of the side console that can be reached without leaning is very dramatic.

SOLUTION

COMBIMAN can compute the intersection of the reach envelope and any surface of the crewstation. This is called the Reach Envelope Analysis and it considers body size, the type of clothing worn by the pilot (more layers and bulk reduce the mobility of the joints), and the type of harness (lap belt, shoulder locked, and shoulder unlocked). Figure 11.4a shows the result of a reach envelope analysis to right side console of a helicopter for a male pilot with a 50th percentile arm length and the seat adjusted to the full down position. By contrast, Figure 11.4b shows the same conditions for the seat adjusted to the full up position. The difference in the part of the side console that can be reached without leaning is very dramatic.

Other types of reach analysis COMBIMAN can perform involve the different types of flight clothing. The type of clothing chosen for the analysis not only affects the appearance of the model, but it also automatically changes the joint mobility limits, for realistic reach analysis. If the analysis described above were repeated with a more restrictive clothing type, the reach curve would have been closer to the body. This capability is very important because crew system designers usually do not have access to different types of pilots' clothing, especially all the sizes of each. So, even if the designer had a mock-up, he or she would still not be able to evaluate the effects of restrictive clothing worn by the population of users of the crew system.

Case study 4

PROBLEM

How much force can a pilot exert on an ejection handle? Now that women are flying trainer aircraft equipped with ejection seats, the question of actuation resistance must be reconsidered. The fact that women, as a group, are not as strong as men, as a group, raised an important design question: how much resistance should ejection handles have? The objective is make the control resistance so great that the ejection controls cannot be activated inadvertently, but low enough so that all pilots can operate it. When the seat was originally designed, there were no female military pilots, so only the male strength characteristics were considered.

SOLUTION

The COMBIMAN program has models to predict the strength a seated operator can exert on eight types of controls: stick-type aileron and elevator, wheel-type aileron and elevator, general-type hand grip (any location in the forward hemisphere reach envelope), side-mounted ejection handles, centre-mounted ejection handles, and rudder pedals. The controls can be operated by left, right, or both hands, and the pedals with left or right foot. The strength models are based on measurements of over 1000 subjects. The strength analysis models provide 1st, 5th, 50th, 95th and 99th percentile forces for the task defined.

The strength available for operating a control depends on the type of control, location of control, orientation of the control, and the direction of force applied. These control characteristics are important because different joint angles and combinations of muscles are involved in exerting forces in different directions and locations. The orientation of the handgrip may produce awkward postures which can reduce the available strength.

In the case of the ejection handles, the side-mounted handles are located in line with the shoulders and the direction of force is toward the shoulders. Pilots can exert a larger force on side-mounted handles than on a centre-mounted handle, which is not in line with the shoulders and requires a force at an angle to the line between the hand and shoulder.

The results of the strength analysis for the centre pull ejection handle are shown in Figure 11.5. The analysis simulates left hand operation of a centre-mounted ejection handle by a female pilot. The results show that 1st percentile female strength value is 29·48 pounds, less than the 49 lb resistance of the controls. The 5th percentile force is 49·25 lb, right at the value of the control resistance. This indicates that almost 5% of the women would not be able to operate the centre-mounted handles with one hand.

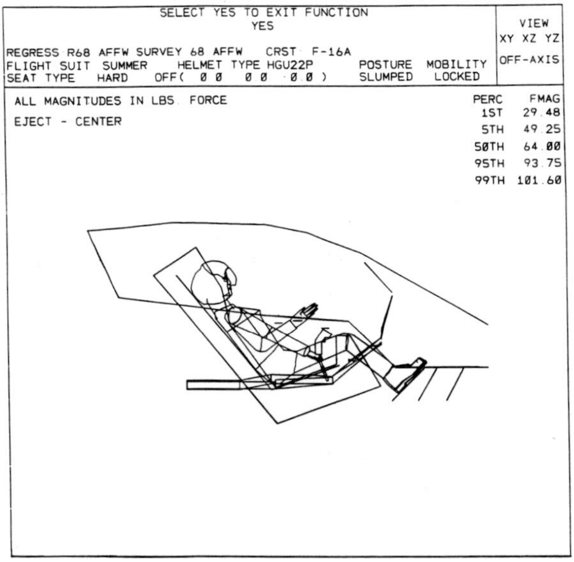

Figure 11.5 The results of the COMBIMAN strength analysis for the centre pull ejection handle with the left hand only. The strength analysis models provide 1st, 5th, 50th, 95th, and 99th percentile forces. The centre-mounted handle is not in line with the shoulders and requires a force at an angle to the line between the hand and shoulder. The results show that 1st percentile female one-hand strength values were 29·48 lb, less than the 49 lb resistance of the controls.

Currently, only the F-16 fighter plane is equipped with the centre-mounted ejection control. Since women are currently prohibited from flying combat aircraft, the situation does not now constitute a deficiency. The seat specification requires one-hand operation, however, because of the possibility of incapacitation of one hand and also because the effective forces may be reduced further by acceleration or buffeting of the aircraft. Trainer and other ejection seat-equipped aircraft have the side-mounted handles.

If the centre-mounted handle were analysed with both hands simultaneously, the 1st percentile values would have been 62·35 lb, well above the control resistance. The two-hand force of 62 lb is more than double the one-hand force of 29 lb because the symmetry of the two hands pulling outward balance each other, causing the sum of the two forces to be aligned with the direction of control movement.

CREW CHIEF applications

Case study 5

PROBLEM

What weight can a worker carry through an access tunnel with a ceiling height of 30 in? There are many passageways inside large systems that require workers to work in other than a standing posture.

SOLUTION

CREW CHIEF Task Analysis evaluates strength and accessibility for eight categories of manual materials handling activities: carry, lift, hold, push, pull, turn, grasp and reach. In all of these tasks, the program considers the posture as well as the location and orientation of the work.

For example, in a carry task, a low ceiling may force the CREW CHIEF human model to bend, stoop, or even crawl while carrying the load. The CREW CHIEF program contains 12 initial postures. The designer chooses the one that is appropriate for the workplace. When the task analysis simulation is run, CREW CHIEF automatically adjusts the posture to accomplish the task. This reposturing is totally automatic, causing the human model to reach around obstacles in the workplace if necessary, and showing graphically whether or not the human model can perform the reach part of the task. The posturing automatically keeps the body's joint angles within realistic limits, adjusted for the additional restriction of the type of clothing chosen for the analysis. CREW CHIEF has four types of clothing: fatigues, cold weather, arctic and chemical defence.

The 12 initial postures are standing, sitting, stooping, squatting, kneeling on one knee,

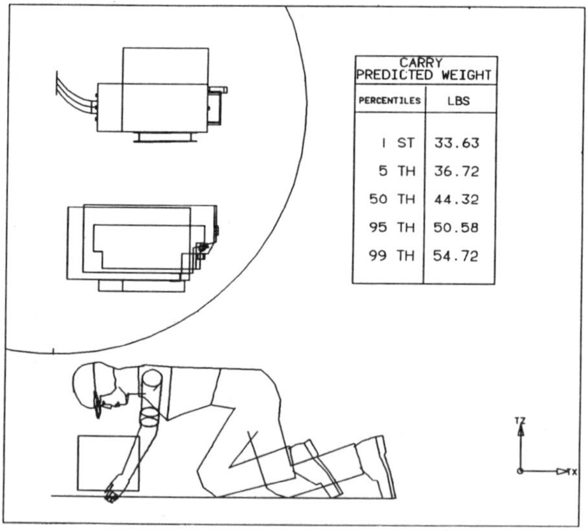

Figure 11.6 After the designer sets up the task to be performed, CREW CHIEF automatically executes the task and displays the available strength for the 1st (weak), 5th, 50th, 95th, and 99th (strong) percentile worker.

kneeling on two knees, lying prone, lying supine, lying on the side, crawling, climbing and walking. Strength analyses are available for all appropriate combinations of tasks and postures. For example, carrying strength is evaluated in the walking, crawling, and stooping postures. Strength for torquing with wrenches is appropriate for all postures except walking, crawling, and climbing. More than 100 000 strength measures were collected to develop the strength models in CREW CHIEF.

For this example, illustrated in Figure 11.6, a 5th percentile female model is chosen, because the small female represents the most limiting condition. (Remember, in CREW CHIEF, a 5th percentile female model has only stature, arm reach, and weight parameters set at 5th percentile population values; the other dimensions are regressed from these three.) If the population of workers includes both men and women, MIL-STD-1472 requires accommodation of a 5th percentile female strength as a minimum. The box carried is medium size, approximately 10 in high and deep and about 18 in wide. The 30 in passage ceiling is denoted by a horizontal line above the floor. To perform the analysis, the designer identifies the location of the work (in this case, the box to be carried), the task (CARRY), and the location of the worker from the work. The CREW CHIEF program will assemble the human model automatically, posture it, and place it in the workplace in the location specified. The available strength for those conditions is then computed and displayed as shown in Figure 11.6. Note the 5th percentile strength value is 36·72 lb. If the object weighs this amount or less, the design criterion is satisfied.

Case study 6

Problem

What is the minimum height of an aircraft fuel tank which will admit the full range of cleaning personnel? A recently discovered design-induced maintainability problem involves the in-wing fuel tanks of a new aircraft. Combat aircraft have self-sealing fuel tanks for survivability. After several years of ageing, the gummy sealant hardens and must be replaced. After the fuel is flushed from the tank, maintenance personnel, wearing protective clothing and breathing apparatus, must enter the tank and remove the old sealant. It was found that larger-than-average fuel maintenance personnel get stuck in wing tanks. When the aircraft was designed, the maintainability and human factors people thought to test the access opening to make sure a large person could get through. However, equal consideration was not given to the height of the tank itself.

Solution

In the case cited, it is too late to redesign or retrofit. The limited accessibility requires small workers; what was originally a design problem now becomes a personnel and recruiting problem. The career field for a major system must now be populated with small men and women.

Had an analysis model like CREW CHIEF been available, this problem could have been prevented, for with CREW CHIEF, the designer can select from a range of body sizes of male and female technicians, and may choose a 1st, 5th, 50th, 95th or 99th percentile model for male and female maintenance technicians. In CREW CHIEF, the human model sizes are much simpler than in COMBIMAN where body proportions are critical. In maintenance, proportions have less significance, and were made simpler to allow the more complicated tasks to be evaluated. The CREW CHIEF human models are based on stature, arm length, and weight, with all other dimensions regressed from these three. It would also be a poor

Figure 11.7 The wing fuel tank problem represented with the CREW CHIEF human model. Here the prone and supine postures for a 99th percentile male are pictured.

design if it considered only gross body fit in the case of this fuel tank, and not the cleaning task which must also be performed.

Figure 11.7 shows the fuel tank problem represented with the CREW CHIEF human model. Here, the prone and supine postures for the 99th percentile male human-model are pictured. The prone posture shows the working posture. A side-lying posture indicates the height necessary to roll over.

Case Study 7

PROBLEM

What is the minimum number of tools needed to maintain a helicopter's jet engine? Maintenance on the flightline is usually done outdoors, in all kinds of weather. Maintainers in this environment have little sympathy for aircraft designs which make their jobs more difficult. While guidelines have recommended no common tools and minimum fasteners for better maintainability for many years they have not been rigorously followed. A notable exception to this is the Army's LHX program (Dornheim, 1989). The T800 engine, to be produced by the Light Helicopter Turbine Engine Co., requires only a literal handful of tools to maintain on the flightline. The engine has only two sizes of bolt heads and can be maintained with a ratchet, extension, two sockets, and two wrenches. Having to use only a few tools lightens the load maintainers must carry to and from the helicopter, and speeds up maintenance because fewer tools changes are necessary to get the job done. This degree of design for maintainability does not just happen; it requires commitment to the MANPRINT goals and work to verify that these goals are met.

SOLUTION

CREW CHIEF can be used to verify that a particular tool can fit into a tight clearance, that there is sufficient range of movement of the tool to remove or replace a fastener, and that the maintainer wearing even arctic or chemical defence clothing has the accessibility to perform the operation. There are 105 types of tools available in the CREW CHIEF tool box. Users can add other tools, if desired. Figure 11.8 shows a female CREW CHIEF human model using a ratchet wrench to remove a bolt from a jet engine. Because the bolt head is in a recessed area, the maintainability engineer must determine if a standard ratchet/socket

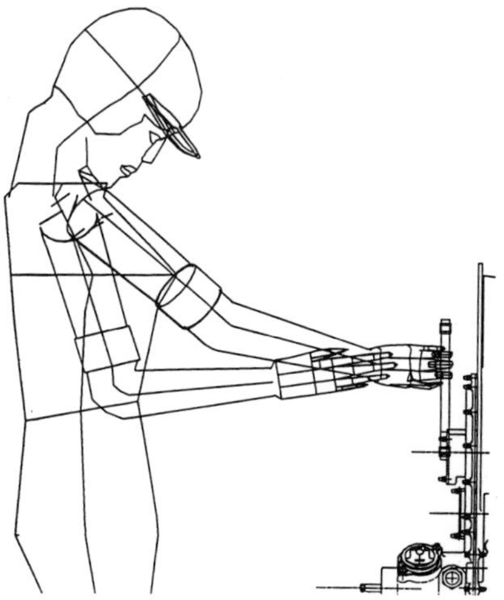

Design torque limits for tool analysis. Predicted torque.

Percentiles	Torque in ft–lbs	
	Tighten	Loosen
1st	8.08	10.35
5th	11.15	12.29
50th	16.97	16.09
95th	27.54	22.96
99th	31.02	25.47

Figure 11.8 Female CREW CHIEF human model uses a ratchet wrench to remove a recessed bolt from a jet engine, demonstrating adequacy of standard ratchet/socket. The table in the figure shows the torque strength available to perform the task.

combination will do the job. By using the CREW CHIEF human model, Figure 10.8 shows not only that the standard tool is adequate, but also displays the range of torque strength available to do the task. In cases in which an unrelated component interferes with the movement of the tool handle, the tool envelope analysis function of CREW CHIEF shows the free range of movement of the tool. In Figure 11.9, the obstructing handles on a box limit the rotation of a ratchet wrench. When an extension rod is added between the ratchet and the socket, the rotation is not obstructed. This tool envelope function works for all of the tools in CREW CHIEF's tool box. The designer could have selected chemical defence clothing or arctic clothing for analysis if hand clearance had been limited.

There are several advantages to using CREW CHIEF instead of real human subjects for this type of analysis. First, CREW CHIEF can be used on the design itself, even before a hardware mock-up is built. Second, there is no need to perform a difficult search for small and large subjects. Three, there is no need to procure the protective clothing. It is expected that analytical models like CREW CHIEF will reduce the need for maintainability demonstrations on prototype equipment. This is particularly important for aircraft where

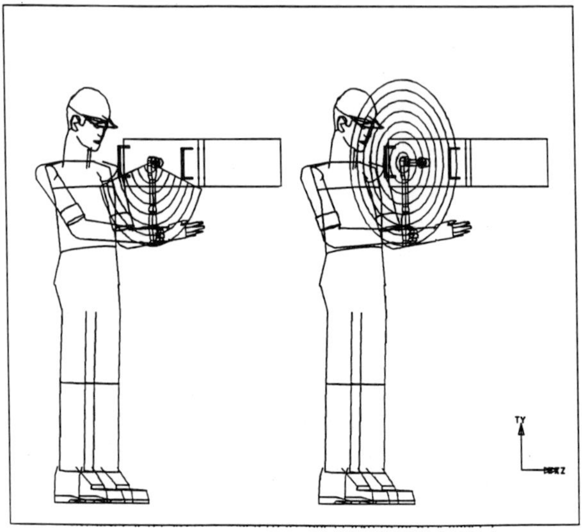

Figure 11.9 The tool envelope analysis of CREW CHIEF (left) shows ratchet wrench interferes with handles on a box. When an extension rod is added between the ratchet and the socket, the rotation is not obstructed.

maintainability demonstrations are concurrent with flight testing. Flight test schedulers are reluctant to allow maintainability demonstrations in which damage to a one-of-a-kind component could delay the flight test until a replacement part is custom made.

Other models

There are many other computer models to help designers (National Research Council, 1988). In addition, even anthropometric data are available on a dial-up computer system at the Center for Anthropometric Research Data (CARD) (see Robinson *et al.*, 1988). The data base contains hundreds of measurements from numerous surveys taken all over the world. As new data become available, they are added to the data base. The dial-up program allows users to extract subsets from the data base and compute statistics on them.

Trends for the future

Someday, we may be able to measure the human body in three dimensions. Already, new technology uses a laser scanner to gather data on head and face shapes (Beecher, 1986; Zehner, 1986). In the near term, such data may be useful for head gear design, such as helmets and protective masks. High density face data are currently being used for face contour comparisons, landmark spatial distribution analyses, and subregion delineations. The volume scanned by current laser mappers is too small to scan an entire body at one time, but work is in progress to develop a system which will.

Models must not only use new technology, but should also anticipate the design problems of the future. Planned enhancements for CREW CHIEF include a model of space suits and space tools. In addition to the currently used 'shuttle suit' which operates at 4.3 psi pressure, NASA is designing a high pressure suit which will operate at 8.3 psi. There are several advantages to the higher pressure suit. First, the high pressure suit will require a much shorter oxygen pre-breathe period (to expel nitrogen from the blood). Second, the high pressure suit will have quick-disconnect interchangeable limb components and size adjusters so fewer suits will be needed in orbit. (The older shuttle suit had sewn adjustment segments, and sizes could not be changed in orbit.)

Development of Predetermined Time Systems (PTS) for the factory environment began in 1911, and is a mature technology. In the factory, however, jobs are designed for efficient operation, with equipment designed to be free of accessibility problems. In aircraft maintenance, limited accessibility is the rule rather than the exception. The Predetermined Time Systems do a very poor job of predicting mean-time-to-repair on the flight line. Driven by procurement policies requiring manufacturers to warrant their products, designers need better tools for determining time to repair. The future CREW CHIEF system of programs will have a task-time estimating capability for remove/replace tasks involving very limited accessibility, as is common in aircraft maintenance. CREW CHIEF will be re-hosted on several other CAD systems to make it available to a greater number of designers. A personal computer version of CREW CHIEF is also being considered.

The CREW CHIEF development program also has by-products supporting equipment design standards and providing research data bases describing the ergonomics attributes of the maintenance technician.

References

Beecher, R. M., 1986, Computer graphics and diagnostics. *Proceedings of the Human Factors Society, 30th Annual Meeting*, 1986, pp. 211–215.

Dornheim, M. A., 1989, McDonnell Douglas/Bell Team Introduces Its LHX Concept. *Aviation Week & Space Technology*, March 6, 56–59.

Korna, M. and McDaniel, J. W., 1985, *User's Guide for COMBIMAN Programs (Computerized Biomechanical Man-Model), Version* 7, Armstrong Aerospace Medical Research Laboratory Technical Report AFAMRL-TR-88-034 (Wright-Patterson, AFB, OH).

Korna, M., Krauskopf, P., Haddox, D., Hardyal, S., Jones, M. Polzinetti, J. and McDaniel, J. W., 1988, *User's Guide for CREW CHIEF: A Computer Graphics Simulation of an Aircraft Maintenance Technician (Version 1 – CD20)*. Armstrong Aerospace Medical Research Laboratory Technical Report AAMRL-TR-88-034 (Wright-Patterson, AFB, OH).

McDaniel, J. W. and Askren, W. B., 1985, *Computer-aided Design Models to Support Ergonomics*, Armstrong Aerospace Medical Research Laboratory Technical Report AAMRL-TR-85-075 (Wright-Patterson AFB, OH).

Moroney, W. F. and Smith, M. J., 1972, *Empirical Reduction of Potential User Population as the Result of Imposed Multivariate Anthropometric Limits*. (Pensacola, FL: Naval Aerospace Medical Research Laboratory).

National Research Council, 1988, *Ergonomic Models of Anthropometry, Human Biomechanics, and Operator-Equipment Interfaces, Proceedings of a Workshop*, (Washington, DC: National Academy Press).

Robinson, J. C., Robinette, K. M. and Zehner, G. F., 1988, *User's Guide to Accessing the Anthropometric Data Base at the Center for Anthropometric Research Data (Card)*. Armstrong Aerospace Medical Research Laboratory Technical Report AAMRL-TR-88-012 (Wright-Patterson AFB, OH).

Zehner, G. F., 1986, Three-dimensional summarization of face shape. *Proceedings of the Human Factors Society, 30th Annual Meeting*, 1986, pp. 206–210.

Chapter 12
SAFEWORK:
A microcomputer-aided workstation design and analysis. New advances and future developments

CHRISTIAN FORTIN, ROBERT GILBERT, ANNE BEUTER[2], FRANCINE LAURENT[1], JEAN SCHIETTEKATTE[1], ROBERT CARRIER and BERNARD DECHAMPLAIN[1]

Department of Industrial Engineering,
Polytechnic School of Montreal

[1]Genicom Consultants Inc.

[2]Department of Kinanthropology,
University of Quebec at Montreal

Abstract. SAFEWORK is a flexible microcomputer-based model that should prove very useful in original machine and workstation design, in accident prevention, and in the study of workstation modification. Its various modules will continuously evaluate potential problems related to postural comfort, motion and posture feasibility, visibility and reachability of controls. Users select critical anthropometric variables for a given population. Then one or several three-dimensional (3-D) manikins are generated, as subjects, from anthropometric data in the data base. SAFEWORK now contains 14 3-D biomechanical models and a small transparent system which leads the user, with simple questions, to choose the appropriate model for the situation. In the anthropometric module, the models are based on multivariate normal distribution and can calculate 200 multiconditional variables for numerical representation of a human being from a minimum of four variables (height, weight, sex, race) to a maximum of 35 (for the user). Finally, we are now implementing a new technique for representation of the volumes of the manikins, using the mathematical theory of dual Kriging. This very recent and efficient technique, based on the mathematical theory of Krige, is a real breakthrough in modelization of forms as complex as the human body. It is much more economic in the number of sampling points, space memory, and execution time than other techniques with the same precision.

Keywords: Workstation; Design; Analysis; Anthropometric models; Three-dimensional biomechanical models; Dual Krigeage; Natural movements models.

Introduction

Several models of manikins have been developed for CAD on Main Frame with complicated user interface such as COMBIMAN (Evans, 1982), BOEMAN (Ryan *et al.*, 1970), SAMMIE (Bonney *et al.*, 1982), CAR (Harris *et al.*, 1980) and others. SAFEWORK is a highly flexible easy-to-use microcomputer program for the ergonomic analysis and design of workstations, especially human–machine interfaces and industrial products. It should also prove very useful in ergonomic task redefinition, workstation modernization, production-line balancing and all problems related to productivity, safety, health and comfort in human–machine systems. It was primarily designed for analysts evaluating potential problems and/or looking for solutions.

The main objectives of SAFEWORK development are to:

(a) function in an IBM AT or similar environment, preferably with a 80386 and an arithmetic coprocessor;
(b) have all the functions necessary to handle the tasks mentioned above with maximum efficiency;
(c) isolate the user from the complexity of its internal models;
(d) be reliable, coherent and robust;
(e) be easy to use;
(f) avoid long periods of dead time during use.

Together these objectives constitute an enormous challenge because there are no programs that meet all these goals. Even more advanced packages only satisfy two criteria, and they use large computers (Stewart, 1985; Pheasant, 1986). To satisfy all the above criteria new developments are required in every aspect of the problem.

SAFEWORK

Using various anthropometric data and models, such as those of NASA, Dempster, Drillis and Contini, McConville and Churchill, and others, as well as original software, the system was built up into three modules based on models published by several researchers (Daniels, 1952; Dempster, 1955; Roebuck, 1957; Hertzberg, 1960; Geoffrey, 1961; Dempster and Gaughren, 1964; Dempster *et al.*, 1964; Laubach and McConville, 1965; Moroney and Smith, 1972; Roebuck *et al.*, 1975; McConville and Churchill, 1976; NASA, 1978; Iwata *et al.*, 1981; Tisserand and Schouller, 1981; Tisserand and Saulnier, 1982; Archambault, 1986; Haslegrave, 1986; Vézina and Gilbert, 1986).

Specific anthropometric models module

This module allows for generation of one or more specific anthropometric models. These models may represent actual individuals, based on their external anthropometric measurements, or a sample population on a boundary hypersurface based on means, standard deviations, correlation coefficients and statistical models and methods.

Models for individuals

The numerical anthropometric module allows for modelling of any human being, normal or not. Every anthropometric model construction for individuals will go through a three-step process:

1. Using the 'generate' function from the menu, the following data (if they are not already in the data base) are entered: (i) general information (e.g. name, occupation, employee number, sex, ethnic origin, age); (ii) anthropometric measurements taken by the user (Figure 12.1a).

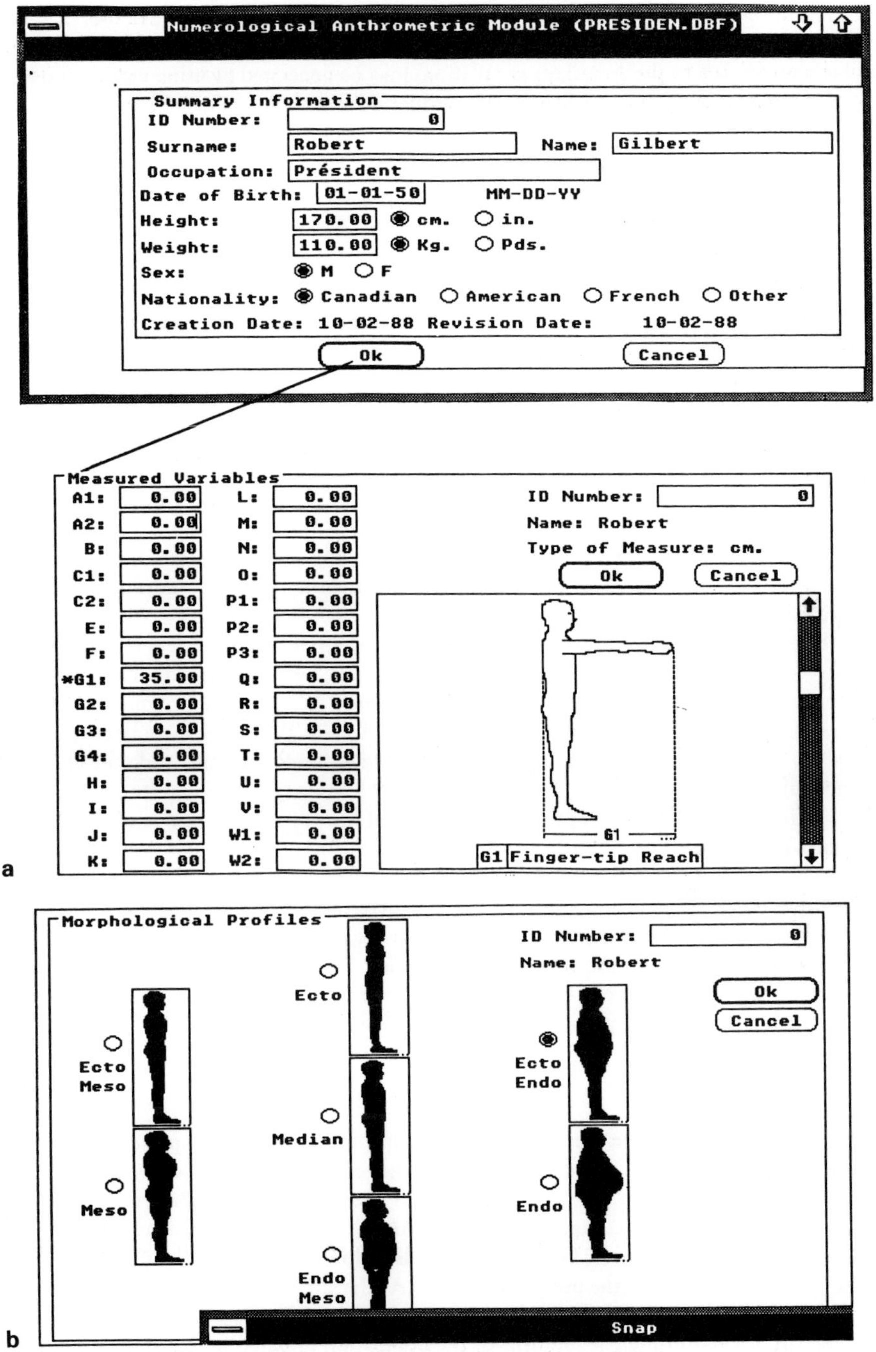

Figure 12.1 *(a) Data input for measured variables (upper two screens).*
(b) Choice of profile (bottom screen).

2. The program will then estimate any variables it needs for which the user has not supplied measurements. These estimates, after automatic population model adjustment with reference to the American population, may be generated by using various models: the multinormal, those of Dempster, Drillis and Contini, McConville and Churchill, and others (Daniels, 1952; Dempster, 1955; Roebuck, 1957; Geoffrey, 1961; Dempster and Gaughren, 1964; Dempster *et al.*, 1964; Moroney and Smith, 1972; Roebuck *et al.*, 1975; McConville and Churchill, 1976; NASA, 1978; Roozbazar *et al.*, 1979).
3. Finally, the program will compute the parameters necessary to build a 35 segment link-man.

The software program uses different correlation matrices from the NASA (1978) data banks for men and women. In the absence of such data on other populations, the same correlation matrices need to approximate the necessary values when the mean values and variances for a given population are available.

To construct links for the manikins, a subset of external measures is generated to which are applied two series of transformations (external measures to length of bones to length of links) whose equations, in part, come from the results of Trotter and Gleser (1952, 1958), Dempster (1955), Dempster *et al.* (1964), Roebuck *et al.* (1975), and Reynolds (1978). Body volumes can be added to the model link at any time (see Figure 12.9). These volumes are built with Bezier equations, which give a more realistic modelling of the human form than the hyperellipse, for example. We will soon be implementing a much more sophisticated representation based on the dual Kriging technique.

Furthermore, since quite different morphologies may yield identical sets of heights and weights for example, a morphological submodule has been developed. The choice of somatotype is based on seven different profiles (Figure 12.1b). This submodule will be modified to allow the user to make continuous choices in the possible somatotype space defined by the three axes in the conventional system (mesomorphic, endomorphic and ectomorphic axes).

To maximize the adaptation of the individual (or of the population) being considered, the user must identify the critical anthropometric variables. It is easy enough in workplaces to identify certain variables which are critical to the accommodation of a population. Thus, the height between the ground and knees is critical to determine the optimum height of a work table at which the work is done sitting down. This height cannot be lower than the highest knees in the population. From these critical variables, SAFEWORK can construct a set of manikins for use in the design process. All the data are stored in dBase III format and can be retrieved and manipulated by the user through this widely-known database manager.

Models for populations

SAFEWORK will allow the user to generate a sample from any given population, based on parameters such as means, standard deviations and correlation coefficients and by using statistical procedures, such as the Monte Carlo method. Missing data on the target population are estimated by comparison with the American population, the one on which we have the most complete data. SAFEWORK will also allow for extraction by similar methods of a sample of individuals in the user's data banks. At the same time, it is possible to generate samples containing only actual or potential marginal individuals. These individuals will be on the hypersurface delimiting the segment of the population targeted in the study. When the objective is to maximize the fraction of a population accommodated, only those individuals corresponding to linear combinations of the first principal components of the data points cloud can be generated on the hypersurface. Transformation to the principal referential axis is

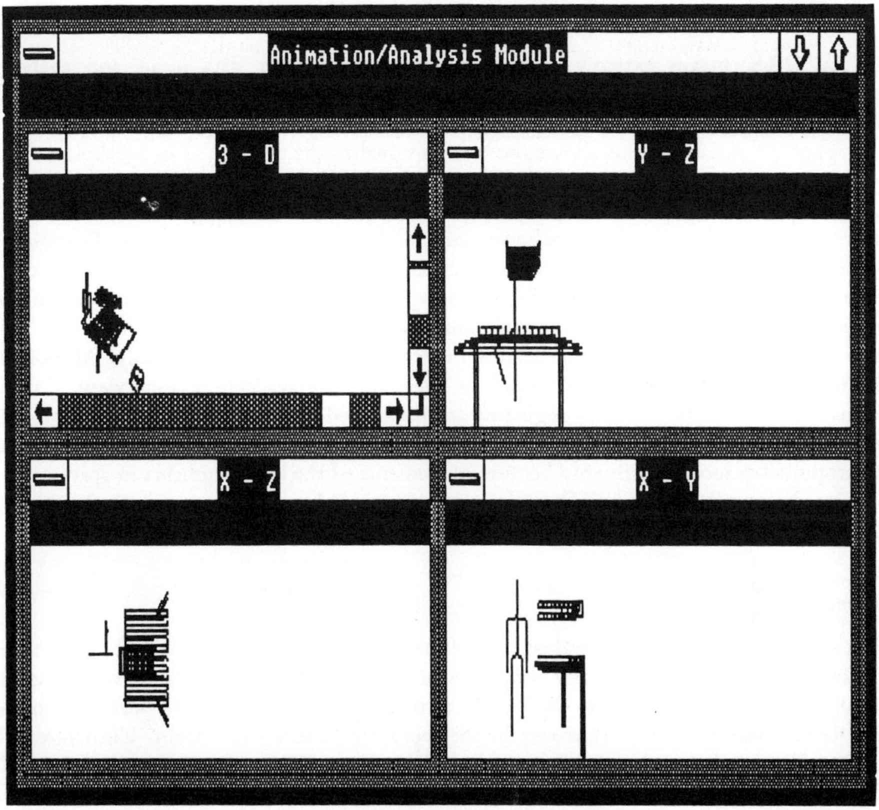

Figure 12.2 *Four types of view in SAFEWORK.*

particularly easy when a multinormal model is used and, after normalization, it is much easier to calculate the volumic integral of the distribution: the integral in *n* dimensions is then replaced by *n* integrals in one dimension. After integration, it is also easy to return to the initial system of variables.

Each of the specific anthropometric models in the sample will then be processed through the two last steps described in the preceding subsection.

Workstation module

The workplace is built using a CAD system, such as AutoCAD, and imported into SAFEWORK. The pre-generated manikins may then be called up for the analysis (see Figures 12.2, 12.3, 12.4, 12.6 and 12.9).

Analysis and animation module

This module integrates the specific anthropometric model into the workstation model. No definite path of use is imposed on the user. The main features follow.

Representation

SAFEWORK makes extensive use of graphics to simplify data input and analysis. Depending on the user's choice, the model links and pseudo-links of SAFEWORK are either represented in orthographic projection or in 3-D. In 3-D, a camera moving along the surface of a sphere of variable radius allows the user to visualize the manikin at its workplace from any angle and distance (Figure 12.5). Three or more active simultaneous views of the workstation and its operator may be represented (Figures 12.2, 12.3, 12.4 and 12.6). The rulers (Figure 12.2) available on each view control displacement in the window. Dragging a corner of a view with the mouse controls the size of the virtual aperture on the workstation. Each new view is generated by specifying a viewpoint and view size. The application can also be temporarily cleared off the desktop to a small icon at the bottom margin of the screen. This user-friendly interface allows non-technical users to master the system rapidly. All normal features (zooming, hiding and semi-hiding lines, etc.) are available on each view. Calculations for volumes of the body segments are performed in the animation-analysis module itself. This feature was incorporated in this module rather than in the numerical anthropometry module primarily because movements of the body segments in space can be calculated reasonably quickly if the volumes are not added to the segments each time, i.e. if the operations are executed on the links.

Animation

First, the user specifies the general posture of the worker in the workstation. There are three choices: standing, sitting or any other posture saved in an earlier session with another manikin. Second, the user drags the specific anthropometric model into the workstation and may then position it very accurately in the workstation using the zoom. Until now, all displacements were made with freezed linked segments. The animation analysis module allows 3-D positioning of every link by shifting in the orthogonal planes. There are two options for adjusting the posture and studying reach and other movements:

1. Manual input of the positions of each segment to be moved. In this option, the user points to the exact (or approximate) target with the mouse. The coordinates of the printed target appear on the screen. The user may also directly enter the coordinates via the keyboard. The system validates the input data at the end.
2. Moving by angular displacements. Here, the user points to a segment, and the proximal angle may be changed within its maximum limits by dragging the segment with the mouse. SAFEWORK then displays the angle and the population percentile capable of achieving the movement. The angular databank comes from several sources, such as NASA (1978), Ryan *et al.* (1970), Houy (1983), Kapandji (1980), Laubach (1970) and Grieve (1981). The value of each angle may appear simultaneously on the screen in a tabular format or near each angle.

A third option is also being planned: moving on target. In this option, the user gives the program a target to reach with a specific segment, and indicates up to which segments the motion will be carried out. SAFEWORK will then use specific algorithms, some found in the literature on graphics and others under development in our laboratory, for the distribution of angular modifications. A special project is to include some natural movements in SAFEWORK (see later section on Natural movement submodule). A specific function will continuously check whether the operator's segments engage other segments or structural components of the workstation, and notify the user. With current methods, this is

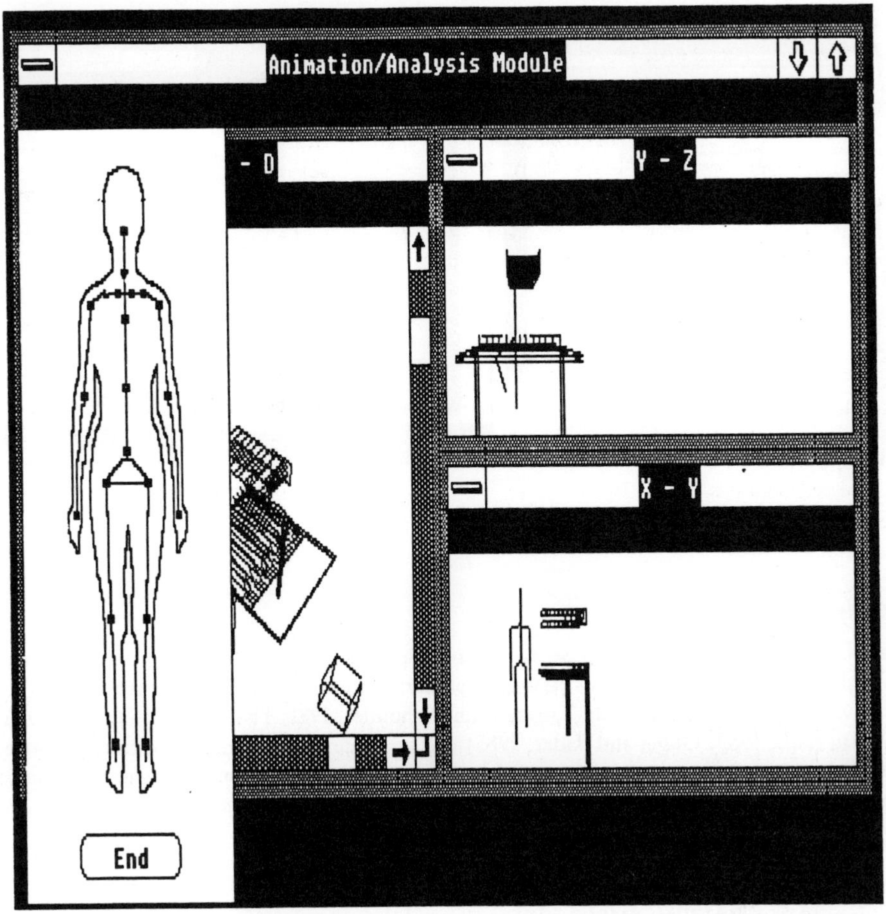

Figure 12.3 Choice of articulation to point.

usually a very tedious problem requiring solid modelling, but it is alleviated considerably with the dual Kriging technique.

EXAMPLE

Figure 12.3 shows the articulations that may be pointed with the mouse. Figure 12.4 shows how to fix the value of a parameter of displacement of a link: the percentage of the population able to realize this displacement appear automatically in the box below the drawing. The user interface for selecting a view angle in 3-D is shown in Figures 12.5 and 12.6. A composite leg movement is shown in Figure 12.6. A set of orthographic and 3-D views are summarized in Figure 12.2. Finally, in Figure 12.9, the representation of the body volumes with the Bezier equations is illustrated. In the same figure, the results of the biomechanical analysis of a specific articulation are displayed graphically.

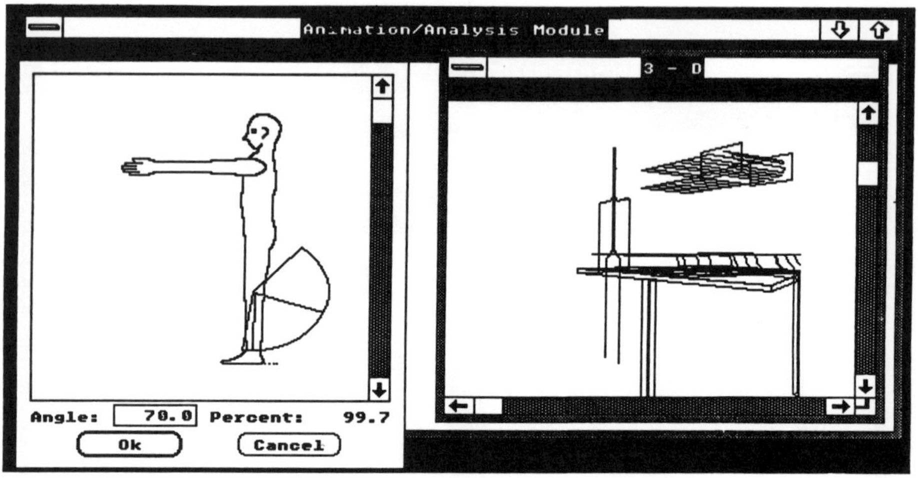

Figure 12.4 Selection of the angular displacement of the leg.

Analysis

Various menu functions (some accessing modules still under development) will evaluate variables and potential problems related to postural comfort, motion and posture feasibility and visibility. Current methods (Tisserand and Schouller, 1981; Tisserand and Saulnier, 1982; Pineau *et al.*, 1983; Unger and Rider, 1985) will generate reach envelopes for testing control reachability. Angular comfort zones, fields of vision and segment centre of mass will also be implemented. Force and torque analysis functions are fully developed but not yet interfaced with other modules.

BIOMECHANICAL ANALYSIS

This module of SAFEWORK has a biomechanical analysis submodule, which is composed of 14 mathematical models managed by a small expert system which, with simple questions, leads the user to choose the appropriate model for the situation. This submodule calculates the over-turning moment (equilibrium) and sliding moment, as well as the resulting forces and torques in 3-D for every articulation. The user can call up this module at any time and easily make the desired choice because the program uses the manikin's anthropometric and biomechanical files and the posture in question. The biomechanical models are based on a set of data and mathematical relations (distribution of the body mass; 'body index'; density and volume, body segments; effect of sex, age, somatotype, etc.) taken from several authors, such as Boyd (1933), Dupertuis *et al.* (1951), Dempster, (1955), Brozek *et al.* (1963), Dempster *et al.* (1964), Dubois *et al.* (1964), Dempster and Gaughren (1965), Drillis and Contini (1966), and Ryan *et al.* (1970).

Reports

Reports available with SAFEWORK are very useful features of the system. The flexibility of 'Windows,' under which the system was developed, allows direct transfer of information to an output device or to a graphic word processor (for example, 'Write' by Microsoft or

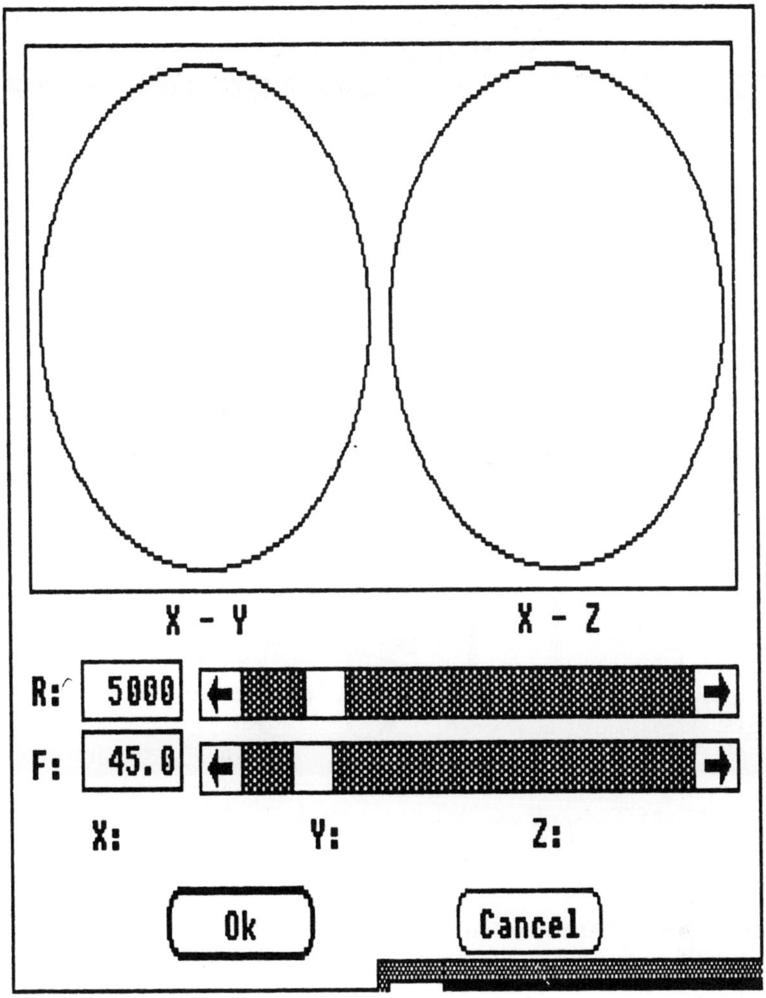

Figure 12.5 3-D view parameters.

'Pagemaker' by Aldus) in a concurrent processing session. Hence, the results displayed in tables or graphics format can be rapidly combined in the same report file. The printout can be obtained on any output device supported by 'Windows'.

Multivariate normal distribution and population accommodation

The numerical anthropometry module uses the multivariate normal distribution mathematical model to calculate, from a miminum of four initial data (height, weight, sex, ethnic origin), 200 multiconditional variables needed for numerical representation of a human being.

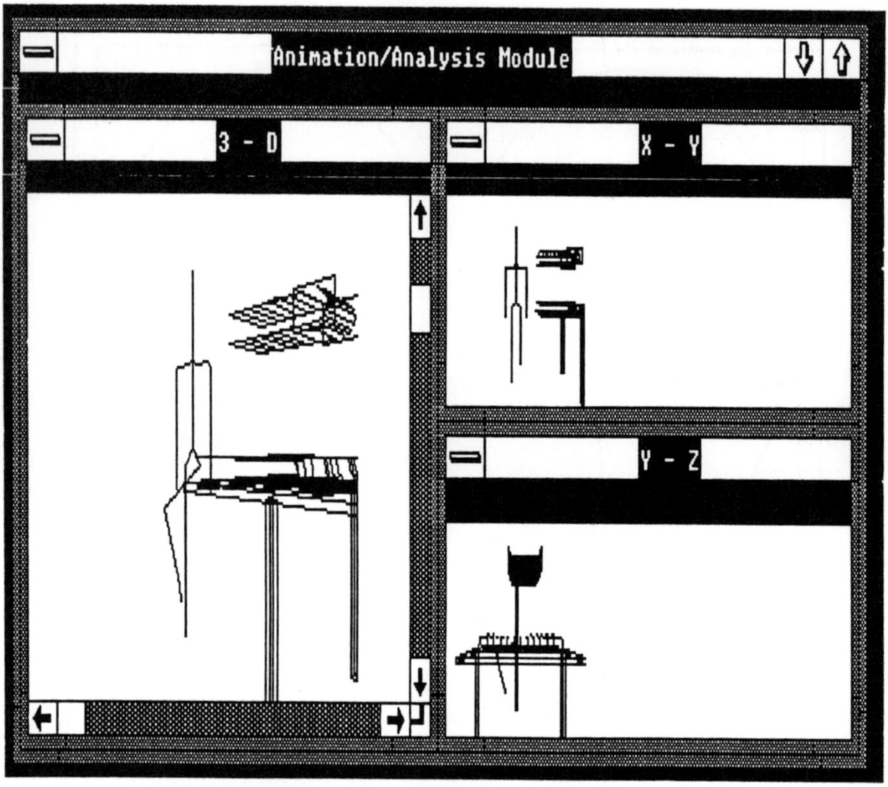

Figure 12.6 Result of a composite movement of the leg. Various angular 3-D views can be obtained by using the arrows around the figure.

Multivariate Normal Distribution

By definition, a vector of variables has a multivariate normal distribution with mean μ and covariance matrix Σ,

$$\mathbf{X} = (X_1, X_2, \ldots X_p) \sim N(\mu, \Sigma)$$

where μ is the vector of the means, and Σ is the covariance matrix defined by

$$\Sigma = \begin{pmatrix} \mathrm{var}(X_1) & \mathrm{cov}(X_1, X_2)\ldots & \mathrm{cov}(X_1, X_p) \\ \cdot & \cdot\ \cdots\ \cdot & \\ \cdot & \cdot\ \cdots\ \cdot & \\ \cdot & \cdot\ \cdots\ \cdot & \\ \mathrm{cov}(X_p, X_1) & \mathrm{cov}(X_p, X_2)\ldots & \mathrm{var}(X_p) \end{pmatrix}$$

if $\mathbf{X}$ has a probability density function given by

$$f(X_1, \ldots, X_p) = \frac{\exp[-0{\cdot}5\,(\mathbf{x}-\mu)'\,\Sigma^{-1}(\mathbf{x}-\mu)]}{(|\Sigma|)^{-1/2}\,(2\pi)^{(p/2)}}$$

where $|\Sigma|$ is the determinant of Σ.

This hypothesis of a multivariate distribution of the anthropometric variables of a population is an approximation because certain variables are not distributed in a Gaussian manner (weight and certain perimetric measures such as the trans-pelvic link). In practice, actual errors made on the estimates of the 5th and the 95th in using this model are generally below 3%, reaching 6% in only very few cases.

The cov $(\mathbf{X}_1, \mathbf{X}_2)$ elements of the matrix Σ are obtained from the correlation matrix with the following relation:

$$\text{CORR}\,(X_i, X_j) = \frac{\text{cov}(X_i, X_j)}{[\text{var}(X_i)\text{var}(X_j)]^{1/2}}$$

The variables to be estimated are based on a given subset of variables and are determined by the following multiconditional calculations.

If $\mathbf{X}_2$ is a vector of variables that has been measured and $\mathbf{X}_1$ a vector of variables to be estimated, then:

(a) partition $\mathbf{X}$ into two subvectors:

$$\mathbf{X} = (\mathbf{X}_1, \mathbf{X}_2);$$

(b) do the same for the vector of means:

$$\boldsymbol{\mu} = (\boldsymbol{\mu}_1, \boldsymbol{\mu}_2)$$

where $\boldsymbol{\mu}_1$ is the mean vector corresponding to $\mathbf{X}_1$ and $\boldsymbol{\mu}_2$ that corresponding to $\mathbf{X}_2$;

(c) partition Σ correspondingly:

$$\Sigma = \begin{pmatrix} \Sigma_{11} & \Sigma_{12} \\ \Sigma_{21} & \Sigma_{22} \end{pmatrix}$$

where Σ_{11} is the covariance matrix of $\mathbf{X}_1$, Σ_{22} is the covariance matrix of $\mathbf{X}_2$, and Σ_{12} and Σ_{21} are the mixed covariances of $\mathbf{X}_1$ and $\mathbf{X}_2$.

(d) lastly, obtain:

$$\mathbf{X}_1 \sim N(\boldsymbol{\mu}_c, \Sigma_c)$$

by calculating:

$$\boldsymbol{\mu}_c = \boldsymbol{\mu}_1 + \Sigma_{12}\,\Sigma_{22}^{-1}(\mathbf{X}_2 - \boldsymbol{\mu}_2)$$

$$\Sigma_c = \Sigma_{11} - \Sigma_{12}\,\Sigma_{22}^{-1}\Sigma_{21}$$

These equations are valid because it is precisely a property of the multivariate distribution that all linear combinations of its variables and all its marginal distributions are normally distributed. In models for individuals, the conditional mean $\boldsymbol{\mu}_c$ of $\mathbf{X}_1$ on $\mathbf{X}_2$ is the best estimate of $\mathbf{X}_1$. The covariance matrix Σ_c is a measure of the sensitivity and accuracy of this estimate, and thus it is an indicator of the amplitude of the somatotype variations of $\mathbf{X}_1$ still possible around $\boldsymbol{\mu}_c$ with the data given in $\mathbf{X}_2$. At the limit, the best model for an individual for whom we have only a few data will be a complete population segment centred on the mean μ_c for $\mathbf{X}_1$ and on the measures of $\mathbf{X}_2$ for $\mathbf{X}_2$. Alternatively, in models for a population, the covariance matrix Σ_c is used directly in the definition of the targeted population segment.

Another important property of the multivariate normal distribution is that if $\Sigma_{12} = \Sigma_{21} = 0$, then $\mathbf{X}_1$ and $\mathbf{X}_2$ are statistically independent. Consequently, if we have a normally distributed vector of data: $\mathbf{X} \sim N(\mu, \Sigma)$, it will always be possible to find a linear transformation that will render the new variables statistically independent. Then, the integral of probabilities will be much easier to calculate because the integral in n dimensions may be replaced by n integrals in one dimension. If, moreover, the new variables are standardized and if we are interested in the segment centred on the population mean, one integral will suffice, and we may then calculate the boundaries corresponding to the particular percentage of the population that we want to accommodate in advance. Various algorithms exist for finding the required linear transformations. They are based on the following considerations.

First, all linear combinations of a normally distributed vector are normally distributed:

if $\mathbf{X} \sim N(\mu, \Sigma)$
and $\mathbf{Y} = \mathbf{DX}$, where $\mathbf{D}$ is a linear transformation matrix,
then $\mathbf{Y} \sim N(\mathbf{D}\mu, \mathbf{D}\Sigma\mathbf{D}')$, where $\mathbf{D}'$ is the transpose of $\mathbf{D}$.

Second, Σ is a symmetric positive definite matrix. Then, we have:

$$\Sigma \phi = \phi \cap$$

where ϕ is the matrix of the eigenvectors of Σ of norm **1**, and

$$\cap = \begin{pmatrix} c_1 & \cdots\cdots\cdots & 0 \\ & \cdots\cdots\cdots & \\ 0 & \cdots\cdots\cdots & c_n \end{pmatrix}$$

where c_i, $i = 1, \ldots, n$ are the corresponding eigenvalues with $c_i \geq \mathbf{10}$, $i = 1, \ldots, n$, because Σ is positive definite, and where $\phi' \phi = \mathbf{I}$, the identity matrix, by the orthogonality of the eigenvectors.

Consequently, the matrix

$$\mathbf{Z} = \phi' \mathbf{X} \text{ is normally distributed:}$$
$$\mathbf{Z} \sim N(\phi\mu, \phi' \Sigma\phi) = N(\phi\mu, \cap)$$

We now have a diagonal covariance matrix. Now, if we look at

$$\sqrt{\cap} = \begin{pmatrix} \sqrt{c_1} & \cdots\cdots\cdots & 0 \\ & \cdots\cdots\cdots & \\ 0 & \cdots\cdots\cdots & \sqrt{c_n} \end{pmatrix}$$

and perform the linear transformation:

$$\mathbf{U} = \sqrt{\cap}\,\mathbf{Z},$$

we obtain:

$$\mathbf{U} \sim N(\sqrt{\cap}\,\phi' \mu, \mathbf{I})$$

and the covariance matrix Σ is replaced at the end by the identity matrix. All the new variables, U_i, $i=1, \ldots n$, the components of **U**, are standardized and statistically independent. We may return into the normal space of **X** by applying the inverse transformations.

Population accommodation

Calculation of the eigenvalues corresponding to the eigenvectors of the covariance matrix Σ is the same as for the principal components of the multivariate normal distribution. In fact, the principal components correspond to the square of the eigenvalues in decreasing order. Each of these squares corresponds to the percentage of the total variance of the data that may be explained or supported by the variance of the corresponding variable. Generally, the sum of the four to seven first squares is sufficient to account for more than 90–95% of the total variance observed. Furthermore, the square roots of the eigenvalues correspond to the lengths of the semiaxes of the hyperellipsoid centred on the population mean and whose surface is a contour of constant density of probability. Thus, if we are interested in the mean population segment, the following procedure applies.

First, given an initial multivariate distribution, the principal components of the distribution are determined and standardized. Next, all the simple linear combinations of the five to seven first principal components are generated. A simple linear combination is one in which all the coefficients are -1 and $+1$. If there are seven variables, there will be 128 combinations each corresponding to a particular individual in the distribution. If we wish to accommodate for $A\%$ of the population and we have n principal variables, we will integrate for each between $Z_{-\alpha}$ and $Z_{+\alpha}$ on the $Z \sim N(0, 1)$ normal univariate distribution, such that:

$$\Pr[Z_{-\alpha} \leq Z \leq Z_{+\alpha}] = 1 - (A\%)^{1/n}$$

This is the same as dilating or contracting the hypercube obtained in the space of **U** by the previous simple linear combinations, to obtain the desired population coverage.

Finally, we may return to the original space of **X** and relate the extreme data points to the concrete problems of adjustment. All that is needed is to find the two extremal points on each original axis. The equivalent operations on the initial distribution are illustrated in Figure 12.7. The effects of the successive linear transformations are shown in Figure 12.8.

In general, it is not advisable to use more than five to seven principal components. First, each is already a linear combination of several variables. Second, it would soon lead to the problem known by mathematicians as the multidimensional explosion of complexity if we wanted to consider additional variables. In trying to accommodate a large percentage of a population on several variables simultaneously, the percentage necessary to accommodate on each variable will increase exponentially with the number of variables and the percentage of simultaneous accommodation. Third, as previously mentioned, the first few principal components are usually sufficient to account for most of the total variance observed in a given distribution of values.

After step 3, the population accommodated will be greater than $A\%$, the objective. In principle, if $A\%$ is greater than say 80%, then the difference will not be significant because of the low density probability outside the ellipsoid, in the hyperrectangle region, containing exactly $A\%$ of the population. If $A\%$ is significantly smaller, a precise evaluation may be required in theory, using for example an n-dimensional integral in the original space of **X** applying the Monte Carlo method of numerical integration. The integral begins close to the mean, and its sample space is gradually increased until the $A\%$ limit is obtained. This method

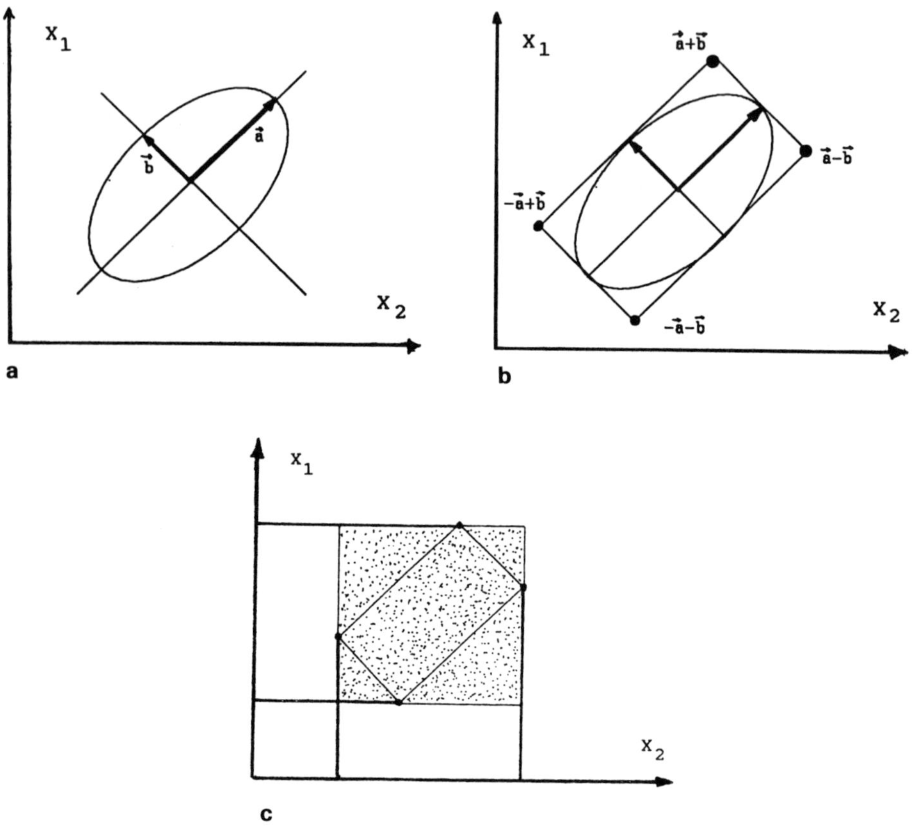

Figure 12.7 (a) STEP 1: Find the principal components.
(b) STEP 2: Find all the simple linear combinations of principal vectors.
(c) STEP 3: Find the extremal points on each axis by finding the minimum and maximum components of the coordinates of all the 2^n points.

is likely to be much more costly in execution time than the others (hours instead of fractions of a second). If an efficient algorithm exists or may be constructed, and if the submatrix of covariance used is sufficiently small, either method may be used. But, in practice A% will be much smaller before a significant difference appears, because the vast majority of the correlation coefficients between anthropometric variables are on the order of 0·2. There are very few exceptions, and these can often be explained (in the case of statures measured by different methods, or to the fact that standing statures included the length of the leg). Consequently, the shape of the cloud of data points is more a hypersphere than an elongated hyperellipsoid. Hence, the hyperrectangle around the hyperellipsoid is more a hypercube and is not very inclined compared with the initial reference axis. Therefore, the differences between the percentages accommodated in the hyperrectangle around the hyperellipsoid and hyperrectangle formed with the minimum and maximum for each variable are very much smaller.

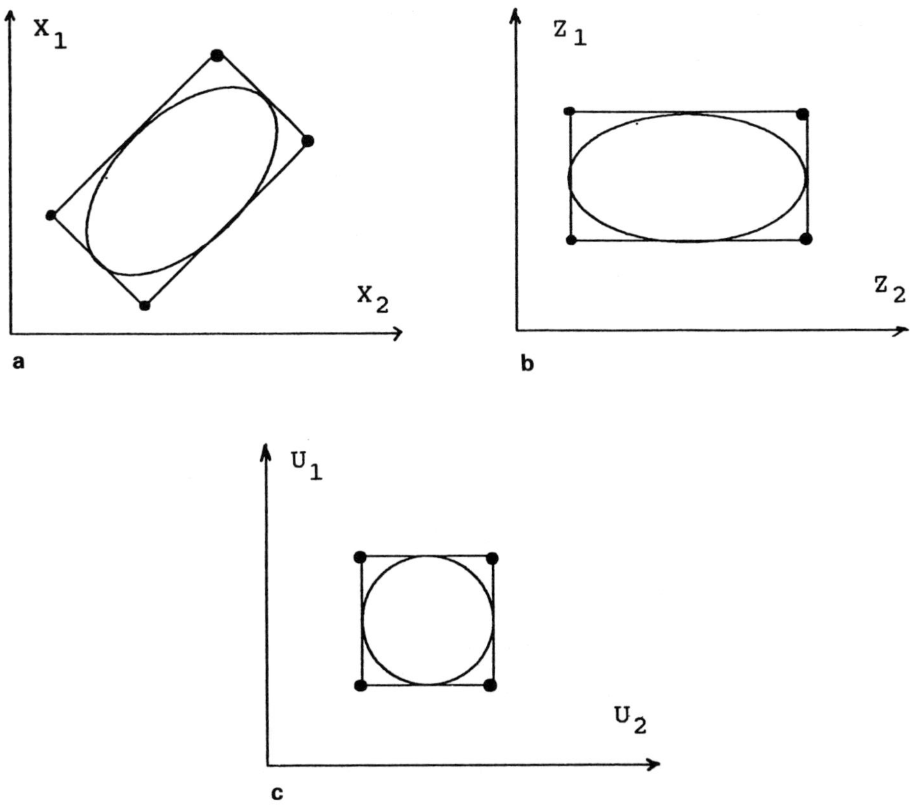

Figure 12.8 (a) Initial distribution.
(b) After diagonalization.
(c) After normalization.

Future developments and dual Kriging technique

Since the latest mathematical development of the Kriging theory made by a colleague, Dr Michel Soulié, further updates of SAFEWORK become of great interest. The Kriging theory is a statistical interpolation method (exact interpolator) based on the generalized increments and covariance theories.

Dr Soulié *et al.* (personal communications) have demonstrated that we can adapt this theory as dual Kriging for geometrical modelling thus, using it in representing the whole human body surface or volume in a single, simple analytical and linear equation. The fidelity of representation is based on the number of points and their particular location taken on the human body. Hence, using the usual anthropometric references is a step and creating new ones is another. Computer memory and processing time are also some important considerations. Kriging is represented using matrix equation. Therefore, using matrix properties, one can considerably reduce the size of the matrix, and most probably look for real-time applications on microcomputer!

A remark made by Stewart illustrates the scope of the anthropometric aspect of the problem:

> "Man modelling with fully integrated workspace design **is not yet available on microcomputers**. The ideal CAD package would permit the user to specify anthropometric criteria of the working population, and then design the workstation in **three-dimensional** space around a representative man-model, **providing instant graphic feedback** to identify obvious incompatibility in the system."

Kriging would not only be used for human body modelling (volume of body segments, connection of volume surfaces of body segments with respect to the body link system, etc.) but also for human body animation thus representing reaching zones, comfort zones, body movements in 3-D, angular limits, etc. One could even determine the spatial trajectory of the eyes in relation with the directional angle of the visual line.

Furthermore, if Kriging can represent such a complex form as a human body, it can certainly represent any workstation using the same methodology. Consequently Kriging is of great interest for 3-D software and more specifically for algorithms testing contact or penetration. Indeed, with the volume surface equation, we can easily check if a point (substituted in the equation) is inside or outside depending if $f(x, y, z)$ is negative or positive. If the value of the function is zero, it is a contact point.

There is a great potential for the applications of the Kriging theory but in order to perform in this area, we must accomplish a certain number of research steps before putting such a powerful tool in application. These steps are concerned with morphometric discontinuities, connection of volume surfaces of the body segments with respect to the body link system, choice of the best function or the best combination of $K(h)$ functions, number of points necessary for the faithful representation, etc.

Biomechanical analysis submodule

When the biomechanical analysis of a posture is presented to the program, the first step is to determine the stability or static equilibrium by calculating the centre of gravity and the overturning and sliding moment. If the position is statically stable, the user may continue the analysis and point to any articulation to obtain the resulting stresses and torques (see Figure 12.9). The biomechanical submodule for calculating the resulting centre of gravity, stresses and torques uses a set of data and mathematical relations taken from several authors as seen earlier in this chapter.

Choice of the biomechanical model

Calculation of the overturning moment (to verify stability of a given posture) can vary with the situation. For example, a slippery floor (low friction coefficient) or a point of support (at the hips, for example) can alter the results. So in order to decrease the uncertainty associated with a single biomechanical model, we have established 14 distinct situations and models. Each situation has different elements, i.e. towing or pushing force, use of one or both hands, lifting a weight, slippery floor, etc. A small interactive interface identifies the situation and calls up the right biomechanical model. The choice is made by means of simple questions asked of the user, for example:

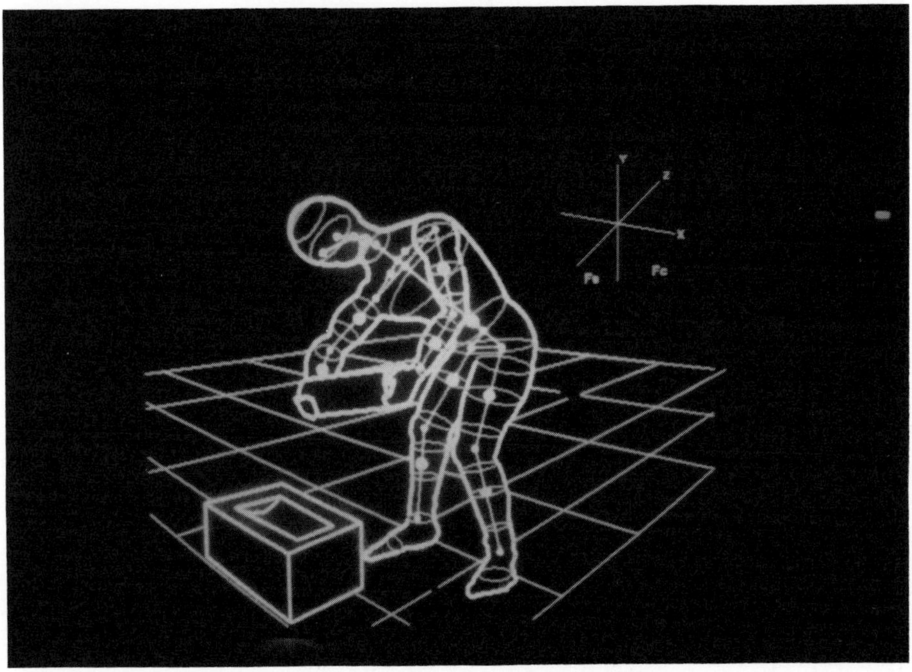

Figure 12.9 Full representation of a workstation with SAFEWORK.

Is there a known pushing or towing force? (Y/N)
Are both hands used to push or tow? (Y/N)
The left and right hand coordinates (*x*, *y*, *z* direction) on which the known pushing or towing force applies?
Case of pushing a car? (Y/N)
Would there be any weight to be lifted? (Y/N)
The magnitude of this weight *Z* (in kg and in *Z*-direction)?

If there is a point of support on the body, it could substantially affect the calculation of the overturning moment, and the system will ask the user to identify the segment at which the point of support is found and its coordinates. When the coordinates (*x*, *y*, *z*) are required, the user will indicate on the screen where the point of application is on the workstation when this submodule will be linked, thereby eliminating the cumbersome manipulation of 3-D coordinates by the user.

Vision analysis submodule

This submodule will distinguish: central and peripheral vision; monocular, binocular and ambinocular fields; stereoscopic vision. It will permit the user to project the various 'conic' fields of vision and to see what the manikins 'see' in a given posture. This submodule will be realized in part with the collaboration of a team of Toronto University in Canada.

Natural movement submodule

Objective

The purpose of this submodule of SAFEWORK will be to allow the user to move several limb segments simultaneously by moving the manikin limb tip in a simple software operation. The user will click the mouse once on the limb tip to be moved and once on the final destination, and the segments involved will move accordingly. For example, reaching for a stationary target with the hand involves the shoulder, arm, forearm and fingers. By moving the hand to the target, the remaining segments of the kinematic chain will automatically be moved in 3-D space according to prescribed trajectories. These prescribed trajectories will be identified through the kinematic analysis and modelling of a series of reaching experiments.

Reaching for the target can be executed with a finger tip (e.g. to depress a knob), with a pinch grasp (e.g. to rotate a knob), and with an open hand (e.g. to manipulate a lever). Thus, hand posture plays a significant role in reaching behaviours. These reaching movements are restricted by task demands such as accuracy or cluttered environment, and by physical contact with external surfaces, such as body posture. In addition, according to Hollerbach (1985), an object can be located in space by six variables (i.e. three for position in Cartesian coordinates and three for orientation expressed in roll, pitch and yaw angles), and we know that the upper limb possesses more than the six degrees of freedom required to grasp an object in space. Therefore, kinematic redundancies will have to be considered in the development of this submodule. Other factors will have to be considered, including: initial/final position of the arm; hand posture in relation to the target; overall body postures; speed of execution; length of the hand path in space; nature of trunk involvement; learning; and so on.

Methodology

In order to develop a representative movement repertoire of reaching behaviours, we plan to study 10 adult subjects with no known neuro-musculo-skeletal disorders of the right upper extremity in a series of reaching tasks in which the aforementioned factors will be systematically varied. Part of this research will be conducted in collaboration with Dr H. Flashner of Mechanical Engineering Department of University of South California in Los Angeles.

A videomotion analysis system (Peak Performance Technologies) will be used to record the kinematics of reaching. Light Emitting Diodes (LEDs) remaining continuously in the field of view of at least two video cameras will be fixed on anatomical landmarks corresponding to the shoulder, elbow, wrist and finger joints. After a calibration procedure, the video signal will be recorded at 60 Hz by three synchronized video cameras for subsequent analysis.

The Automatic Digitizing Module (Peak Performance Technologies) will be used to locate and track the LEDs placed on the moving limb. This module will search, detect and then calculate the position of the centroid of contrasting markers for each frame of a videotaped sequence. The first three frames of this sequence will be digitized by an operator using an overlaid crosshair cursor. Once these initial frames are digitized, the computer (a Compaq 386-20) will take over and capture the remainder of the frames in the sequence. If an LED disappears, the software will request the operator to intervene and make the correct location decision. The coordinates of the centroids will be stored for subsequent data processing, including 3-D coordinate calculations using a Direct Linear Transformation, smoothing with a fourth order, zero lag Butterworth digital filter, calculating the segmental and total body centre of mass, linear and angular velocity and acceleration, etc.

Model

Human movements appear to be organized in discrete units and it has been shown that motor skills are composed of elementary procedural units (Johnson, 1984). In fact, slips of actions occur at specific moments of the performance (i.e. transitions) corresponding to more than one possible ensuing course of action (Reason, 1979). Furthermore, a hardwire element of motion has been proposed with the concept of scaling of movement load, condition, path, and even intersubject variability following a normalization procedure for distance and time (Hollerbach and Flash, 1982; Atkeson and Hollerbach, 1985).

We have recently proposed to represent experimental kinematic data by a finite number of predetermined functions and a small number of task-dependent undetermined coefficients using a least squares approximation described in the next section (Flashner *et al.*, 1988, 1989). We showed that the different basic functions investigated such as cosines, even polynomials, and their associated coefficients reflected the qualitative behaviour of joint excursions. This kinematic approach provides an adequate, simple and flexible analytic description of limb movements executed at natural cadence and we intend to use it to describe reaching movements in the natural movements submodule.

Least squares approximation

The task-dependent coefficients will be calculated and associated with a small number of basis functions (Flashner *et al.*, 1988). In brief, suppose we are given n data points:

$$(t_i, Y_i), i = 1, \ldots, n$$

Let t be the dependent variable and y be the independent variable. Then, if there is a functional relationship $y(t)$ between the variables, the following relationship must be satisfied:

$$Y_i = y(t_i), i = 1, \ldots, n$$

Since the exact relationship is not known, we approximate it by linear combination of m given basis functions:

$$y(t) = c_1 f_1(t) + \ldots + c_m f_m(t),$$

where $c_j, j = 1, \ldots, m$ are undetermined coefficients.

To find the coefficients, c_i, the least squares method minimizes the sum of the squares of the deviation at the measurement points:

$$r^2 = \sum_{i=1}^{n} (y(t_i) - y_i)^2$$

Therefore the problem of best approximation is defined by:

$$\min_{c_j} \sum_{i=1}^{n} \left(\sum_{j=1}^{m} c_j f_j(t_i) - y_i\right)^2$$

The solution of this problem exists and is always unique (Davies, 1975). Furthermore, if the set f_1, f_2, . . . is closed, the data can be approximated with arbitrary accuracy by a linear combination of f_i, $i = 1, \ldots, m$. Define:

$$A_{ij} = f_j(t_i), \ \mathbf{b} = \begin{pmatrix} y_1 \\ \cdot \\ \cdot \\ \cdot \\ y_n \end{pmatrix}, \ \mathbf{x} = \begin{pmatrix} c_1 \\ \cdot \\ \cdot \\ \cdot \\ c_m \end{pmatrix}$$

The least squares problem can be reformulated as the following minimization problem:

$$\min \|\mathbf{Ax} - \mathbf{b}\|$$

where $\|\cdot\|$ is the Euclidean norm and $A \in R^{nxm}$

It can be shown that this minimization problem is equivalent to finding the minimum norm (length) solution to the set of linear equations:

$$\mathbf{Ax} = \mathbf{b}$$

The solution is given by: $\mathbf{x} = \mathbf{A}^+ \mathbf{b}$, where $\mathbf{A}^+$ is the pseudo-inverse of $\mathbf{A}$ and is given by:

$$\mathbf{A}^+ = (\mathbf{A}^t\mathbf{A})^{-1}\mathbf{A}^t$$

and evaluated by singular value decomposition algorithm.

For each experimental condition, joint kinematics of the upper extremity will be characterized by a small number of analytical functions and task-dependent coefficients. Thus, a collection of prescribed trajectories will be developed as pertinent factors are manipulated. We expect that, in some cases, changes in coefficient values alone will be sufficient to approximate the data without altering the basis functions. In other cases, different basis functions may have to be added or substituted. This approach will be used to implement the natural movements submodule. We also intend to use this approach to enable the user to simulate work-related movements executed, for example, with minimum effort and maximum efficiency.

Conclusion and future developments

We hope that our model will be very useful in original machine and workstation design or modification, accident prevention, and in the assessment of problems, actual or potential, related to postural comfort, motion, visibility and control reachability. We intend to develop and test on the field as soon as possible the following developments:

1. The development of muscular models, in particular for the back, is in progress. Presently, some 8000 hours have been dedicated to fine and ultra-fine 3-D biomechanical modelization of the back (dissecting, analysing and measuring) by our work physician and our physiologist. These models are already anticipated in the programming of biomechanical models.
2. A vision module will be developed.
3. The module dealing with morphological profiles will be modified to allow the user to make continuous choices between the somatotype extremes.

4. A 3-D pointing animation module will be developed with forced and freely reached distances models conditional to the freezing of a given body segment. This submodule is interesting not only because of its utility but also because the user will no longer need to manipulate every body segment in 3-D separately. Instead, the mouse will be used to point to the member and the destination, taking into account the nature of the contact and a fixed body link previously specified by the user.
5. Clothing will be represented with selective dilatation techniques.
6. We will develop a way to represent likely natural movement and to choose that most likely in any given setting. As in the works of Flashner *et al.* (1988, 1989), the characteristics of the likely paths will be a function of individual traits as age, sex, experience and particular pathologies (Parkinsonism, cerebral paralysis, partial paralysis, etc).
7. In the more distant future, we will establish the neurokinetic basis of therbligs and ways to simulate them.

All these developments and the representation of body volumes will use the most recent mathematical advances in the theory of Kriging, which constitutes one of the most powerful geometric modelling tools available. Furthermore, we plan to implement detailed models, specific to certain needs of analysis as, for example, in the studies of the hand or the head and neck, in the field of vibration control and ergonomics of handtools and personal protectors.

Acknowledgements

Part of this research was supported under CRSNG (CANADA) Grant CRD-0038986 to R. Gilbert. Microsoft C is a trademark of Microsoft Corporation. Microsoft Windows is a trademark of Microsoft Corporation. dBase III is a trademark of Ashton-Tate.

References

Archambault, H., 1986, Etablissement d'une procédure informatique pour l'étude du dimensionnement des postes de travail. P.F.E., avril, 1986. Industrial Engineering Department. Polytechnic School of Montreal.

Atkeson, C. G. and Hollerbach, J. M., 1985, Kinematic features of unrestrained vertical arm movements. *Journal of Neuroscience*, **5**, 2318–2330.

Bonney, M. C., Case, K. and Porter, J. M., 1982, User needs in computerized man models. In: *Anthropometry and Biomechanics, Theory and Application*. edited by R. Easterby, K. H. E. Kroemer and D. B. Chaffin. (New York: Plenum Press), pp.97–101.

Boyd W., 1933, The specific gravity of the human body. *Human Biology*, **5**, 646–672.

Brozek, J., Grande F., Anderson J. T. and Keys, A. 1963, Densitometric analysis of body composition: Revision of some quantitative assumptions. *Annals of the New York Academy of Science*, **110**, 113–140.

Chaffin, D. B., 1969, A computerized biomechanical model—development of and use in studying gross body actions. *Journal of Biomechanics*, **2**, 429–441.

Daniels, G. S., 1952, *The 'Average Man'?* Technical Note No. CRD 53-7. (Aero Medical Laboratory, Wright Air Development Center).

Davis, P. J., 1975, *Interpolation and Approximation* (Mineda, NY: Dover).

Dempster, W. T., 1955, *Space Requirements of the Seated Operator. Geometrical, Kinematic and Mechanical aspects of the Body with Special Reference to the Limb*. WADC Technical Report No. 55–159 (Wright Air Development Center).

Dempster, W. T. and Gaughren, R. L., 1965, Properties of body segments based on size and weight. *American Journal of Anatomy*, **120**, 33–54.

Dempster, W. T., Sherr, L. A. and Priest, J. T., 1964, Conversion scales for estimating humeral and femoral lengths and the lengths of function segments in the limbs of American caucasoid males. *Human Biology*, **36**, 246–262.

Drillis, R. and Contini, R., 1966, *Body Segments Parameters*, Technical Report No. 1166.03. (New York University, School of Engineering and Science, NY).

Dubois, J., Santschi, W. R., Walton, D. M., Scott, C. O., and Hazy, F. W., 1964, *Moment of Inertia and Centers of Gravity of the Living Human Body Encumbered by a Full Pressure Suit*. Technical Report AMRL-TR-64-110, Wright-Patterson Air Force Base, Ohio.

Dupertuis, C. W., Pitts, G. C., Osserman, E. F., Welham W. C. and Behnke, A. R., 1951, Relationship of specific gravity to body build in a group of healthy men. *Journal of Applied Physiology*, **3**(1), 676–680.

Evans, S. M., 1982, Some computational problems in developing computerized man-models. In: *Anthropometry and Biomechanics. Theory and Application*, edited by R. Easterby, K. H. E. Kroemer and D. E. Chaffin. (New York; Plenum Press).

Flashner, H., Beuter, A. and Arabyan, A., 1988, Fitting mathematical functions to joint kinematics during stepping: implications for motor control. *Biological Cybernetics*, **58**, 91–99.

Flashner, H., Beuter, A. and Boettger, C., 1989, Parameter optimization model for learning in stepping motion. *Biological Cybernetics*, **60**, 277–284.

Frisch, F. D. and D'Aulerio, L. A., 1980, Bioman – An improved occupants/crew station compliance modelling system. *Aviation, Space and Environmental Medicine*, **51**(2), 160–167.

Geoffrey, S. P., 1961, *A 2-D Manikin – The Inside Story. X-rays Used to Determine a New Standard for a Basic Design Tool*. SAE Preprints, for presentation at the 1961 SAE International Congress and Exposition of automotive engineering. Society of Automotive Engineering, New York.

Grieve, G. P., 1981, *Common Vertebral Joint Problems* (New York: Churchill Livingstone).

Harris, R., Bennet, J. and Dow, L., 1980, Car-II: A revised model for crewstation assessment of reach. Tech. Dept. 1400. 06b. Willow Grave P. A. Analystic Inc.

Haslegrave, C. M., 1986, Characterizing the anthropometric extremes of the population. *Ergonomics*, **29** (2), 281–301.

Hertzberg, H. T. E., 1960, Dynamic anthropometry of working positions. *Human Factors* **2**, 147–155.

Hollerbach, J. M. and Flash T., 1982, Dynamic interaction between limb segments during planar arm movements. *Biological Cybernetics*, **44**, 67–77.

Hollerbach, J. M., 1985, Levels of analysis in biological and robot limb movements: Kinematics and control. In: *Motor Control: From Movement Trajectories to Neural Mechanisms*. Short Course Syllabus. (Dallas, TX: Society for Neuroscience). pp. 2–13.

Houy, D. R. 1983, Range of joint movement in college males. *Proceedings of the Human Factors Society*, **1**, 374–378.

Iwata, K., Moriwaki, T., Kawano, T., Misaki, N., Nomura, H. and Yanagida, Y., 1981, Modelling and dynamic analysis of three-dimensional human body motion via computer graphics. *Memoirs of the Faculty of Engineering, Kobe University*, **27**, 15–23.

Johnson, P., 1984, The acquisition of skill. In: *The Psychology of Human Movement*, edited by M. M. Smyth and A. M. Wing (London: Academic Press).

Kapandji, I. A., 1980, *Physiologie Articulaire*, (Paris, Librairie Maloine).

Laubach, L. L. and McConville, J. T., 1965, Relationships between flexibility anthropometry and the somatotype of college man. *The Research Quarterly*, **37** (2) 241–251.

Laubach, L. L., 1970, Characteristics of the range of joint motion and its relationships to selected anthropometric dimensions and somatotype components, Ph.D. Dissertation, Ohio State University, 108 pp.

McConville, J. T., and Churchill, E., 1976, *Statistical Concepts in Design*. Technical Report No. AMRL-TR-76-29. Aerospace Medical Division Air Force Systems Command (Wright-Patterson Air Force Base, OH), 57 pp.

Moroney, W. F. and Smith, M. J., 1972, Empirical Reduction of Potential User Population as the Result of Imposed Multivariate Anthropometric Limits. Naval Aerospace Medical Research Laboratory. Naval Aerospace and Regional Medical Center. Florida.

NASA, 1978, *Anthropometric Source Book*, Vol. I, *Anthropometry for Designers*, Vol. II, *A Handbook of Anthropometric Data*; Vol. III, *Annotated Bibliography of Anthropometry*, edited by Webb Associates NASA Reference Publication 1024. National Technical Information Service (NTIS). US Department of Commerce.

Pheasant, S., 1986, *Bodyspace* (London: Taylor & Francis).

Pineau, J. C., 1986, Mollard, R. Sauvignon, M. and Amphoux, M., 1983, Biostereometric data processing in Ergodata *Biostereometrics '82. Proceedings of the SPIE*. International Society of Optical Engineers, **361**, 169–171.

Reason, J. T., 1979, Actions not as planned. In: *Aspects of Consciousness*, edited by G. Underwood and R. Stevens (London: Academic Press).

Reynolds, H. M., 1978, Inertial properties of body and its segments. In NASA, 1978, *Anthropometric source book, Vol I, Anthropometry for designers*, NASA reference publication 1024.

Roebuck, J. A., 1957, Anthropometry in aircraft engineering design. *Journal of Aviation Medicine*, **28**, 41–56.

Roebuck, J. A., Kroemer, K. H. E. and Thomson W. G.,1975, *Engineering Anthropometry Methods*. (New York: John Wiley).

Rohmert, W., Wangenheim, M., Mainzer, J., Zipp, P. and Lesser, W., 1986, A study stressing the need for a static postural force model for work analysis. *Ergonomics*, **29**(10), 1235–1249.

Roozbazar, A., 1973, Biomechanical modelling of the human body. *Proceedings of the 17th annual meeting of the Human Factors Society*, 16–18 October 1973, 181–191.

Roozbazar, A., Gordon, W. B. and Richerson, M. E., 1979, A theoretical model to estimate some ergonomic parameters from age, height and weight. *Ergonomics*, **22**(1), 43–58.

Ryan, P. W., Springer, W. E. and Hlastala, M. P., 1970, *Cockpit Geometry Evaluation*. JANAIR REPORT 700202 (Boeing Military Airplane Systems Division).

Stewart, L., 1985, Microcomputer applications in human factors research. *Communiqué* **15**(– 3), 1–3.

Tisserand, M. and Saulnier, H., 1982, *Dimensionnement Des Postes de Travail. Application Sur Terminal Écran-clavier, 2.-Conception du Poste*. Cahiers de notes documentaires, ND 1389–108–82. (Paris: INRS).

Tisserand, M. and Schouller, J.-F., 1981, *Dimensionnements Des Postes de Travail. Application Sur Terminal Écran-clavier, 1.-Le Diagnostic*. Cahier de notes documentsaires, ND 1347–105–81. (Paris: INRS).

Trotter, M. and Gleser, G. C., 1952, Estimation of stature from long bones of American Whites and Negros. *American Journal of Physical Anthropology*, **10**, 463–514.

Trotter, M. and Gleser, G. C., 1958, A re-evaluation of estimation of stature based on measurements of stature taken during life and of long bones after death. *American Journal of Physical Anthropology*, **16**, 79–123.

Unger, R. L. and Rider, J. P., 1985, A computer based model of machine operator compartments. *IEEE 1985 Proceedings of the International Conference on Cybernetics and Society*, pp. 147–153.

Vézina, L. and Gilbert, R., 1986, The first version of specific anthropometric model on IPC-HP. Unpublished report. Department of Industrial Engineering, Polytechnic School of Montreal.

Chapter 13
Computer-aided workplace layout and line balancing

HO NAI CHOON and LIM TEOW EK[1]
GINTIC CAD/CAM Centre
[1]Nanyang Technological Institute,
Nanyang Avenue,
Republic of Singapore

Abstract. With the advent of the computer-aided design system, it is now feasible to optimize workplace layout and balance a production line by simulation using a personal computer. This paper describes a new software package for the optimization of workplace and line balancing. The package uses computer graphics to assist the user in visualizing the operation sequence. Each motion is simulated with a man model's arm movement with a menu containing the basic motions. The cycle time is automatically calculated. Finally an optimization routine determines the optimum workplace layout based on the activities defined by the user.

Keywords: Line balance; Workplace optimization.

Introduction

The production line is used by most manufacturing firms to assemble large quantities of standardized products at low costs. Ideally the production line should have workstations whose subsequent operations are located adjacent to each other, allowing the product to move at the most efficient rate through a series of balanced operations. The traditional manual methods used by industrial engineers are no longer effective in meeting the challenge of rapid product change and most firms spend many industrial engineering man-hours in planning the production line to cater for the introduction of new product models. The advent of computer-aided design (CAD) systems has enabled workplace design packages to be developed. Work study packages such as 4M DATA (Karger and Hancock, 1982), TIMELINK (Schofield, 1982) and MOST (Zandin, 1980) can be implemented on a wide range of mini and mainframe computers. However, the cost is high and their use may not be economical for smaller firms. Many computer programs have also been developed for line balancing, with procedures varying from simple list processing prioritizing schemes to optimal seeking procedures of the optimal seeking methods. To overcome the problem of high hardware cost, a software package using a personal computer for the optimization of workplace layout and line balancing was developed.

Description of the software package

The package consists of (i) a Work Study Analysis Program, (ii) a Workplace Layout Optimization Module and (iii) a Line Balancing Subsystem. The work study analysis program uses computer graphics to generate a menu-driven interactive man–workplace model which can simulate the motions of an assembly operator. It allows the work study analyst to draw the workplace layout on the computer screen; plan out the elemental basic motions of the operation; compute the total operation time and compile a sequence chart of all the operations performed. The Workplace Layout Optimization Module determines the optimum placement of materials and tools based on load–distance analysis. Finally, the Line Balancing Subsystem uses the standard time established for each work element and the precedence constraints defined by the users to determine the allocation of work elements to each station to achieve minimum balance loss.

System requirements

The software has been developed with an extensive use of computer graphics on a personal computer. The hardware requirements are:

1. An IBM AT personal computer (or a compatible computer) equipped with a fixed disk and at least 640 Kbytes of main memory;
2. A maths co-processor compatible with the computer;
3. An Enhanced Graphics Adapter (EGA) card; and
4. A high resolution colour monitor.

The software requirements are:

1. IBM Personal computer disk operating disk system (DOS), Version 2.1 or later;
2. IBM personal computer graphical kernel system (GKS);
3. IBM virtual device interface (VDI);
4. IBM enhanced graphics adapter colour display 640 × 350 16-colour device driver, VDIDY010.SYS; and
5. IBM extended video and keyboard driver, ANSI.SYS.

Work Study Analysis Program

Workplace layout and other industrial engineering techniques have been practised in the USA and other industrialized European countries as early as the 1900s. The findings of notable pioneers like Frederick Taylor, Frank and Lillian Gilbreth, and other researchers have resulted in the availability of time standards for use in workplace layout. The best known predetermined motion time systems are motion time analysis (MTA), work-factor (WF), basic motion time study (BMT), and methods time measurement (MTM). Among these, MTM has become very widely used throughout the world (International Labour Organization, 1978). Because of its wide acceptance, MTM time standards are used in this program.

The main features of the program developed are:

1. Generation of a two-dimensional man model based on local anthropometric data;
2. Facility for the analyst to define a desired workplace layout with a library of fixtures and tools;
3. Facility for selecting standard MTM-2 elements;
4. Work element assembly simulation using computer graphics;
5. Data retrieval and storage routines;
6. Editing features to facilitate modifications to previously simulated sequence of operations.

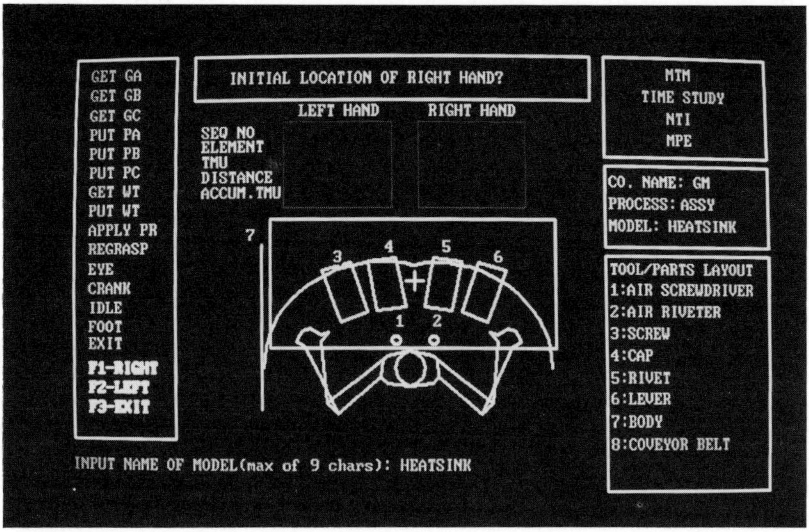

Figure 13.1 Simulation of man–workplace model and MTM motions.

The basic motions are based on the 39 time standards of the MTM-2 system. MTM-2 elements used in the package are: GET(GA, GB, GC), PUT(PA, PB, PC), PUT, WEIGHT, APPLY PRESSURE, REGRASP, EYE, CRANK, FOOT and STEP. Process time and resilience time are also included.

The analyst first needs to define a layout of the component parts and the location of the man model. Motions of the assembly operator can then be simulated by selecting the hand to be moved and the associated MTM elements (Figure 13.1). The hand's destination must be digitized. The distance travelled by the hand will be automatically generated by the computer. Upon completion of the simulation, a process sequence chart is compiled to give the total cycle time and motion assigned indices (MAI), where

$$\text{MAI} = \frac{\text{(accumulated time of hand) less (waiting time)}}{\text{cycle time}}$$

This ratio shows how completely the hand has been assigned work. If a low ratio is found when compared with a standard for that type of work, the analyst can try to improve the work methods by utilizing both hands more effectively. Incorporated with the simulation program are the editing features which enable the analyst to insert and amend elements in the sequence chart.

Workplace Layout Optimization Module

Upon completion of the simulation, the Workplace Layout Optimization Module can be used to carry out a load-distance analysis. It uses the proximity criterion (Muther, 1974) to locate the optimum placement of materials and tools. The load-distance analysis considers not only the number of inter-workplace movements of a product but also the distances over which the movements are made. Using this method, the optimization criterion (c) which considers both the number of movements and distance moved is minimized:

$$c = \sum_{i=1}^{N} \sum_{j=1}^{N} L_{ij} D_{ij}$$

where N = the number of work centres; L = the number of loads or movement of work element between work centres i and j, and D = the distance between centres i and j.

During the simulation, the relationship matrix based on the load distance criterion is established. The total closeness rating (TCR) for each tray is computed, and the components to be placed are ranked according to their TCR in a decreasing order. This establishes the sequence in which the components will enter the layout. The dimensions of the trays are converted to modular units (specified by the user) and the size of layout matrix is established. The trays are placed one by one in modular units into the layout matrix according to the previously established sequence. Each time a new tray is placed, a sub-optimization problem is solved to determine the optimum position of this box which minimizes the sum criterion of the load-distances between the new tray and those previously located. Constraints (such as certain trays to be placed at fixed location) can be specified and taken into consideration during the location process. The flow chart for the optimization routine is depicted in Figure 13.2.

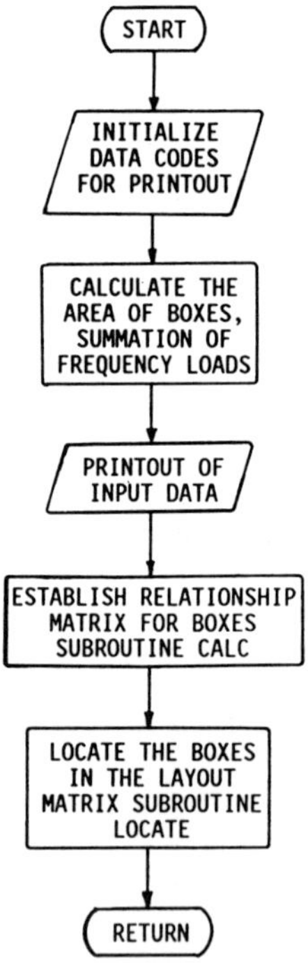

Figure 13.2 Optimization routine.

Line Balancing Subsystem

The objective of line balancing is to minimize the idle time on the assembly line for different combinations of workstations and work elements, subject to certain restrictions. The first of these restrictions is the production volume that is to be achieved. Frequently, it is the production volume that determines the pace at which the line should operate. If the demand for the product changes, it will require the line to be re-balanced. A number of variable factors, such as lack of materials, a change in the design of the product, labour shortages, or a change in production method, can influence the line balance.

Thus, there are two approaches to assembly line balancing. The first assumes a fixed cycle time and finds the optimal number of workstations. The second assumes the number of workstations to be fixed and seeks to minimize the total delay time by minimizing the cycle time, that is the time allowed at the station with the largest work content. In the latter, several line balancing solutions, both manual and computerized methods are available.

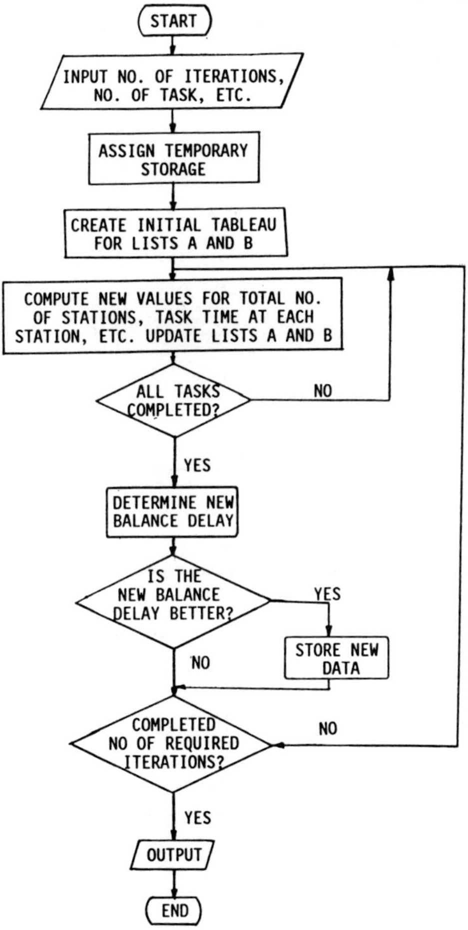

Figure 13.3 Flowchart for line balancing program.

The line balancing subsystem in this module was developed using COMSOAL (Computer Method of Sequencing Operations for Assembly Lines) (Arcus, 1966). It includes precedence constraints and enables the user to minimize balance loss by determining the optimum number of work elements to be assigned to each workstation. The time for each work element can be established using the workplace simulation package described in the previous section. A summary of the COMSOAL program for the optimization of line balancing is depicted in the flowchart (Figure 13.3).

Table 13.1 List of work elements.

No.	Element description	Te*	Immediate predecessor
1	Place frame on work holder and clamp	0·15	–
2	Assemble fan to motor	0·37	–
3	Assemble bracket 1 to frame	0·21	1
4	Assemble bracket 2 to frame	0·21	1
5	Assemble motor to frame	0·58	1,2
6	Affix insulation to bracket 1	0·12	3
7	Assemble angle plate to bracket 1	0·29	3
8	Affix insulation to bracket 2	0·12	4
9	Attach link bar to motor and bracket 2	0·30	4,5
10	Assemble 3 wires to motor	0·45	5
11	Assemble nameplate to housing	0·18	–
12	Assemble light fixture to housing	0·20	11
13	Assemble blade mechanism to frame	0·65	6,7,8,9
14	Wire switch, motor and light	0·72	10,12
15	Wire blade mechanism to switch	0·25	13
16	Attach housing over motor	0·35	14
17	Test blade mechanism, light, etc.	0·16	15,16
18	Affix instruction label to cover plate	0·12	–
19	Assemble grommet to power cord	0·10	–
20	Assemble cord and grommet to cover plate	0·23	18,19
21	Assemble power leads to switch	0·40	17,20
22	Assemble cover plate to frame	0·33	21
23	Final inspect and remove from workholder	0·25	22

*Te, time (minutes) for each work element.

Table 13.2 Sample of the input format.

How many iterations? 20
How many tasks? 23
What is the maximum cycle time? 1·0
Enter the starting value for
Random no. generation: 555555
How long does task 1 take? 0·15
Total # of pre-task for task 1? 0
How long does task 2 take? 0·37
Total # of pre-task for task 2? 0
.
.
.
.
How long does task 23 take? 0·25
Total # of pre-task for task 23? 1
What is pre-task 1 for task 23? 22

For example, consider a company which has broken down the tasks to be assigned to various assembly line workstations, and where certain tasks must be performed before others. The relationships between the tasks are shown in Table 13.1, where *Te* represents the time in minutes for each work element, and the precedence diagram is depicted in Figure 13.4. The input and output formats are shown in Tables 13.2 and 13.3, respectively.

The maximum cycle time is estimated by determining how many units need to be produced per hour. It is assumed that the maximum cycle time is equal to 1 min in this case. The data entry for this program is as shown in Table 13.2.

Table 13.3 Sample of the output format.

STATION	Element	Te
# 1	18	0·12
	1	0·15
	3	0·21
	2	0·37
Total time at station: 0·85		
Total idle time: 0·15		
# 2	4	0·21
	5	0·58
	11	0·18
Total time at station: 0·97		
Total idle time: 0·03		
# 3	10	0·45
	19	0·10
	20	0·23
	8	0·12
Total time at station: 0·90		
Total idle time: 0·01		
# 4	7	0·29
	9	0·30
	6	0·12
Total time at station: 0·71		
Total idle time: 0·29		
# 5	12	0·20
	13	0·65
Total time at station: 0·85		
Total idle time: 0·15		
# 6	14	0·72
Total time at station: 0·72		
Total idle time: 0·28		
# 7	16	0·35
	15	0·25
	17	0·16
Total time at station: 0·76		
Total idle time: 0·24		
# 8	21	0·40
	22	0·33
	23	0·25
Total time at station: 0·98		
Total idle time: 0·02		
	Balance delay = 15·75%	

After trying different solutions, the program reports the 'best solution'. This process continues until all eight stations have been displayed. The output format is shown in Table 13.3.

Although the line balancing program does not generate an optimal solution, it gives good answers when compared with other heuristic approaches. It is found that COMSOAL is an attractive approach to solving line balancing problems on microcomputers because it is easy and simple to program. However, COMSOAL can take excessive computer time for large problems.

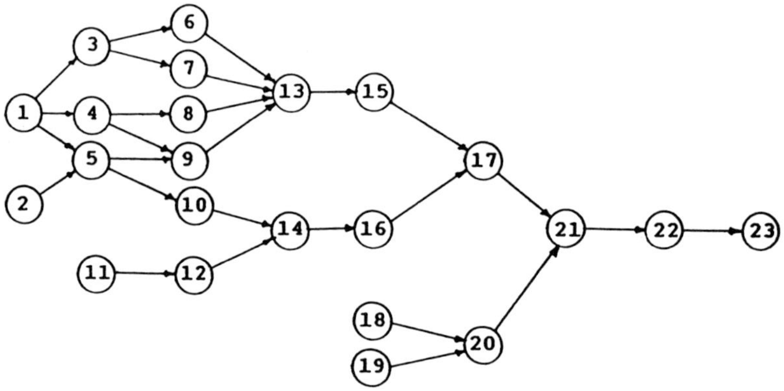

Figure 13.4 The precedence diagram for the example in the text.

Interfacing of line balancing with workplace layout

The minimum number of workstations required for an assembly line is first obtained by running the COMSOAL program. The data obtained is then used by the workplace layout program to optimize the layout so that each individual station time can be reduced. The workplace layout optimization results in a reduction of the cycle time whenever it is possible. This new cycle time in turn can be used by the line balancing program to compute a better line balance. In this way the user can run both line balancing and workplace layout programs to obtain an optimized solution.

Case study

Description of operation

This case study was taken from a manufacturer of vehicle stereo equipment. The operation was known as 'Heatsink assembly'. Its objective was to install a heatsink unit in a power transistor. The procedure is given in Table 13.4.

Table 13.4. 'Heatsink Assembly' procedure.

1. Obtain unit insulator from bin and position onto fixture (LH)
2. Obtain transistor from bin and position onto nut insulator (RH)
3. Bring compound dispenser to assembly (LH)
4. Press to release compound (LH)
5. Obtain insulator strip from bin and position onto assembly (RH)
6. Press to release compound (LH)
7. Put aside compound dispenser (LH)
8. Obtain heatsink from bin and position onto assembly
9. Obtain screw from bin and position onto assembly (RH)
10. Bring and position air screwdriver onto screw (RH)
11. Screw (RH with LH assisting)
12. Put aside air screwdriver (RH)
13. Take complete assembly from fixture to bin (RH)
14. Actuate counter to indicate one more complete heatsink (RH)

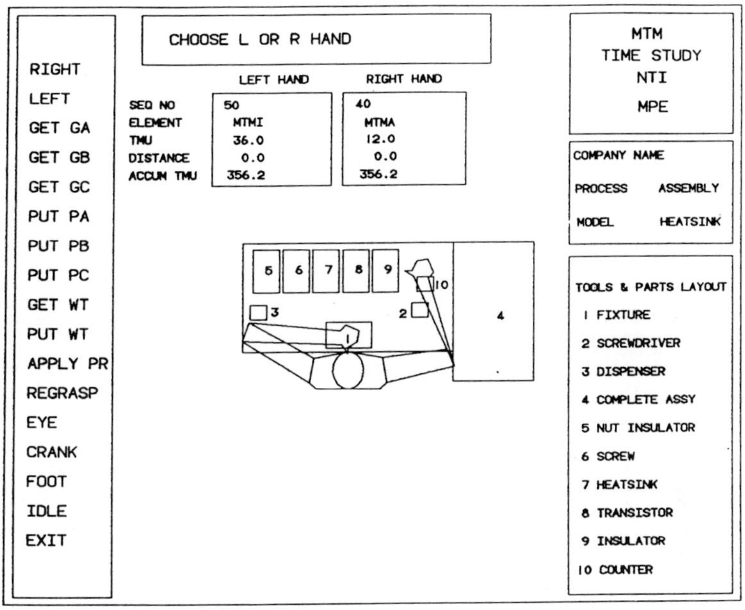

Figure 13.5 Existing work layout.

Layout constraints

In the existing layout (see Figure 13.5) the tray containing completed assemblies was actually a huge carton resting on a side table. Therefore, due to its bulky size, this box was represented as a fixed installation in the layout optimization exercise. Similarly, the compound dispenser and air-screwdriver (hanging from overhead springs) were regarded as fixed installations, with their positions predetermined by the industrial engineer. The rest of the trays could be located anywhere on the main table. The fixture should be placed at the centre edge of the table, nearest the operator's body. It was desirable to specify 100 mm as minimum distance between the centre box and the rest of the boxes to prevent cluttering of boxes around the work centre.

Simulation and optimization results

The following were the results and outputs generated in the case study:

(a) A computer hardcopy of the existing layout (Figure 13.5).
(b) A computer printout of the process sequence chart using MTM-2 based on existing layout (Figure 13.6).
(c) Relationship chart between the 10 boxes (Figure 13.7a and b).
(d) Ordered sequence of the boxes according to decreasing TCR (Figure 13.8).
(e) A computer hardcopy of the completed optimized layout (Figure 13.9).

COMPANY NAME :
PROCESS :ASSEMBLY
MODEL :HEATSINK

PROCESS SEQUENCE CHART

LEFT HAND							RIGHT HAND			
MTMU	DIST	TMU	SUM	LOC	SEQ	LOC	MTMU	DIST	TMU	SUM
START	0.00	0.00	0.00				START	0.00	0.0	0.00
					1		MTMI	0.00	8.0	8.00
MTGC	343.36	27.0	27.00		2					
MTMI	0.00	4.0	31.00		3		MTGC	240.63	23.0	31.00
					4		MTPA	241.69	11.0	42.00
MTPA	342.93	15.0	46.00		5					
MTMR	0.00	6.0	52.00		6					
MTPC	24.44	21.0	73.00		7					
MTGB	31.19	7.0	80.00		8		MTMI	0.00	38.0	80.00
MTPB	23.46	10.0	90.00		9					
MTMI	0.00	11.0	101.00		10		MTPC	26.03	21.0	101.00
MTGB	276.50	14.0	115.00		11		MTGB	277.36	14.0	115.00
					12		MTPA	287.64	11.0	126.00
MTPB	302.86	19.0	134.00		13					
MTMA	0.00	14.0	148.00		14					
MTPA	29.54	3.0	151.00		15		MTMI	0.00	25.0	151.00
MTMI	0.00	21.0	172.00		16		MTPC	34.07	21.0	172.00
MTPA	27.04	3.0	175.00		17					
MTMA	0.00	14.0	189.00		18					
					19		MTMI	0.00	27.0	199.00
MTGC	263.89	23.0	212.00		20					
MTMI	0.00	10.0	222.00		21		MTGC	231.03	23.0	222.00
MTPB	263.47	19.0	241.00		22		MTPB	221.04	19.0	241.00
					23		MTPC	31.19	21.0	262.00
					24		MTGB	23.46	7.0	269.00
MTMI	0.00	38.0	279.00		25		MTPB	21.36	10.0	279.00
MTPB	20.88	10.0	289.00		26		MTMI	0.00	10.0	289.00
MTGB	28.49	7.0	296.00		27					
					28		MTGB	284.68	14.0	303.00
					29		MTPB	315.39	24.0	327.00
					30		MTMA	0.00	14.0	341.00
MTMI	0.00	66.7	362.70		31		MTMI	0.00	21.7	362.70
MTPA	23.46	3.0	365.70		32					
					33		MTGB	31.08	7.0	369.70
					34		MTPA	349.95	15.0	384.70
					35		MTGA	170.77	9.0	393.70
MTMI	0.00	42.0	407.70		36		MTMA	0.00	14.0	407.70

TOTAL CYCLE TIME = 407.70
LEFT HAND MOTION ASSIGNED INDEX,MAI = 0.53
RIGHT HAND MOTION ASSIGNED INDEX,MAI = 0.68

Figure 13.6 Process sequence chart using MTM–2 (existing layout).

OPTIMUM PLAN LAYOUT USING THE NEIGHBORING TECHNIQUE

****

NUMBER OF BOXES TO BE LOCATED .... = 10

TABLE LENGTH .... = 700,000 MM

TABLE WIDTH .... = 375,000 MM

INITIAL SPATIAL CONFIGURATION OF EACH BOX

NO.	NAME	LENGTH (MM)	WIDTH (MM)	AREA
1	FIXTURE	125.00	100.00	12500.00
2	SCREWDRIVER	50.00	50.00	2500.00
3	DISPENSER	50.00	50.00	2500.00
4	COMPLETE ASS	250.00	500.00	125000.00
5	NUT INSULATO	100.00	150.00	15000.00
6	SCREW	100.00	150.00	15000.00
7	HEATSINK	100.00	150.00	15000.00
8	TRANSISTOR	100.00	150.00	15000.00
9	INSULATOR	100.00	150.00	15000.00
10	COUNTER	50.00	50.00	2500.00

a

	1	2	3	4	5	6	7	9	10
1									
2	2.								
3	2.	0.							
4	1.	0.	0.						
5	2.	0.	0.	0.					
6	2.	0.	0.	0.	0.				
7	2.	0.	0.	0.	0.	0.			
8	1.	0.	0.	0.	0.	0.	0.		
9	2.	0.	0.	0.	0.	0.	0.	0.	
10	1.	0.	0.	1.	0.	0.	0.	1.	0.

b

Figure 13.7 (a) Initial spatial configuration. (b) Relationship chart.

BOX NO.	AREA	LENGTH	WIDTH	TCR
1	20	5	4	15.
10	4	2	2	3.
2	4	2	2	2.
3	4	2	2	2.
4	200	10	20	2.
5	24	4	6	2.
6	24	4	6	2.
7	24	4	6	2.
8	24	4	6	2.
9	24	4	6	2.

** NOTE BOTH LENGTH AND WIDTH ARE REPRESENTED IN MODULES
ONE MODULE = 25 MM

Figure 13.8 Ordered sequence of the boxes according to decreasing TCR.

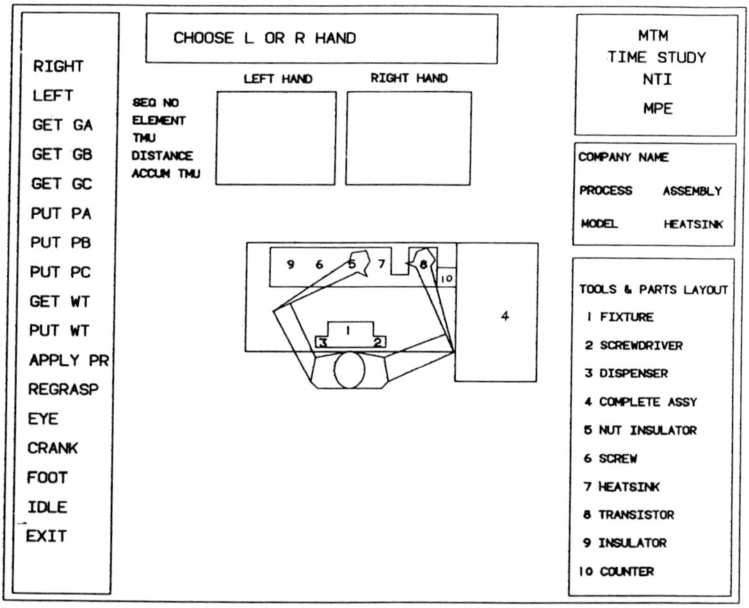

Figure 13.9 Completed optimized layout.

Optimization results:

$$\text{Total cycle time before optimization} = 407{\cdot}70\ \text{TMU}$$

$$\text{Total cycle time after optimization} = 378{\cdot}70\ \text{TMU}$$

$$\text{Percentage saving} = \frac{407{\cdot}70 - 378{\cdot}70}{407{\cdot}70} \times 100\% = 7{\cdot}1\%$$

The savings of 7% could have been more if the frequency load between the boxes were higher.

Conclusions

Among the advantages of using a computerized system for generating time standards and balancing lines are:

1. Reduction in time required by the time study analyst to set the standard. (In application, a 50% reduction in clerical effort was reported as compared with manual methods of applying standard data.)
2. Greater accuracy and uniformity in the time standards.
3. Ease of maintaining the methods and standards file when engineering and methods changes occur.
4. Time standards can often be set before the start of production.

The package described in this paper provides a low cost and user-friendly tool to achieve the above advantages. The software package can be used to eliminate the tedious task of designing workplace layout and balancing a production line using conventional methods. An additional advantage is that the computer graphics are produced in colour and apart from providing a better vision, they allow the user to make a sketch of any workplace layout at any workstation. It is expected that the optimization of workplace layout and balancing of production line using such a package may lead to higher productivity in mass production factories.

References

Arcus, A. L., 1966, COMSOAL: A computer method of sequencing operation for assembly lines. *International Journal of Production Research*, **4** (4), 259–277.

International Labour Organization (ILO), 1978, *Introduction to Work Study* (Geneva, Switzerland), p. 314.

Karger, D. W. and Hancock, W. M., 1982, *Advanced Work Measurement* (New York: Industrial Press), 208 pp.

Muther, R., 1974, *Systematic Layout Planning*, 2nd edn (Mass: Cahners Books).

Schofield, N. A., 1982, Workstudy Management System. *Computers and Industrial Engineering*, **6** (2), 75.

Zandin, K. B., 1980, *Works Measurement Systems* (New York: Marcel Dekker), 133 pp.

PART IV
WORKLOAD AND WORK MEASUREMENT

Chapter 14
Job analysis and work design using ergonomic databases

KURT LANDAU, REGINA BRAUCHLER, WOLFGANG BRAUCHLER, ANNA LANDAU[1] and RAINER BOKRANZ[2]

Ergonomics Department (530),
University of Hohenheim,
D-7000 Stuttgart 70, Federal Republic of Germany

[1]*ERGON Engineering Inc.,*
D-7432 Bad Urach,
Teckstraße 1, Federal Republic of Germany

[2]*WAB Consulting Inc.,*
D-6157 Pfungstadt,
von Ketteler Str. 5b,
Federal Republic of Germany

Abstract. This chapter concentrates on the procedures of ergonomic analysis and the description of working systems using ergonomic databases. First, objectives and applications of job analysis procedures are discussed and then concepts of epidemiological expert systems are developed. Work design checklists are reviewed in order to evaluate design alternatives for suitability and to obtain information on deficiencies and constraints from reliable actual state data for corrective and for conceptual design. Applications of cluster analysis techniques for evaluation of ergonomic design checklists are shown. Ergonomic design programmes for companies are developed. A multi-level job analysis model is derived, and computer software is discussed.

Keywords: Job analysis; Task profiles; Work sphere design; Cluster analysis; Design checklist.

Introduction

With the introduction of new legal standards for industrial safety in a number of European countries, the need arises to analyze the structural state of working systems in existence or in preparation with respect to stress and strain in the employees. In general, the exorbitantly expensive physical and physiological measurements, which seem to be necessary, exceed the possibilities of small and medium enterprises and often the material and personnel conditions of large-scale enterprises as well. The question arises if there is a less expensive procedure of ergonomic analysis and description of working systems which enables examinations of work physiology to be completely replaced, or limited to small 'bottleneck' areas.

The analysis of *industrial safety*, the research for *accident causes*, and the *rehabilitation* of handicapped persons – subjects which represent focal points of research of the European Communities – suggest the possibility of elaborating and applying a procedure for the ergonomic analysis of working systems which can be used at numerous workplaces, safeguarding an acceptable cost–utility ratio.

In order to further the humanization and rationalization of working systems a great amount of field research has recently been carried out in industry, motivated on one hand by the initiative of individual enterprises and on the other hand by financial aids from various government institutions. The aim of this field of research is the improvement of existing working systems (by corrective ergonomics) and the design of new systems (by conceptive ergonomics). Thus it is desirable to compare the selected possibilities of designing human work with other existing alternatives of design or to evaluate them absolutely. Therefore, new procedures of job analysis and job description had to be developed. The last 15 years have been marked by a creative phase of invention and testing of new job analysis procedures (e.g. the AET: Rohmert and Landau, 1983a; Landau and Rohmert, 1981, 1989).

The origins of the AET (Arbeitswissenschaftliches Erhebungsverfahren zur Tätigkeitsanalyse; ergonomic job analysis procedure) date back to a study (Rohmert *et al.*, 1975) ordered by the German government to investigate discrimination against women at work with respect to pay. A job analysis procedure was required that allowed a detailed investigation of workload and strain within a given work system. At that time there was no job analysis procedure that could readily be used, although the Position Analysis Questionnaire (PAQ) (McCormick *et al.*, 1969) seemed to provide a basis for the psychological items.

Starting from that particular research project, the AET has been continuously developed over the last few years and applied in different industrial situations to analyze a wide range of shop-floor and management jobs (a total of about 5000 job analyses).

Objective and applications of job analysis procedures

There are different ways to analyze the structure of the working system and the interaction of its elements. Thus, one cannot speak of the work or job analysis, or of the activity-oriented or the task-related job analysis. How analyses of working systems are designed and performed depends largely on the underlying theoretical foundation and the related object of the work analysis, i.e. the working system, the job, the individual, and so on.

The following criteria can be used to evaluate job analysis procedures (Frieling, 1975; Hoyos, 1974). The procedure should:

1. Be based on a theoretical model that allows a practical interpretation of the results obtained with the work analysis procedure,
2. Offer a complete coverage of all demands that are present on a specific working system,
3. Offer maximum cost-effectiveness with regard to application, data processing, and data evaluation; the application of the procedure should allow standardization,
4. Go beyond a merely verbal work description and allow quantitative statements at least at the ordinal scale level.

When applying the procedures, statements should be possible as to: (i) the reliability with which several raters analyze a working system at the same time (inter-rater reliability), and

Table 14.1. Selected fields of application of work analysis procedures.

Field of application	Actual application and possible result
Ergonomic and technical design of working systems	
Documentation and diagnosis of working systems	Description and possibly quantification of system elements and their characteristics, e.g. as to the stresses exerted by them; deduction of design needs; formation and verification of design priorities; prevention of possible impairments by detection of unsupportable stresses/strains; purposeful reduction of stress
Product design	Description and possibly quantifications of product characteristics as to their human-related impact during their later utilization by the consumer or investor
Management organization	
Preparation of reorganization measures	Analysis of demand profiles of different jobs before and after corporate reorganization measures of before and after technical/technological changes; forecasting of shifts in demands through technical change
Work process design	Task and demand sequences over working hours
Design of shifts and free time	Determination of free times based on stress rankings for working systems; work distribution for specific shifts according to types and extent of demands; prevention of absences through stress-related cause analysis
Personnel administration	
Personnel recruitment	Preparation of personnel ads according to existing job analyses
Selection and positioning	Selection of job applicants according to the degree of congruency between demand and aptitude profiles; determination of aptitude/fitness criteria
Instruction/training/retraining	Deduction of training needs/training measures according to the task/demand profile of working systems
Integration of disabled persons and other groups of persons into the organization	Search for suitable working systems with specific task/demand profiles to ensure tolerability of the work
Qualitative personnel planning or personnel appointment planning	Comparison of aptitude and demand features (long-term and short-term view)
Personnel representation	Job analysis as discussion basis in talks or disputes between employers and employees
Wage fixing	Review of classifications by comparison of the demand structure of critical working systems with corporate working systems
Diagnosis of payment policy	Comparison of corporate demands and industry-typical demands with regard to the adequateness of a payment policy
Vocational guidance and research	
Job classification	Deduction of a demand- or task-related job classification and comparison with conventional job classifications; health status in different occupational groups; analysis of jobs or occupations as to required psychical and physical aptitude; analysis of seriousness and frequency of accidents according to job/occupational groups; job classification according to socio-demographic factors
Vocational guidance and information	Description of job spectra when providing guidance for school leavers

(ii) the reliability with which all items of the work analysis procedure can be rated (item reliability).

In addition to these criteria, it is required that the procedure be applicable to all possible physical and non-physical forms of work, including work contents that range from the generation of forces to the generation of information.

Obtaining an overview of the different strategies in job analysis is made easier when starting the discussion with the fields of applications and the objectives of such procedures. Possible applications of job analysis lie in the fields of demand analysis and work design. The evaluation of the job analysis data allows conclusions on the demands that a specific job imposes on the working person. From the job analysis data, the design status of working systems can be analyzed, classified, and compared with each other; the analysis also provides information on situations of extreme stress and their possible elimination through corrective design measures. Often job analysis procedures can be used as a forecasting tool, referring to non-existent working systems that are still in the planning phase.

In personnel management, a job analysis that aims at a demand analysis is a valuable tool when it is combined with the assessment of the required psychical and physical qualifications.

Other possible applications such as job and professional classification, investigations into the cause of accidents, or the deduction of education and training objectives show the accent on the universal nature of work analysis procedures. Table 14.1 summarizes some major fields of application and objectives of analysis procedures (Landau and Rohmert, 1989).

This contribution concentrates on applications of ergonomic databases on problems of epidemiological stress research and on work design projects. An example of a multipurpose ergonomic database follows these.

Epidemiological expert system

In epidemiological socio-medical research, the concept has been applied for many years relating stress caused by job conditions to the employee's diseases. In this context, the job is characterized by the employee's occupation, whereas the disease is characterized by clinical findings. However, the indicative value of the data has been limited by the fact that, besides the occupation designation, there was no detailed information available concerning the job done by the employee and the stress and strain caused by the job. In several selected publications (e.g., Medical Research Council, 1982; Caplan *et al.*, 1975; Occupational Mortality, 1978), it could be pointed out that certain diseases cause early retirement or death with a significantly higher frequency as compared with the average population. In these cases, the early retirement or even death of the employee were often caused by toxic and carcinogenic working situations.

For most health impairments an employee incurs throughout his working life, however, this method of relating job and impairment is inappropriate. Since many factors resulting from working and private life are neglected in this simple relation model, there remains an enormous erratic value which makes it impossible effectively to compare work stress and health condition.

By directly relating work stress on the one hand and disease or impairment data on the other hand, the results of so-called socio-epidemiological investigations would be neglected. The results of these investigations proved to have an inverse relationship between social class and disease (Kessler and Cleary, 1980; Dohrenwend and Dohrenwend, 1981). In the course of several socio-epidemiological research projects, the relationship between the variables of social class, stressing events in the life of a person, problem management behaviour, social support, and disease were comprehensively analyzed. In this context, there exist three conceptual hypotheses: the *simple stress hypothesis* traces the outbreak of a disease back to life events while it is assumed that both the number of events and the extent of the stress caused by these events is of crucial importance; the *interactive stress hypothesis* introduces social support and/or problem management behaviour as additional variables which help to explain the relationship between stressing life events and disease; finally, the *additive stress hypothesis* assumes that social support and problem management behaviour have an effect on the disease which is independent of stressing life events (see Biefang, 1985).

The established (expanded) stress–strain concept according to Rohmert *et al.* (1975) links 'objective' workplace/job/stress data with subjective strain data in order to predict possible impairments caused by the job. When this concept is taken into account it becomes obvious that the appropriate analysis is merely based on stress factors caused by working life.

The concept considers neither stress factors resulting from the habitual situation (nutrition, way of life, use of drugs, etc.) nor the employee's social situation nor the employee's approach to the given habitual and social situation (i.e. the 'redefinition'). Consequently, the concept is an extreme simplification of matters. The influence of private life – which can be crucial, and might cause or hinder the outbreak of a disease – might well be responsible for much of the erratic values of clinical findings. By simply comparing the influencing factors of the job to which the employee is exposed for eight hours a day on the one hand with the 24 hours of the entire day on the other hand, it becomes obvious that the concept considers only one third of the variables in question. Furthermore, the stress–strain concept neglects the numerous retroactive actions between the individual determinants.

Therefore, the first analysis should focus on the question as to what extent the expanded stress–strain concept should *additionally* be expanded by including the determinants mentioned above in order to be able to create an appropriate description of the relationship between work stress and job environment. These relationships are represented in Figure 14.1.

A few exemplary retroactions have been included in Figure 14.1. It becomes obvious that fairly complex calculations or simulation computations will be required to explain clinical

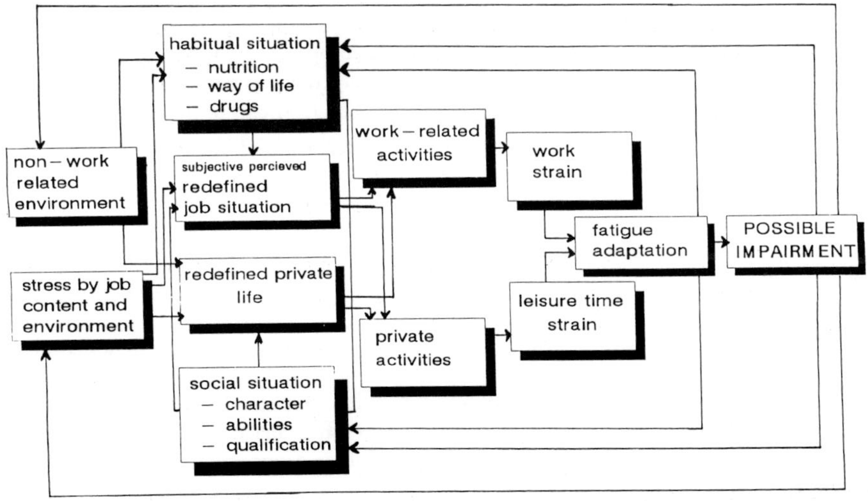

Figure 14.1 Hypothetical relationship between stress/strain determinants and clinical findings.

findings on the basis of the relations indicated. In addition, the stress–strain concept is designed for short-time observation of, e.g. a single shift, but not for modelling stress and strain factors which affect the employee throughout his entire working life.

In our opinion, it is not permissible to predict that an existing working situation together with certain stress and relief factors which result from private life will cause specific impairments of an employee's health. (In this context and with respect to the working situation it is assumed that necessary industrial safety measures have been adopted and that established work study findings as to specific levels of work stress are observed.) In view of the high mobility between occupations and jobs occurring in the course of a working life, clinical findings cannot be directly linked to current stress factors resulting from the present working and private life. For example, an employee's health may have been impaired even before he took up his present job.

Consequently, the above concept should as well be made *dynamic*. This means that the concept should *retrospectively* integrate former stress factors resulting from the employee's job and private life by means of a suitable questionnaire. Of course, no employee can be expected to be able reliably to apply a questionnaire designed according to ergonomical findings (e.g. the AET, see Rohmert *et al.*, 1975; Landau and Rohmert, 1981) with respect to his former positions and jobs. Therefore a simplified method was developed, which has to be applied by the company physician – the Standardised Preventive Examination (SPE) procedure.

The SPE covers about 250 items and contains a retrospective stress analysis together with a job, family and social anamnesis, that is intended to minimize the residual variance in establishing the correlation between the medical findings (diagnoses and laboratory results) and the stress data. The stress data of the actual job has to be assessed by company ergonomists using the Stress Evaluation System (SES).

The SES was designed on the theoretical basis of an objective bottleneck-oriented job analysis using the model of the working system and the stress–strain concept. Its structure and questionnaire has been derived from the existing job analysis method AET (Rohmert *et al.*, 1975; Landau and Rohmert, 1981; Rohmert and Landau, 1983b). When designing the SES, basic problems were encountered with item selection and item scaling. Finally the evaluation of stress height and duration were made using largely standardized rating indices. On assessing the bridging examples the results of physiological and psychological strain analyses documented in the literature were used as appropriate guidelines.

The selection of items for the SES was ensured by the co-operation of ergonomists, physicians, work safety specialists and psychologists. The 130 items of the SES are marked in an assessment interview and /or in an analysis of documents. This assessment technique offers the advantage of highly cost-effective application, data processing, and data evaluation. From the ordinal rating of the 130 items stress factors are derived that allow detailed information about the work content and work environment. To give a quick overview of the height and duration of stresses the stress factors are represented graphically as profile diagrams. Finally, SPE and SES records are linked together using an epidemiological expert system (EES) (see Figure 14.2).

EES is designed to assess the indicated data items from the professional and private spheres and to evaluate them with regard to their possible integration into an epidemiological early warning system. Studies of EES validity and reliability are presently done in coal mining (Brauchler and Landau, 1989). After the application of EES analysis modules on a selected population of the company concerned, it will be possible to determine the predictive value of the individual types of data.

Consequently, the results of these projects will allow a statement as to whether an early warning system for potential impairments caused by the job should contain all data elements

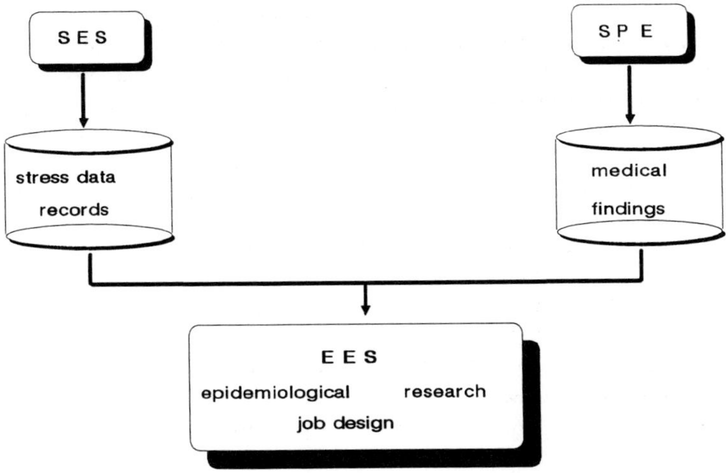

Figure 14.2 Structure of the Epidemiological Expert System (EES).

discussed in this paper or whether it can be justified to reduce the number of data sources. If the latter were possible, the survey expenditure for ongoing epidemiological monitoring could be limited to a responsible level.

Work design using ergonomic databases

Work design in companies can take place in different situations:

1. A work system is newly designed, for example in the case of new investments, so that it cannot be compared with existing work systems, i.e. actual states cannot be assessed.
2. A work system is designed, for example in the case of expansion investments, so that it replaces an existing work system while offering so many innovations that it can hardly be compared with its predecessor. In this case, an analysis of the actual state may be desirable.
3. A work system is designed in a corrective manner only, for example in the case of rationalization investments. In this case, an analysis of the actual state is indispensable.

Expansion and rationalization investments are typical applications of work design measures which require an analysis of the actual state. The quality of actual state analyses is key to the quality of the work design result.

Actual state analyses are performed when information is needed: (i) to evaluate design alternatives for suitability, (ii) to obtain information on deficiencies and constraints from reliable actual state data for corrective design, or (iii) to obtain actual state data from similar, comparable work systems for conceptual design.

In order to be able to perform systematic ergonomic actual state analyses, ergonomic databases are generated which can be used to detect deficiencies or possible improvements for individual jobs. Such actual state analyses are usually performed using ergonomic checklists which can be developed for different target groups, e.g. for work design professionals, security professionals or industrial engineers. Figure 14.3 shows an extract of such a checklist

CONSOLIDATION LEVEL	LEVEL (description see report)					
	0	1	2	3	4	5
1. Ergonomically insufficient body support		X				
2. Parts container unsuitable for gripping			X			
3. Improper use of body supports and parts containers			X			
4. Improper arrangement of containers actuators, devices, equipment			X			
5. Neglect of parts suitibility to assembly in product design		X				
6. Unergonomic and inefficient working methods		X				
7. Considerable amount of outside work			X			
8. Excessive division of labor, i.e. work content per operation too low			X			
9. No clear distinction between program com-pilation, pre-assembly and main assembly			X			

Figure 14.3 Design checklist (extract).

which has been developed for the automotive industry and the automotive suppliers industry (Bokranz and Landau, 1986).

It is applied by shop managers who want to detect deficiencies in ergonomic work design at individual workplaces in their own field of responsibility. For this purpose, the shop managers are trained and are allowed to participate in pilot projects in order to acquire the necessary skills. Design rules were derived from the training material, which were summarized in a checklist. In this way, continuity was obtained between the training material and the checklist. It has been proved in controlled applications that most of these managers were able to detect major design deficiencies and to develop suitable improvements. Using this concept, working system design could broadly be applied in a short time both in an assembly plant of a German automotive manufacturer and with an automotive supplier. This success was due to the fact that the concept has not been implemented by a few 'professionals' but by many 'amateurs' who had been made aware of ergonomic matters.

However, actual state analysis at the individual workplace using an ergonomic checklist is not sufficient because only the specific deficiencies which are present at the workplace can be detected and corrected. The question of whether detected deficiencies also exist at other workplaces or even throughout entire operating units of an organization, cannot be resolved without further examination. Findings derived from individual workplaces cannot be transferred to working spheres unless statistical methods are used to minimize the risk of faulty conclusions.

The necessity of work sphere analyses is supported by the fact that problems which are present in an entire division of a company (e.g. low technical level of working equipment) usually cannot be detected from findings of individual workplaces. It can also be concluded from results of pilot projects that work sphere analyses are the best cure for the popular 'company blindness', i.e. a reduced awareness of deficiencies. In addition, work sphere analyses are necessary in all cases where design measures for individual workplaces are to be complemented by design programs for entire companies or company divisions.

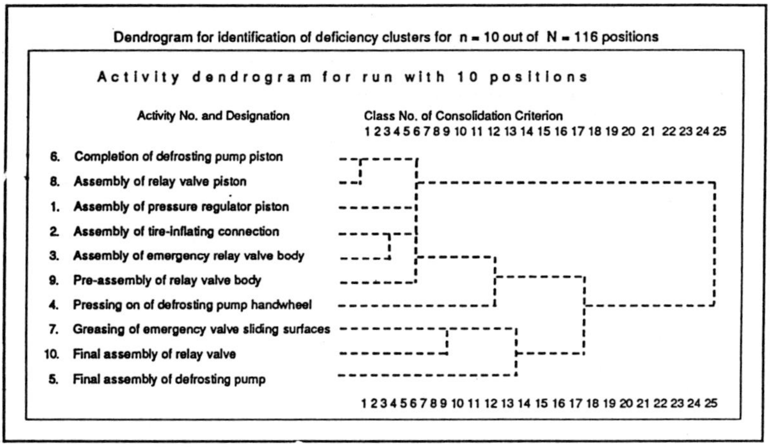

Figure 14.4 Deficiency dendrogram.

A description of the analysis method follows. It will make clear that the analysis is based on conventional job analysis and uses statistical methods (cluster or profile analysis), and that common major deficiencies are detected with a very high probability.

The definition the work spheres is usually oriented at existing company and department boundaries and summarizes, for example, pressroom, raw parts assembly and paint shop, rather than raw parts assembly and mechanical manufacture or mechanical manufacture and assembly. The evaluation technique of cluster analysis requires a minimum of 10 representative workplaces. At these workplaces, obvious deficiencies are assessed by means of a checklist using an ordinal scale. Assessment can be made by (i) checking items in the preprinted checklist form, (ii) checking an OCR document, or (iii) data input into a portable front-end computer.

The data records are subjected to a cluster analysis. The resulting dendrogram shows the workplaces with comparable deficiency structures. Figure 14.4 contains an example of such a deficiency dendrogram.

The lower the consolidation level, the higher is the similarity between workplaces with regard to assessed deficiencies. This makes workplaces # 6 and # 8 virtually identical, because a consolidation has already been made in the first consolidation class. Workplaces # 6, # 8 and # 1 have very similar deficiencies, but are generally different from all other workplaces because a consolidation with these is forced only in the last consolidation class. Experience with this kind of cluster analysis for sample sizes of 10–20 workplaces has shown that useful similarities exist with cluster sets up to consolidation class 15. For this reason, three deficiency clusters should be used in this example:

1. Workplaces # 1, # 6 and # 8,
2. Workplaces # 2, # 3, # 4 and # 9,
3. Workplaces # 5, # 7 and # 10.

The deficiency clusters are used to identify workplaces which are comparable with regard to assessed deficiencies (or advantages). However, this does not provide any information about nature, extent and cause of these deficiencies. Profile comparisons can be used to detect similarities with regard to assessed deficiencies for each cluster. Figure 14.5 also shows the six level ordinal scale used in cluster analysis, which shows the degree at which the objective is met for each of the 95 target or design criteria of the checklist used in this example:

Level 0: Does not apply, objective cannot or even should not be met.
Level 1: Objective could be met, but there is no indication that it will be met.
Level 2: There is some attempt to meet the objective, but this is only a trend 'in the right direction'.
Level 3: Objective is met to a certain degree, but the result is not too impressive.
Level 4: Objective is nearly met but there still is room for improvements.
Level 5: Objective is met or (applies only for some design criteria) even exceeded.

The objective or design criteria are divided into nine groups:

1. Body position, posture
2. Field of movement
3. Field of vision
4. Work surfaces
5. Body supports
6. Actuators
7. Indicators
8. Sequence of movements
9. Stress

Figure 14.5 shows an extract of the profile comparison for the first cluster (workplaces # 1, # 6 and # 8) for the case described in this example.

A list of relevant deficiencies can be derived from the individual deficiencies per cluster which have been detected by the profile comparison.

Experience has shown that an increasing degree of non-homogeneity of the workplaces within a work sphere with regard to the work system items 'task', 'input', 'output', 'work

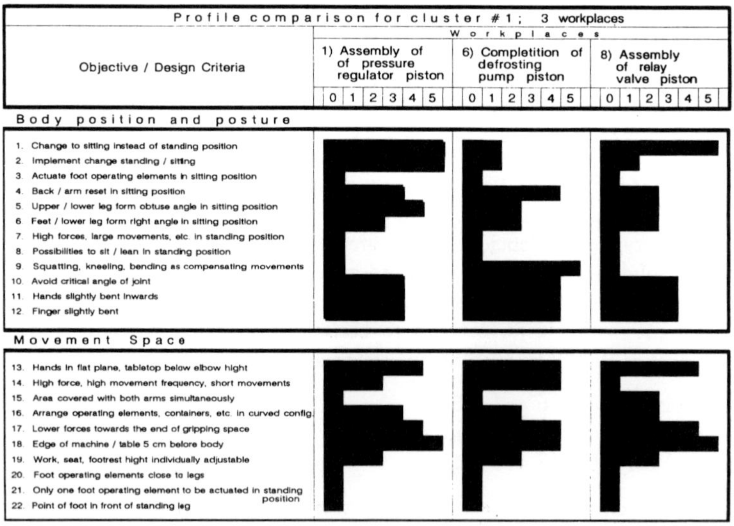

Figure 14.5 Profile comparison for detection of similar deficiencies within a deficiency cluster (extract).

DEFICIENCY	Work-place design	Work-organization	Product and program planning	Lack of qualification	Work-discipline
1. Ergonomically insufficient body support	X				X
2. Parts container unsuitable for gripping				X	
3. Improper use of body supports and parts containers	X				X
4. Improper arrangement of containers actuators, devices, equipment	X			X	X
5. Neglect of parts suitibility to assembly in product design			X	X	X
6. Unergonomic and inefficient working methods	X			X	X
7. Considerable amount of outside work		X	X	X	
8. Excessive division of labor, i.e. work content per operation too low		X		X	
9. No clear distinction between program compilation, pre-assembly and main assembly		X	X	X	

Figure 14.6 Work sphere deficiencies and their possible causes (extract).

flow', 'equipment' and 'environment' leads to an increased diversity of the deficiencies assessed per cluster; and, vice versa, the more homogeneous a work sphere is, the more similar are the deficiencies assessed per cluster.

In formulating work sphere deficiencies, the (individual) deficiencies of all clusters can be summed up if the work sphere shows highly homogeneous deficiencies as in the case described. Otherwise, it might make more sense to distinguish between cluster-specific and generic work sphere deficiencies. The formulated (individual) deficiencies will have to be generalized so that their causes can be identified. The result obtained for the described example can be seen from Figure 14.6.

A distinction should be made between technical and personal causes. Technical causes should be split up into causes relating to (i) work place design, (ii) work organization, and (iii) product and program planning. This makes it possible to classify planning deficiencies in order to implement specific revision projects. With personal causes, however, it may be difficult to decide whether a deficiency is caused by lack of qualification (especially in the areas of shop management, job planning, product and equipment construction, safety professionals, and work council members), i.e. by lack of knowledge, or by lack of work discipline, i.e. by lack of willingness.

The analytical–inductive approach described above can be summed up in four theses:

1. The quality of planning and design of human work is the result of a learning process. Therefore, planning and design deficiencies should be assessed systematically and generic findings should be derived.
2. Planning and design of human work becomes more time-consuming due to the increasing knowledge about human work and due to increasing design restrictions. Therefore, any measures applied in modification of existing work systems (i.e. rationalization or expansion investments) should be checked for universality.
3. In planning and designing human work, more capital-intensive workplaces have been created, and the costs per time unit have increased steadily. Therefore, when modifying existing work systems, related risks and problems to be solved should be identified.

4. The real problems (and thus the primary organization and investment requirements) are only revealed by work sphere analysis rather than by the analysis of individual workplaces. Therefore, before modifying existing work systems, not only workplace-specific deficiencies should be detected but also the main reasons for major deficiencies in a work sphere.

The programme system described in the following section allows work sphere analyses to be executed and supports the creation of company ergonomic databases.

ERGON job profiles – a cost-effective job analysis instrument

Extensive workplace analyses require great effort in terms of time and qualified specially trained analysts. This is why the use of methods like the AET or the SES is limited to selected workplaces or workplace collectives. Small and medium companies often want to avoid the expenditure related to job analysis. In many cases, there is no qualified personnel who could perform the job analyses. For this reason, there is a need for instruments which are able to gather sufficient information about a workplace so that this workplace can be evaluated for work hardness and difficulty, while keeping expenditure low. This workplace supervision can be implemented with a multi-level job analysis model (see Figure 14.7).

On the first level, a complete workplace documentation system with minimized assessment effort is introduced. The items assessed with this method are used as indicators for the selective work analysis which will be performed on the next levels. These consist of a job analysis (AET, Design Checklist, SES, see previous sections) and at level 3 of special methods of psychophysiological strain examination or physicochemical environmental measurements at the workplace (e.g. noise level measurement, vibration measurement, as required). Owing to the low assessment effort on the first level, workplace observations can be made at regular intervals. This makes it possible to replace the static snapshot of a workplace analysis with dynamic workplace supervision.

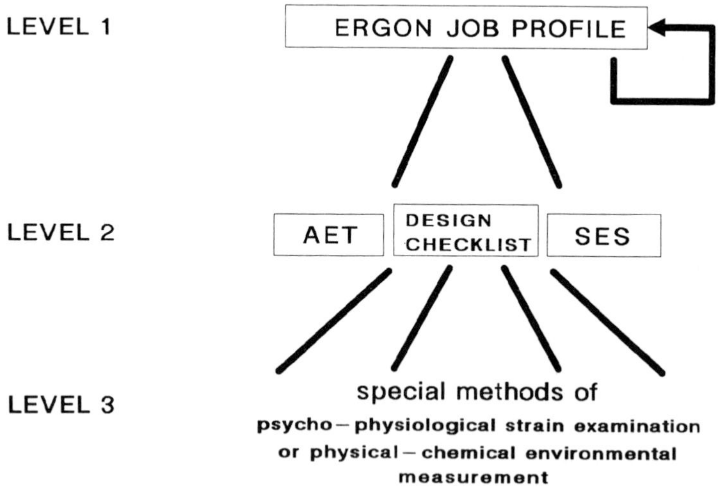

Figure 14.7 Multi-level job analysis model.

A complete workplace documentation system must meet the following requirements:

All relevant workplace information must be assessed.
The documentation must provide a realistic picture of the workplace.
The assessed data must lend itself to standardization so that comparisons and electronic processing are possible.
The assessment effort must be minimal.
The assessed data must be objective and repeatable.
The assessment procedure must be easy so that it can be applied independent of the presence of a trained job analyst.

The 'ERGON Job Profile' procedure designed for this purpose is structured as follows.

1. Assessment log

First, the following details are recorded: (i) time of assessment, (ii) workplace being assessed, (iii) person making the assessment, (iv) conditions of assessment.

2. Work system description

The work system is assessed as described by Nadler (1967). The work task is accurately described in verbal form. If several jobs are performed at one workplace, each of them has to be analyzed. The following details are assessed for characterization of the working system.

a. Task analysis

The job being analyzed is classified according to task-specific characteristics. There are 19 predefined task groups which are split up according to tasks that are performed: (a) on material work objects (eight task groups); (b) on abstract work objects (seven task groups); (c) with relation to persons (four task groups).

Tasks mainly related to material work objects

- adjusting/preparing
- assembling/disassembling
- equipping/inserting
- transporting
- adjusting/setting up/measuring
- operating
- checking
- supervising

Tasks mainly related to information – processing and planning

- planning/organizing
- coding/decoding information
- transcribing information
- arranging/classifying
- combining information
- analysing information or data
- using mathematics

Tasks mainly related to other persons

- Public speaking/representing
- instructing/training
- serving/supplying/attending
- judging human behavior

Figure 14.8 Task analysis.

This allows for a structured task description and the comparison of workplaces (see Figure 14.8).

b. Work flow

The work flow is to be described in the form of keywords which are stored as plain text.

c. Work object

The work object is designated and specified as being material, energetical, informatory or biological.

d. Working equipment/auxiliary equipment

Working equipment and auxiliary equipment is designated and characterized for its degree of automation, mobility and importance.

e. Individual

The individual as a part of the work system is characterized by his/her age, qualification, and experience.

f. Working environment

Negative influences of the working environment, such as illumination, climate, vibration, noise and dangerous substances, are assessed.

g. Temporal organization of work

Working hours and breaks are assessed.

The assessed data are stored in a database supported on a personal computer. Flexible access techniques allow for selective evaluation. Task profiles can be created for individual jobs or a group of jobs (see Figure 14.9). Using special requirement profiles, job classes (clusters) can be formed. Further processing of these selections can then be made with the detailed procedures of our multi-level model (Figure 14.7).

Used as a stand-alone system, the 'ERGON Job Profile' lends itself to task and requirement analyses. It can therefore be used as a tool for personnel assignment planning. It allows jobs to be classified according to work tasks. The knowledge of the special requirements of individual jobs can contribute to decisions for special personnel training. In addition, the 'ERGON Job Profile' can be integrated into a more comprehensive job information system. In this context, it can be used as a front-end system for the AET, ergonomic design checklists or the SES. It can generate an analysis log for a more comprehensive workplace analysis. The selective access capabilities allow the extraction of workplace groups which can directly be processed in the AET and SES.

The 'ERGON Job Profile' data can be used as a basis for decisions on more comprehensive job analyses. The complete assessment of data of all workplaces in a company provides an early warning system for stress bottlenecks at the workplace.

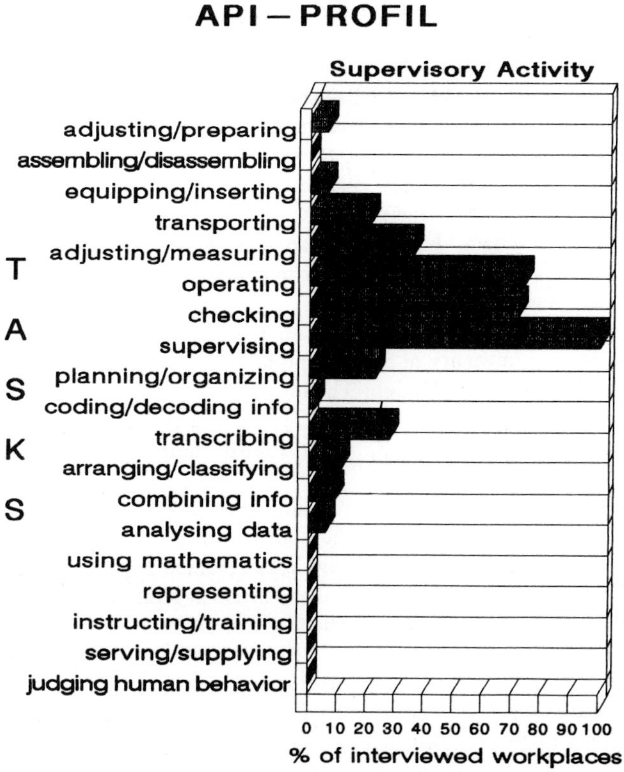

Figure 14.9 Sample task profile.

References

Biefang, S., 1985, Messung psychosozialer Erfolgs- und Prädiktorvariablen, Habilitationsschrift eingereicht bei der theoretischen Fakultät für Medizin der Universität Ulm.

Bokranz, R. and Landau, K., 1986, Mängelanalyse in Arbeitsfeldern – ein strategisches Instrument zur Arbeitsgestaltung. *Fortschrittliche Betriebsführung und Industrial Engineering*, **35** (2), 70–74.

Brauchler, R. and Landau, K., 1989, Epidemiological analysis of workload data using ergonomic databasis. In: *Recent Developments of Job Analysis. Proceedings of an International Symposium on Job Analysis*, University of Hohenheim, 14–15 March 1989, edited by K. Landau and W. Rohmert. (London: Taylor & Francis).

Caplan, R. D., Cobb, S., French, J. R. P., van Harrison, R. and Pineau, S. R., 1975, *Job Demands and Workers' Health*. NIOSH Research Report No. 75–160.

Dohrenwend, B. P. and Dohrenwend, B. S., 1981, Socio-environmental factors, stress and psychopathology. *American Journal of Community Psychology*, **9**, 128–164.

Frieling, E., 1975, *Psychologische Arbeitsanalyse* (Stuttgart: Verlag W. Kohlhammer).

Hoyos, G., 1974, *Arbeitspsychologie* (Stuttgart: Verlag W. Kohlhammer).

Kessler, R. C. and Cleary, P. D., 1980, Social class and psychological distress. *American Sociological Review*, **45**, 463–478.

Landau, K. and Rohmert, W., 1981, AET – a new job analysis method. In: *Proceedings 1981 Spring Annual Conference and World Productivity Congress*, 17–20 May 1981, pp. 751–760.

Landau, K. and Rohmert, W., (Eds), 1989, *Recent Developments in Job Analysis, Proceedings of an International Symposium on Job Analysis*, University of Hohenheim, 14–15 March 1989. (London: Taylor & Francis).

McCormick, W. T., Mecham, R. C. and Jeanneret, P. R., 1969, *The Development and Background of the Position Analysis Questionnaire*. (Purdue University, Lafayette: Occupational Research Center).

Medical Research Council, 1982, *Job Exposure Matrices*. Proceedings of a conference held in April 1982 at the University of Southampton.

Nadler, G., 1967, Work System Design: The Ideals Concept (Homewood, Ill: Richard D. Irwin).

Office of Population Censuses, 1978, Occupational mortality: decennial supplement. (London: HMSO).

Rohmert, W., 1979, Determination of stress and strain at real work places. Methods and results of field studies with air traffic control officers. In *Workload, its theory and measurement*, edited by N. Moray, (New York: Plenum Press), pp. 423–443.

Rohmert, W., 1983a, Determination of Stress and Strain at Real Work Places: Methods and Results of Field Studies with Air Traffic Control Officers.

Rohmert, W., 1983b, Formen menschlicher Arbeit. In: *Praktische Arbeitsphysiologie*, edited by W. Rohmert and J. Rutenfranz (Stuttgart: Thieme).

Rohmert, W., Rutenfranz, J., Luczak, H., Landau, K. and Wucherpfennig, D., 1975, Arbeitswissenschaftliche Beurteilung der Belastung und Beanspruchung an unterschiedlichen industriellen Arbeitsplätzen. In: *Arbeitswissenschaftliche Beurteilung der Belastung und Beanspruchung an unterschiedlichen Arbeitsplätzen*, edited by W. Rohmert and J. Rutenfranz (Bonn: Der Bundesminister für Arbeit und Sozialordnung) pp. 15–250.

Rohmert, W. and Landau, K., 1983, *A New Technique for Job Analysis* (London: Taylor & Francis).

Chapter 15

OWLKNEST: a knowledge-based expert system for selecting operator workload techniques

SUSAN G. HILL and REGINA M. HARRIS[1]

Idaho National Engineering Laboratory,
Idaho Falls, Idaho, U.S.A.

[1]Analytics, Inc.,
Willow Grove, PA, U.S.A.

Abstract. The Operator Workload Knowledge-based Expert System Tool (OWLKNEST) is an interactive, microcomputer-based tool that provides guidance in selecting the most appropriate technique(s) to use for estimating or predicting Operator Workload (OWL) for a particular set of conditions. This chapter presents background material concerning operator workload measurement and describes why an expert system was determined to be the best approach. The OWLKNEST design and design parameters are also presented. Several potential system applications are described which illustrate how OWLKNEST can be used for a variety of needs.

Keywords: Operator workload; Expert systems; Knowledge-based expert systems; Workload measurement techniques; OWLKNEST.

Introduction

In the past, the term 'workload' has generally referred to how hard a person is working physically. However, with the advent of less physical and more cognitive, decision-making, monitoring demands, the term 'operator workload' (OWL) has come to include how hard a person is working *both* mentally and physically. The major application area of such interest has been in the aviation community where the mental workload associated with flying aircraft, particularly advanced tactical military aircraft, has been of great concern. Mental workload assessments in military applications other than aviation have been increasing (see Hill *et al.*, 1987, for a discussion of OWL requirements in the US Army). This concern about mental workload has also been reported in industrial settings. For example, Beare and Dorris (1984) examined the workload of nuclear power control room crews. With increasing focus on mental workload, the question of how to measure that workload has received widespread attention. The obvious difficulty in assessing *mental* workload is that the behaviours of interest – cognitive process – are not very observable.

A variety of techniques for estimating OWL have been developed and most have been documented in published papers and reports. Many applications of these techniques have been similarly documented. One way to illustrate how much information is available is by noting the number of papers and reports that have been cited in major reviews of the workload literature. For example, Wierwille and Williges (1980) compiled an annotated bibliography on operator mental workload assessment methods. The bibliography contains over 600 citations. In this review, workload methods are divided into 28 specific techniques which are classified into four major categories: opinion, spare mental capacity, primary task and physiological. More recently, O'Donnell and Eggemeier (1986) reviewed workload assessment methodologies in a chapter with more than 245 citations. Lysaght *et al.* (1988) reviewed more than 475 references to obtain information about both analytic and empirical OWL methodologies. It is easy to conclude that there is an overwhelming body of literature (that keeps growing) on techniques and applications of techniques to measure operator workload.

Along with the wealth of information, such reviews as those just mentioned also provide some guidelines as to how to choose the most appropriate technique to use. O'Donnell and Eggemeier (1986), for example, divide the OWL methods into four major classes (primary task measures, secondary task methods, physiological techniques and subjective techniques) and discuss each of the classes in terms of the criteria of sensitivity, diagnosticity, intrusiveness, implementation requirements and operator acceptance. Depending on the particular circumstances, the resulting capabilities of the various methods can be used to choose the most appropriate class or classes of workload methods.

Wierwille *et al.* (1980) took a similar approach. However, they devised a matrix of critical criteria (physical space required, portability, intrusion-safety, data transmission and recording, experimental control, integration into aircraft system and crew acceptance) and workload measures and made judgments as to whether there were any potential problems in implementing the workload technique based on the seven criteria. This matrix represents the feasibility of the workload technique for use during aircraft flight. Another matrix was devised that provided relative rankings of the applicability (based on published research results) of the workload techniques for investigating different behavioural activities (i.e. the universal operator behaviours proposed by Berliner *et al.*, 1964). The two matrices together provide a structured paper-and-pencil method for deciding which workload assessment techniques are most appropriate for specific circumstances.

A more sophisticated approach was taken by the Human Performance Research Group (1988) at the NASA Ames Research Center. Using a commercially available expert system shell, a decision support tool, Workload Consultant for Field Evaluation (WC FIELDE), was developed to assist researchers in choosing the appropriate techniques for assessing operator workload (Casper *et al.*, 1986). The system includes a number of rules that are used to rank 24 workload measurement techniques in terms of their appropriateness for the particular circumstances of the proposed study (Casper *et al.*, 1987). The focus of the current version of WC FIELDE is on empirical techniques.

Greater interest in OWL assessment has resulted from the increasingly cognitive nature of system operation tasks. A variety of techniques for estimating OWL are available, but information about these techniques and their appropriateness may be very difficult to assimilate. Furthermore, information on practical issues concerning the applicability of a particular tool is often unavailable, for example time, cost, personnel issues. The selection of the appropriate technique(s) for assessing OWL is complex and beyond the scope of many practitioners. Therefore the Operator Workload Knowledge-based Expert System Tool (OWLKNEST) was developed as an automated tool to provide guidance in selecting the most appropriate technique(s) to use for estimating or predicting operator workload.

System design and development

OWLKNEST design

An expert system approach was determined to meet the needs of OWL practitioners best. A design goal of OWLKNEST was to be maximally flexible and adaptable for new applications and highly usable in terms of both operation/execution and analysis of results. OWLKNEST utilizes a question–answer dialogue which facilitates use by inexperienced analysts. It incorporates an hierarchical logic structure that attempts to focus quickly on the most applicable technique(s) and minimize the number of questions posed by the system. The result is a suggested list of the appropriate analytical and/or empirical techniques. The user is also provided with the capability to obtain brief descriptions of each recommended technique including references and descriptions of prior applications.

OWLKNEST incorporates a rule-based knowledge representation system. The knowledge base has been developed from a comprehensive review and evaluation of current OWL assessment techniques (Lysaght *et al.*, 1988) as well as laboratory and field test experiences culled from research literature and individuals who have applied the various techniques.

Figure 15.1 illustrates the OWLKNEST flow of information. The user inputs are fed into the expert system, which applies rules and knowledge dependent on the various system, operator and other characteristics input by the user. A list of suggested workload techniques is the system output.

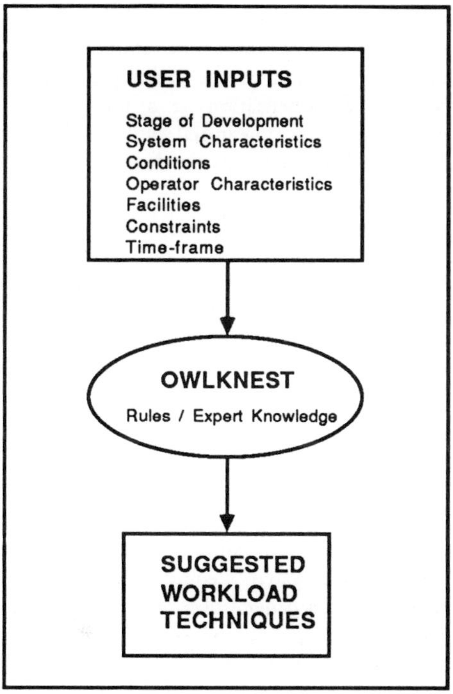

Figure 15.1 *Flow of user inputs through OWLKNEST to produce suggested workload techniques.*

User characteristics and requirements

The targeted users for OWLKNEST are analysts responsible for assessing operator workload for a system, for example, the Army MANPRINT analyst or the human factors specialist in a commercial setting. The user may also be the analyst who suggests the most appropriate method to examine OWL (e.g. research psychologists in field support positions). OWLKNEST assumes that users have at least a fundamental knowledge of OWL concepts. This implementation of OWLKNEST is targeted at novice computer and expert system users.

Although OWLKNEST focuses on Army users who can be assumed to possess knowledge of the Army acquisition cycle, other potential users include other Department of Defence agencies, NASA and industry. The OWLKNEST dialogue terminology was chosen to maximize understanding by both Army and other users. An on-line help and explanation feature are provided.

Knowledge requirements

As noted earlier, workload techniques can be divided into several different categories. The taxonomy implemented in OWLKNEST is that suggested by Lysaght *et al.* (1988) which divides OWL techniques into analytical and empirical techniques. Analytical techniques can be thought as predictive techniques that may be applied early in system design without an 'operator in the loop'. Empirical techniques are those that are taken with an operator in the loop during simulator, prototype, or system evaluations. Within each of these major categories, other subcategories of workload techniques can be identified. Table 15.1 lists all the workload techniques that have been included in OWLKNEST.

Only those OWL techniques that meet the following criteria are included in OWLKNEST: (i) satisfactory robustness of the technique as applied to Army systems, (ii) the existence of published documentation at a fine enough level of detail, and (iii) sufficient validation of the technique. The initial implementation focuses on a set of well-documented techniques that are clearly relevant for Army applications.

Software and hardware environment

OWLKNEST requires an 80286 microprocessor-based microcomputer equipped with a minimum of 640 Kbyte memory and two floppy diskette drives of at least 360 Kbyte (or one floppy diskette and a hard disk) (e.g. IBM PC/AT). It also requires DOS 2.0 or higher. EXSYS Professional, a rule-based, backward-chaining system, is the expert system shell selected for OWLKNEST.

System application

To illustrate the use of OWLKNEST, two sample applications are presented. The first case represents early system design. The system's equipment may not yet be fully designed, although the operator functions and a general idea of how the tasks are accomplished have been determined. The second case is an example of a more mature system in which the equipment and operator tasks exist in a well-defined form. There is to be form sort of test and evaluation of the system.

The dialogues between OWLKNEST and the user required to characterize each OWL assessment problem are illustrated below. The user selects one or more of the numbered options in response to each question. For illustration purposes, the user's response is shown in bold. The last two responses for each question are used when none of the options are applicable to the particular OWL assessment. A response of 'I don't know at this time' is used when the user does not possess suffiicient information at this time to answer the question, while a response of 'This question is not applicable to my problem' is used to indicate that the question is not pertinent to the problem. These last two responses are used to prevent the expert system from forcing the user to select a clearly inappropriate option and subsequently using that response in ranking the choices of appropriate OWL techniques.

Table 15.1. OWL techniques included in OWLKNEST.

- Analytical operator workload techniques
  - Expert Opinion
    - Interviews
    - Questionnaires
    - Delphi
    - Prospective ratings—SWAT, TLX and OW
  - Comparability analysis
  - Simulations
    - HOS
    - MicroSaint
    - TAWL
    - SIMWAM
  - Task analysis
    - McCracken–Aldrich
    - Zaklad/Zachary cognitive analysis
    - Timeline analysis
- Empirical operator workload techniques
  - Primary Measures
    - Type 1—System
    - Type 2—Operator
  - Subjective measures
    - Interviews
    - Questionnaires
    - Modified Cooper–Harper
    - NASA-TLX
    - SWAT
    - OW
    - Bedford
    - AHP
  - Secondary Tasks
    - Embedded
    - Choice RT
    - Sternberg memory task
    - Time estimation task
  - Physiological measures
    - Heart rate
    - Heart rate variability
    - Evoked potentials
    - Eye movements and scanning
    - Pupil measures
    - Blink rate

Case 1: Early system development

The dialogue between OWLKNEST and the user is illustrated for the case of assessing workload in early system design and development where operators and equipment are not yet available for use.

The current cycle in system development is:

1. **Pre-concept exploration**
2. Concept exploration
3. Demonstration and validation
4. Full scale development
5. Production and deployment
6. I don't know at this time
7. This question is not applicable to my problem

The total time available for implementation, performing and analyzing the OWL assessment is:

1. **Less than 1 week**
2. Less than 1 month
3. Less than 3 months
4. Within 6 months
5. Within 1 year
6. More than a year
7. I don't know at this time
8. This question is not applicable to my problem

I have access to the indicated special equipment:

1. Polygraph
2. EKG
3. Oculometer
4. Biomedical Laboratory Analysis
5. Video
6. Audio
7. **None of the above**
8. I don't know at this time
9. This question is not applicable to my problem

The system data that are currently available include:

1. **Missions scenario**
2. **Equipment descriptions**
3. Task list descriptions
4. Comparable system performance data descriptions
5. Estimated task times
6. Task analysis
7. None of the above
8. I don't know at this time
9. This question is not applicable to my problem

I have access to the indicated computers for performing the OWL assessment:

1. Any microcomputer
2. **IBM or DOS compatible microcomputer**
3. Any minicomputer
4. VAX minicomputer
5. Any mainframe computer
6. None of the above
7. I don't know at this time
8. This question is not applicable to my problem

A subject matter expert is available for the OWL assessment:

1. **Yes**
2. No
3. I don't know at this time
4. This question is not applicable to my problem

The personnel resources available for the OWL assessment include:

1. Technician
2. **Human factors specialist**
3. Engineer
4. Computer Scientist/Programmer
5. Biomedical
6. None of the above
7. I don't know at this time
8. This question is not applicable to my problem

The availability of a mock-up, simulator, or comparable system for OWL assessments is:

1. Yes
2. **No**
3. I don't know at this time
4. This question is not applicable to my problem

The immediate concerns that need to be addressed include the following OWL assessments:

1. Overall system performance
2. Ability to handle multiple tasks
3. Identify mission specific OWL chokepoints
4. Shoot-off contest
5. **Impact of new operator tasks**
6. Test and evaluation
7. None of the above
8. I don't know at this time
9. This question is not applicable to my problem

Based upon these responses, the following OWL techniques and associated rankings are recommended by OWLKNEST:

Analytical – Expert opinion	8
Analytical – Task Analysis	8
Analytical – PRO-TLX	5
Analytical – PRO-OW	5
Analytical – PRO-SWAT	5
Analytical – Interviews	5

Case 2: Field Test

The dialogue between OWLKNEST and the user is illustrated below for the case of a test and evaluation field test where the equipment and operators exist.

The current cycle in system development is:

1. Pre-concept exploration
2. Concept exploration
3. **Demonstration and validation**
4. Full scale development
5. Production and deployment
6. I don't know at this time
7. This question is not applicable to my problem

The total time available for implementation, performing and analyzing the OWL assessment is:

1. Less than 1 week
2. **Less than 1 month**
3. Less than 3 months
4. Within 6 months
5. Within 1 year
6. More than a year
7. I don't know at this time
8. This question is not applicable to my problem

I have access to the indicated special equipment:

1. Polygraph
2. **EKG**
3. Oculometer
4. Biomedical laboratory analysis
5. **Video**
6. **Audio**
7. None of the above
8. I don't know at this time
9. This question is not applicable to my problem

The system data that are currently available include:

1. **Missions scenario**
2. **Equipment descriptions**
3. Task list descriptions
4. Comparable system performance data/descriptions
5. **Estimated task times**
6. Task analysis
7. None of the above
8. I don't know at this time
9. This question is not applicable to my problem

I have access to the indicated computers for performing the OWL assessment:

1. **Any Microcomputer**
2. IBM or DOS compatible microcomputer
3. Any minicomputer
4. VAX minicomputer
5. Any mainframe computer
6. None of the above
7. I don't know at this time
8. This question is not applicable to my problem

A subject matter expert is available for the OWL assessment:

1. **Yes**
2. No
3. I don't know at this time
4. This question is not applicable to my problem

The availability of a mock-up, simulator or comparable system for OWL assessments is:

1. **Yes**
2. No
3. I don't know at this time
4. This question is not applicable to my problem

An operator is available for the OWL assessment:

1. **Yes**
2. No
3. I don't know at this time
4. This question is not applicable to my problem

Can the operator be safely interrupted during the performance of system tasks:

1. **Yes**
2. No
3. I don't know at this time
4. This question is not applicable to my problem

Would the attachment of sensors to the operator interfere with the safe use of the system:

1. **Yes**
2. No
3. I don't know at this time
4. This question is not applicable to my problem

The immediate concerns that need to be addressed include the following OWL assessments:

1. **Overall system performance**
2. Ability to handle multiple tasks
3. **Identify mission specific OWL chokepoints**
4. Shoot-off contest
5. **Impact of new operator tasks**
6. **Test and evaluation**
7. None of the above
8. I don't know at this time
9. This question is not applicable to my problem

Based upon these responses, the following OWL techniques and associated rankings are recommended by OWLKNEST:

Empirical primary measures	8
Empirical – Embedded secondary task	8
Empirical – NASA-TLX	5
Empirical – OW	5
Empirical – SWAT	5
Empirical – Interviews	5
Empirical – Questionnaires	5

Additional information is available on all the workload techniques included in OWLKNEST. Once a suggested list of techniques has been given, the analyst can obtain further information concerning the characteristics and application of the techniques as illustrated in Table 15.2 for NASA-TLX.

Table 15.2 Example of additional information available for suggested techniques.

NASA-TASK LOAD INDEX (TLX)

DESCRIPTION: A multi-dimensional scale derived from the NASA-Bipolar scales that uses an individual weighting procedure to reduce between-subject variability. It is comprised of two parts: (1) six rating scales covering different dimensions of workload used to rate OWL; and (2) the 'Sources of Workload Evaluation' uses paired comparisons of the six dimensions to obtain individual weightings of the dimension importance to workload for any task. The ratings and weightings are combined to produce a composite workload rating between 0 and 100.

SENSITIVITY: Sensitive to differences in task loading in a number of different types of tasks.

DIAGNOSTICITY: Gives a global composite rating of workload. However, the six subscales can be examined individually and used for diagnostic purposes

INTRUSION: Very little, although it does require a judgement. There was concern that the judgement might interfere with flight duties, but ratings were able to be obtained real-time.

IMPLEMENTATION REQUIREMENTS:

Data Collection: A 'Sources of Workload Evaluation' will be obtained for each task under study. The procedure uses only 15 paired comparisons and does not require much time to accomplish. The six scales are used to obtain ratings either real-time or after the session via visual recreation.

Operator Training: Practice in using and understanding the operational scales definitions is helpful.

OPERATOR ACCEPTANCE: Has been used successfully in real-time and post-flight aviation applications.

SAFETY: Plans must be made as to what to do if the operator is too busy to give a rating. Ratings should be secondary to the primary concern with operational safety (e.g. flying a plane or controlling a land vehicle).

PHYSICAL SPACE REQUIRED: n/a

PORTABILITY: n/a

INTEGRATION INTO SYSTEM: n/a

RESTRICTIONS: n/a

RELATIVE COST OF USE:

Testing time: The 'Sources of Workload Evaluation' takes about 10 min to make paired comparisons. The six ratings would not take significant time if the operators were familiar with the scale descriptions.

Equipment: Can be obtained via paper and pencil, or via computer. Video recording equipment is necessary in order to tape operator activity for use in post-test visual recreation.

Setup and support: Minimal.

Data analysis: The weighting and global measure computation can be done by hand, although a computer would be helpful. Descriptive and inferential statistics can be applied. Parametric and non-parametric statistics have been used to examine significant differences between mission segments or task variables.

Table 15.2 Continued.

References:
Hart, S. G. and Staveland, L. E., (1988, Development of NASA-TLX (Task Load Index): Results of empirical and theoretical research. In: *Human Mental Workload,* edited by P. A. Hancock and N. Meshkati (Amsterdam: Elsevier), pp. 139–183.
Human Performance Research Group, 1986, *Collecting NASA Workload Ratings: A Paper-and-pencil Package (Version 2.1)* (Moffet Field, CA: NASA Ames Research Center).

AVAILABILITY:
Human Performance Group, National Aeronautics and Space Administration
Ames Research Center
Moffet Field, CA 94035
Indicate version: Pencil and paper or Computer

Discussion of the two cases

Several comments can be made regarding the two example cases. First, the output of the system is a list of appropriate workload techniques with associated numeric rankings. The higher the ranking, the more appropriate the technique is to the particular situation. OWLKNEST uses the following rankings: 8, high applicability; 5, average applicability; 2, low applicability. These rankings are based on probabilities that are built into the expert system rules. Since the rankings are based upon informed judgement, they serve as a guide to indicate the order in which the user should consider applying the technique. It is incumbent upon the analyst using OWLKNEST to use judgement in choosing which of the listed techniques will be used to assess workload.

A second comment is that OWLKNEST can be used in a sensitivity analysis mode. The system can be used several times for the same application by changing one or more of the responses given. For example, in the first run, the analyst may choose to respond that no special equipment is available and obtain results based on that answer as one of the inputs. In the next run, however, the analyst may want to see what other techniques would be available if audio and video recording equipment were available. In this case, the suggested list might include different techniques. In this way, the analyst will be provided with some information on which to base decisions as to whether additional resources should be allocated to the workload assessment effort.

The system also is intended to be used iteratively across time. A workload analysis may be desired early in system design (e.g. Case 1). Then after the initial development is complete and a prototype is available, OWLKNEST might be used to provide suggested workload techniques based on the currently available information and resources. Hence, the expert system can be used throughout the development cycle of the system.

It should be emphasized that OWLKNEST will continue to be refined as more information about the existing techniques is found. Further research, validation and application of the techniques will be included as needed. Similarly, as new ways of assessing workload are validated, OWLKNEST can be expanded to include them.

Conclusion

The increasing criticality of performing OWL assessments throughout the system development cycle has led to the recognition that a comprehensive, easy-to-use tool to recommend the appropriate battery of OWL techniques is needed. OWLKNEST begins to provide this capability and supplies information on the appropriate OWL techniques based on individual needs and resources. Refinement of the expert system will continue as more information and experience with the included techniques is obtained and other workload techniques are identified for inclusion.

Although originally designed for military application, there appears to be a clear opportunity to transfer the technology to more industrial, commercial applications. Although terms may be different, concern with an operator's ability to control systems that use complex technology is the same. Similarly, the workload techniques are not application specific, but can be used in a variety of domains. OWLKNEST can provide the information and guidance of an appropriate selection of workload techniques for a broad range of applications.

Acknowledgement

This work was supported by the US Army Research Institute (ARI) under Contract No. MDA903-86-C-0384. The views, opinions and/or findings contained in this paper are those of the authors and should not be construed as an Official Department of the Army position, policy or decision.

References

Beare, A. and Dorris, R., 1984, The effects of supervisor experience and the presence of a shift technical advisor on the performance of two-man crews in a nuclear power plant simulator. In: *Proceedings of the Human Factors Society 28th Annual Meeting* (Santa Monica, CA: Human Factors Society), pp. 242–246.

Berliner, C., Angell, D. and Shearer, D., 1964, Behaviors, measures, and instruments for performance evaluation in simulated environments. In: *Proceedings, Symposium and Workshop on the Quantification of Human Performance* (Albuquerque, NM).

Casper, P., Shively, R. and Hart, S., 1986, Workload consultant: A microprocessor-based system for selecting workload assessment procedures. In: *Proceedings of the IEEE Conference on Systems, Man, and Cybernetics* (Piscataway, NJ: IEEE Service Center), pp. 1054–1059.

Casper, P., Shively, R. and Hart, S., 1987, Decision support for workload assessment: Introducing WC FIELDE. In: *Proceedings of the Human Factors Society 31st Annual Meeting* (Santa Monica, CA: Human Factors Society), pp. 72–76.

Hart, S. G. and Staveland, L. E., 1988, Development of NASA-TLX (Task Load Index): Results of empirical and theoretical research. In: *Human Mental Workload*, edited by P. A. Hancock and N. Meshkati (Amsterdam: Elsevier), pp. 139–183.

Hill, S., Lysaght, R., Bittner, A., Jr., Bulger, J., Plamondon, B., Linton, P. and Dick, A., 1987, *Operator Workload Assessment Program for the Army: Results from Requirements Document Review and User Interview Analysis* (DTR 2075-2) (Willow Grove, PA: Analytics).

Human Performance Research Group, 1986, *Collecting NASA Workload Ratings: A Paper-and-pencil Package (Version 2.1)* (Moffet Field, CA: NASA Ames Research Center).

Human Performance Research Group, 1988, *NASA Workload Consultant for Field Evaluation (WC FIELDE), Version 2.0* (Moffet Field, CA: NASA Ames Research Center).

Lysaght, R., Hill, S., Dick, A., Plamondon, B., Wherry, R., Jr., Zaklad, A. and Bittner, A., Jr., 1988, *Operator Workload: Comprehensive Review and Evaluation of Operator Workload Methodologies* (Willow Grove, PA: Analytics).

O'Donnell, R. and Eggemeier, T., 1986, Workload assessment methodology. In: *Handbook of Perception and Human Performance*, Vol. II, edited by K. Boff, L. Kaufman and J. Thomas (NY: John Wiley), pp. 42-1–42-49.

Wierwille, W. and Williges, B., 1980, *An Annotated Bibliography on Operator Mental Workload Assessment* (SY-27R-80) (Patuxent River, MD: Naval Air Test Center).

Wierwille, W., Williges, R. and Schiflett, S., 1980, Aircrew workload assessment techniques. In: *Survey of Methods to Assess Workload, Proceedings of the NATO Advisory Group for Aerospace Research and Development (AGARD)*, AGARDograph No. 246, (Neuilly sur Seine, FR: AGARD), pp. 19–54.

Chapter 16
Computer-aided workload analysis using MTM

FRANK M. KÜHN AND WOLFGANG LAURIG[1]

Klöckner-Moeller GmbH
Technical Services Department (TD)
Hein-Moeller-Str. 7-11
D-5300 Bonn 1, Federal Republic of Germany

[1]Institute for Occupational Health
Department of Ergonomics
University of Dortmund
Ardeystr. 67
D-4600 Dortmund 1, Federal Republic of Germany

Abstract. Users of predetermined motion time systems are interested in more efficient human movements. They seldom consider the risk of mismatching the task features to the workers' abilities. Based on this approach, an increase in load may lead to fatigue which in turn can result in productivity losses. Starting from the problem of identifying motor workload in assembly tasks, it is explained that the concept of assessing load can be understood axiomatically. A pragmatic approach is proposed in order to derive indices of load using the precise description of movements in the MTM system. The practicality of such indices is exemplified. A computer-aided procedure for the application of this concept is introduced. It is currently being tested in practice for objectifying the ergonomic evaluation of different designs of manual operations.

Keywords: Motor workload; Indices of load; Predetermined motion times; Assembly tasks; Computer-aided procedure.

Introduction

The analytical concept underlying the various predetermined motion time systems (PMTS) can be traced back to the beginning of time and motion studies (Gilbreth and Gilbreth, 1921). When applying this analytical concept for predicting and controlling human movements and movement times, users of PMTS such as 'methods time measurement' (MTM) (Maynard *et al.*, 1948; Antis *et al.*, 1969) or 'work factor' (WF) (Quick *et al.*, 1965) are interested in the rationalization of manual operations, i.e. in saving time by means of more efficient motions. Consequently, PMTS have often been discussed with regard to the introduction and application of incentive wage agreements in particular.

Work on this project was done at the Institute for Occupational Health at the University of Dortmund. The concept was developed in collaboration with Klöckner-Moeller GmbH.

However, increasing the efficiency of movements involves the risk of creating fatigue (Kerkhoven, 1962) and reducing productivity if variations in the skill of the operators are ignored. Minter (1977) identifies this problem in connection with the application of PMTS since they are based on time values for an 'average operator', 'average conditions' and the 'average nature of the process'. A decrease in the motivation level resulting from a mismatch between the skill of the operator and the features of a task (described in the following as components of load) should also be taken into account. The effect may be the same as for tasks demanding highly efficient motions: a mismatch between skill and workload involves the risk of fatigue and productivity losses. Practical experience has shown that workloads are sometimes so high that workers are even unable to achieve the predetermined times previously regarded as optimum times.

For these reasons, the task features must be taken into account in order to avoid a mismatch with the abilities of the working persons. Such risks can often be predicted by carrying out an ergonomic workload analysis at the planning stage. This should help to avoid productivity losses resulting from subsequent corrections and design modifications.

Rationalization by means of PMTS also involves work simplification. As with work measurement in general, a method is required for the assessment of attempts at simplification. Finally, practitioners are interested in identifying the variation in workload in the iterate process of work design. Consequently, there is a clear need for both a method and a tool for identifying, describing and even assessing workload, specific components of load and differences between the loads in various tasks as early as the planning stage. Such a tool called MAID, employing a pragmatic approach in order to derive indices of load on the worker's motor system on the basis of MTM, is proposed in this chapter.

MAID: both ergonomic concept and computer program

MAID is the abbreviation for 'Motion Analysis and Index Derivation', i.e. an ergonomic concept and a program system permitting computer-aided workload analyses using the precise description of movements found in the MTM-1 system and the simplified systems MTM-UAS and MTM standard data (basic values).

The MTM system has been chosen because it is widespread and reflects the true work situation reasonably well (Minter, 1977). The utilization of a PMTS already employed in industry should enable planning engineers to execute workload analyses and determine indices of load almost as a by-product of predetermining motion times. This permits consideration of ergonomic aspects (Casey and Wilson, 1980) in spite of the time pressure often encountered during time measurements. Additional computations of indices of load at the planning stage, or even during the evaluation of previous analyses, are accepted only reluctantly. If proposed methods cannot be used simply in practice, they will be too expensive to apply and will not be accepted by planning engineers. For these reasons, a computer-aided procedure was developed for both the calculation of indices of load and the application of MTM (Kühn, 1984). It may be assumed that a concept for rating workload based upon a PMTS such as MTM will be more acceptable to practitioners than any sophisticated concept. The concept presented in this chapter permits an exact and comprehensible description of different manual working methods. Such descriptions help to identify the load on the respective workers and to derive, according to ergonomic rules, indices of load which facilitate the evaluation of a workplace design.

Comparison with the cost of an empirical rating for the assessment of motor workload reveals the advantages of using the MTM system. The effort involved in analyzing manual operations is mainly justified by the expected 'improvement potential' (Salwiczek, 1982) through rationalization. The additional cost of checking the workload is comparatively low. Thus, good arguments for a humane design of work systems result as a side effect of attempts to rationalize assembly tasks.

MAID was developed both to help, for example, planning engineers to evaluate and design working operations and workplaces according to the concept of 'prospective ergonomics' (Laurig, 1986) and to identify realistic planning quantities (hitherto often calculated manually). An approach with pragmatically defined indices of load is therefore proposed. It has a number of advantages: (i) little effort is required to determine the indices, (ii) the indices are derivable from all MTM analyses, and (iii) their objectivity is determined only by the quality of the available analyses.

Ergonomic hypothesis and MTM-based model

An empirical rating of workload based on shop-floor measurements is expensive and not always practicable in industry. An axiomatic concept (Laurig *et al.*, 1983) based upon the analytical concept of PMTS is therefore preferred. On account of this analytical concept, the PMTS contain a possible solution to all of the problems mentioned in the introduction.

The concept underlying PMTS can be regarded as axiomatic since the application of these systems according to well-defined rules always leads to the same results regardless of the user. This axiomatic objectivity is one reason for the success of PMTS in practice.

These considerations suggest that it should be possible to derive pragmatically defined indices of load from PMTS and to use them axiomatically (Laurig *et al.*, 1983). Even simple approaches such as classifying and counting the basic motions within one assembly cycle or adding up the lengths of the motions given in PMTS-analyses permit surprising insights into differences between various designs of manual operations (Laurig *et al.*, 1985). Similar attempts to derive indices of load are discussed by various authors (Minter, 1977; Heinz and Salwiczek, 1983; Kühn, 1984).

There are, however, doubts about whether such pragmatic approaches can be justified in terms of ergonomics. As a consequence, such arbitrarily constructed indices seem to be of limited value. It might therefore be useful to base the derivation of such indices on more scientific ergonomic criteria.

Accordingly, the MAID concept provides for a more detailed workload analysis using load-related features of the MTM system described as 'conditions' or 'factors of influence' (Schwab, 1956). Factors used in the MTM system to characterize the difficulty of a work task are, for example, motion length, degree of control (accuracy of motion) and forces (see Table 16.1).

It is due to such factors that the MTM system permits an exact and comprehensible description of different manual working methods and helps to derive, according to ergonomic rules, indices of load. There is, for example, the 'degree of control' factor in MTM when considering "precise motions which demand sight . . . physical control and mental concentration" (Schwab, 1956). Such factors illustrate the system's implicit conception of motor workload. In accordance with the axiomatic concept of determining load, they can be understood as 'factors of load'.

Mapping the findings for the factors on to one scale would provide an overview in spite of the differentiation according to many different factors and their various scales. Following the concept of the 'degree of control', the motion elements can, for example, be classified as mainly 'muscular' (coarse movements requiring mainly muscular energy) or 'sensory motor

Table 16.1. Excerpt from the MTM table (Antis et al., 1969).

	Standard time table			Supplementary weight factor table	
	Time values (TMU)				
	degree of control (cases)				
Motion length (cm)	Low M-A-	Medium M-B-	High M-C-	Weight or force (kg or dN)	Factor
26	11·5	12·3	13·7	4	1·07
28	12·1	12·8	14·4	6	1·12
30	12·7	13·3	15·1	8	1·17
35	14·3	14·5	16·8	10	1·22
40	15·8	15·6	18·5	12	1·27

MTM motion element 'Move', abbreviated to 'M-', e.g. M30C10: 15·1 TMU * 1·22 = 18·4 TMU (1 TMU = 0·036 s).

work' (fine movements requiring a high degree of sensory activity for motor control) (Rohmert and Laurig, 1971). Consequently, it is possible to classify, for example, the basic motion 'reach' as muscular load and 'grasp' as sensory motor load. If the introduction of such criteria for differentiating load is assumed to be plausible in ergonomic terms, basic work motions can be classified pragmatically.

A greater variety of indices is possible when factors from the MTM system are utilized in a more differentiated way. However, these indices do not describe the factors mentioned previously themselves, but instead ergonomically defined components of load, derived from these factors as well as from supplementary items in the MTM system:

1. Load on the finger/hand system due to movements,
2 Load on the hand/arm system due to movements,
3 Load on the body system due to movements,
4 Increased degree of control of movements,
5 Visual control of objects or processes,
6 Increased forces to be applied.

The first three components of load describe the demands particularly related to mobility and influenced by motion lengths. The last three describe supplementary features.

The indices of load are computed with reference to an ergonomic hypothesis consisting of a simple model of load. This model states that the motor workload, i.e. the difficulty of a special motion element, is reflected by the execution time of the element. A correlation between the duration of motion elements derived from MTM and the load itself is assumed: the more difficult the operation is, the longer the execution time.

It can be concluded from this that the predetermined motion times themselves can be utilized to quantify load (Kühn, 1984; Laurig *et al.*, 1985). In order to allow examination of the components of load, the proposed concept includes a detailed analysis of all of the basic motion elements employed in a MTM analysis. Thus, using this model, each MTM analysis can be translated into a list of portions of the total cycle time corresponding to the aforementioned components of load. These portions of predetermined times can be added up within each category of components of load and, for purposes of comparing the load in various sequences of motions, related to the total cycle time.

The calculation of percentages of components of load is exemplified in Table 16.2 for a single element of motion called M30C10 (see also Table 16.1) which is taken from a real

Table 16.2. Load analysis of the MTM motion element M30C10 (see also Table 16.1.)

Components of load	Motion elements	Standard time and differences (TMU)	Percentage (%)
Load on the hand/arm system	M30A	12·7	69
Increased degree of control	M30C-M30A	2·4	13
Increased forces	M30C10-M30C	3·3	18
Total (standard time)	M30C20	18·4	100

MTM analysis (Kühn and Laurig, 1985a): the simplest bringing action over a distance of 30 cm is described by M30A and calculated for 12·7 TMU. This element only requires basic mobility of the hand/arm system, e.g. no increased degree of control and no particular forces. In addition to this load on the hand/arm system, the M30C element requires an increased degree of control. This supplementary component of load takes 2·4 TMU (the difference between the standard time values for M30C (15·1 TMU) and M30A (12·7 TMU). Increased forces represent a further component of load and take an additional 3·3 TMU (the difference between the standard time values for M30C10 (18·4 TMU) and M30C (15·1 TMU)). Comparability is facilitated by converting these time values into percentages of the total execution time of the motion. They are used in the following as indices of load.

In order to achieve an optimum adaptation of the workload to individual characteristics of workers, this calculation must refer to a total operation or working cycle which is constantly repeated by the workers throughout the work period. Such a total operation consists of numerous basic motion elements. One example for a load analysis of a short sequence of motions (reach something (R30A), grasp it (G1A), move it (M30C10) and, finally, release it (RL1)) for one hand, e.g. right hand, is computed in Table 16.3.

In addition to the procedure outlined here, supplementary rules help to increase the accuracy and reliability of the MTM and the load analyses.

In contrast to the conventional MTM analysis, the load-oriented analysis also includes operations which are not critical in time (contrary to the 'principle of the limiting motion'; Schwab, 1956) since these still represent a load. Thus, complete analyses of the left and the right hands are required, i.e. all non-limiting, combined, overlapping and simultaneous

Table 16.3. Load analysis of a sequence of motions (R30A–G1A–M30C10–RL1).

	Time values (TMU)					
	Motion elements					
Components of load	R30A	G1A	M30C10	RL1	Total	Percentage (%)
Load on the finger/hand system	—	2·0	—	2·0	4·0	12·5
Load on the hand/arm system	9·5	—	12·7	—	22·2	69·6
Increased degree of control	—	—	2·4	—	2·4	7·5
Increased forces	—	—	3·3	—	3·3	10·4
Total	9·5	2·0	18·4	2·0	Σ 31·9	Σ 100·0

motions must also be determined. This applies to all operations to be executed during process time as well. Thus, when the part of the load due only to the limiting motions is set at 100%, the real load of complex tasks for both hands together may even exceed 100%.

A further problem as regards a complete description of a sequence of operations is that the MTM method can describe only those operations which are fully influenced by the workers. Other operations, such as holding a hand drill or observing an instrument, are not considered. For this reason, MAID supports the analysis of the additional elements XV for visual control and XS for static forces. These supplements are defined and explained with reference to the MTM basic method, i.e. similar to the conventional MTM motion elements.

Where appropriate, allowances should also be analyzed using the MTM method. They are thereby considered as sequences of motions which occasionally have to be performed. Such a sequence of motions, probably appearing every hundred cycles (e.g. material procurement), can later be assigned a relative factor. No other possibility for considering the load resulting from the allowances is provided since, in fact, the load analysis is based exclusively on the MTM standard times. However, since such accurate allowance analyses cannot always be realized, the MAID program also allows simplified allowance analyses to be conducted.

MTM-1 or simplified MTM systems

Using data from the MTM-1 scale is time consuming. Consequently, two further MTM systems were considered as a basis for the application of the proposed approach (Kühn and Laurig, 1985b). These systems 'MTM-UAS' and 'MTM standard data' are second-generation systems, i.e. condensed or simplified versions of the MTM-1 system. These procedures, which are similar to MTM-2, are also widespread in German industry. They are used to obtain weighted average times for common combinations of movements and are therefore advantageous for the quick and easy application of PMTS. On the other hand, the description of the sequences of motions under analysis is not as precise as with the basic MTM-1 system, even in those cases where movements of the left and right hands can be differentiated. The use of such systems for predetermining motor workload therefore has to be discussed critically. It may be assumed that, in fact, the load analyses based on basic systems are more reliable. Results from the various MTM systems are discussed in connection with the application of computer-aided procedure.

Computer-aided procedure

The use of computer-aided methods for the application of PMTS (Salwiczek, 1982; Kühn, 1984) has been shown to reduce substantially both the time needed for an analysis and the number of errors when applying the PMTS. Since routine operations, such as, for example, the assignment of standard time values in tables and the addition of time values, are taken over by the program the analysis speed can be increased. The acceptability of computer-aided methods as cost-effective solutions has increased as a result of the development of the personal computer market. MAID runs on personal computers such as the IBM-PC or AT.

The MAID program (see Figures 16.1 and 16.2) has been developed to reduce the expense involved in applying MTM systems and executing load analyses. Special attention was paid to the user interface in order to improve system acceptability. The description of sequences of motions composed of the MTM basic motion features and total operations is, however, not performed by the computer, i.e. the user must be familiar with the MTM system. One

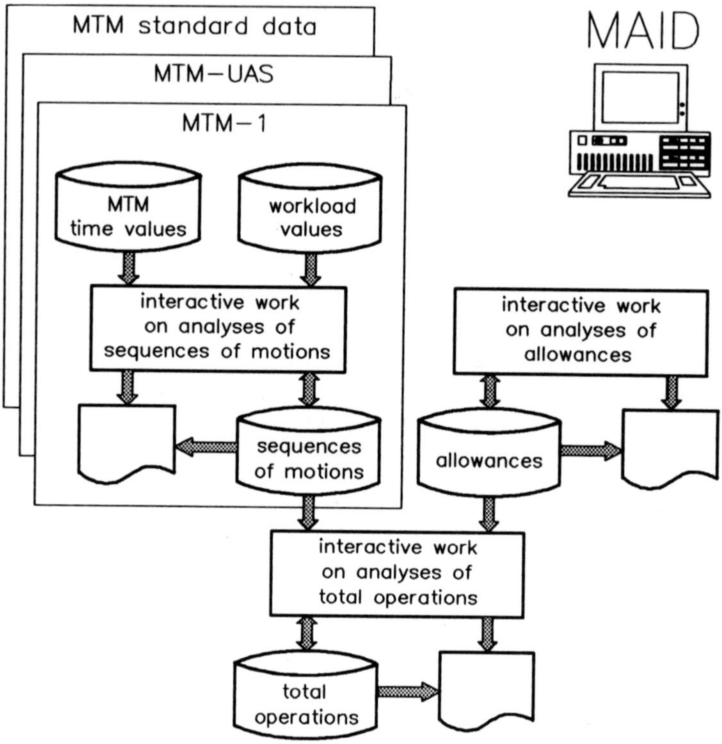

Figure 16.1 MAID program structure and data flow.

prerequisite for the introduction of this computer program was therefore correspondence of its input and output formats to the forms normally used in a MTM analysis. Paper and pencil are replaced by a keyboard and a screen (see Figure 16.3) where the user can freely move and insert, delete and copy characters and lines in a similar manner to text editors and other worksheet-based programs. In the case of larger errors the user no longer has to transcribe the sheet.

The real benefit from introducing the computer becomes apparent in the automatic assignment of predetermined times to each basic motion. In addition, the program adds the standard time values for the sequences of motions and total operations, indices of load and planning quantities.

The program also provides for the addition of allowances to the computed predetermined times. MAID permits use of the MTM method even in the analysis of allowances, i.e. sequences of motions that occasionally have to be performed. Such a sequence of motions, probably occurring every hundred cycles (e.g. material procurement), can later be assigned a relative factor (in this case 0·01, see also Figure 16.6, line 11). There is no other possibility in the load analysis for considering the workload resulting from the allowances. However, MAID also permits simplified allowance analyses in cases where accurate allowance analyses based on MTM standard times cannot be realized.

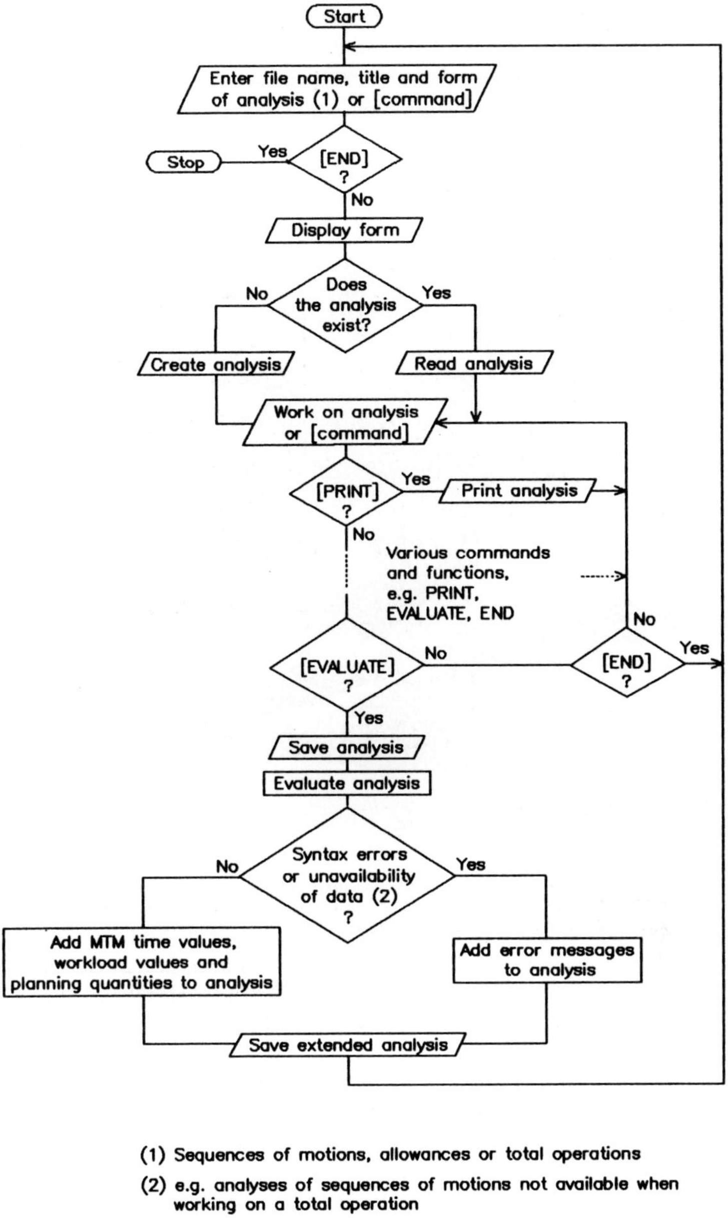

Figure 16.2. MAID flow chart.

Another advantage of the computer-aided MTM and load analysis lies in the interactive modification of the manual operation task on the computer screen. The system also allows arbitrary merging of the existing sequences of motion into new operations and computations based on these operations. Each modification input is subsequently used to calculate new standard time values, indices of load and planning quantities. In this way, the user works interactively on sequences of motions and allowance analyses, and, finally, combines these modules into total operations which form complete working cycles. The user's entries and

Figure 16.3. Application of MAID in practice.

File : ASSEM_A3.MTM
Title: Exchange enclosure in equipment
Form : MTM

No	Text l.h.	Code l.h.	(TMU)	(TMU)	Code r.h.	Text r.h.
1	reach enclosure					move enclosure
2	in box	R20C			M4B	out of equipment
3					M20C	to box
4					(G2	regrasp
5	grasp enclosure	G4A				
6					P1NSE	position enclosure
7	move enclosure					in box
8	to equipment	M20C			RL1	release enclosure
9	regrasp	(G2				
10	position enclosure					
11	in equipment	P2NSE			R10E	reach screws
12	release enclosure	RL1				
13						
14						
15						
16						
17						

Figure 16.4. MTM analysis of a sequence of motions, user's entries.

the results of his analyses as calculated by the program can be printed directly in the form represented on the screen.

Analyses of a sequence of motions and of a total operation are exemplified in the following. Figure 16.4 shows the entries for a sequence of motions, including the file name and the title of the analysis as well as the textual description of the sequence of motions and the appropriate MTM motion element codes. Figure 16.5 displays the MTM analysis supplemented by MAID. On the first screen page of the analysis (lines 1 to 17) the program adds up the results, i.e. standard time, sum of the motion lengths and indices of load for the left and right hands. On the second screen page, the user's entries are supplemented by the individual standard times of the motion elements.

Screen page 1

File : ASSEM_A3.MTM
Title: Exchange enclosure in equipment
Form : MTM

No	Text l.h.	Code l.h.	(TMU)	(TMU)	Code r.h.	Text r.h.
1						
2						
3	standard time 75.5 TMU					
4						
5						
6	workload				left right	
7						
8	motion lengths				40.0 34.0	cm
9						
10	load on the finger/hand system				18.7 13.4	%
11	load on the hand/arm system				28.6 28.5	%
12	load on the body system				0.0 0.0	%
13	increased degree of control				31.0 11.8	%
14	visual control				0.0 0.0	%
15	increased forces				0.0 0.0	%
16						
17						

Screen page 2

File : ASSEM_A3.MTM
Title: Exchange enclosure in equipment
Form : MTM

No	Text l.h.	Code l.h.	(TMU)	(TMU)	Code r.h.	Text r.h.
18	reach enclosure					move enclosure
19	in box	R20C	11.4	4.0	-M4B	out of equipment
20				11.7	M20C	to box
21				5.6	(G2	regrasp
22	grasp enclosure	G4A	7.3			
23				10.4	P1NSE	position enclosure
24	move enclosure					in box
25	to equipment	M20C	11.7	2.0	RL1	release enclosure
26	regrasp	(G2				
27	position enclosure					
28	in equipment	P2NSE	21.0	6.8	R10E	reach screws
29	release enclosure	RL1	2.0			
30						
31						
32						
33						
34						

Figure 16.5. MTM and load analysis of a sequence of motions supplemented by MAID.

Once all of the sequences of motions and allowances (analysis of the latter is not demonstrated here) are analyzed and defined by file names, the modules can be combined into a total operation (see Figure 16.6, cf. program structure shown in Figure 16.1). The file names of sequences of motions analyzed according to MTM-1 are recognizable by the extension MTM, the file names of analyses of allowances by the extension ALL. Since MAID permits the combination of sequences of motions which have been analyzed using different methods (MTM-1 and simplified methods), the program also interprets the extensions UAS and STD (standing for MTM-UAS and MTM standard data).

```
File : ASSEM_A.TOP
Title: Assemble enclosure
Form : TOP

No | Total operation                       | (TMU)
 1 | ASSEM_A1.MTM
 2 |
 3 | 'two screws
 4 | 2 * ASSEM_A2.MTM
 5 |
 6 | ASSEM_A3.MTM
 7 |
 8 | ASSEM_A4.MTM
 9 |
10 | 'repeat nearly every five times
11 | 0.2 * ASSEM_A4 MTM
12 |
13 | 'allowances for this operation
14 | ASSEM_A.ALL
15 |
16 |
17 |
```

Figure 16.6. Total operation entered by the user, including arbitrary remarks.

Screen page 1

```
File : ASSEM_A.TOP
Title: Assemble enclosure
Form : TOP

No | Total operation                                                          | (TMU)
 1 |
 2 |      planning quantitities
 3 |
 4 |
 5 |      standard time              781.60   TMU
 6 |
 7 |    = standard time                0.4690 min
 8 |    + allowances (11.0%)           0.0516 min
 9 |    = total operation time         0.5206 min
10 |
11 |
12 |      number of workers            1     1      1     4     4      4
13 |      quantity/cycle/worker        1     1      1     1     1      1
14 |      production period           60   480   9600    60   480·  9600
15 |      quantity                   115   922  18440   461  3688  73761
16 |
17 |
```

```
Screen page 2

File : ASSEM_A.TOP
Title: Assemble enclosure
Form : TOP

No | Total operation                                              (TMU)
---+------------------------------------------------------------------
18 |
19 |
20 |
21 |
22 |      workload                          left  right
23 |
24 |      motion lengths                   136.5  330.0  cm
25 |
26 |      load on the finger/hand system     9.6   16.1  %
27 |      load on the hand/arm system        8.6   22.7  %
28 |      load on the body system            0.0    0.0  %
29 |      increased degree of control       12.1   21.5  %
30 |      visual control                     0.0    0.0  %
31 |      increased forces                   0.0   21.1  %
32 |
33 |
34 |
```

```
Screen page 3

File : ASSEM_A.TOP
Title: Assemble enclosure
Form : TOP

No | Total operation                                      | (TMU)
---+------------------------------------------------------+------
35 | ASSEM_A1.MTM                                         | 167.3
36 | "Prepare fitting mechanism"                          |
37 | 'two screws                                          |
38 | 2 * ASSEM_A2.MTM                                     | 322.8
39 | ""                                                   |
40 | ASSEM_A3.MTM                                         |  75.5
41 | "Exchange enclosure in equipment"                    |
42 | ASSEM_A4.MTM                                         | 180.0
43 | "Fit enclosure"                                      |
44 | 'repeat nearly every five times                      |
45 | 0.2 * ASSEM_A4 MTM                                   |  36.0
46 | "Fit enclosure"                                      |
47 | 'allowances for this operation                       |
48 | ASSEM_A.ALL                                          | %11.0
49 | "Allowances for assembling enclosure"                |
50 |                                                      |
51 |                                                      |
```

Figure 16.7. Total operation supplemented by MAID.

The analysis of the total operation supplemented by MAID is shown in Figure 16.7. For the calculation of the planning quantities the program is based on a 40-h week, one worker and one quantity per cycle per worker. These default statements can be modified by the user. In this example the user defined four workers for the operation. The first screen page of the analysis completed by MAID (lines 1 to 17) shows the most important operational planning quantities. Line 15 gives the quantities for different production times, here 60, 480 (i.e. one day in a 40-h week) and 9600 min (i.e. 1 month or 4 weeks or 20 days). The indices of load are provided on the second screen page (lines 18 to 34). From the third page onwards (from line 34) the user will again find the total operation with the individual sequences of motions supplemented by standard times, percentages of allowances and the titles of the modules.

A major advantage of computer-aided procedures is the possibility of storing the results and the raw data in files and on a storage medium such as a floppy disk. Thus, in addition to facilitating and systematizing the analysis, the program system aims at simplifying the data exchange between the MAID users and its developers. By computing and comparing the various analyses the latter are then able to derive or to supplement recommendations for the evaluation of indices of load. Data exchange between different users permits the development of standard analyses which can subsequently be used as standards within the company organization.

Evaluation and experiences

The indices of load (i.e. percentage values previously described) computed by MAID can be evaluated and interpreted in different ways. However, this always presupposes that the operations have been completely analyzed.

The complexity of a work task is indicated by the sum of all indices of load for one operation. Many simultaneous or non-limiting motion elements result in indices of load which in total exceed the 100% value since, in contrast to the calculation of the standard time, the load analysis has to consider all basic motion elements. Values far exceeding 100% indicate that the total operation is very concentrated and that difficulties may arise in the practical execution of the task if a worker's skills and abilities are not sufficiently considered.

The arrangement of the indices of load in a profile of load allows more detailed statements to be made. A profile of load consists of a bar chart in which the different bar lengths represent the indices of load for the left and the right hands (see Figure 16.8). Such a profile of load is an easy means of distinguishing between workplaces with differing demands on the workers. Analyses of about a hundred assembly tasks reveal both basic differences in the

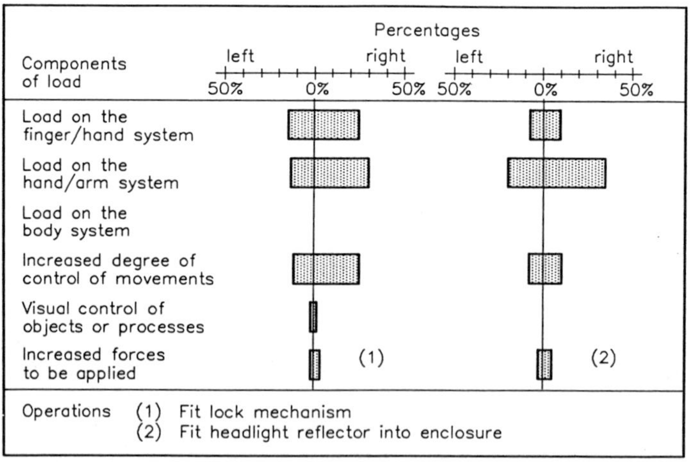

Figure 16.8. Profiles of load for two assembly tasks 'Fit lock mechanism' and 'Fit headlight reflector into enclosure'.

Table 16.4. Comparison of the load indices based on MTM-1 and simplified MTM systems for the same assembly task (calculation of indices is exemplified in Tables 16.2 and 16.3).

	MTM-1		MTM-UAS		MTM standard data	
Components of load	Left (%)	Right (%)	Left (%)	Right (%)	Left (%)	Right (%)
Load on the finger/hand system	9	16	13	20	8	13
Load on the hand/arm system	18	28	19	35	18	27
Load on the body system	3		3		3	
Increased degree of control	10	14	7	11	12	17
Visual control of objects or processes	—	—	—	—	—	—
Increased forces	1	5	—	2	1	5

profiles of load for different workplaces and different plants and similar effects, e.g. asymmetric sequences of motions (the load on the right hand exceeds that on the left hand). Two results provided in Figure 16.8, in which the loads on the left and right hands are compared, clearly exemplify asymmetric sequences of motions. In these cases the load on the right hand exceeds that of the left hand. This tendency has also been found in most MTM analyses in a variety of plants. It can be concluded from this that these sequences of motions have been planned on the assumption that the tolerance for motor load is higher for the right hand than for the left. This is surprising since such a preference for one hand is not included in the instructions for applying PMTS. The tendency to prefer the right hand when planning workplaces represents an ergonomic problem since these workplaces are not always manned exclusively by right-handed persons.

High values of load in the graphics display indicate the risk of local muscle fatigue (Laurig, 1974) even when the effects of the aforementioned asymmetry are disregarded. Differences between the predetermined times and the real cycle time may therefore be explained by work-related fatigue. Similar effects can be expected for sensory motor load. Deterioration in the quality of work may also result. Considerations such as these show how load profiles should be applied in order to avoid a mismatch between the task features and the characteristics of the working persons.

Table 16.4 provides a comparison of load indices (as used in Figure 16.8 to create a load profile) based on different MTM methods, MTM-1 and simplified methods. The most precise description of the load is given by MTM-1. Nevertheless, assuming that movements of the left and right hands are differentiated, the profiles computed on the basis of the simplified methods are similar to the indices computed with MTM-1. However, the indices of the simplified methods must be interpreted carefully since the methods are based upon average times for common combinations of movements.

The aspects discussed above concern individual workplaces. Attention must also be paid to the considerable variance in load at different workplaces even within the same production process. In practice, workplaces and working conditions can be systematically designed to take this into account. Large differences in the load at various workplaces can be equalized as early as the planning stage. However, a given profile of load should still be matched, as far as

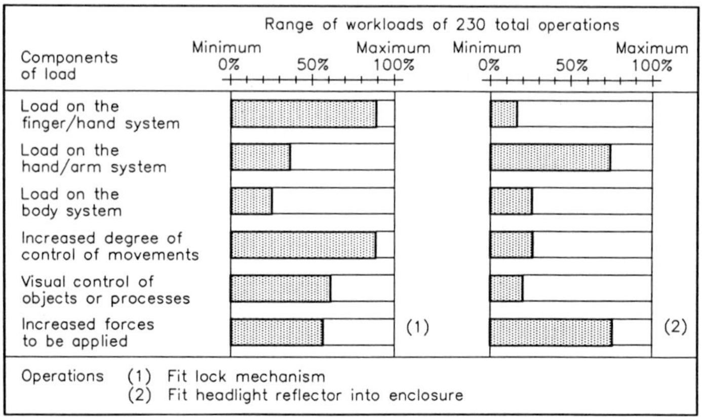

Figure 16.9. Comparative evaluation of two assembly tasks 'Fit lock mechanism' and 'Fit headlight reflector into enclosure' based on the profiles of load shown in Figure 16.8.

possible, with the capabilities of the working persons (Wilcock and Mink, 1982) in order, for example, to equalize large differences in the strain of different persons. A comparison of the results in Figure 16.8 points to the potential for improvements in the design of manual operations.

Comparative evaluation of an individual work task can often be facilitated by classifying the task within a certain group or sample of tasks, for example, all tasks within a particular division in a factory. The result of such an evaluation permits relative statements to be made such as, for example, that the load of a task involving 'increased degree of control' is higher than in 90% of all the tasks included in the group under comparison. Appropriate bar charts (see Figure 16.9) may be used to match work tasks and workers as well as possible.

Apart from their utilization in the matching process, the indices must be examined to establish which ones exceed some limit values and should therefore be judged as critical even at the planning stage of a workplace. When defining limit values for the individual indices of load, interest centres on those workplaces where (according to the experience of the enterprise) the planning quantities calculated by MTM can hardly be achieved by some or the greater part of the workers. The evaluation of such a critical workplace ('Fit lock mechanism' as shown in Figure 16.9) demonstrates the reliability limits of the MTM system for tasks with a considerable sensory motor load. The simplified calculation for this load is the sum of the four indices of load on the left and right finger/hand systems and the increased degree of control. The sum for the workplace in question produced a value of more than 70%. The two indices of load on the finger/hand system alone accounted for nearly 40%. Other analyses with similar values also appear questionable on account of their unrealizable time default statements. It would seem that some limit values have already been exceeded. However, this has still to be proved in detail by the evaluation of further analyses.

Conclusion

A pragmatic ergonomic approach seems most appropriate for solving the problems of work measurement in practice which are outlined in the introduction. There are several advantages to the proposed computer-aided procedure: (i) computer-aided application of the widespread MTM system, (ii) calculation of workload analyses as by-products of predetermining motion times and planning quantities, and (iii) an easily understandable procedure for load analysis.

The proposed procedure therefore provides pragmatic and reproducible indications of the load-related features of the task and, consequently, allows the matching of these features to the abilities of the working persons.

The computer-aided procedure for the assessment of load when applying MTM interactively is well accepted in practice. It is already applied by research institutes and manufacturing plants. A large sample of real data which can act as a basis for evaluating the further application of the system is thereby created.

The application currently concentrates on objectifying the state of different designs of manual operations, workplaces and working conditions. However, it is also being tested for predetermining workload as early as the planning stage. The examples discussed above and the many possibilities for applying the proposed approaches simultaneously indicate the limitations of current knowledge. The 'predetermination of load' has to be validated in a similar way to the 'predetermined motion-times' (Laurig *et al.*, 1983). It could be stated in simple terms that findings about load can only be validated by findings about the person's reaction to the load, i.e. strain. A further challenge for industrial ergonomics will therefore be to discuss a 'system for predetermining strain' instead of a system for predetermining the load itself.

Despite these problems, the proposed approach provides answers which are acceptable because of their reproducibility. In addition, it is based upon a PMTS which is assumed to reflect the real working situation reasonably well. The significance of further scientific results relating to the application of the proposed approaches will have to be discussed along these lines.

References

Antis, W., Honeycutt, J. M. and Koch, E. N., 1969, *Die MTM-Grundbewegungen* (Düsseldorf: H. B. Maynard).

Casey, F. and Wilson, A., 1980, Applied ergonomics results in improved methods, morale and reduced costs. *MTM-Journal*, **7** (3), 2–4.

Gilbreth, F. B. and Gilbreth, L. M., 1921, Process Charts. *Transactions of the ASME*, **43** (1818), 1029–1050.

Heinz, K. and Salwiczek, P., 1983, Analysis and evaluation system using MTM-1 (Anabes). *MTM-Journal*, **10** (1), 2–14.

Kerkhoven, C. L., 1962, Kenelly's Law. *Work Study and Industrial Engineering*, 48–60.

Kühn, F. M., 1984, Belastungsanalyse mit Mikrocomputer. *REFA-Nachrichten*, **4**, 41–46.

Kühn, F. M. and Laurig, W., 1985a, Computer-aided assessment of work load using predetermined motion-time-systems. In: *Ergonomics International '85*, edited by I. D. Brown, R. Goldsmith, K. Coombes and M. A. Sinclair (London: Taylor & Francis), pp. 502–504.

Kühn, F. M. and Laurig, W., 1985b, Computer-aided rating of work load when applying MTM. In: *Toward the Factory of the Future*, edited by H. J. Bullinger and H. J. Warnecke (Berlin: Springer), pp. 892–896.

Laurig, W., 1974, *Beurteilung einseitig dynamischer Muskelarbeit* (Köln: Beuth).

Laurig, W., 1986, Prospective ergonomics. In: *Trends in Ergonomics/Human Factors V*, edited by W. Karwowski (Amsterdam: Elsevier–North Holland), pp. 41–50.

Laurig, W., Gencoglu, M. and Luttmann, A., 1983, Der axiomatische Ansatz als Möglichkeit der Lösung des methodologischen Problems der ergonomischen Belastungsprognose. *Zentralblatt für Arbeitsmedizin, Arbeitsschutz, Prophylaxe und Ergonomie*, **33** (4), 130–136.

Laurig, W., Kühn, F. M. and Schoo, K. C., 1985, An approach to assessing motor workload in assembly tasks by the use of predetermined-motion-time-systems. *Applied Ergonomics*, **16** (2), 119–125.

Maynard, H. B., Stegemerten, G. J. and Schwab, J. L., 1948, *Methods-Time Measurement* (New York: McGraw–Hill).

Minter, A. L., 1977, The measure of activity of manual work. Ph.D. Thesis for the University of London.

Quick, J. H., Duncan, J. H. and Malcolm, J. A., 1965, *Das Work-Factor-Buch* (München: Hanser).

Rohmert, W. and Laurig, W., 1971, Work Measurement. *International Journal of Production Research*, **9** (1), 157–168.

Salwiczek, P., 1982, Rechnergestützte Planung und Gestaltung manueller Arbeitsmethoden auf der Basis eines Systems vorbestimmter Zeiten. *Fortschrittsberichte VDI-Z*, **16**, 14.

Schwab, J. L., 1956, Methods-time measurement. In: *Industrial Engineering Handbook*, edited by H. B. Maynard (New York: McGraw–Hill).

Wilcock, R. and Mink, J. A., 1982, The International MTM Directorate rehabilitation project 'A study of man and work'. *MTM-Journal*, **9** (2), 3–11.

Chapter 17

Easy Task Analysis: a prototype expert system for computerized work measurement

DAVID J. RAMCHARAN and LOUIS A. MARTIN-VEGA

Department of Industrial and Systems Engineering
University of Florida
Gainesville, Florida 32611, U.S.A.

Abstract. The objective of this paper is to apply expert systems technology to computerized work measurement. First a brief overview of Computerized MOST is presented along with a discussion of the application of expert systems technology to the MOST analysis process. This is followed by a description of a prototype expert system that implements this technology. Finally, some examples based on the prototype are developed and discussed.

Keywords: Computerized work measurement; Expert system; Computerized MOST.

Introduction

Probably the first technique ever used to measure work was intuitive estimation based on experience (Barnes, 1980). This extremely subjective method was improved upon through direct observation time studies which added the concept of performance rating to stopwatch readings in an attempt to gauge better the time taken by 'an average-skilled worker, working at a normal pace, to complete a clearly defined task'. Even though this approach is still widely used today, the subjectivity inherent in performance rating led eventually to the development of predetermined motion time systems.

The concept of predetermined motion time systems had its origin in the pioneering work of the Gilbreths and Taylor (Barnes, 1980). The Gilbreths' classification of manual basic motions into 'therbligs' coupled with Taylor's efforts in time study laid the foundation for Motion Time Analysis (MTA), a predetermined motion time system introduced by Segur in 1920 (Brisley and Eady, 1982). Segur used hundreds of empirical observations to estimate the normal times of small work elements similar to the manual basic motions. His approach then consisted of breaking down any task into these small work elements and using the tabulated estimates to reconstruct the normal time of the original task.

The evolution of these systems was basically industry specific throughout the years 1920–40. In 1948, H. B. Maynard Co. introduced Methods Time Measurement (MTM), a generic system designed for the study of highly repetitive short cycle tasks (Barnes, 1980). MTM remained an industry standard through the early 1970s with subsequent versions added to either reduce the application time (MTM-2, MTM-3), or adapt the concept to specific work environments such as machine shops (MTM-V). The concept has also been extended to clerical and indirect labour environments (MTM-C) (Karger and Hancock, 1982).

Systems such as MTM and Modapts (Brisley, 1978) consist of tables with predetermined time values for different body moves, allowances for body positions during the move, and factors that take the difficulty of controlling an object into consideration. While these systems can provide very accurate and consistent normal times, their use requires an exact description of the workplace and a very detailed listing of all the motions involved in the task. Moreover, the analyst must also translate the detailed motion descriptions into specified formats, using specialized phrases, before the normal times can be determined. In order to try to alleviate some of these problems, other predetermined time systems using larger data blocks have been designed so that tasks will not have to be described in such detail.

Computerized MOST

During 1967–72, the Swedish Division of H.B. Maynard, Co. developed Maynard Operating Sequence Technique (MOST) which was derived from the structure of MTM. This manual predetermined motion time system is applicable to tasks of any cycle length and repetitiveness as long as there are no variations in the method sequence from one cycle to another. Variations in the human performance are compensated through a balancing time which is a calculated part of the MOST system design (Zandin, 1980). MOST systems were first introduced in the United States in 1974 and have been applied to a broad spectrum of tasks from fabrication and assembly to clerical activities.

The primary work units of MOST are fundamental activities (collections of basic motions) which are used to develop a small number of sequence models that cover practically all aspects of manual activity. The most commonly used sequence models are the following:

1. General Move Sequence: this sequence model describes the spatial movement of an object freely through the air. This can account for as much as 35% of the work of a machine operator and even more of an assembly operator.
2. Controlled Move Sequence: this sequence model describes work associated with the movement of an object when it remains in contact with a surface or is attached to another object during the movement.
3. Tool Use Sequence: this sequence model is used for operations that make use of conventional hand tools such as wrenches, gauges and writing tools.

Further details on these and other MOST sequence models are given in Zandin (1980).

Training required

The computerized version of MOST was first introduced in 1987. A principal feature of Computerized MOST the development of an activity keyword structure that allows the sequence models to be constructed by describing a task in a sentence format. In order to use Computerized MOST (Maynard Computer Services, 1987), the user must be trained to create a Work Description file and a Methods Description file and to enter the data correctly into the program. Even though the calculations and much of the bookkeeping of the manual version have been automated, the analyst still has to translate the steps of the task into the given format before they can be entered into the system. Moreover, development of the Methods Description or MOST Analysis is very difficult without knowledge of the underlying parameters that make up each of the previously mentioned sequence models.

Input required

Creation of a Work description file requires that the user first select that option through a series of menus. The system will then prompt him for information about all the following:

Name of Workplace(s)
Coordinates or Location of Workplace(s)
Dimension of Workplace(s)
Name and Location of Operator(s)
Name and Location of Carrier(s)
Tools used to carry out the task
Objects that are moved
Equipment used to carry out the task
Distances (in steps) between every possible pair of workplaces

The MOST Analysis or Methods Description consists of a list of steps which require the use of certain Location words, Activity Keywords and Supplementary words. The keywords and format for the General Move Method Step as well as the 14-step procedure for Method Step Construction for the General Move Format are detailed in Maynard Computer Services (1987). The procedure involves filling the 14 positions of the step format with action keywords and those location words that were defined in the Work Area Description. Some examples of completed General Move steps are (Maynard Computer Services, 1987):

1. TURN DEPLETED ELECTRODE AT SELF
2. HOLD + THROW DEPLETED ELECTRODE FROM FITTER TO BUCKET
3. PLACE NEW ELECTRODE FROM FITTER TO ELECTRODE HOLD ON FITTER

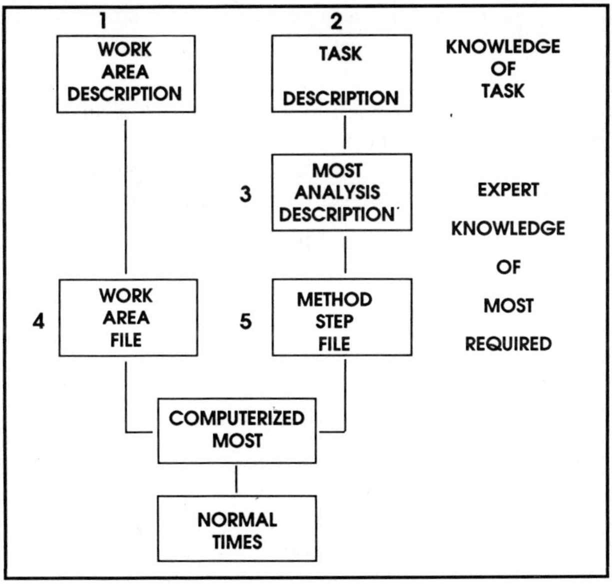

Figure 17.1 Steps in a MOST analysis.

Analysis

These Method Steps are then used by the system along with the data from the Work Area description file to generate normal times for the task under study. A summary of the MOST analysis process is shown in Figure 17.1.

Steps 1 and 2 can easily be carried out by someone who is familiar with the task being analyzed. Steps 3, 4, and 5, however, require a thorough and working knowledge of MOST as a prerequisite for using the computer system (Maynard Computer Services, 1987). The bottom line is that in order for someone to use Computerized MOST, that person either has to be specifically trained in MOST or must be able to use expert knowledge from some readily available source.

Easy Task Analysis

In this section we will describe the development of a prototype expert system that addresses many of the aforementioned problems involved in applying Computerized MOST. The basic idea is to require as little input and data manipulation by the user as possible. As illustrated in Figure 17.2, the trend towards increased automation in work measurement systems has generally been accompanied by an increase in the training required to use the systems effectively. The use of expert system technology provides a means of reversing this trend since the objective is to reduce the training required by the user and simplify the analysis process.

An expert system is a computer program specifically designed to represent expertise in a particular domain (Texas Instruments Learning Center, 1986). Expert systems act as intelligent assistants to human experts as well as assisting people who otherwise might not have access to expertise. Some characteristics of an expert system are (Texas Instruments Learning Center, 1986) (i) it should be useful, (ii) it should be usable, (iii) it should be able to learn new knowledge, and (iv) its knowledge should be easily modifiable.

The prototype described in this paper, which we call Easy Task Analysis (ETA), incorporates these characteristics through the extensive use of menus and instructional prompts as well as a data base that grows as more information is provided. Another goal

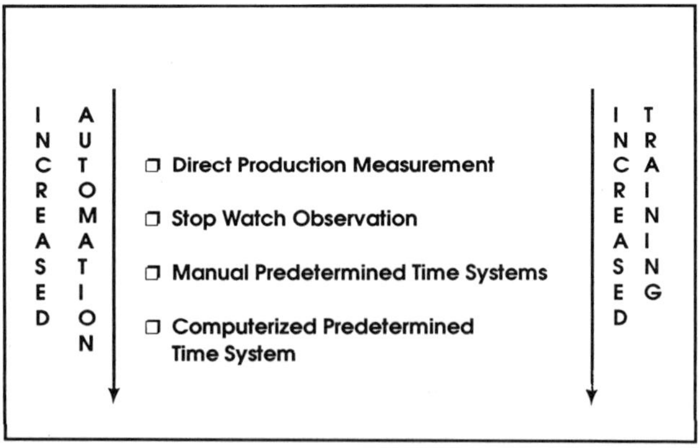

Figure 17.2 Trends in work measurement.

pursued in the development of ETA was to keep user set-up time to a minimum. This is accomplished by limiting the input requirements to method steps.

Information related to the Work Area Description is generated through a prompt-driven question and answer session with the user. Currently, the prototype only develops normal time estimates for tasks that can be described through General Moves or movement of objects through free space. The prototype is written in Turbo Prolog and can be run on any IBM compatible machine with 384K of memory.

Training required

By having the expert knowledge of MOST built into the ETA, the analyst is only required to have knowledge of the task being analyzed. The program guides the analyst through the session with prompts and messages for any required information.

Input required

Input into the system consists of sentences in English that describe the steps or method required to carry out the task being analyzed. The constraints on the input are as follows:

1. Each sentence must contain a verb previously defined in the system. The prototype currently has six verbs in its vocabulary, but users may define verbs specific to their application if desired or deemed necessary.
2. An object must be specified directly after the verb.
3. The preposition 'from' must immediately precede any word specifying the location* of an object.
4. Any word specifying the destination* of an object must appear after the word specifying the location and must be immediately preceded by the preposition 'to'.

Other features that have been included to allow the use of more free-form sentences include:

1. The pronoun 'it' can be used in a sentence to refer to the most recently mentioned object.
2. The conjunction 'and' can be used to string two or more sentences together.
3. Except for the positions specified in the constraints, supplementary words can be added to make the sentence more readable.

Some examples of valid inputs are:

GET BOLT
GET BOLT FROM BIN
PUT BOLT TO FIXTURE
GET BOLT FROM BIN AND PUT IT TO FIXTURE

From the previous examples, it can be seen that a method step input can be as simple as just specifying a verb followed by the object to be acted upon.

---

* The specification of a location and a destination are optional. If they are not provided in the method sentences, the system will prompt the user for the necessary information.

Analysis

After the sentences are entered into the system, ETA scans the first sentence looking for the keywords specified in the constraints. The system then checks through its data base to retrieve any information it might have associated with the size, location and/or destination of the object(s) specified in this sentence. Any information about body positioning and difficulty involved in the move is also retrieved if available. If no information is found, it then prompts the user for that information and adds it to its data base.

The expert knowledge about MOST has been coded into ETA as a set of logic rules. Once the analysis of a sentence is completed, the system matches the information generated with the appropriate rule to arrive at the correct TMU value or normal time for this step. This is continued sentence by sentence until the complete method description has been analyzed by ETA and a normal time for the complete task has been determined. The concept is also illustrated in Figure 17.3.

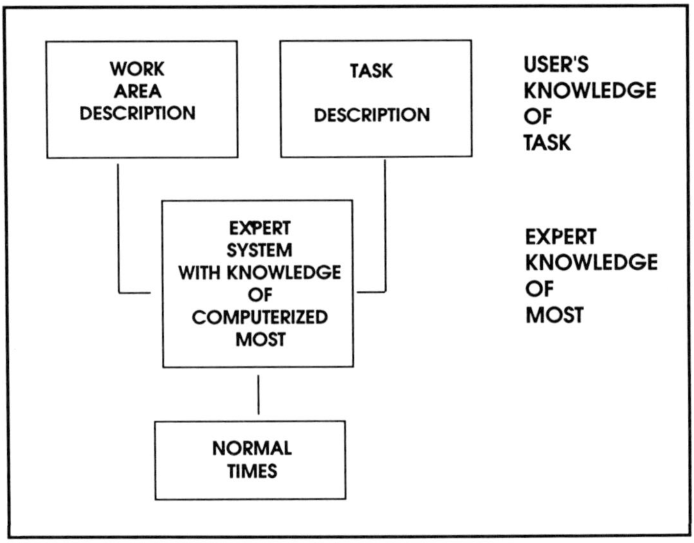

Figure 17.3 Steps in Easy Task Analysis (ETA).

Example 1

Consider the following task. A worker is using a hydraulic press to bend pieces of metal into flanges. He places the piece of metal into the press, activates the press, and after the press is done, he removes the pressed metal (the flange) and puts it on the stack of finished flanges. Since ETA currently only analyzes General Moves, we will examine the last element of that task's cycle.

Let us assume that the user input for that element consisted of the following task description:

GET FLANGE FROM PRESS AND PUT IT TO STACK

ETA would then break off the first sentence from this input, 'GET FLANGE FROM

PRESS', and attempt to build an estimate of its normal time. ETA would recognize GET as a verb, FLANGE as an object, and PRESS as a location. Assuming that it has no knowledge of the task in its data base, it would then prompt the user as follows:

Is flange:

Light
Heavy

To get flange from press:

0. No body movement
3. Bend 50% of the time
6. Bend all of the time
10. Sit or stand
16. Use step or door

How far is the flange at press from operator at self:

0. Less than 2 inches
1. Within reach
3. One or two steps
6. Three or four steps
10. Five, six or seven steps
16. Eight, nine or ten steps.

Assuming the responses were: light, bend 50% of the time, one or two steps; then ETA would respond with a normal time estimate of 70 TMUs (Zandin, 1980).

The second part of the input, 'AND PUT IT TO STACK', is analyzed in a similar fashion. Since ETA has already learned that FLANGE (i.e. IT) is light, it does not need to ask the user for this information. It does not, however, have any knowledge of 'PUTTING FLANGE TO STACK' and so it will prompt the user for this information as follows:

In putting flange to stack, which of the following is necessary:

0. Hold or toss it
1. Lay aside or loose fit
3. Small adjustments or light pressure
6. Care, heavy pressure or blind moves

To put flange to stack:

0. No body movement
3. Bend 50% of the time
6. Bend all of the time
10. Sit or stand
16. Use step or door

How far is stack from operator at press:

0. Less than 2 inches
1. Within reach
3. One or two steps
6. Three or four steps
10. Five, six or seven steps
16. Eight, nine or ten steps

If we assume that the answers were: small adjustments, no body movement and one or two steps; then the normal time that would be presented to the user for the second sentence would be 60 TMUs and the time for the total task would be 130 TMUs.

Finally, let us assume that at this point the user chooses to describe another task:

GET FLANGE FROM PRESS AND PUT IT TO STACK2

Since ETA already has all the information it needs concerning 'GET FLANGE FROM PRESS', it will just present the normal time for this part of the input and go on to analyze the second sentence 'AND PUT IT TO STACK2'. Since STACK2 is a different destination from STACK, ETA will ask for information concerning this destination and then complete the normal time, after presenting the normal time for getting the flange, as follows:

["get","flange","from","press"] 70 TMUs

In putting flange to stack2 which of the following is necessary:

0. Hold or toss
1. Lay aside or loose fit
3. Small adjustments or light pressure
6. Care, heavy pressure, or blind moves

To put flange to stack2:

0. No body movement
3. Bend 50% of the time
6. Bend all of the time
10. Sit or stand
16. Use step or door

How far is stack2 from the operator at press:

0. Less than two inches
1. Within reach
3. One or two steps
6. Three or four steps
10. Five, six, or seven steps
16. Eight, nine, or ten steps

Assuming the responses were: lay aside, no body movement and within reach; the normal time for this new sentence would be 20 TMUs.

The following two examples provide more detailed information from sample sessions that were carried out using ETA. Each session is made up of two input sentences and demonstrates how ETA questions the user for required information during an analysis, and how it learns information about the objects and the workplace as the session progresses.

Example 2

In this example, ETA is used to analyze general moves involved in an assembly process. The first input involves getting a part from somewhere on the floor and putting it on the bench. This is entered as 'Get part from floor and put it to bench'. ETA first displays some general information and prompts for the task description. It then proceeds to analyze the task.

As seen in the following, ETA requests information about the object, body motion, distances, and difficulty involved in carrying out this move. It uses the responses of those questions (in italics) to come up with 150 TMUs for the getting part, and 70 TMUs for the putting part.

Enter your task description in simple English.
The verbs recognized are:

To Gain control	: get, pick_up
To Release	: put, toss
Gain control then Release	: carry, move

Sentences can be joined by the conjunction, and. The pronoun, it, can be used.

Task: *get part from floor and put it to bench*

[''get'',''part'',''from'',''floor'',''and'',''put'',''it'',''to'',''bench'']

Is part light or heavy? *heavy*

Motion of operator's body
To get part from floor:

0. No body movement
3. Bend 50% of the time
6. Bend all of the time
10. Sit or stand
16. Use step or door

CHOOSE NUMBER: *6*

How far is the part at floor from operator at self:

0. Less than 2 inches
1. Within reach
3. One or two steps
6. Three or four steps
10. Five, six, or seven steps
16. Eight, nine, or ten steps

CHOOSE NUMBER: *6*

A6 B6 G3 A0 B0 P0 A0
TMU = 150

In putting part to bench, which of the following is necessary:

0. Hold or toss it
1. Lay aside or loose fit
3. Small adjustments or light pressure
6. Care, heavy pressure or blind moves

CHOOSE NUMBER: *1*

Motion of operator's body
To put part to bench:

0. No body movement
3. Bend 50% of the time
6. Bend all of the time
10. Sit or stand
16. Use step or door

CHOOSE NUMBER: *0*

How far is bench from operator at floor:

0. Less than 2 inches
1. Within reach
3. One or two steps
6. Three or four steps
10. Five, six, or seven steps
16. Eight, nine, or ten steps
CHOOSE NUMBER: *6*

A0 B0 G0 A6 B0 P1 A0
TMU = 70

The second input describes the operator moving a finished part from the work table to a storage pan: 'Move part1 from table to pan'. ETA questions the user for similar information as for the first input. One difference is that it remembers that the operator was at the bench at the end of the previous move, and so requests the distance from the bench to the table. After analyzing this input, ETA returns a value of 200 TMUs for this move.

Do you want to continue (Y/N): *y*
Enter your task description in simple English.
The verbs recognized are:

To Gain control	: get, pick_up
To Release	: put, toss
Gain control then Release	: carry, move

Sentences can be joined by the conjunction, and. The pronoun, it, can be used.

Task: *move part1 from table to pan*
[''move'',''part1'',''from'',''table'',''to'',''pan'']

How far is the part1 at table from operator at bench:

0. Less than 2 inches
1. Within reach
3. One or two steps
6. Three or four steps
10. Five, six, or seven steps
16. Eight, nine, or ten steps
CHOOSE NUMBER: *3*

Motion of operator's body
To move part1 from table:

0. No body movement
3. Bend 50% of the time
6. Bend all of the time
10. Sit or stand
16. Use step or door
CHOOSE NUMBER: *0*

Is part1 light or heavy? *heavy*
How far is the pan from operator at table:

0. Less than 2 inches
1. Within reach
3. One or two steps
6. Three or four steps
10. Five, six, or seven steps
16. Eight, nine or ten steps
CHOOSE NUMBER: *10*

Motion of operator's body
To move part1 to pan:

0. No body movement
3. Bend 50% of the time
6. Bend all of the time
10. Sit or stand
16. Use step or door
CHOOSE NUMBER: *3*

In moving part1 to pan, which of the following is necessary:

0. Hold or toss it
1. Lay aside or loose fit
3. Small adjustments or light pressure
6. Care, heavy pressure or blind moves
CHOOSE NUMBER: *1*

A3 B0 G3 A10 B3 P1 A0
TMU = 200

Example 3

During this example, a general move in a clerical environment is analyzed. In this move, the clerk picks up a pencil and puts it into a drawer of the desk at which he is sitting. The input for this move is 'Pick_up pencil and put it to drawer'.

ETA accepts this input and prompts the user for information about the object, distances, body motion, and difficulty involved in performing this move. It then compiles that information and presents 20 TMUs for the picking up and 20 TMUs for the putting down of the pencil.

Enter your task description in simple English.
The verbs recognized are:

To Gain control	: get, pick_up
To Release	: put, toss
Gain control then Release	: carry, move

Sentences can be joined by the conjunction, and. The pronoun, it, can be used.

Task: *pick_up pencil and put it to drawer*
[''pick_up'',''pencil'',''and'',''put'',''it'',''to'',''drawer'']

Where does operator pick_up pencil from? *desk*
Is pencil light or heavy? *light*

Motion of operator's body
To pick_up pencil from desk:

0. No body movement
3. Bend 50% of the time
6. Bend all of the time
10. Sit or stand
16. Use step or door

CHOOSE NUMBER: *0*

How far is the pencil at desk from operator at self:

0. Less than 2 inches
1. Within reach
3. One or two steps
6. Three or four steps
10. Five, six, or seven steps
16. Eight, nine, or ten steps

CHOOSE NUMBER: *1*

A1 B0 G1 A0 B0 P0 A0
TMU = 20
In putting pencil to drawer, which of the following is necessary:

0. Hold or toss it
1. Lay aside or loose fit
3. Small adjustments or light pressure
6. Care, heavy pressure or blind moves

CHOOSE NUMBER: *1*

Motion of operator's body
To put pencil to drawer:

0. No body movement
3. Bend 50% of the time
6. Bend all of the time
10. Sit or stand
16. Use step or door

CHOOSE NUMBER: *0*

How far is drawer from operator at desk:

0. Less than 2 inches
1. Within reach
3. One or two steps
6. Three or four steps
10. Five, six, or seven steps
16. Eight, nine, or ten steps

CHOOSE NUMBER: *1*

A0 B0 G0 A1 B0 P1 A0
TMU = 20

In the second part of this example, the user inputs 'Move pencil', in order to check the previous result. Since the user did not specify the location or destination of this move, ETA prompts for that information. Since a different verb is used (move instead of pick_up and put), ETA requests information about the body motion and difficulty involved in this movement. Since the user had entered the distances to the pencil, and from the desk to the drawer, ETA already knows that information and simply retrieves it from its data base. Thus, after this analysis, it returns the same value of 40 TMUs as it did previously.

Do you want to continue (Y/N): *y*
Enter your task description in simple English.
The verbs recognized are:

To Gain control : get, pick_up
To Release : put, toss
Gain control then Release : carry, move

Sentences can be joined by the conjunction, and. The pronoun, it, can be used.

Task: *move pencil*
[''move'',''pencil'']

Where does operator move pencil from? *desk*
Where does operator move pencil from? *drawer*
Motion of operator's body
To move pencil from desk:

0. No body movement
3. Bend 50% of the time
6. Bend all of the time
10. Sit or stand
16. Use step or door
CHOOSE NUMBER: *0*

Motion of operator's body
To move pencil to drawer:

0. No body movement
3. Bend 50% of the time
6. Bend all of the time
10. Sit or stand
16. Use step or door
CHOOSE NUMBER: *0*

In moving pencil to drawer, which of the following is necessary:

0. Hold or toss it
1. Lay aside or loose fit
3. Small adjustments or light pressure
6. Care, heavy pressure or blind moves
CHOOSE NUMBER: *1*

A1 B0 G1 A1 B0 P1 A0
TMU = 40

Summary

The objective of this paper has been to describe a prototype expert system (ETA or Easy Task Analysis) that implements expert knowledge of MOST to generate normal times for tasks that are described to the system in simple English. The main differences between this concept and version 2.4 of Computerized MOST are:

1. Training required. Prior to using the computerized version of MOST, an analyst is required to have a thorough, working knowledge of MOST. In ETA, expert knowledge of MOST is built into the system so that the user is only required to have knowledge of the task being analyzed.
2. Input required. Implementation of Computerized MOST requires the analyst to create two files: a Work Area file and a Methods Description file. Creation of the Work Area file requires entering a work area description into the system through a very structured set of system prompts. Each sentence in the Method Description file requires implementation of a 14-step procedure for its construction in adherence with a specified format. For all practical purposes, it is very difficult to construct a methods description sentence without prior knowledge of MOST. If there are any mistakes in either of the files, no analysis is carried out until all mistakes have been corrected. The input required by ETA consists of a simple sentence in English for each step in the task. The sentence structure has only four constraints and information concerning the work area description is obtained from the user through prompts as needed.
3. Analysis. Computerized MOST requires that a Work Area Description file be created before it begins its analysis. This file contains information about all of the workplaces, objects, and their relationships with each other. Therefore, to process each Method Step, Computerized MOST has to search through all of the information in this file, some of which may never be used, in order to get the information needed. Alternatively, ETA builds a data base with information about workplaces and objects one step at a time. When it encounters a method step which needs new information, ETA prompts the analyst for that information and adds it to the data base. If the method step needs information that has already been submitted to ETA, that information will be accessed from the data base. Since ETA builds this data base throughout the analysis of the method steps, the data base will contain only necessary information and the data searched will be, on the average, much smaller than that in Computerized MOST. Finally, it should be noted that the prototype moves contrary to the historical trend shown in Figure 17.2, since in this case increased automation of the work measurement process results in a requirement of less training of the user.

References

Barnes, R., 1980, *Motion and Time Study: Design and Measurement of Work* (New York: John Wiley).

Brisley, C. L., 1978, Comparison of Predetermined Time Systems (PTS), *Proceedings of AIIE Conference*.

Brisley, C. L. and Eady, K., 1982, Predetermined Motion Time Systems. In: *Handbook of Industrial Engineering*, edited by G. Salvendy (New York: John Wiley), pp. 4.5.1.–4.5.31.

Gadre, M. and Martin-Vega, L. A., 1986, *MOST-like Sequence Models for Robotic Work Measurement*, Research Report No. 86–5, ISE Department, University of Florida.
Karger, D. W. and Hancock, W. M., 1982, *Advanced Work Measurement* (New York: Industrial Press).
Maynard Computer Services, 1987, *MOST Computer System, User's Manual*.
Smith, G. L., 1978, *Work Measurement: A Systems Approach* (Grid Publishing).
Texas Instruments Learning Center, 1986, *Understanding Artificial Intelligence*.
Zandin, K. B., 1980, *MOST Work Measurement Systems* (New York: M. Dekker).

PART V
POSTURE AND MANUAL HANDLING OF LOADS

Chapter 18
Computer-aided posture analysis of the trunk, neck, shoulders and lower extremities

W. MONROE KEYSERLING
Department of Industrial and Operations Engineering,
The University of Michigan,
1205 Beal Avenue,
Ann Arbor, Michigan 48109, U.S.A.

Abstract. A computer-aided system for analysing and describing the position of the trunk, neck, shoulders and lower extremities has been developed to evaluate postural activity during work. The system uses a videotape to create a permanent record of work activities and a personal computer to perform the clerical and time-keeping tasks associated with posture analysis. This system can be used to evaluate the postural demands of a specific job or to generate exposure data for epidemiological studies of postural stress.

Keywords: Posture; Ergonomics; Computer-aided analysis; Trunk; Shoulders; Neck; Lower extremities.

Introduction

Working posture is determined by the interaction of many factors in the work place. Features of workstation layout (e.g. the height, orientation, and location of benches, conveyors, equipment, and storage containers) determine how a worker must position his/her body when performing a task (Grandjean, 1980). The design of hand tools, particularly the location and orientation of handles, interact with the workstation layout to determine the position of the shoulders and the upper extremities (Armstrong, 1986). Visual demands interact with workstation layout to determine the posture of the neck and trunk (Grandjean, 1980; Chaffin and Andersson, 1984). Individual work methods such as lifting technique determine the extent to which a worker bends and/or twists the trunk during manual materials handling activities (NIOSH, 1981). The anthropometric characteristics of a worker interact with all of the above factors to determine the specific postures used to perform a job (Van Cott and Kincade, 1972). The association between awkward work posture and the development of fatigue and/or musculoskeletal disorders has been widely discussed in recent ergonomic literature (Van Wely, 1970; Magora, 1972; Corlett and Bishop, 1976; Karhu *et al.*, 1977; Grandjean and Hunting, 1977; Tichauer, 1978; Kilbom *et al.*, 1986; Keyserling *et al.*, 1988).

A positive relationship between awkward trunk postures and the development of fatigue and musculoskeletal disorders has been demonstrated in several recent investigations. Laboratory studies have shown that trunk flexion, lateral bending, or twisting increases mechanical stresses on the spinal muscles and intervertebral discs (Andersson *et al.*, 1977;

Schultz *et al.*, 1982) and that prolonged trunk flexion causes extreme levels of muscle fatigue (Chaffin, 1973). Epidemiologic studies have shown that sustained static postures of the trunk such as prolonged sitting or forward bending result in increased risk of low back pain (Magora, 1972; Kelsey and Hochberg, 1988). Periodic or repetitive bending and/or twisting of the trunk have also been cited as factors in the development of back pain (Keyserling *et al.*, 1988).

Laboratory studies of shoulder posture have shown that prolonged elevation of the arms (shoulder flexion or abduction) causes extreme levels of muscle fatigue, and in some cases, acute tendinitis (Chaffin, 1973; Hagberg, 1982). The relationship between shoulder elevation and increased risk of tendinitis has been demonstrated in a cross-sectional field study (Hagberg, 1984). Shoulder elevation and extension have been associated with increased risk of a variety of cervicobrachial disorders including thoracic outlet syndrome (Feldman *et al.*, 1983; Armstrong, 1986; Jonsson *et al.*, 1988).

Most of the literature on neck posture has emphasized discomfort and disorders related to the head inclination angle. Kumar and Scaife (1979) showed biomechanically that small neck flexion angles cause significant muscle contractions. Subjective discomfort rating methods have demonstrated a relationship between forward neck flexion and localized pain (Hunting *et al.*, 1980). Epidemiological studies have found a relationship between awkward neck posture and cervicobrachial disorders (Jonsson *et al.*, 1988).

Outside of documented cases of lower extremity disorders described in the medical literature, there have been relatively few studies which relate lower extremity postures and activities to pain, discomfort or injuries. Biomechanical models have been developed to describe joint moments at the knee and the ankle during walking. These are primarily descriptive and have not been used to relate the posture of the lower extremity joints to injuries and illnesses. Grandjean and Hunting (1977) reviewed various problems in standing and sitting posture and found relationships among static loads, awkward postures and pain. Standing produced pains in the feet and legs; excessive seat height produced pains in the knees, lower legs, and feet; while inadequate seat height produced pains in shoulders and neck.

The measurement of work posture has two important occupational health applications. One application is to obtain exposure data for use in epidemiologic studies of posture-related injury. The second application is to evaluate a particular job to quantify postural stress and identify specific causes of awkward posture. Control programmes can then be developed to either eliminate or reduce the stress to an acceptable level. In order to study the relationship between work activities and work posture (and ultimately to improve the workplace design), a system for observing and recording posture on a continuous basis is needed. To be practical, such a system should be reasonably easy to learn and use, reliable and easily interfaced with a computer for data reduction, storage, and analysis (Keyserling, 1986a,b).

Several systems have been recently developed for measuring work posture. The 'Posturegram' system, developed by Priel (1974), described the position of 16 links relative to three reference planes. This system required the analyst to estimate 48 limb angles; therefore, several minutes were required to document a single posture. Because of this time requirement, the system was best suited for jobs with few and infrequent posture changes.

A more time-efficient system called the Ovako Working Posture Analysis System (OWAS) was developed in the late 1970s for workplace posture analysis (Karhu *et al.*, 1977). This system utilized pre-defined standard postures and a three-digit code to classify the position of the trunk (four choices), shoulders (three choices), and legs (seven choices). While the OWAS method was extremely time-efficient (i.e. only a few seconds were required to observe, code and record a posture), it could not be used to develop a continuous record of posture changes on a highly dynamic job because several seconds were required to observe and record a single work posture.

Corlett *et al.* (1979) developed a system called Posture Targeting. With this system, the analyst observed the worker and recorded the position of the head, trunk, upper arms, lower arms, upper legs and lower legs by making 10 marks on a chart. Because the procedure required about 30 s to complete, it was best suited for jobs with relatively static work postures.

A posture analysis system called VIRA was developed during the early 1980s to allow continuous recording of neck and shoulder posture during dynamic work (Kilbom *et al.*, 1986). A videotape of the job, recorded at the workplace, was played back in the laboratory in order to classify the position of the neck and shoulders into predetermined posture categories. A personal computer was used to assist the analyst in measuring and recording the time spent in each posture. VIRA produced an accurate description of the time spent in each posture, as well as the number and nature of posture changes.

In order to identify specific causes of postural stress (e.g. poor workstation layout, improper tool selection, inappropriate work methods), it is necessary to measure work activities and working posture on a common, continuous time scale. A computer-aided system with this capability was developed at the University of Michigan to analyse the working posture of the trunk and shoulders on repetitive, assembly line jobs (Keyserling, 1986a,b). This system was later enhanced to analyse the posture of the neck and lower extremities (Leonard and Keyserling, 1989). Because the computer simplifies data recording activities and reduces analysis time, this system can be used to analyse postural stresses quickly in the workplace.

This paper defines a standard scheme for classifying working postures of the trunk, neck, shoulders and lower extremities and describes a computer-aided system for recording the postural activity of these body segments. Problems associated with using the system are discussed and recommendations are presented for improving system performance.

Description and methods

Similarity to time study

The new system for observing and recording work posture is developed from the time study methods used by industrial engineers. Time study and posture analysis are conceptually similar; the goal of the time study is to measure the amount of time required to complete a work element while the goal of posture analysis is to measure the amount of time that the worker spends in certain postures (Keyserling, 1986a; Niebel, 1988). To make posture analysis feasible, it is first necessary to define standard postures. Given a menu of standard postures, time study methods can be used to measure the time spent in each posture.

Posture classification

The posture classification system presented in Figure 18.1 defines a menu of standard postures for the trunk, neck, shoulders and lower extremities. This menu was developed by observing videotapes of a variety of work activities in several manufacturing facilities to establish a taxonomy of common work postures and by reviewing the literature to identify specific work postures associated with the development of fatigue and/or musculoskeletal disorders (Keyserling, 1986a; Leonard and Keyserling, 1989). For each joint of interest, the neutral posture was considered to be a position which placed relatively low levels of stress on the musculoskeletal system and which could be maintained for prolonged periods without unusual fatigue or discomfort. Non-neutral postures were associated with increased levels of muscular activity (and associated fatigue) and/or strain on connective tissues. Finally, posture categories were selected to facilitate the need for the analyst to make rapid classification decisions while viewing a videotape. (See the Equipment and Procedures section for additional discussion of this issue.)

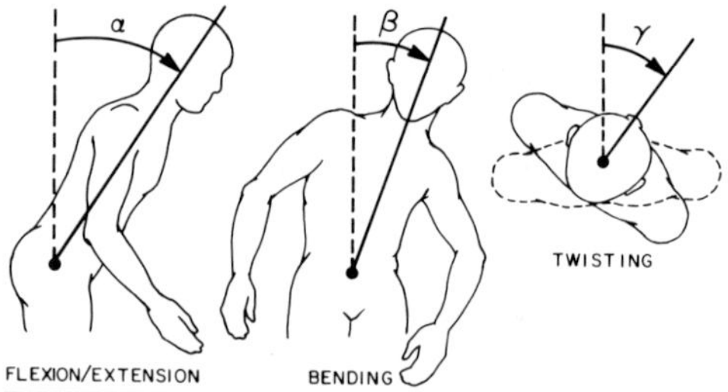

STANDARD TRUNK POSTURES	
1. Stand - Extension (α< -20 degrees)	6. Lie - On Back or Side
2. Stand - Neutral	7. Sit - Neutral
3. Stand - Mild Flexion (α> 20 degrees)	8. Sit - Flexion
4. Stand - Severe Flexion (α > 45 degrees)	9. Sit - Bend/Twist
5. Bend/Twist (β or γ> 20 degrees)	

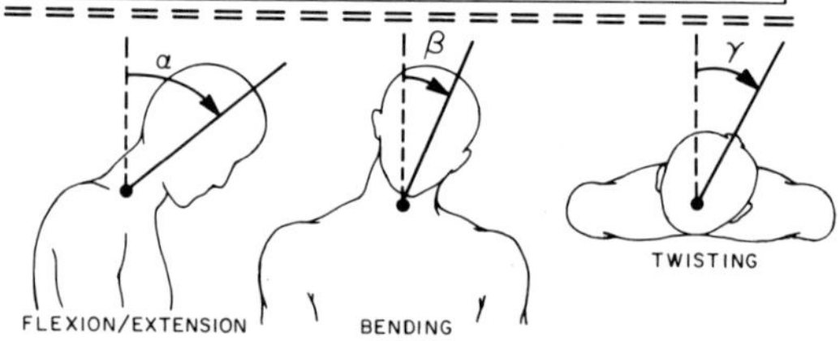

STANDARD NECK POSTURES
1. Extension (α< -20 degrees)
2. Neutral
3. Mild Flexion (α> 20 degrees)
4. Severe Flexion (α > 45 degrees)
5. Bend/Twist (β or γ> 20 degrees)

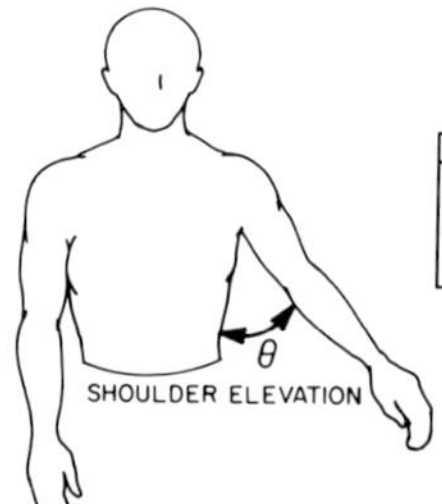

STANDARD SHOULDER POSTURES
1. Neutral (θ < 45 degrees)
2. Mild Elevation (θ > 45 degrees)
3. Severe Elevation (θ > 45 degrees)

STANDARD LOWER EXTREMITY ACTIVITIES
1. Walk
2. Stand
3. Lean (body weight supported by external object)
4. Squat (included knee angle 90-180 degrees)
5. Deep Squat (included knee angle less than 90 degrees)
6. Kneel (one or both knees touch the floor)
7. Sit

Figure 18.1 Standard posture categories for the trunk, neck, shoulders and lower extremities.

For a standing worker, the trunk is classified as deviated from a neutral posture (vertical with no axial twisting) if it is:

Extended (bent backward more than 20°),
Mildly flexed (bent forward between 20 and 45°),
Severely flexed (bent forward more than 45°),
Bent sideways or twisted more than 20° from the neutral position.

Work that requires prolonged or highly repetitive use of these non-neutral postures may result in fatigue or musculoskeletal disorders (Magora, 1972; Habes, 1984; Keyserling *et al.*, 1988). A similar rationale was used to classify trunk posture for a seated worker (Kelsey *et al.*, 1988).

The neck posture is defined relative to the position of the trunk. (If the head and trunk move as a unit, no posture change occurs at the neck.) The neck is classified as deviated from neutral if it is:

Extended (bent backward more than 20°),
Mildly flexed (bent forward between 20 and 45°),
Severely flexed (bent forward more than 90°),
Bent sideways or twisted more than 20°.

Prolonged or highly repetitive use of these postures may result in fatigue or musculoskeletal disorders (Hunting *et al.*, 1980; Chaffin and Andersson, 1984; Jonsson *et al.*, 1988).

Shoulder posture is measured as the included angle between the trunk and the humerus. For either a standing or seated worker, the shoulder is classified as:

Neutral (included angle less than 45°),
Mild elevation (included angle between 45 and 90°),
Severe elevation (included angle more than 90°).

Elevated rates of fatigue or musculoskeletal disorders at the shoulder have been associated with these non-neutral postures (Chaffin, 1973; Hagberg, 1982; Feldman *et al.*, 1983).

The standard postures for the lower extremities are based on various activities such as walking, standing and leaning rather than on specific appendage positions. The following lower extremity activities are recognized:

Walk (locomotion while maintaining an upright posture),
Stand (free standing with no support other than the feet),
Lean (weight partially supported by an external object such as a wall or railing),
Squat (knees bent with the included angle between thigh and calf 90–180°),
Deep squat (included angle between thigh and calf less than 90°),
Kneel (one or both knees contact the floor and support part of the body weight),
Sit (body weight primarily supported by the ischial tuberosities).

The primary objective in defining lower extremity standard 'postures' is to cover the range of lower extremity activities as completely as possible with the fewest categories (Leonard and Keyserling, 1989).

When using the system, it is necessary to describe independently the posture of the trunk, neck, left shoulder, right shoulder and lower extremities using these standard positions. Trunk posture is classified with respect to an absolute neutral (i.e. vertical orientation with no twisting). Neck and shoulder postures are classified relative to the position of the trunk. Lower extremity posture is classified based on activities and/or the method of supporting body weight.

Equipment and procedures

The major steps required to perform a postural analysis are presented schematically in Figure 18.2 and described below.

The first step in the analysis is to obtain a continuous videotape record of the job. Any portable video camera and recorder can be used for this purpose. A colour system is recommended, however, because this increases the detail and quality of the playback. When making the tape, it is essential that the camera angle be chosen so that all joints of interest (i.e. the trunk, neck, shoulders and lower extremities) are not obstructed. In some cases, it is advisable to use more than one camera angle to record the job. For a highly repetitive job (e.g. an assembly line job with a cycle time of less than 5 min) several work cycles should be recorded continuously from each camera angle. This assures that all major work activities will be documented. A dimensioned workplace sketch, showing the location of the operator, equipment, tools, materials handling devices, etc. should be made at the time of videotaping. This sketch is used in conjunction with the results of the posture analysis to evaluate the layout of the workstation (Keyserling, 1986b).

The second step of the analysis process is to develop a sequential description of the major tasks required to perform the job. This is done in the laboratory and requires a videotape playback unit and monitor. The analyst views the entire tape of the job and selects one or more work cycles for subsequent analyses. The job is then broken down into fundamental work elements. [Note: The term 'element' is used by industrial engineers to describe a basic work action or objective (such as getting and placing a part) or performance of a specific operation (such as securing a flange with a pneumatic nut-runner). A work cycle is the sequence of elements performed by a worker on one unit of product. For repetitive jobs, such as assembly line work, many work cycles can be performed during a single shift (Niebel, 1988).]

After the elements have been identified and listed sequentially, a stopwatch or other timing device is used to measure the duration of each element and its end-point relative to the start of

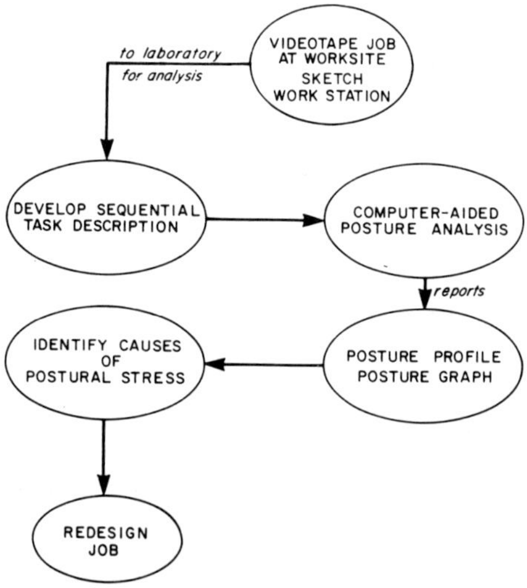

Figure 18.2 Principal steps in the postural analysis of a job (from Keyserling, 1986b).

the work cycle. This procedure establishes a common time scale so that work postures can be associated with specific work activities (Keyserling, 1986b).

The third step is to collect the postural data. To do this, the tape must be played back at the same speed that it was recorded. Furthermore, in order to keep all analyses on a common time scale, it is necessary to analyse the same sequence of videotape that was used to develop the task description. A personal computer (IBM or compatible, equipped with a printer and two floppy disk drives) is used to assist the analyst in collecting and recording time and posture data (Keyserling, 1985, 1986a). The computer keyboard is used to enter data while the tape is played; and the computer's internal clock is used to measure time. To do this, each of the standard postures is assigned a key. Whenever the worker changes posture, the analyst hits the key corresponding to the new posture. The value of the new posture and the time of the posture change (measured by the internal clock relative to the start of the work cycle) are stored on diskette for subsequent analysis and archiving. Because the computer performs all required time-keeping and clerical functions, the analyst can devote uninterrupted attention to the videotape. This is an essential feature due to the very short time (sometimes less than 0·5 s) between posture changes on highly dynamic jobs.

To eliminate the need to observe multiple joints simultaneously, the tape is played one time for each joint of interest (e.g. trunk, neck, left shoulder, right shoulder, and lower extremities). The joints can be analysed in any order. Each analysis is performed on a common timescale, with zero corresponding to the start of the work cycle.

Results and system reports

Following the final data entry, the computer generates a posture profile for each joint. These profiles provide basic descriptive statistics for postural activity (e.g. the total time spent in each standard posture during the work cycle, the minimum and maximum times spent in each standard posture, the number of times the posture was entered, etc.) Posture profiles generated from a case study job are presented in Table 18.1.

The system also has the optional capability to generate a graph which shows posture changes over time for any joint of interest. Figure 18.3 presents an example of this graph for an analysis of right shoulder posture. For the purposes of job redesign, it is useful to show postural activity and task activities on a common time scale. This graph can be used in conjunction with the results of the task analysis to identify specific work activities associated with awkward postures.

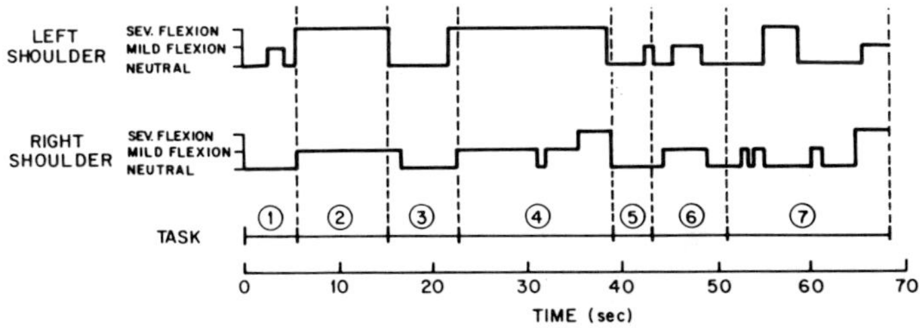

Figure 18.3 Example of graphical output showing posture changes at the shoulders. Dotted lines indicate break points between work elements (from Keyserling, 1986b).

Table18.1 Sample posture profiles (adapted from Keyserling 1986b).

Posture	Frequency	Min (s)	Max (s)	Mean (s)	S.D. (s)	Total time	% of cycle
Trunk							
Neutral	4	2·3	31·4	14·9	13·6	59·5	87
Mild Flexion	3	1·6	3·7	2·7	1·1	8·0	12
Bend/Twist	1	0·5	0·5	0·5	—	0·5	1
Number of postures			= 8				
Average time per posture			= 8·5 s				
Work cycle duration			= 68·1 s				
Neck							
Neutral	3	4·7	31·1	21·4	14·5	64·2	94
Bend/Twist	2	1·5	2·4	1·9	0·6	3·9	6
Number of postures			= 5				
Average time per posture			= 13·6 s				
Work cycle duration			= 68·1 s				
Left shoulder							
Neutral	7	0·9	7·3	4·2	2·5	29·4	43
Mild elevation	4	1·1	3·2	2·1	0·9	8·4	12
Severe elevation	3	3·0	16·9	10·8	6·9	30·2	45
Number of postures			= 14				
Average time per posture			= 4·7 s				
Work cycle duration			= 68·1 s				
Right shoulder							
Neutral	8	0·5	5·9	3·4	2·2	27·5	40
Mild elevation	8	0·7	11·2	4·2	3·8	33·5	49
Severe elevation	2	3·4	3·8	3·6	0·3	7·2	11
Number of postures			= 18				
Average time per posture			= 3·8 s				
Work cycle duration			= 68·1 s				
Lower extremities							
Walk	1	68·1	68·1	68·1	—	68·1	100
Number of postures			= 1				
Average time per posture			= 68·1 s				
Work cycle duration			= 68·1 s				

Discussion

Human learning and subjective judgement

It is important to recognize the importance of the human analyst in the system described above. The computer is strictly used to assist the analyst in recording data, reducing data, and summarizing and displaying results. The analyst must observe the videotape, determine when posture changes occur, and correctly classify the observed posture into one of the standard posture categories. As discussed below, this is not a trivial exercise and a well-trained, experienced analyst is essential in order to obtain good results.

In time study, job elements usually follow a consistent pattern. Each element has well defined beginning and end-points (often with audible cues). Furthermore, each element is usually of sufficient duration (typically at least 2 s, frequently 5 s or longer) that the analyst can momentarily redirect his/her attention away from the job in order to write observations and/or comments to a recording form. Furthermore, after observing several cycles, the analyst can learn the sequence of work activities and anticipate break points between elements.

Work posture does not always follow a consistent pattern from cycle to cycle. There are rarely audible cues to indicate posture changes. Posture can change very frequently, and a given posture may be held for only a fraction of a second. For these reasons, the posture analyst must constantly focus his/her attention on the video playback monitor and cannot be distracted in any way. Furthermore, he/she must learn the appropriate key to strike when the worker's posture changes. (Note: The analyst cannot look at the keyboard during the analysis, even momentarily, because a posture change could be missed. In effect, the analyst must respond to posture changes in a manner similar to a musician 'sight-reading' music. This implies that psychomotor learning must occur in order to perform accurate posture analyses efficiently.)

Experience has shown that the basic concepts of posture analysis and computer data entry can be learned in a short period of time. Once the concepts are learned, the analyst must learn the motor skills required for data entry. This requires 'hands-on' practice with the computer while viewing videotape. For most new analysts, the first 30 min of practice are associated with a large number of keying errors. It is not uncommon for the analyst to become quite frustrated during this early learning phase. In most cases, the rate of data-entry errors decreases substantially during the next 30 min of practice. Most individuals require between 5 and 10 h of practice to develop competence and confidence in using the system.

Data entry is only part of the learning process. Analysts must also learn the standard posture definitions presented in Figure 18.1 and use this system to classify working postures displayed on a video monitor. In order to do this, the analysis must mentally process information presented on a two-dimensional display (i.e. the video screen) to develop a three-dimensional mental image of posture. This is not always a trivial exercise, particularly in situations where the video camera is not orthogonal to the joint of interest. Experience has shown that certain posture categories can be confused in the translation process from two to three dimensions (Keyserling, 1986a). For example, under certain conditions one analyst may classify a trunk deviation as 'flexion' (i.e. movement in the worker's sagittal plane) while a second analyst viewing the same screen may classify the deviation as 'bent/twisted' (i.e. movement in the worker's frontal plane with axial twisting). Regardless of the postural analysis system used, inter-analyst discrepancies of this nature will occur as long as two-dimensional displays are used to represent three-dimensional postures. These discrepancies can be reduced, however, by using one of the following approaches:

1. Having the analyst participate in the taping of the job (Keyserling, 1986b). This allows the analyst to observe the job and required work postures in a true three-dimensional environment. The two-dimensional image on the video screen serves as a cue which helps the analyst recall the three-dimensional posture.
2. Simultaneously taping the job from several camera angles (Demko, 1986; Yang, 1987). By viewing multiple tapes, the analyst is better prepared to formulate a three-dimensional mental image.
3. Using stationary two-dimensional images (e.g. 35 mm slides) as a training tool for learning standard posture classifications. The training process should include rapid feedback of the 'true' posture and time for discussion of classification errors (Yang, 1987).

Classification errors also occur at the intersection between two standard postures. For example, if the worker being analysed flexes his/her trunk at an angle in the vicinity of 45°, the analyst must decide between the 'mild flexion' and 'severe flexion' categories. Inter-analyst and intra-analyst (i.e. repeated measures) consistency is poor in situations when the joint of interest is at a boundary between two standard postures (Yang, 1987). This problem is inherent with any posture analysis system that requires subjective judgements. In situations where 'clean' boundaries are required, it may be necessary to use an instrumented system. Nordin (1982) have developed a portable, battery-powered instrument for measuring trunk flexion angles. This device utilizes five posture categories ranging from neutral (upright) to full flexion (90°). Each category is 18° wide. This instrument is limited to motion occurring in the sagittal plane and does not measure lateral bending or twisting.

Summary

It has been frequently hypothesized that awkward work posture can cause a variety of musculoskeletal pains and disorders (Van Wely, 1970; Corlett and Bishop, 1976; Grandjean and Hunting, 1977). Until recently, it has not been feasible to perform epidemiological studies of the relationship between working posture and health problems due to the time required to record, code, and analyze postural data when using traditional analysis methods. Because the system described above substantially reduces the time required to analyse working posture (for example, it takes less than 30 min to perform posture analyses of the trunk, neck, shoulders and lower extremities for a job with a 75-s work cycle), it is now feasible to undertake the epidemiological studies of the relationship between working posture and health. On a smaller scale, the system can also be used by ergonomists, engineers, and safety and health professionals to evaluate quickly the postural stresses on a single job. The results of the analysis can be used to improve tool design and workstation layout (Keyserling, 1986b).

The system described above was developed to analyse only the posture of the trunk, neck, shoulders and lower extremities. It can be easily enhanced however to develop posture classification schemes for other joints. This would require the development of standard postures to describe the position of each joint of interest. It would also require that the videotape provide an unobstructed, high-resolution view of the joint. (Each new joint added to the system would also increase the analysis time, since an additional replay of the videotape would be required.)

References

Andersson, G., Ortengren, R. and Herberts, P., 1977, Quantitative electromyographic studies of back muscle activity related to posture and loading. *Orthopedic Clinics North Amer.*, **8**, 85–96.

Armstrong, T., 1986, Ergonomics and cumulative trauma disorders. *Hand Clinics*, **2**, 553–565.

Chaffin, D. B., 1973, Localized muscle fatigue – definition and measurement. *Journal of Occuaptional Medicine*, **15**, 346–354.

Chaffin, D. B. and Andersson, G., 1984, *Occupational Biomechanics* (New York: Wiley-Interscience).

Corlett, E. N. and Bishop, R. P., 1976, A technique for assessing postural discomfort. *Ergonomics*, **19**, 175–182.

Corlett, E. N., Madeley, S. J. and Manenica, I., 1979, Posture targetting: a technique for recording working postures. *Ergonomics*, **22**, 357–366.

Demko, D., 1986, *Evaluation of Subject Accuracy Using a Computer-aided System to Analysis Postural Stress*. Technical Report (Ann Arbor, MI: University of Michigan, Center for Ergonomics).

Feldman, R., Goldman, R., and Keyserling, W., 1983, Peripheral Nerve Entrapment Syndromes and Ergonomic Factors. *American Journal of Industrial Medicine*, **4**, 661–681.

Grandjean, E., 1980, *Fitting The Task To The Man* (London: Taylor & Francis), pp. 41–62.

Grandjean, E. and Hunting, W., 1977, Ergonomics of posture; review of various problems of standing and sitting posture. *Applied Ergonomics*, **8**, 135–140.

Habes, D. J., 1984, Use of EMG in a kinesiological study in industry. *Applied Ergonomics*, **15**, 297–301.

Hagberg, M., 1982, Local shoulder muscular strain – symptoms and disorders. *Journal of Human Ergology*, **11**, 99–108.

Hagberg, M., 1984, Occupational musculoskeletal stress and disorders of the neck and shoulder: a review of possible pathophysiology. *International Archives of Occupational and Environmental Health*, **53**, 269–278.

Hunting, W., Grandjean, E. and Maeda, K., 1980, Constrained postures in accounting machine operators. *Applied Ergonomics*, **11**, 145–149.

Jonsson, B., Persson, J. and Kilbom, A., 1988, Disorders of the cervicobrachial region among female workers in the electronics industry. *International Journal of Industrial Ergonomics*, **3**, 1–12.

Karhu, O., Kansi, P. and Kuorinka, I., 1977, Correcting working postures in industry: a practical method for analysis. *Applied Ergonomics*, **8**, 199–201.

Kelsey, J. and Hochberg, M., 1988, Epidemiology of chronic musculoskeletal disorders. *Annual Review of Public Health*, **9**, 379–401.

Keyserling, W., 1985, *Posture Analysis Program (Software)* (Ann Arbor, MI: University of Michigan, Center for Ergonomics).

Keyserling, W. M., 1986a, Postural analysis of the trunk and shoulders in simulated real time. *Ergonomics*, **29**, 569–583.

Keyserling, W. M., 1986b, A computer-aided system to evaluate postural stress in the workplace. *American Industrial Hygiene Association Journal*, **47**, 417–425.

Keyserling, W., Punnett, L. and Fine, L., 1988, Trunk posture and low back pain: identification and control of occupational risk factors. *Applied Industrial Hygiene*, **3**, 87–92.

Kilbom, A., Persson, J. and Jonsson, B., 1986, Risk factors for work-related disorders of the neck and shoulder – with special emphasis on working postures and movements. In *The Ergonomics of Working Postures*, edited by E. N. Corlett, J. R. Wilson and I. Manenica (London: Taylor & Francis), pp. 44–53.

Kumar, S. and Scaife, W. G. S., 1979, A precision task, posture, and strain. *Journal of Safety Research*, **11**, 28–36.

Leonard, J. and Keyserling, W. M., 1989, A method to evaluate neck and lower extremity postures using simulated real time analysis. In *Advances in Industrial Ergonomics and Safety I*, edited by A. Mital (London: Taylor & Francis), pp. 593–600.

Magora, A., 1972, Investigation of the relation between low back pain and occupation; physical requirements: sitting, standing, and weight lifting, *Industrial Medicine and Surgery*, **41**, 5.

NIOSH (National Institute for Occupation Safety and Health), 1981, *Work Practices Guide for Manual Lifting*, Technical Report No. 81–122 (Cincinnati, OH: NIOSH).

Niebel, B. W., 1988, *Motion and Time Study*, 8th edn (Homewood, IL: Richard D. Irwin).
Nordin, M., 1982, *Methods for Studying Work Load with Special Reference to the Lumbar Spine* (Goteborg, Sweden: Department of Orthopaedic Surgery, University of Goteborg).
Priel, V. Z., 1974, A numerical definition of posture. *Human Factors*, **16,** 576–584.
Schultz, A., Andersson, G., Ortengren, R., Nachemson, A. and Haderspeck, K., 1982, Loads on the lumbar spine: validation of a biomechanical analysis: measurements of intradiscal pressures and myoelectric signals. *Journal of Bone and Joint Surgery*, **64A,** 713–720.
Tichauer, E. R., 1978, *The Biomechanical Basis of Ergonomics* (New York: Wiley-Interscience).
Van Cott, H. and Kincade R., 1972, *Human Engineering Guide to Equipment Design* (Washington: United States Government Printing Office).
Van Wely, P., 1970, Design and disease. *Applied Ergonomics*, **1,** 262–269.
Yang, M., 1987, *Comparison of a Computer-aided System Used to Analyze Postural Stress to a Trunk Inclinometer System*, Technical Report (Ann Arbor, MI: The University of Michigan Center for Ergonomics).

Chapter 19

OWASCA: computer-aided visualizing and training software for work posture analysis

SEPPO VÄYRYNEN, KARI OJANEN[1], MATTI PYYKKÖNEN[2], AARO PEURANIEMI[3], TIMO SUURNÄKKI[3], MAURI KEMPPAINEN[2]

University of Oulu, Institute of Work Science, Linnanmaa, SF-90570 Oulu, Finland

[1]Kuopio Regional Institute of Occupational Health, PO Box 93, SF-70701 Kuopio, Finland

[2]CaeMatic Oy (Inc.), PO Box 36, SF-87400 Kajaani, Finland

[3]Centre for Occupational Safety, Kalevankatu 12, SF-00100 Helsinki, Finland

Abstract. OWAS is a system for observing, analysing and developing work postures. It is based on thorough planning carried out in the Finnish steel industry and research institutes. OWASCA is OWAS training software. It is also suitable for visualizing the work postures and for the basic analysis of the postures and their loads. The posture is presented with parametric vector 2-D graphics, OWAS codes and texts. The OWAS posture of the back, arms and legs, posture combination, force or effort used, additional postures and action categories can be studied interactively step by step. OWAS skills can be tested by OWASCA. The program shows a random work posture, the user gives his/her answer and OWASCA tells whether the answer is right or wrong. It also describes the errors in detail and gives the numbers of test postures and right answers. OWASCA runs both on IBM PC/XT AT or compatibles and on Apple Macintosh.

Keywords: Work postures; Ergonomics; Computer-aided design.

Introduction

When ergonomic aspects are studied or analysed using a computer-aided design system, the procedure can be characterized by the concept 'ergoCAD'. A review of computer-aided workplace design has recently been presented (Kern and Bauer, 1988).

Effects of accessibility on postures in forest machinery maintenance have been studied in Finland. The first ergoCAD application of the Kuopio Regional Institute of Occupational Health was developed in collaboration with the University of Oulu for the Computervision CAD system (Väyrynen *et al.*, 1985; Väyrynen, 1988). Both in the study mentioned above and in the ergoCAD application, the work postures in forest machinery maintenance were analysed and rated according to the Finnish method called OWAS.

Two additional ergoCAD applications have been developed in Finland, one for Computervision in Tampere (e.g. Leppänen and Mattila, 1987) and one for PCs in Helsinki (Launis *et al.*, 1988).

Work postures are one of the most suitable targets that can be controlled more effectively with computer-aided ergonomic design and analysis systems. Strenuous work postures load the musculoskeletal system, increase energy consumption, cause other unwanted physiological reactions, and may be linked to accident occurrence (Chaffin and Andersson, 1984; Corlett *et al.*, 1986). In manual materials handling, the postures required for lifting, pushing or pulling, greatly affect the low back muscle force and disc compression force. For example, in Finland in 1986, overexertion was often cited as the most common occupational accident (20·2%) and physical effort was often (17·1%) the cause of the accident (Työsuojeluhallitus, 1987).

OWAS method

OWAS, Ovako Working posture Analysing System, was developed according to the experience obtained in the steel industry (Karhu *et al.*, 1977, 1981; Heinsalmi, 1986; Suurnäkki *et al.*, 1988). The features of OWAS are also described elsewhere (Chaffin and Andersson, 1984; Corlett *et al.*, 1986; Kilbom *et al.*, 1986). The OWAS system identifies the most common work postures for the back, arms and legs, and estimates the weight of the loads handled or the extent of the strength (load/effort) used (Figure 19.1).

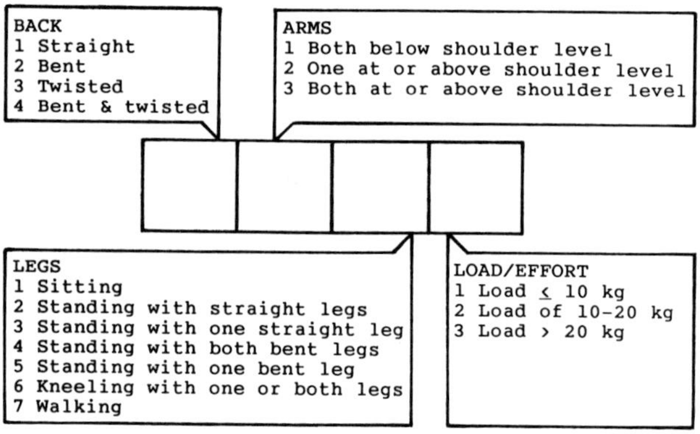

Figure 19.1 Coding of OWAS work postures (Suurnäkki et al.*, 1988).*

The observed posture combinations are classified according to the OWAS method into ordinal scale action categories. The four action categories described here are based on experts' estimates on the health hazards of each work posture or posture combination in the OWAS method on the musculoskeletal system.

Action category 1. Work postures are considered usual with no particular harmful effect on the musculoskeletal system. No actions are needed to change work postures.

Action category 2. Work postures have some harmful effect on the musculoskeletal system. Light stress, no immediate action is necessary, but changes should be considered in future planning.

Action category 3. Work postures have a distinctly harmful effect on the musculoskeletal system. The working methods involved should be changed as soon as possible.

Action category 4. Work postures with an extremely harmful effect on the musculoskeletal system. Immediate solutions should be found to reduce these postures.

The qualitative and quantitative determination of work postures creates a basis for adjusting posture load towards the optimum, as regards the worker's health and safety. This is the basic idea underlying the development of the OWAS method. It is a system consisting of the work postures of various parts of the body and of the combination of postures, whose quantitative proportions in work can be calculated reliably by the observation technique, and whose degree of hazard and urgency of correction measures can be estimated.

OWASCA

OWASCA (OWAS Computer-Aided) is intended to visualize work postures and to train posture analysis according to the OWAS system (Figure 19.2). OWASCA is a two-dimensional graphic program for personal microcomputers (both IBM PC XT/AT or compatible and Apple Macintosh). The program is written with Turbo-Pascal. Both Finnish and English versions are available on a single disk.

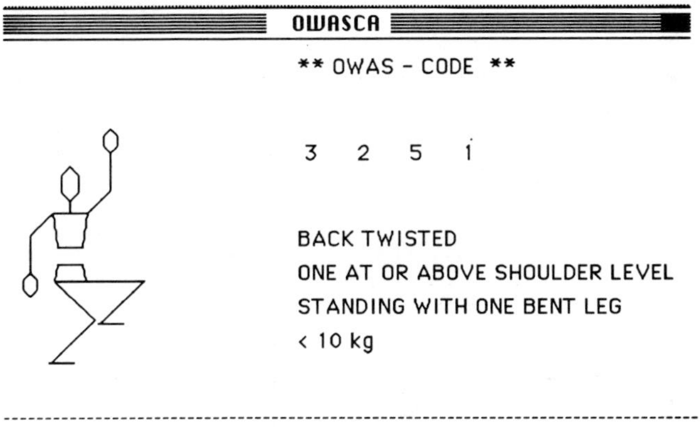

Figure 19.2 OWASCA is a microcomputer-based software including man model facilities.

The program operates with posture figures, external load, and the OWAS action category, which ranks the strenuousness of each posture combination on an ordinal scale (urgency of correction). The man model (stick diagram) has 14 links (Figure 19.3). Ten of them (the upper torso, the head, the upper arms, the lower arms, the lower legs and the feet) have three degrees of freedom (df). The middle torso and the upper legs have 2 df and the lower torso has 1 df. The control of the model is based on parametric vector graphics. The natural human moving limitations have been taken into consideration.

The use of OWASCA

OWAS posture codes are a good basis for computer programs and computer-based training and education. The interactive structure of the program is based on menus. The first menu

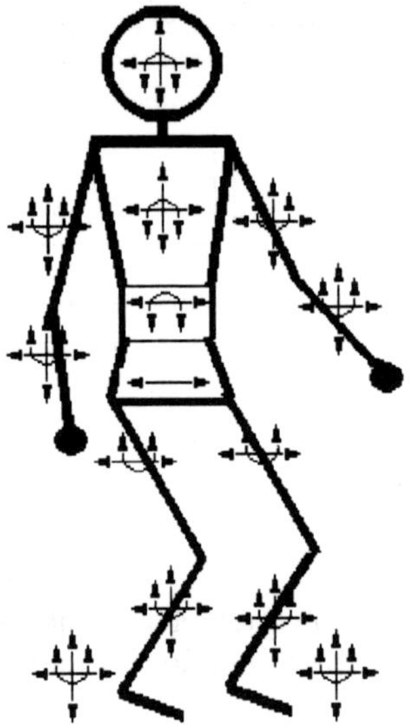

Figure 19.3 The structure of the OWASCA man model and the degrees of freedom of the links.

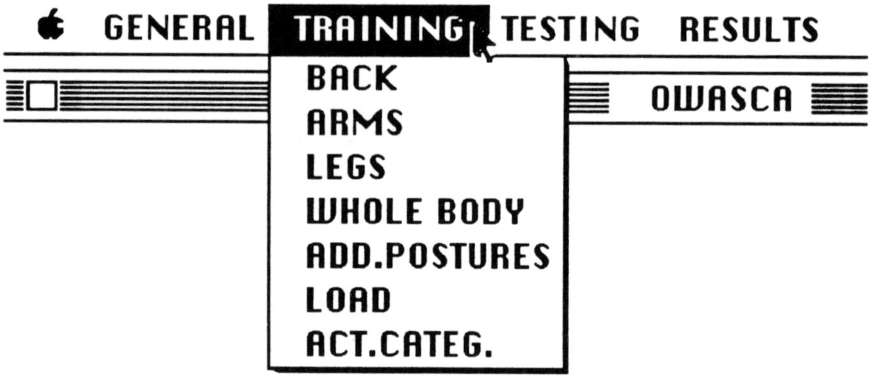

Figure 19.4 Training menu in the Macintosh version.

contains the explanation of the OWAS system and the program. The second menu contains the interactive training of the OWAS method (Figure 19.4). The user can learn step by step the postures of the back, the arms and the legs, entire posture, force classes, additional postures and action categories.

After the user has learned the OWAS method, OWAS skills can be tested with OWASCA program. The test can be run from a simple level to a more difficult one according to the user's skills. The testing goes through as many random work posture figures as the user or the teacher wishes.

The answer is given in the answer window through the keyboard (PC), or using the mouse (Macintosh). In the feedback window the program indicates whether the answer was

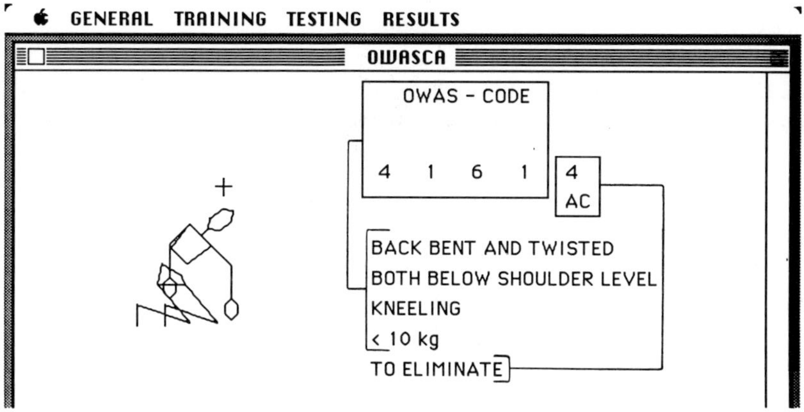

Figure 19.5 An example of the screen of OWASCA containing the OWAS code, action category of the posture combination and corresponding explanatory texts.

right or wrong. All the answers are saved for later checking. The user can check his/her own results from the fourth menu. OWASCA tells both the given and the right answer. These answers can be compared in detail. The numbers of right answers and total test postures can be seen on the screen (see Figure 19.5).

With the program one can study, for example, the effect of the load or effort on the degree of hazard of the work posture. It is also possible to illustrate by examples the alternatives for doing the same work by resorting to different work postures, either worse or better. The screen text of the computer can also be printed. The combination of a microcomputer, a projection pad and OWASCA is the most useful equipment to demonstrate the effects of work postures on the musculoskeletal system.

Discussion

Safety and health personnel, in addition to designers, are the potential users of OWASCA. The training of OWAS specialists is one of the main applications of this system. The OWASCA software is a basic instrument for redesigning work methods towards decreasing postural load. However effective redesigning of work spaces and designing of new work spaces need a graphic CAD program to work with OWASCA. At the present, OWASCA software is not compatible with any commercial CAD software.

Controlling the length of the links of the OWASCA model, various body measurements could be taken into account, and the model could be scaled according to the work environment. The user-friendly interface, like in OWASCA, also makes it easy for inexperienced microcomputer users to start the work with computerized ergonomic software. The compatibility of the OWASCA with the great majority of microcomputers makes it possible to utilize OWASCA in most workplaces and organizations.

Acknowledgements

OWASCA was sponsored by the Finnish Work Environment Fund.

References

Chaffin, D. B. and Andersson, G., 1984, *Occupational Biomechanics* (New York: John Wiley).

Corlett, N., Wilson, J. and Manenica, I. (Eds), 1986, *The Ergonomics of Working Postures* (London: Taylor & Francis).

Heinsalmi, P., 1986, Method to measure working posture loads at working sites (OWAS). In: *The Ergonomics of Working Postures*, edited by N. Corlett *et al.* (London: Taylor & Francis), pp. 100–104.

Karhu, O., Härkönen, R., Sorvali, P. and Vepsäläinen, P., 1981, Observing working postures in industry: Examples of OWAS application. *Applied Ergonomics*, **12**, 13–17.

Karhu, O., Kansi, P. and Kuorinka, I., 1977, Correcting working postures in industry: A practical method for analysis. *Applied Ergonomics*, **8**, 199–201.

Kern, P. and Bauer, W., 1988, Computer aided workplace design. In: *Ergonomics International 88. Proceedings of the Tenth Congress of the International Ergonomics Association*, 1–5 August 1988, Sydney, edited by A. S. Adams *et al.*, pp. 99–101.

Kilbom, Å., Horst, D., Kemmlert, K. and Richter, A., 1986, Observationsmetoder för registrering av belastningar på rörelseapparaten – en litteraturstudie (Observation methods for registration of load and strain on the human body). *Arbete och Hälsa*, **21**, (Swedish with an English summary).

Launis, M., Lehtelä, J. and Kuusisto, A., 1988, Two-dimensional anthropometric human model system for computer-aided design (CAD) of work places. *Työ ja ihminen*, **6**, 68–83, (Finnish with an English summary).

Leppänen, M. and Mattila, M., 1987, Computer aided ergonomics design. In: *New Methods in Applied Ergonomics. The Proceedings of the Second International Ergonomics Symposium*, 14–16 April 1987, Zadar, edited by J. R. Wilson *et al.*, pp. 81–87.

Suurnäkki, T., Louhevaara, V., Karhu, O., Kuorinka, I., Kansi, P. and Peuraniemi, A., 1988, Standardised observation method for assessment of working postures; the OWAS-method. In: *Ergonomics International 88. Proceedings of the Tenth Congress of the International Ergonomics Association*, 1–5 August 1988, Sydney, edited by A. S. Adams *et al.*, 281–283.

Työsuojeluhallitus, 1987, Työtapaturmat: Suomen virallinen tilasto XXVI: 38 (Industrial accidents in Finland), (Tampere) (Finnish with an English summary).

Väyrynen, S., 1988, The computer as an aid in the analysis of ergonomic and safety features of maintenance (description of two graphic applications). *Scandinavian Journal of Work Environment and Health*, **14** (suppl 1), 105–107.

Väyrynen, S., Ojanen, K. and Kuusisto, A., 1985, Computer aided design of a workplace – a survey and an experiment. *Työterveyslaitoksen tutkimuksia*, **3**, 332–336, (Finnish with an English summary).

Chapter 20
Development of an expert system for designing workplaces in manual materials handling jobs

EUI S. JUNG and ANDRIS FREIVALDS[1]

Department of Industrial Engineering,
Pohang Institute of Science and Technology,
Pohang, Korea

[1]Department of Industrial and Management Systems Engineering,
The Pennsylvania State University, PA,
U.S.A.

Abstract. An expert system was developed as a framework of integrating diverse and multi-factored ergonomic knowledge for workplace design. Though numerous computer-assisted approaches have been made to overcome the lack of integrated design principles, those models being used require very specific information of various design activities that may not be available in the early stage of design. On the other hand, an expert system would be an effective design aid that is capable of guiding the designer to solve a problem. However, most expert systems lack detailed evaluation capabilities due to the qualitative nature of inference mechanisms. Furthermore, these approaches were independently developed, focusing mainly on a single aspect such as biomechanics or physiology. In this research, the architecture of the expert system that suits the data-intensive and multi-objective nature of ergonomic workplace design is discussed. Also discussed are the ways in which ergonomic knowledge is integrated into a modularized knowledge base.

Keywords: Expert systems; Manual materials handling; Multiple criteria decision making.

Ergonomic workplace design and its evaluation

Introduction

The design and evaluation of the workplace in industrial and manufacturing environments has long been the focus of attention of many specialists, each viewing the problem from a diverse viewpoint. In real applications, workplace design is the outcome of a complex system of interrelated disciplines and techniques for which no-one is solely responsible. Workplace design requires a multi-factored process requiring a systems approach to consider alternative factors and to achieve the optimal solution (Evans, 1985). In general, the group of workplace designers include specialists in manufacturing, ergonomists in biomechanics and work physiology, motion and time-study engineers, plant layout specialists, occupational safety engineers, industrial psychologists and behavioural scientists. This list is by no means exhaustive. They are all concerned with the improvement of productivity and worker performance (Huchingson, 1981).

As the design process becomes more complex, the number of factors to be considered also increases, as does the need for more interaction between the above-mentioned specialists. However, as the factors increase in diversity and complexity, the limitation of such interaction and the deficiency of integrated design and evaluation techniques often force designers to overlook many parameters involving human workers (Chaffin and Andersson, 1984). Since those designers have traditionally been too concerned with the engineering aspect of the task in the workplace, human operators or workers are considered as another unit which can be adjusted to meet the systems' needs. The role of ergonomics has often been to improve the resulting designs by considering human workers in the implementation process. Indeed, Alexander (1986) pointed out that a critical element in workplace design is the use of data about human dimensions. From the design standpoint, it is certainly quicker to design for a limited range of workers rather than ensure more universal operability. Yet the cost saved on faster design later becomes as burden to the operating cost.

Ergonomists should be able to apply ergonomics to the particular workplace design throughout the whole design and evaluation process, by integrating their knowledge with the techniques and methodologies employed in these diverse areas. Considerable work has been done on these factors and the information listed can be obtained from various handbooks, journals and textbooks (Maynard, 1971; NIOSH, 1981). It is, however, apparent that such information has not been well integrated into a design framework that enables the resulting design to improve productivity and performance. From the literature review, it is evident that industrial workplace design fails to consider the human worker throughout all stages of design and evaluation. It may be partly because it is difficult economically to justify the expense of the proposed changes made by ergonomic considerations. Lost productivity due to inefficient work methods, increasing worker compensation costs due to overexertion and musculoskeletal injuries reveal such failure. In spite of OSHA regulations for safe and healthful working conditions, overexertion injuries, specifically, low back disabilities, have increased in both frequency and compensation costs. More than half of the working population in both the USA and European countries suffer from low back pain at least once during their working careers (Pope *et al.*, 1984). Specifically, it is obvious that large numbers and types of injuries and illnesses have been related to manual materials handling jobs. It is generally agreed that manual materials handling (MMH) jobs appear to be a major source of most of the musculoskeletal injuries in industry (Jiang and Ayoub, 1987).

Although the exact causation of overexertion injuries in MMH jobs is not conclusive, the prevention has traditionally been attempted by three approaches: (i) selection of workers with sufficient capacity, (ii) training of workers in proper work methods, and (iii) ergonomic design of the job to fit workers. Worker selection is to detect those people who are at high risk of a potential low back injury. Common techniques of medical history and physical examination have not been so effective and often of sceptical value for this purpose (Snook, 1978; NIOSH, 1981). It also involves complex legal matters. A great deal of accident prevention education has also been directed toward the reduction of low back injury. However, there has not been any significant reduction in injuries while the straight back, bent knees lifting method has been favoured.

It appears that a structural and effective improvement can be realized through ergonomic design of the tasks at the workplace. This has been considered as the most promising method for the prevention of manual handling injuries. Yu *et al.*, (1984) also stated 'Preemployment strength testing and ergonomic job design offer the greatest promise. Job design alone should not be expected to be completely effective in preventing low-back injuries, but it would appear to be somewhat more effective than selecting the worker for the job or training the worker to lift properly.' Furthermore, the complexity of the problem, the great need for reliable and valid guidelines, and the serious impact of injuries on workers justify more research in workplace design.

The reason why it is difficult to incorporate ergonomic information into a workplace design and evaluation may be the multidisciplinary nature of ergonomic principles and workplace design itself. First, there are many ergonomic aspects that directly affect physical stresses, such as biomechanical strength, physiological endurance, workplace-related posture and anthropometric reach. Ergonomic problems are, therefore, characterized by multiple objectives and criteria which lead an ergonomic workplace design and its evaluation to formulate a multiple-criteria, decision-making problem having multiple objectives which are possibly conflicting. The second issue would be the process of design itself. The design process is by nature iterative and highly interactive. A myriad of input information is required, including specialists' ball-park judgements, reference table look-up, computer simulation graphics, experimental analysis, and mock-up. Such complexity of the design process in which ergonomics is considered limit the number of alternatives without properly weighing the importance of each alternative. Consequently, once the first design is set up, it continues until deficiencies lead to redesign (Evans, 1985).

Numerous computer-assisted approaches have been made to overcome the lack of integrated design principles. Some of these are SAMMIE for posture and reach analysis, The University of Michigan Biomechanical Strength Prediction Model which estimates physical stresses imposed on a human operator's musculoskeletal system, Garg's metabolic prediction model which predicts metabolic energy expenditure of the tasks. In fact, such computerized models overcome many restrictions that the traditional approaches previously had. Although those models offer sufficient flexibility to adapt the range of applications, these were independently developed, focusing mainly on a single ergonomic aspect. It requires a number of demands on the designer to refine the parameters based on his experience. Ergonomic design often requires several areas to be considered simultaneously in order to reach the optimal or satisfactory solution across possibly conflicting areas. Specifically, for ergonomics to be considered from the early stage, the information must be available in integrated formats. A more effective framework is therefore needed, packaging the existing models into an integrated system, containing a common data base of human dimensions and performance, and efficiently communicating between the models and the designer.

Expert systems as a framework for ergonomic workplace design

Applying computer-aided ergonomic design principles to the workplace is not new. A problem with existing computer-aided design models is their specificity. In other words, they require specific information of various design activities, and can be regarded as being designed for use in the later stages of design in which comprehensive specifications are available. At the early stage of design, designer or user may not have all the specifications required but, often, should provide necessary information or advice throughout the design process (Coury, 1983). Another problem stems from the fact that an accumulation of ergonomic knowledge is vast and massive, so that the expertise is required both in selecting relevant classes of knowledge and in combining them with some sort of heuristic judgements according to their importance.

One approach for solving these problems is to introduce the expert systems approach in integrating existing models and expertises into a framework which guides the designer through necessary steps to reach a conclusion, and to explain how it is reached. With the increasing popularity of artificial intelligence and expert systems, there have been many attempts to build expert systems, without keeping in mind the basic question as to whether an expert system really helps the problem-solving process. In many cases it may be possible to solve the problems related to ergonomic design by rather direct and often simple analysis. When we adopt this relatively new concept to deal with ergonomic design problems, its validity to the ergonomic design area should be discussed to see whether an expert system is able to surpass other computer models in their performance. Tirupathikumara *et al.*, (1985) summarized engineering application problems which ideally suit an expert systems approach as follows:

1. Problems in the domain cannot be well defined analytically.
2. Problems can be formulated analytically but the number of alternative solutions is large.
3. The domain knowledge is vast and relevant knowledge needs to be used selectively.

As discussed, problems in ergonomic workplace design well match those summarized here. It has been also recently suggested that ergonomic evaluation of manual workplace designs should incorporate concepts from knowledge-based engineering and artificial intelligence to capture the expertise in solving the specific ergonomic problem (Coury, 1983; Evans, 1985; Karwowski *et al.*, 1986). An integrated framework should be adaptive to the problem specificity in both the initial design stage where more general and global information is needed to formulate the design, and the evaluation stage where comprehensive specifications become available and alternative designs are being evaluated. In this regard, an expert systems approach is worthwhile for solving ergonomic problems having subjective, qualitative information, and multiple criteria.

The proposed expert system as a framework for integrating ergonomic information into workplace design should satisfy the following needs. First, it should include a data base of human dimensions which are commonly used in ergonomic design and analysis, and provide basic inputs to the designer. Second, it should also include a separate knowledge base of ergonomic expertise, consisting of the so-called declarative knowledge about available facts and relations of ergonomic variables, and the procedural knowledge about rules and methods to find relevant facts. Then, it should, in essence, incorporate an inference mechanism to draw inferences from whatever information is available to reach a solution and to provide its reasoning, resembling what an expert does. Inference mechanisms which include heuristics, search mechanisms, control strategies, guide the user to reach the optimal solution or at least a satisfactory solution by way of either user-provided information or default reasoning.

System scope

It is common practice for designers to select design guidelines from textbooks, handbooks, scientific articles and experiments. The question then arises as to what specific guidelines can contribute to the prevention of injuries at the workplace, and to which groups of workers these can be applied. However, there are still considerable discrepancies among diverse sources. According to the literature survey by Dul and Hildebrandt (1987), the maximum acceptable weight of lift, for instance, varies from 27 to 80 kg for male workers and from 10–12 to 64 kg for female workers in the lifting task. One of the reasons for such differences among the guidelines is probably the difference in criteria that are applied. In other words, the back compressive forces on L5/S1 as a biomechanical criterion played a role in one source, while other criteria such as energy expenditure did in another source. In fact, such a situation is fairly representative for most ergonomic guidelines. Yu *et al.* (1984) indicated that three criteria (or factors) have been advocated to determine the maximum permissible workload that an individual can handle. These are psychophysical, biomechanical, and physiological criteria. Psychophysical studies usually provide recommendations on maximum acceptable weight of the load for infrequent lifting as well as for repetitive lifting tasks. Biomechanical studies, in general, are for infrequent lifting tasks, whereas physiological criteria are more applicable to repetitive lifting.

The development of the proposed ergonomic expert system is restricted to the manual materials handling tasks in the manufacturing environment in terms of evaluating physical stresses governed by biomechanical, physiological, and psychophysical factors. By sufficiently limiting the initial scope of the system, an expert system can be reasonably justified in solving specific problems. This research focuses on organizing information regarding the role of ergonomics, and developing a framework that enables the user systematically to utilize the information by way of following expert reasoning processes.

An expert system is a promising problem-solving approach that uses domain-specific knowledge and an inference mechanism to derive an appropriate solution or solutions in such an ill-defined domain. However, not much work has been done in applying an expert systems approach to an ergonomic domain. Several prototype expert systems have been developed in the area of ergonomics (Coury, 1983; DeGreve and Ayoub, 1987; Chen, 1987; Karwowski *et al.*, 1987). An expert system, by its definition, has to be capable of solving problems that require expert knowledge in a particular domain. The expert systems mentioned above possess that knowledge in some form. These can be, therefore, classified as knowledge-based systems. However, not every knowledge-based system can be considered an expert system. An expert system also has to be capable, in some way, of (i) efficiently selecting appropriate pieces of knowledge like a data base system, (ii) obtaining the best inference(s) from the given situation, possibly, from an uncertain situation, (iii) explaining its behaviour and decisions to the user, as human experts do. Such features enhance the user's confidence in the system's advice or enable the user to detect a possible flaw in the system's reasoning.

In addition, the proposed expert system must be adaptive to the characteristics of an ergonomic domain. First, since ergonomic information such as anthropometric data or table look-ups requires the capability of handling massive ergonomic data in the form of tables, equations, etc., it should have an interface to the external data base system to store and retrieve necessary data efficiently. Second, multiple, possibly conflicting factors (e.g. biomechanical factors, physiological factors) are involved at different levels of the design process for drawing solutions to the problem. This implies that the system should also incorporate a decision-making scheme to weigh the importance of multiple factors in a more objective way. Third, those factors often require using not only IF-THEN type of deductions useful in the design stage, but complex analytical models whenever specific information is sufficient to use those models.

In order to meet these criteria, this research is aiming at creating a formal architecture of the proposed ergonomic expert system. Emphasis is, therefore, placed on developing the following:

1. Construct an effective framework that combines artificial intelligence and expert systems methodology for design and evaluation of an ergonomic workplace in MMH jobs.
2. Build a comprehensive and modularized knowledge base that incorporates knowledge and expertise on work physiology, biomechanics and psychophysics.
3. Construct a problem-solving scheme or inference mechanism capable of using such domain-specific knowledge, coupled with user-interaction facilities which include explanation of the decisions made during the problem-solving process.
4. Link the knowledge base to the complex analytic models and the data base that efficiently manipulates massive ergonomic information.

Artificial intelligence and expert systems

An overview

Expert systems originated from the idea of artificial intelligence in the field of computer science that can be traced back to the 1950s. Although the term, artificial intelligence, has been recently spread out and it is still thought to be in its infancy, the seeds were sown about three decades ago since the idea of a general problem solver. There have been numerous arguments from the beginning as to what is intelligence. The commonly accepted notion of intelligence is that intelligence is the ability to perceive, understand and learn about new situations, resembling what a human brain does. The idea of artificial intelligence is that if a computer system can duplicate this reasoning ability in some way, then, the system can also be classified as being intelligent. That is where the term, artificial intelligence, came from, in a sense that human intelligence is captured by an external system. As it means different things to different people, there could be as many formal definitions as potential application fields. Feigenbaum stated that artificial intelligence is the field of computer science concerned with designing intelligent computer systems, that is, a computer system that exhibits the characteristics we associate with intelligence in human behaviour (Barr and Feigenbaum, 1982).

More direct concern is using AI techniques through expert systems, which are of immense value to engineering domains. In relation to ergonomic workplace design, an expert system can be defined as a tool that can understand design specifications and intelligently use domain knowledge to suggest possible alternatives and desired paths of planning actions to be taken. Expert systems researches have concentrated on the high performance software development in specialized domains and therefore put an emphasis on constructing knowledge and expertise in an organized way, as well as implementing necessary techniques and formal reasoning methods. This is why expert systems are also called knowledge-based systems (KBS).

Then, what makes expert systems differ from conventional user-friendly systems or data base systems? A common misconception about expert systems is due to considering it as a collection of IF–THEN constructs along with some method of transferring knowledge from experts to non-experts. Expert systems not only use techniques to transfer knowledge but also use analytical tools to evaluate it and techniques to learn. First, expert systems separate the declarative knowledge about general facts, theories, etc., from the procedural knowledge which includes subjective experiences, heuristic rules and knowledge about knowledge, so called, meta-knowledge. Such a separation has distinct advantages because changes in either the declarative knowledge base or the procedural knowledge base do not affect each other, so that the system can keep growing without too much patching. Second, expert systems employ general reasoning mechanisms to draw their inferences and to provide the user with justifications of conclusions reached. Finally, expert systems possibly incorporate some sort of learning mechanism which is one of the important aspects of human intelligence.

Although an expert system is a highly experimental field far from general theory, it is promising that there has been a record of successful applications in many fields, for instance, MYCIN for medical diagnosis, PROSPECTOR for mineral deposit analysis, and R1 for VAX system configurations, and so forth (Hayes-Roth *et al.*, 1983).

Architecture of an expert system

Since many techniques have been developed in artificial intelligence research, several types of expert systems have been built. As mentioned in the overview, they are termed knowledge-based because their performance depends critically on utilizing domain-specific knowledge such as facts and heuristics used by experts. Some authors discuss principles and guidances of building expert systems based on different applications. However, there is no firm guide to aspiring expert systems architects through the issues and choices faced in designing and evaluating a system. In general, expert systems can be classified depending on their applications or areas where artificial intelligence methods have been applied. The types of possible applications are well summarized in Waterman (1986).

These applications largely fall into two distinct area of problems. Among these, interpretation, analysis, or diagnosis problems require inferring the abstract state as a potential solution to the problem (the highest level of problem representation), given the detailed state or description (lowest level of problem representation). For example, the most well-known application area of expert systems is that of medicine. Such an area requires an expert system to extract a possible disease from the given symptoms. Whereas, expert systems that perform design or planning require a process that produces a sequence of actions to transform a problem world from abstract state to detailed state. Process planning or manufacturing, in general, is such a task domain, that of generation problems which need to generate the lowest level representation (the sequence of intended actions), when the highest level of representation is presented. Therefore, interpretation problems can be best solved by a bottom-up approach, while generation problems tend to be better solved by a top-down approach (Tsatsoulis and Kashyap, 1987).

When we look into an application of expert systems to the domain of ergonomic workplace design for MMH tasks, the system developer must consider both of the above-mentioned aspects, according to the problem specificity and the user needs. If the user intends to design a general ergonomic workplace in the initial design stage, the system should be capable of developing task descriptions and parameters that satisfy the constraints of the problem. It should be, therefore, equipped with an internal control structure (or possibly, generic solution plans) to help the user reach possible solutions. Second, if the user wants to analyze potential designs or has sufficient information to evaluate the problem using analytical models, the system should lead the user to formulate the problem in formats appropriate for these models. In addition, the system should attempt to optimize decision functions that measure the desirable properties of potential designs. To achieve these requirements, the structure of expert systems needs to be highly modular and expandable.

As seen in Figure 20.1, architecture of an expert system is, therefore, a method of designing the following components:

1. Knowledge base:
   (i) knowledge consisting of domain related facts,
   (ii) knowledge embodying domain related rules for drawing inferences.
2. Inference mechanism:
   (i) an interpreter that applies the rules,
   (ii) a scheduler that orders the application of rules,
   (iii) a consistency enforcer keeping the inference consistent when knowledge base is updated,
   (iv) a justifier that explains the system's reasoning.
3. User-friendly interface.

Existing expert systems seldom contain all the components, but one or more components occur in every expert system. The interpreter and the scheduler act as an intermediate decision maker which schedules and executes the chosen actions by applying the corresponding rules in the knowledge base. Existing expert systems employ a lot of

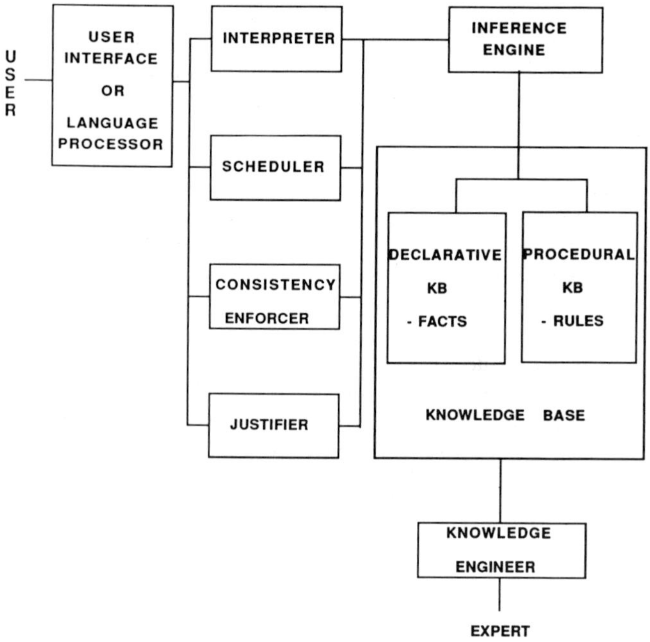

Figure 20.1 An architecture of expert systems.

variations which play such a rôle. The consistency enforcer attempts to keep the solution consistent. The need to implement such a consistency enforcer arises from the inherent characteristics of data acquisition, for instance, incomplete data, unreliable data, or imprecision of data. The justifier traces the chain of reasoning in the system to give explanations to the user. It generally gives answers to question as to why a specific conclusion was reached and what alternatives were considered. The core and structure of an expert system would be the knowledge base coupled with an inference engine. As mentioned before, the knowledge base consists of two parts. A declarative knowledge base provides facts related to the problem domain, whereas a procedural knowledge base contains domain-specific, problem-solving knowledge. An inference engine defines the problem-solving approach and domain independent control structure to manipulate the knowledge base.

Knowledge representation schemes and the inference mechanism

The ultimate goal of expert systems research is to build a system that has various knowledge modules to produce an intelligent output. The language operated on various modules we call the representation. The word 'representation' not only refers to conventions for structuring data in a computer implementable form like the data base, but the abstract structuring of expressions denoting facts (Charniak and McDermott, 1985). The predicate logic was chosen to represent the knowledge since it has so far served as the most flexible vehicle. Kowalski (1979) supported this argument, stating that there is only one language suitable for representing information, whether declarative or procedural, and that is first-order predicate logic. Predicate logic is a formal language in which any statement or belief as well as propositions can be expressed as formula.

A few well-known representation schemes will be examined in this section. These schemes are, however, considered as a part of inference mechanisms, since the advantages and the disadvantages are closely related to its inference capabilities. The most widely used representation schemes are rule-based deduction systems which are also called production systems, frame-based abduction systems and semantic networks. The use of production rules for knowledge representation has led to systems that have achieved high levels of performance in their particular domains. The advantage of this approach stems from the fact that the representation is modular and declarative, so that it provides conceptual clarity, encourages incremental development, and makes the knowledge base directly accessible. Production rules are also well suited to accept the declarative and procedural knowledge transformed into predicate calculus. MYCIN and PROSPECTOR are two well-known examples. In frame-based abduction systems, the domain-specific, problem-solving knowledge is represented in descriptive 'frames' of information. Inference is typically based on hypothesize-and-test cycles which model human reasoning as follows. Given one or more initial problem features, the expert system generates a set of potential hypotheses. These hypotheses are then tested by various procedures which measure their feasibilities and plausibilities. DENDRAL and INTERNIST are the typical examples (Barr and Feigenbaum, 1982).

Another widely used representation scheme is semantic networks developed by Quillian (1968). A semantic network can be defined as a labelled directed graph in which nodes represent concepts, arc labels represent binary relations. Several advantages can be cited for this semantic network representation. Among these are the ability to represent subset and element taxonomic information, the ability to include the same relation in several different premises and conclusions. There are also a number of expert systems based on the statistical decision theory. Bayesian approach is the most widely used statistical approach. Domain specific problem-solving knowledge in Bayesian systems is represented as tables of probabilities. This approach seems to be the most applicable to a medical domain, since medical diagnosis is based on an accumulation of doctors' experiences and statistics of pathological cases. However, it is very difficult to estimate the conditional probability of all the possible outcomes.

However, very little has yet been written comparing the relative advantages and disadvantages of the intrinsically different schemes. In fact, a number of AI researchers have favoured the clear superiority of rule-based deductive approaches. For example, Hayes-Roth *et al.* (1978) stated "As a result, pattern-directed inference systems based on antecedent-consequent rules make a strong claim to being the best available scheme for knowledge representation . . ." However, Ramsey *et al.* (1986) argued that there is no single 'best method' for building expert systems. The real issue for consideration is which methods are best suited for a specific application problem. As noted, if we develop an expert system to interpret or analyze the problem situations (e.g. medical diagnosis), the rule-based production systems have been known as the best system, due to the versatility and expressive power. However, if the system is needed to perform primarily design and planning functions which require generating a sequence of actions, some system developers prefer the frame-based system such as that in STRIPS (Tsatsoulis and Kashyap, 1987). In general, frame-based systems require a larger size inference engine and are more difficult to implement than production systems, while production systems have shortcomings in implementing the control structure. However, both systems showed a reasonable performance in terms of run time and inference time (Niwa *et al.*, 1984).

The problems we face in the domain of ergonomic workplace design and evaluation include both interpretation and generation problems. The system should not only provide appropriate information to formulate the problem and to find the potential design, but it should also evaluate and analyze the given design alternatives. It should be highly flexible to the problem specificity and modularized to accommodate multi-factored knowledge. In the prototype expert systems, a pattern-directed rule-based inference mechanism was chosen to meet these requirements. Several advantages can be cited. First, its representation retains the desirable modularity. Second, it permits an explicit, structured description of the problem domain. Also, it can easily propagate measures of uncertainty (Duda *et al.*, 1978).

A pattern-directed inference system is a program that focuses on situations occurring as patterns. These patterns lead the system to select pieces of code (or rules) to be executed. It responds directly to a wide range of unanticipated events rather than operating on expected data in known formats using a prespecified and inflexible control structure. Such an organization can be characterized as pattern-directed (Waterman and Hayes-Roth, 1978). Production systems and rule-based systems can be, in general, characterized as pattern-directed systems, in which the pattern-directed modules consist of antecedent pairs as rules in the following:

IF conditions THEN actions

Each pair in the pattern-directed module tests the condition part and executes the action part only when all the conditions have been asserted. Here, much judgemental knowledge can be represented in the form of rules such as a & b & $c = > h$ in the following:

'Maximum acceptable weight which 90% of men can lift from floor to knuckle is 37·3 lbs'
or
'If the gender and population of operator is known and the task variables are known, use Snook's results for the maximum weight of the load.'

The pattern-directed architectures does not require all the connections between modules to be planned and defined in advance. Consequently, each module can be designed and implemented relatively autonomously. This renders a high degree of modularity. It is especially desirable in systems with complex knowledge bases because it is difficult to predict in advance all the interactions between pieces of knowledge in the base (Bratko, 1986).

Expert systems in ergonomics

Applying the expert systems approach to the field of ergonomics has recently begun and has not yet matured. Coury (1983) suggested an expert systems framework to provide useful information and the means for assessing the ergonomic impact on workplace design. His system features include the knowledge-based system coupled with CAD. Another attempt was made by Karwowski *et al.* (1986, 1987) to build a conceptual framework of an expert system for designing manual material handling jobs. The early system called LIFTAN adopted fuzzy logics to characterize linguistic variables involved in the design. Later, they developed a microcomputer version of LIFTAN called M-LIFTAN which was expanded to incorporate a comprehensive analysis and design of manual materials handling tasks. The system seems to be more useful in the early stage of design mainly to aid non-experts. DeGreve and Ayoub (1987) developed a knowledge-based system to assist design engineers in achieving an ergonomically sound workplace design. The present prototype system is confined to reach analysis based on anthropometric data. Chen (1987) developed an interactive knowledge-based system for physical work stress analysis. The system calculates the physical work stress index in order to measure the workload. It also has the capability of predicting lifting capacity and providing general ergonomic guidelines. However, the system does not employ a formal inference mechanism. A potential solution or suggestion is rather drawn by IF–THEN statements written in BASIC.

A more mature system was developed by Kabuka *et al.* (1988) for the design of new and existing repetitive manual materials handling (RMMH) tasks. It provides maximum acceptable weight for new RMMH tasks and recommendations for the redesign of existing RMMH tasks. Quantitative psychophysical study results were incorporated as rules and classified into subframes according to the type of manual handling activities; lifting, lowering, carrying, and pushing/pulling. Now, the system scope is lmited to the ergonomic knowledge in psychophysics. In general, applying an expert system to an ergonomic domain is still in its infancy.

The prototype expert system implementation

By building an expert system we mean the development of three main modules of an expert system: a knowledge base, an inference mechanism, and a user interface. Here, constructing a comprehensive knowledge base would be the most important task, since the essential purpose of building an expert system is to capture the expertise to solve ergonomic problems. As mentioned, the declarative knowledge includes facts related to the ergonomic domain, such as those found in the NIOSH guidelines (NIOSH, 1981). The procedural knowledge, usually represented in the form of rules, interacts with the declarative knowledge and contains the procedures for deriving possible solutions. One of the important characteristics of ergonomic workplace design reflects a multi-factored iterative process. Ergonomic information, which is classified into operator, task, workplace information, is further categorized based on three ergonomic factors; biomechanical, psychophysical, and work physiological factors.

As described in Figure 20.2, each module of the procedural knowledge base generates a working rule module to the specific case by the inference mechanism including meta-knowledge. The declarative knowledge is stored in different submodules according to the type of information. When it contains massive data like anthropometric data, it is separately stored in the data base so that it is efficiently retrieved and updated without bothering the main inference chain. When it requires mathematical computations, it is written in FORTRAN and placed outside the memory. This information is loaded through the system interpreter only when it is necessary (IBM, 1985). If it is only to be displayed to the user, it is put into the separate files and displayed through the user interface. The main modules shown in Figure 20.2 will be discussed later.

Programming language selection

Since the system architecture is greatly affected by the data structure of the language chosen, an appropriate programming language should be selected. LISP or PROLOG are the two possible general-purpose languages in artificial intelligence and expert systems. LISP has been the primary language of artificial intelligence for many years, since it is very flexible for manipulating arbitrary objects in a list structure. It is also a low-level language which brings more flexibility than PROLOG. However, at the same time, LISP is harder to understand because it is too low for most programmers. LISP also has a hard-to-read syntax unlike that of any other programming language. PROLOG which has steadily gained in popularity, has two advantages over LISP. First, its syntax and semantics are much closer to predicate logic, the way of representing facts and reasoning mechanism chosen for the proposed expert system. Second, it provides automatic backtracking, making search easier (Rowe, 1988).

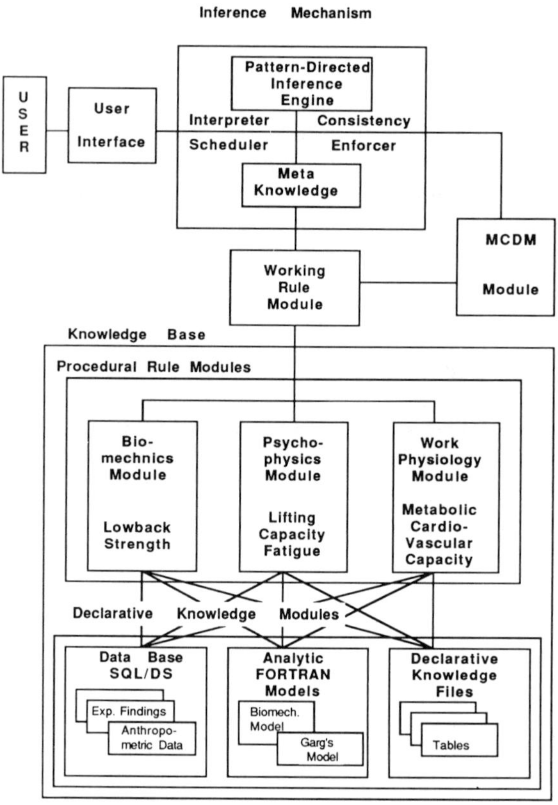

Figure 20.2 The architecture of the proposed expert system.

There is another aspect in both languages which should be discussed: the capability of interfacing to other languages. Both languages were primarily built for natural language processing. In fact, they support neither graphic functions nor efficient computational facilities, which are very useful for workplace design and user interface. Also, they are not appropriate for dealing with large amounts of ergonomic data in terms of efficiency. For the purpose of linking to the system analytic models written in FORTRAN such as biomechanical strength models, these languages should be used in conjunction with FORTRAN and possibly PASCAL. In general, PROLOG retains a higher modularity and better I/O predicates than LISP, which make the system possible to interface. VM/PROLOG, which is an IBM implementation of the PROLOG language, was selected for the prototype expert system based on the reasons above. In addition to the common features of PROLOG, VM/PROLOG has three positive features. First, it is capable of accessing a relational data base called SQL/DS, a standard relational data base concept. Second, it can communicate with the system interpreter from which other languages can be concurrently loaded. Finally, it is highly expandable, since it employs a concept, a so-called nucleus extension, which allows other modules to be loaded and run concurrently in the virtual machine (IBM, 1985).

Inference mechanism

In the prototype expert system, the inference mechanism is referred to as a more general scheme within which the inference engine with meta-knowledge play such roles of preprocessors. It interprets the user descriptions of the specific problem prepared through the user interface and schedules, firing appropriate rules to be asserted. When asserting the rules fired, it also keeps the record of those rules throughout the entire session. As discussed, the prototype system adopted a pattern-directed, rule-based inference system, since the system gives a system a high degree of modularity and expandability. The rules in the knowledge base take the form of

IF Condition_Part THEN Action_Part_1 ELSE Action_Part_2

The inference mechanism examines each rule in the working rule module in a breadth-first search (BFS) fashion, by testing Condition_Part which is a list of preconditions to be satisfied. If those are satisfied from the global parameter stack that represents the system status, then Action_Part_1 is executed, otherwise Action_Part_2 is executed. Execution of Action_Part involves

1. Flagging the global parameters to update the system status.
2. Asserting new facts which are to be used for further chaining.
3. Directing the user to find appropriate knowledge through the data base or the menus.
4. Suggesting general guidelines for workplace design.
5. Preparing inputs to the models available.

Use of an external database

Ergonomic knowledge consists of a variety of massive data ranging from psychophysical lifting guidelines to human body segment data, since the use of data about human dimensions and performance is a critical element in workplace design. In such a domain, it is desirable that a PROLOG program supports a large number of assertions efficiently. However, there are usually system limits on speed and main memory, which can force an extensive swapping of data. The concurrent use of a relational data base system avoids this extra burden. The prototype system using VM/PROLOG has an interface to Standard Query Language/Data System (SQL/DS). It has the functions required for data management, including indexing, data sharing and recovery control. Basically, the interface to SQL/DS is done by specifying SQL statements in PROLOG predicates.

There are three types of SQL statements being used: data definition statement, data insertion statement, and data retrieval statement (Chang and Walker, 1986). Data definition statements are used to create or drop a table as follows;

```
CREATE TABLE AYOUB79
  (LIFTHT CHAR(6), SEX CHAR(6), POP SMALLINT, MAWLB DECIMAL
  (5,2) IN SQLDBA.E4J)
```

The first statement makes the system to create in the data base the table named AYOUB79 that has four columns of data in appropriate data types; lifting height, sex, population, and maximum acceptable weight in pounds. For most of the time during the reasoning process, only the data retrieval statements will be used to retrieve specific values from the table.

```
SELECT MAWLB FROM AYOUB79
  WHERE SEX = 'MALE' AND POP = 75 AND LIFTHT = 'FLOOR-TO-
  KNUCKLE'
```

This statement retrieves the maximum acceptable weight from the table, when the lifting height ranges from floor to knuckle and 75% of the male workers should be capable of performing the task safely. The above statements are drawn from the example of utilizing the experimental findings obtained by Ayoub *et al.* (1979) for determining the maximum acceptable weight.

Analytic FORTRAN Models

While many existing expert systems have excellent capability to draw general guidances and advices, they often lack analytical and mathematical problem-solving capabilities mainly due to the limitations of AI languages to deal with numbers and the natural language-oriented properties. A number of analytic models developed for the ergonomic analysis of workplace design can serve to compensate for this deficiency. These models provide the analysts with an accurate estimate of physical stresses imposed on the human operator for a given ergonomic factor. The prototype expert system encompasses the performance of the existing expert system by guiding the user to formulate the input to these models. The inference mechanism keeps track of the user specifications and the system's own expertise, prepares the input to utilize appropriate models in the analytic model module, such as metabolic energy cost prediction models and biomechanical models.

Multiple Criteria Decision Making (MCDM) module

Ergonomic design is itself a decision-making process to derive a solution among multiple factors. Through the derivation of an appropriate design and possible alternatives, we always face a decision-making process. Since the ergonomic knowledge base is modular and hierarchical, each module which is being invoked may have a different weight due to its importance. Importance of each module can be determined by the designer. Therefore, the general decision-making scheme is needed when combining different modules. In this research for developing a general ergonomic expert system, a formal multiple-criteria, decision-making scheme was adopted for considering multi-factored ergonomic principles. Suppose we fired three different rules from the rule base, with attribute values calculated from each rule respectively $\{X_1, X_2, X_3\}$. The attribute values are, of course, normalized. If the user or the system by default estimates the importance of each rule by assigning w_i, then, one of the schemes in MCDM will be the simple additive weighting method. The output confidence value is given by

$$D = \frac{\Sigma(x_i\, w_i)}{\Sigma w_i}$$

In fact, the rule-based deduction as an inference mechanism can be more complicated in that a combination of antecedent rules consists of both conjunctive and disjunctive rules. A simple approach based on classical logic representation has been used in several systems such as MYCIN; if these rules act as a conjunctive antecedents, then the consequent can be simply asserted with $\min\{x_i\}$ or if they are disjunctive, $\max\{x_i\}$. More generally, suppose we have three rules or alternatives $X = \{X_1, X_2, X_3\}$, and we evaluate these alternatives with respect to the two different attributes $\{A1, A2\}$. These attributes can be, for example, two representative ergonomic factors such as physiological demands, low back compressive forces. If we equally deal with these attributes or factors, the decision can then be made by

$$D = \max_{A_i} \min\,\{X_i\}$$

If the designer feels that one factor is more important than another factor, we assign the importance to each attribute A_i. As Yager (1977) suggested, each attribute can be weakened or strengthened by using the operation similar to concentration and dilation operations given by

$$D = \max_{A_i} \min \{X_i^{w_i}\}$$

$$\Sigma w_i / n = 1$$

where w_i is the importance of each attribute. The general decision-making scheme of an expert system depends not only on the designer's needs but also on the system's prior knowledge accumulated by different experts or knowledge engineers, as the system develops.

Expert system development details

Typical design tasks of manual handling jobs that the designer often faces include various activities such as lifting, lowering, pushing and load carrying, which impose physical and mental stresses on the human operator. Ergonomic information which should be incorporated involves operator variables such as gender, age, strength, and population of the operators, task variables such as the weight, vertical and horizontal location, vertical height of the object, moving distance, walking speed and grade, and environmental variables such as relative humidity and temperature, and so on. The designer must either determine the value of these variables which are of interest by using the knowledge that the system and the designer possess.

The prototype system starts the dialogue by asking the user of the task he wants to design. It asks what type of job he is interested in, what factors should be considered. When the user is so naive that he is not able to provide the set of input parameters, the system either shows all the options or its reasonings about the relationships between the job and the factors involved. Next, it invokes appropriate modules (e.g. work physiology module, biomechanics module, etc.) to evaluate. While the control of the system passes throughout all the modules, the inference mechanism gathers all the inferences and evidences from the rules fired. When the system is able to gather sufficient input parameters for the analytic models available, it prepares the input file for these models and performs an evaluation of a possible solution and/or alternative designs.

Ergonomists have compiled a massive body of research results over decades. These results have been published in various forms; tables, graphs, formulae, and regression equations. One of the common characteristics of the results is that none of these fits all the facets of the problems or can even be generalized. Each research has a certain type of precondition in which numerous assumptions are used, and deals with only a part of specified variables. Therefore, the object-level rules are to be made on a basis of each researcher's results. Preconditions and assumptions of each research (condition_part of the rules) are then utilized to find the appropriate pieces of knowledge that help the user design an ergonomically sound workplace.

For example, the knowledge in psychophysics module will be helpful in determining the weight of the load to be handled in the initial design stage. The prototype system suggests to the user appropriate models according to the variables available and the experimental conditions of those models. When little information is available, the system provides a simple guideline like the ILO suggested limits for weight lifting (ILO, 1962). Here, the ILO guideline is stored in the data base and retrieved through the interface between the PROLOG and the data base.

Rule base: IF only age and sex are known,

THEN use ILO62

Fact base: ILO62: Maximum acceptable weight in kg

Age	Male	Female
14–16	14·6	9·8
16–18	18·5	11·7
18–20	22·6	13·7
20–35	24·5	14·6
35–50	20·6	12·7
50–	15·6	9·8

More detailed research results become available as the user specifies the job conditions. The system possesses the results from Snook and his colleagues (Snook *et al.*, 1970; Snook, 1978), Ayoub *et al.* (1979), Ergonomics Guides (AIHA, 1970), NIOSH guidelines (NIOSH, 1981), etc. One of the common characteristics is that most of their findings are provided in tables and graphs, which are somewhat difficult and inefficient to implement. Thus, these tables and graphs were converted into either regression equations residing in the fact base or the tables stored in the data base. The system, then, provides the user with the appropriate maximum acceptable weight or workload. For example, the following rules which require a different set of variables in the psychophysical module show the use of tabularized knowledge.

IF vertical distance, horizontal distance, gender, population are known
THEN use Eguide70
IF horizontal distance, population are known and gender is male
THEN use Snook70
IF horizontal distance, frequency, gender are known
THEN use Ciriello83
IF horizontal distance, population, gender, frequency, box size are known
THEN use Ayoub79
IF horizontal distance, vertical distance, frequency, moving distance are known
THEN use NIOSH81

Of course, rules are actually implemented in more PROLOG-like expressions. Here, it is quite plausible to have multiple potential solutions given the design specifications. So, it is natural to assume that those alternatives are next evaluated by physiological and biomechanical criteria. It is, of course, possible to run those modules independently. The main flowchart is shown in Figure 20.3.

The modules in the prototype system consist of three main parts; a rule base, a fact base, and supporting functions. The rule base consists of IF–THEN–ELSE type of rules and guides the user to find the appropriate design specifications. The fact base is, basically, a collection of research results in the specific field. It contains two types of knowledge; informative facts such as suggestions and simple guidelines, and computational facts such as regression equations and graphs. Here, it should be noted that table look-ups are separately stored in the data base.

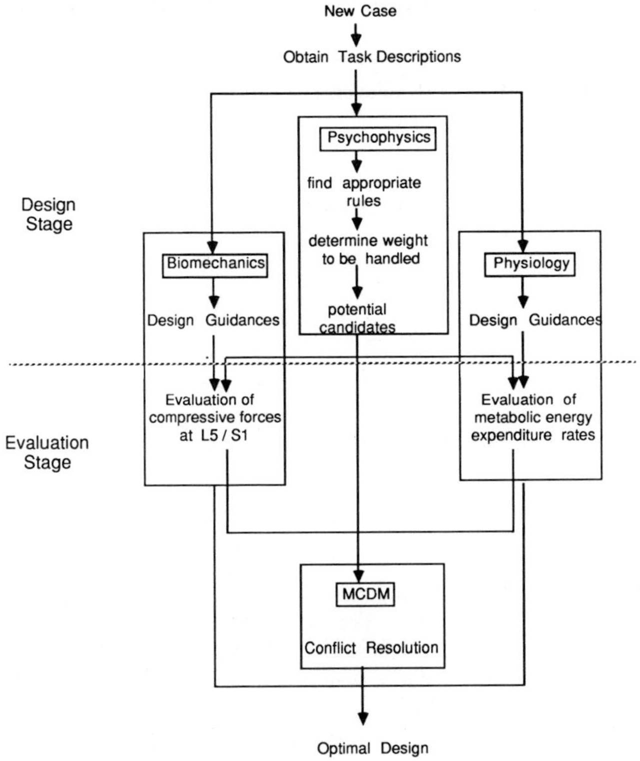

Figure 20.3 *The system's flowchart.*

For instance, the work physiology module concerns the energistic effects of the work. In order to assure that the demands do not exceed worker's capability, the system first tries to obtain the physical work capacity (PWC) of the group of operators in the workplace. If not available, the system is capable of providing appropriate guidelines like the experimental result by Bink (1962). Next, the PWC or V_{O2}max should be corrected to establish the limit for a daily work, or to accommodate the age of the operator.

Lehmann58: metabolic energy expenditure rate

5·2 kcal/min for male, 8 h work
3·5 kcal/min for female, 8 h work

Bink62M: metabolic energy expenditure rate

$$\text{Limit} = 17{\cdot}9 - 4{\cdot}77^{*}\text{LOG(TIME(min))}$$
$$\text{Limit} = 20{\cdot}4 - 0{\cdot}0688^{*}\text{AGE} - 4{\cdot}75^{*}\text{LOG(TIME(min))}$$

Once the physiological limit allowed for an operator is set up, the next step would be to evaluate the metabolic requirements of the particular task or job. In the early stages of design, the user may have no specific information, but a general idea about the type of tasks and it severity. In this case, the system finds the rule related to a general metabolic estimation according to the task severity. It displays the file which has the experimental results by Passmore and Durnin (1955) for metabolic energy expenditure rates of the tasks depending on their severities as below. More specifically, the user can pick up the activities involved in the job, estimate the proportion of time for each activity, and calculate the weighted metabolic requirement of the job. Rough estimations of operator and task variables are very useful especially in the early stages of design. The values of parameters may be better represented by linguistic values such as light, moderate, and heavy, etc. It is believed that fuzzy logics may play an important role in drawing appropriate inferences in this stage. Here, for example, heavy metabolic requirements can be represented as

$$\text{heavy} = 0{\cdot}5/5{\cdot}0 + 0{\cdot}8/7{\cdot}0 + 1{\cdot}0/9{\cdot}0 + 0{\cdot}8/11{\cdot}0$$

Finally, when the user is capable of providing the necessary information, the system prepares input files to run the analytic models such as Garg's model (Garg *et al.*, 1978), displays the output back to the user, and returns the control back to the main system.

Conclusion

The system is being tested by experts and various ergonomic design cases available from the literature to refine the system parameters. The use of an external data base made it possible to manipulate massive ergonomic data efficiently. The system is also capable of drawing inferences from quantitative knowledge as well as qualitative knowledge by using analytical models. No significant increase in inference time and run time showed that the system architecture is highly modularized and expandable. Further refinements are being made to validate its use.

References

AIHA, 1970, Ergonomics guide to manual lifting. *American Industrial Hygiene Association Journal*, **31**, 510–546.

Alexander, D. C., 1986, *The Practice and Management of Industrial Ergonomics* (Englewood Cliffs, NJ: Prentice-Hall).

Ayoub, M. M., Dryden, R., McDaniel, J., Knipper, R. and Dixon, D., 1979, Predicting lifting capacity. *American Industrial Hygiene Association Journal*, **40**, 1075–1084.

Barr, A. and Feigenbaum, E. A., 1982, *The Handbook of Artificial Intelligence*, Vol. 1. (Los Altos, CA: William Kaufmann).

Bink, B., 1962, The physical work capacity in relation to working time and age. *Ergonomics*, **5**(25), 25–28.

Bratko, I., 1986, *PROLOG Programming for Artificial Intelligence*, (Reading, MA: Addison-Wesley).

Chang, C. L. and Walker, A., 1986, PROSQL: A PROLOG programming interface with SQL/DS. In: *Expert Data Base Systems*, edited by Larry Kerschberg (Benjamin/Cimmings) pp. 233–246.

Charniak, E. and McDermott, D., 1985, *Introduction of Artificial Intelligence*, (Reading, MA: Addison-Wesley).

Chaffin, D. B. and Andersson, G. B. J., 1984, *Occupational Biomechanics*, (New York: John Wiley).

Chen, J., 1987, Prototype expert system for physical work stress analysis. Ph.D. Dissertation, Norman, OK, School of Industrial Engineering, The University of Oklahoma.

Coury, B. G., 1983, *A Specification for an Ergonomic Expert System* Nottingham University, Department of Production Engineering and Production Management.

DeGreve. T. B and Ayoub, M. M., 1987, A workplace design expert system. *International Journal of Industrial Ergonomics*, **2**(1), 37–48.

Duda, R. O., Hart, P. E., Nilsson, N. J. and Sutherland, G. L., 1978, Semantic network representation in rule-based inference systems. In: *Pattern-Directed Inference Systems*, edited by D. A. Waterman and F. Hayes-Roth (New York: Academic Press), pp. 203–221.

Dul, J. and Hildebrandt, V. H., 1987, Ergonomic guidelines for the prevention of low back pain at the workplace. *Ergonomics*, **30**(2), 419–429.

Evans, S. M. R., 1985, Ergonomics in manual workplace designs: current practices and an alternative computer-assisted approach. Ph.D. Dissertation, Ann Arbor, MI, Department of Industrial Engineering, The University of Michigan.

Garg, A., Chaffin, D. B. and Herrin, G. D., 1978, Prediction of metabolic rates for manual materials handling jobs. *American Industrial Hygiene Association Journal*, **39**, 661–674.

Hayes-Roth, F., Waterman, D. A. and Lenat, D. B., 1978, Principles of pattern-directed inference systems. In *Pattern-directed Inference Systems*, edited by D. A. Waterman and F. Hayes-Roth (New York: Academic Press) pp. 577–601.

Hayes-Roth, F., Waterman, D. A. and Lenat, D. B., 1983, *Building Expert Systems*, (Reading, MA: Addison-Wesley).

Huchingson, R. D., 1981, *New Horizons for Human Factors in Design*, (New York: McGraw-Hill).

IBM, 1985, *VM/Programming in Logic*, Program No. 5785-ABH.

ILO, 1962, *Maximum Permissible Weight to Be Carried by One Worker*, (Geneva: ILO Office).

Jiang, B. C. and Ayoub, M. M., 1987, Modelling of maximum acceptable load of lifting by physical factors. *Ergonomics*, **30**(3), 529–538.

Kabuka, M., Genaidy, A. M. and Asfour, S. S., 1988, A knowledge-based system for the design of manual materials handling. *Applied Ergonomics*, **19**(2), 147–155.

Karwowski, W., Mulholland, N. O. and Ward, T. L., 1986, LIFTAN: an experimental expert system for analysis of manual lifting tasks. *Ergonomics*, **29**(10), 1213–1234.

Karwowski, W., Mital, A., Palenque, L. E., and Ward, T. L., 1987, Development of a microcomputer-based expert system for the analysis of manual materials handling tasks in industrial settings. *International Journal of Industrial Ergonomics*, **2**(1), 49–59.

Kowalski, R., 1979, Algorithm = Logic + Control. *ACM Communications*, **22**(7), 424–436.

Maynard, H. B. (Ed.), 1971, *Industrial Engineering Handbook*, 3rd edn. (New York: McGraw-Hill).

NIOSH, 1981, *Work Practices Guide for Manual Lifting*, No. 81–122, Doc US HE20 7108 (Washington DC: NIOSH).

Niwa, K., Sasaki, K. and Ihara, H., 1984, An experimental comparison of knowledge representation schemes. *The AI Magazine*, 29–36, Summer.

Passmore, R. and Durnin, J. V. G. A., 1955, Human energy expenditure. *Physiological Reviews*, **35**(4), 801–840.

Pope, M. H., Frymoyer, J. W. and Andersson, G. B. J., 1984, *Occupational Low Back Pain* (New York: Praeger Scientific).

Quillian, R., 1968, Semantic memory. In *Semantic Information Processing*, edited by M. Minsky (Cambridge, MA: MIT Press).

Ramsey, C. L. Reggia, J. A., Nau, D. S. and Ferrentino, A., 1986, A comparative analysis of methods for expert systems. *International Journal of Man–Machine Studies*, **24**, 475–499.

Rowe, N. C., 1988, *Artificial Intelligence through PROLOG* (Englewood Cliffs, NJ: Prentice-Hall).

Snook, S. H., 1978, The design of material handling tasks. *Ergonomics*, **21**(12), 963–985.

Snook, S. H., Irvine, C. H. and Bass, S. F., 1970, Maximum weights and workloads acceptable to male industrial workers. *American Industrial Hygiene Association Journal*, **31**, 579–586.

Tirupathikumara, S. R., Kashyap, R. L. and Moodie, C. L., 1985, *Artificial Intelligence Techniques in Facilities Layout Planning: The Development of An Expert System*. TR-ERC 86–1 (West Lafayette, IN: Department of Industrial Engineering, Purdue University).

Tsatsoulis, C. and Kashyap, R. L., 1987, *Using Dynamic Memory Structures in Planning and Its Applications to Manufacturing*. TR-ERC 87–9 (West Lafayette, IN: Department of Industrial Engineering, Purdue University).

Waterman, D. A., 1986, *A Guide to Expert Systems*, (Reading, MA: Addison-Wesley).

Waterman, D. A. and Hayes-Roth, F., 1978, An overview of pattern directed inference system. In: *Pattern-Directed Inference Systems*, edited by D. A. Waterman and F. Hayes-Roth (New York: Academic Press), pp. 3–22.

Yager, R. R., 1977, Multiple objective decision-making using fuzzy sets. *International Journal of Man–Machine Studies*, **9**, 375–382.

Yu, T., Roht, L. H., Wise, R. A., Killan, D. J. and Weir, F. W., 1984, Low-back pain in industry: an old problem revisited. *Journal of Occupational Medicine*, **26**(7), 517–524.

Chapter 21
ERGON-EXPERT: a modular knowledge-based approach to reduce health and safety hazards in manual materials handling tasks

VOLKER ROMBACH and WOLFGANG LAURIG

Institut für Arbeitsphysiologie an der Universität Dortmund Abteilung Ergonomie, Ardeystraße 67, D-4600 Dortmund 1, Federal Republic of Germany

Abstract. A deficiency in the practical use of ergonomic knowledge is derived from discussions about its development. Applying the methodology of expert systems in ergonomics is one possibility for reducing this deficiency. The stage of development of an expert system for assessing lifting and carrying tasks is presented. Special attention was paid to the user interface in order to improve the acceptance and the educational value of the system. The knowledge base of the system contains basic knowledge for dealing with ergonomic problems relating to manual materials handling.

Keywords: Development of ergonomic knowledge; Expert systems; Metabolic load; Assessment and evaluation of work design.

Introduction

The Ergonomics Information Analysis Centre has been publishing the *Ergonomics Abstracts* since 1969. They contain such a wealth of insights and results relating to specific problems that it is hardly possible to retain an overall view. Although these abstracts are of considerable use to the ergonomics scientist, the practitioner is rarely qualified enough to benefit from the knowledge they document. As far as the problems encountered on the shop floor are concerned, there is an imbalance between the supply of knowledge from the field of industrial ergonomics and the demand for it.

The application of ergonomic knowledge thereby seems to have reached a critical stage, one at which no essential methodological progress in terms of the translation of the knowledge into practice can be expected (Marshall, 1989a, b). A number of computerized approaches have been developed to improve the application of ergonomic knowledge in industry (Dooley, 1982; Chaffin and Evans, 1986; Karwowski, 1987a). Taking as an example the traditional problem area of manual materials handling tasks, computer systems have been developed for the (i) analysis (e.g. Chaffin and Anderson, 1984), (ii) evaluation (e.g. Laurig *et al.*, 1985), (iii) simulation (cf. Kroemer *et al.*, 1988), (iv) energy and cardiac cost prediction, (e.g. Asfour *et al.*, 1985), and (v) design (e.g. Genaidy *et al.*, 1987) of workplaces and work sequences. However, it must be emphasized here that many of these systems have remained prototypes, scientific studies or special-purpose tools of limited availability or applicability.

Some recent approaches to develop tools for the special tasks of 'industrial ergonomics' have used results from research in the field of 'knowledge-based expert systems' (Karwowski, 1987b).

Using current understanding of the concept of expert systems for orientation purposes, it can be stated, following Jackson (1986), that:

> Expert systems are computer programs which use rules and facts about a particular problem area to enable the user to solve problems from that area.

Consequently, expert systems must be well-suited to dealing with the aforementioned imbalance between the supply of ergonomic knowledge and the demand for it. However, in practice a variety of misunderstandings exist concerning the concept of expert systems, these occurring particularly in conjunction with misunderstandings about the aims of AI research. It should therefore be noted before proceeding that expert systems do not involve 'artificial experts'. In order to prevent further misunderstandings, it would be advisable in future to use the term 'expert-support systems' instead of 'expert systems'.

Accordingly, the development of expert systems in ergonomics could help to support the shop-floor expert as well as, indeed, the ergonomics expert himself.

The development of expert systems in ergonomics involves two different problem areas. First, the technical requirements for the development of expert systems must be clarified taking into account the interests and demands of the future users (Laurig, 1987). Using experience from knowledge engineering it is also necessary to examine the procedures used by ergonomics in the solution of design tasks. Only in this way will it be possible to represent in a generally valid form the methods employed by ergonomics to solve particular problems. The knowledge required for the solution of these problems must also be structured in such a way that it can be processed and utilized according to the technical requirements (= ergonomic knowledge engineering).

Similar suggestions, from the point of view of human factors, were presented and discussed by Chignell and Peterson (1988). One of the first applications of these concepts for manual materials handling was an experimental system for the analysis of lifting tasks (LIFTAN) as presented by Karwowski *et al.* (1986).

The following considerations refer to the development of a system named ERGON-EXPERT (Rombach and Laurig, 1988). In accordance with the explanation of the concept of expert systems provided in the foregoing, this development is limited for the present to the 'particular problem area' of manual materials handling (MMH) activities.

Concept of ERGON-EXPERT

ERGON-EXPERT is intended as a tool for identifying health hazards by means of a comprehensive analysis and assessment of work sequences of manual materials handling tasks. To achieve this goal, the analysis and assessment process requires knowledge from various domains of occupational health. The sequence of the analysis and assessment, determined according to the importance in practice, is:

1. Compliance of the work sequence with laws, regulations, rules, guidelines and standards of occupational safety and health.
2. Compliance with constraints related to practice.
3. Compliance with constraints related to man.
4. Compliance with ergonomic principles of work design.

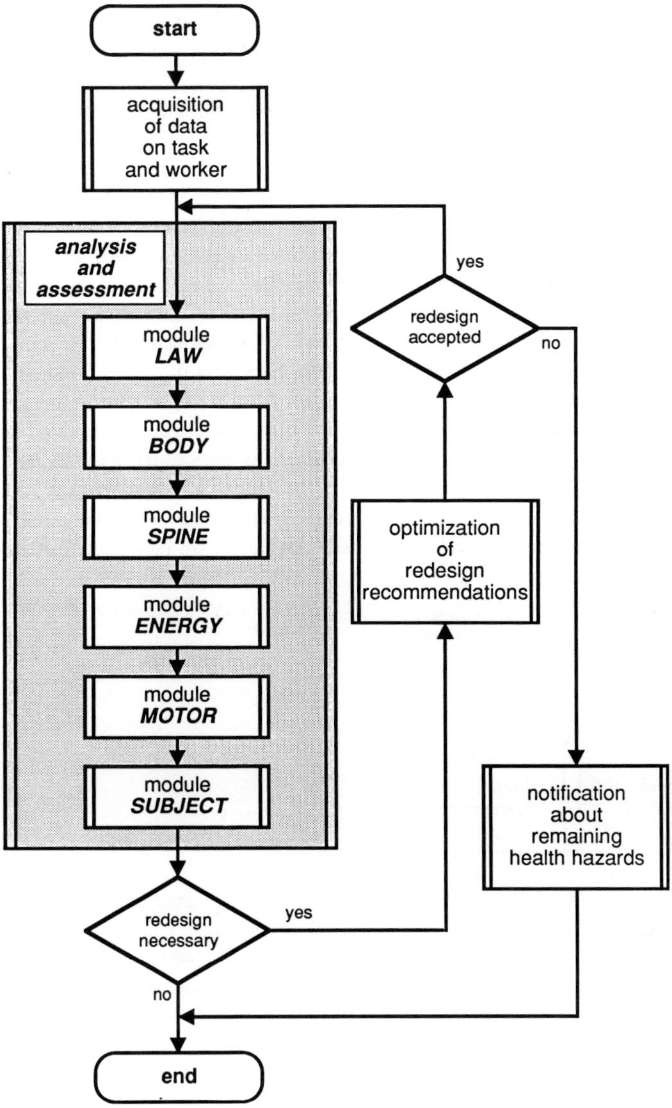

Figure 21.1 Principal structure of ERGON-EXPERT.

An example of the regulations mentioned above are the Occupational Safety and Health Regulations issued by the Australian Minister for Labour (1988). Examples of typical constraints related to practice would be special task requirements such as the type and shape of packing materials or the priority of operation sequences during a manual materials handling task. Cost and benefit considerations are also of major importance for the actual design to be translated into practice. Similar to the above sequence, the order of the constraints related to man is determined by their importance for the design of a work sequence. The order is as follows: (i) anthropometric factors, (ii) skeletal abilities, (iii) muscular abilities (maximum strength), (iv) metabolic capacity, (v) cardiovascular capacity, (vi) motor skills, and (vii) subjectively perceived workload.

Anthropometric demands result from the specific task and workplace. Demands on musculoskeletal abilities can be deduced from the attributes describing the load to be handled. Comparison of the anthropometric data of the worker with the demands at the workplace and the assessment of the musculoskeletal abilities against these demands determine the 'practicability' of the task. This term is found in the hierarchic ergonomic evaluation concept proposed by Rohmert (1972). In accordance with this concept, comparison of the demands of the work sequence with the metabolic and cardiovascular capacity conforms to the evaluation goal 'tolerability'. The motor skills of the specific worker and the subjectively perceived workload are used for evaluating the 'acceptability' of the workplace. The general principle of the analysis and assessment process is shown in Figure 21.1.

In the first module (LAW) the work sequence is checked for compliance with laws, regulations, rules, guidelines and standards as found, for example, in the NIOSH *Work Practice Guide for Manual Lifting* (NIOSH, 1981). The procedure used in the CAD tool WERNER (Kloke, 1989) is to be adapted for the second module (BODY). The module SPINE is developed along the lines recommended by Troup *et al.* (1984) and uses a simplified version of the biomechanical calculation model developed by Jäger and Luttmann (1987).

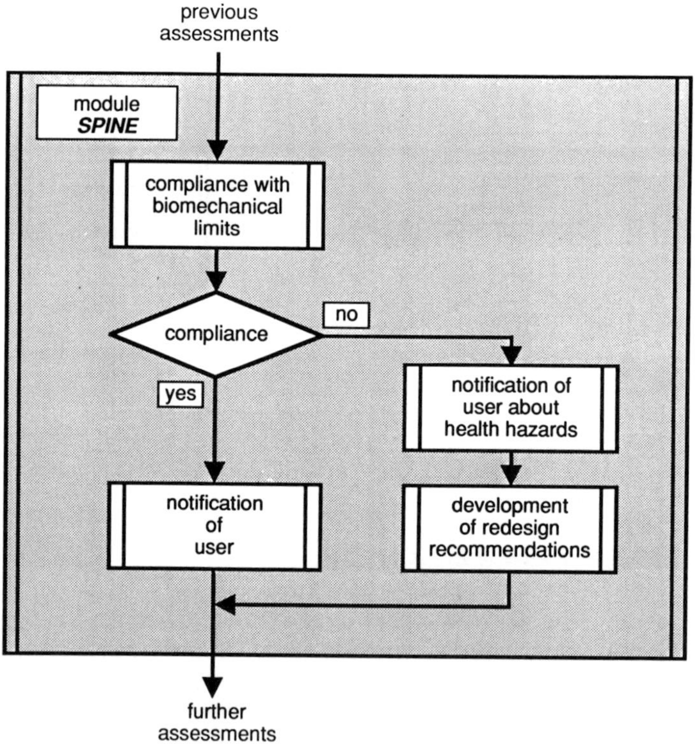

Figure 21.2 Schematic representation of the procedure in the analysis and assessment module "BODY".

This module is used for the calculation of the forces and torques acting at the lumbo-sacral joint (L5-S1). Since the limit for the maximum acceptable compressive force is still under discussion, a limit value of 4 kN is used for present assessment purposes (cf. NIOSH, 1981 or Jäger and Luttmann, 1986). The module ENERGY, employed in the assessment of the metabolic load resulting from the work sequence, is developed following considerations published, for example, by Garg *et al.* (1978) and Genaidy and Asfour (1987). The techniques used in the product MAID (Laurig *et al.*, 1985) are being adapted for module 5 (MOTOR). The module SUBJECT uses the limit values published by Mital (1984a, b). The modular structure of ERGON-EXPERT permits the easy extension of the list of modules and assessment methods.

All of the constraints are checked to ensure comprehensive results. The results found in the various analysis and assessment modules are used, where necessary, for the development of redesign recommendations. Since an evaluation using biomechanical, physiological and psychophysical methods can lead to contradicting redesign recommendations (Garg and Ayoub, 1980), an optimizing function is being considered. This function takes into account the boundaries of the applicability of the evaluation methods, the capabilities of the specific worker, specific task requirements and ergonomic job design principles (cf., for example, Garg and Ayoub, 1980; NIOSH, 1981; Konz, 1983). If one of the redesign recommendations is accepted by the user, the assessment process can be repeated to illustrate the effect of the new design. Figure 21.2 provides a detailed example of the structure of one of the analysis and assessment modules. The other modules are structured similarly.

Implementation

Since ERGON-EXPERT is intended for use by various user groups, it is impossible to predict the hardware environment at the place of application. Special attention must therefore be paid to the portability of the system. A development tool which is found on a wide variety of hardware platforms is NEXPERT OBJECT by NEURON DATA, Inc. of Palo Alto, Ca. The expert system shell was chosen for the development of ERGON-EXPERT because it combines an easy-to-use developer interface with powerful knowledge representation and knowledge processing mechanisms. The knowledge contained in a knowledge base

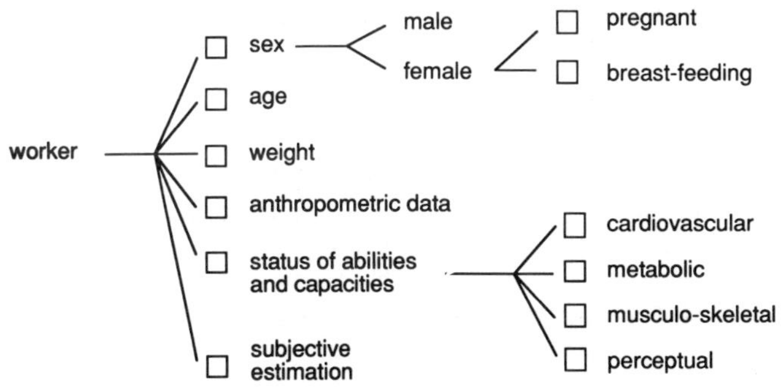

Figure 21.3 Example for the description of elements of the work system via objects and their attributes.

belonging to a system built with NEXPERT OBJECT is represented using both object descriptions as in the SMALLTALK programming environment (Goldberg and Kay, 1976) and IF . . . THEN . . . rules similar to those used in the well-known expert system MYCIN (Jackson, 1986). An example of the description of a worker is to be seen in Figure 21.3.

Since ERGON-EXPERT's main purpose is to detect and to minimize health hazards in manual materials handling tasks, a number of descriptions of standard tasks are included in the system. These are, for example: (i) lifting a load, (ii) lowering a load, (iii) transferring a load, (iv) standing with/without load, and (v) walking with/without load.

The load to be lifted is described by attributes such as size, weight, dimensions and filling. The results acquired during the analysis and assessment process contribute to the determination of redesign requirements. These requirements are then transformed into redesign recommendations using ergonomic knowledge. Since the knowledge for the analysis and assessment is derived from different knowledge domains it is only sensible to use the same structure when building ERGON-EXPERT. The knowledge required for each analysis and assessment module is contained in a separate knowledge base. This allows easy modification and extension of the knowledge bases. They are linked together at run time by meta-level rules describing the relationship and hierarchy of the modules. An example of such a meta-level rule is shown in Figure 21.4.

If there is no evidence of

violation of regulations and standards

then hypothesis

workplace complies with

regulations and standards

is confirmed

and load knowledge base "**body**"

Figure 21.4 Example of a meta-level rule.

Such a design facilitates the extension of the areas of application of ERGON-EXPERT through the addition of further modules, e.g. for the assessment of heat stress factors (cf., Taylor and Corlett, 1987 or Das, 1988). The calculation models used by the analysis and assessment modules are programmed in C and PASCAL and are also separated from unrelated knowledge, so allowing flexible modification and exchange.

The user interface

Since the users' knowledge about ergonomics, computer systems, and the expert system itself varies, the user interface of the expert system has to fulfil a number of demands. These include, among other things, (i) a high degree of self-explainability, (ii) robustness against

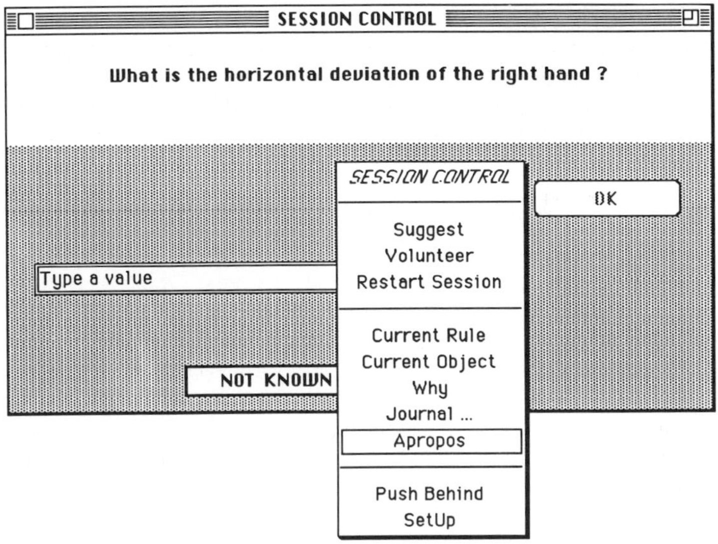

Figure 21.5 Example for the design of the user interface of ERGON-EXPERT.

incorrect input, and (iii) adaptability of the user interface to the needs of experienced and lay users.

The above-mentioned demands led to the decision to use a highly interactive user interface with special attention being paid to the explanation facilities. The main development guidelines are to:

1. Allow, wherever possible, the user to choose between possible answers to a question,
2. Give the user access to detailed background information on the problems under investigation,
3. Inform the user about incorrect or illogical input as soon as possible.

The provision of some of these features in the expert system shell further justifies its choice. The ability to display at all times the status of the inference process by investigating the current values of both hypotheses and variables, together with information about reasons for the confirmation or rejection of any hypothesis, gives valuable help when analysing a workplace. These facilities are complemented by the tracing facility of the shell and a network display technique which provide an elegant means of investigating the conclusions in their contexts. The user interface of ERGON-EXPERT is designed for various groups of persons. The potential usership of the system includes OSHA-inspectors, company physicians, industrial engineers, workplace designers, human factors experts, rehabilitation experts, and students of industrial engineering and workplace design.

The provision of the above-mentioned information therefore contributes to the extension of the knowledge of non-experts in ergonomics.

An example of the interface of the current version of ERGON-EXPERT is shown in Figure 21.5. Using the 'Apropos'-function of the system leads to the display of additional information as, for example, in Figure 21.6. Typical details concerning the results of the analysis and assessment process are provided in Figures 21.7 and 21.8.

Since not all of the modules in Figure 21.1 are complete, ERGON-EXPERT is currently at the prototype stage.

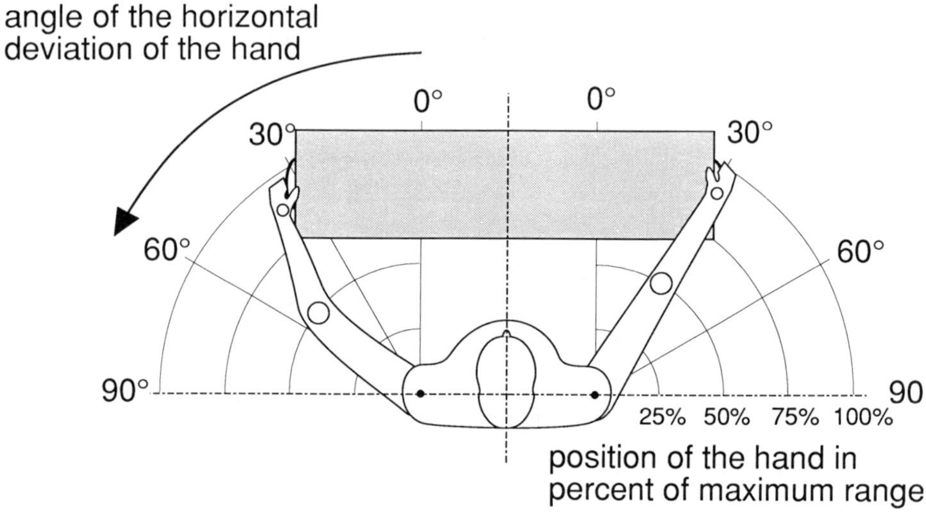

Figure 21.6 Display of additional information about system requests via the "Apropos"-function.

!

End of
biomechanical assessment

In all of the analysed body postures the calculated compressive forces on the intervertebral disc at L5–S1 **do not exceed the limit value** of 4 kN.

There seems to be no risk of damage to the intervertebral disc.

For exact values of the calculated compressive forces in the various postures refer to the object network.

Figure 21.7 Notification of the user about the result of the biomechanical assessment.

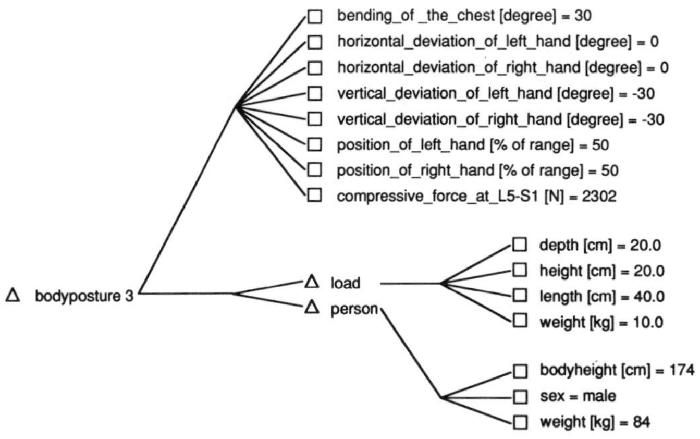

Figure 21.8 Comprehensive display of detailed results.

Concluding remarks

The development of expert systems for ergonomics provides an opportunity for scientific justifications for the various assessment approaches and their results to be compared with each other. This has hardly been possible previously. In order to exploit this new aspect in the methodology of knowledge-based systems, a distinction was made in the development of ERGON-EXPERT between analysis and assessment modules containing knowledge from different domains.

Using the prototype it could be demonstrated in principle that this methodology enables progress in ergonomics in terms of systematic work design and evaluation. The encountered difficulties were due less to the adaptation of the specialized software to problems relating to ergonomics than to the inconsistency of the raw data. This became evident during the establishment of the knowledge base. It is hardly possible, for example, to standardize the results for body forces to compensate for the differences in measurement methodology between the various investigations.

In developing ERGON-EXPERT attention has been focused particularly on the user interface of the system in order to reduce time-consuming redesign processing during development. Since the user interface is the key to the acceptance of a system, it plays an essential role in translating the knowledge contained in a knowledge-based system into a usable form. This translation process therefore has a major influence on the benefits of a system for the ergonomic design of manual materials handling tasks.

Acknowledgements

The research project ERGON-EXPERT is funded in part by the German Ministry for Research and Technology (BMFT) under 01HK8263 within the framework of the research programme 'Humanization of Work'.

References

Asfour, S. S., Ayoub, M. M. and Genaidy, A. M., 1985, Computer models for energy cost prediction of manual lifting and lowering. *Computers and Industrial Engineering*, **9** (Suppl 1), 94–98.

Australian Minister for Labour, 1988, *Occupational Health and Safety (Manual Handling) Regulations 1988* (Department of Labour: Victoria, Australia).

Chaffin, D. B. and Anderson, G. B. J., 1984, *Occupational Biomechanics* (New York: John Wiley).

Chaffin, D. B. and Evans, S., 1986, Computerized biomechanical models in manual work design. In: *A Cradle for Human Factors: Proceedings of the Human Factors Society 30th Annual Meeting*. (Dayton: The Human Factors Society Inc.).

Chignell, M. H. and Peterson, J. G., 1988, Strategic issues in knowledge engineering. *Human Factors*, **30**, 381–394.

Das, B., 1988, Ergonomics and new technology implementation. In: *Success Factors for Implementing Change*, edited by K. M. Blache (Dearborn, MI: Society of Manufacturing Engineers), pp. 87–107.

Dooley, M., 1982, Anthropometric modeling programs – a survey. *IEEE Computer Graphics and Applications*, **2** (9), 17–25.

Garg, A., Chaffin, D. B. and Herrin, G. D., 1978, Prediction of metabolic rates for manual material handling jobs. *American Industrial Hygiene Association Journal*, **39**, 661–674.

Garg, A. and Ayoub, M. M., 1980, What criteria exist for determining how much load can be lifted safely? *Human Factors*, **22**, 475–486.

Genaidy, A. M. and Asfour, S. S., 1987, Review and evaluation of physiological cost prediction models for manual materials handling. *Human Factors*, **29**, 465–476.

Genaidy, A. M., Asfour, S. S. and Kabuka, M., 1987, An expert system for the design of manual material handling tasks. In: *Trends in Ergonomics/Human Factors IV*, edited by S. S. Asfour (Amsterdam: Elsevier – North Holland), pp. 559–566.

Goldberg, A. and Kay, A. (eds), 1976, *SMALLTALK-, Instruction Manual* (Palo Alto: The learning research group, XEROX-PARC).

Jackson, P., 1986, *Introduction to Expert Systems* (Wokingham: Addison-Wesley).

Jäger, M. and Luttmann, A., 1986, Biomechanical model calculations of spinal stress for different working postures in various workload situations. In: *The Ergonomics of Working Postures. Models, Methods and Cases* edited by E. N. Corlett, J. R. Wilson and I. Manenica (London: Taylor & Francis), pp. 144–154, ref. 399–423.

Jäger, M. and Luttmann, A., 1987, Determination of spinal stress by biomechanical model calculations and comparison with spinal mechanical strength. In: *Biomechanics, Basic and Applied Research*, edited by G. Bergmann, R. Kölbel, and A. Rohlmann (Dordrecht: Nijhoff), pp. 473–478.

Karwowski, W. (Ed.)., 1987a, Special Issue: Application of knowledge-based expert systems in ergonomics. *International Journal of Industrial Ergonomics*, **2**.

Karwowski, W., 1987b, Preface to Special Issue: Application of knowledge-based expert systems in ergonomics. *International Journal of Industrial Ergonomics*, **2**, 3–5.

Karwowski, W., Mulholland, N. O., Ward, T. L., Jagannathan, V. and Kirchner, R. L., Jr., 1986, LIFTAN: An experimental expert system for the analysis of manual lifting tasks. *Ergonomics*, **29**, 1213–1234.

Kloke, W. B., 1989, WERNER – A small computer based tool for ergonomic design. In: *Application of Human Performance Models to System Design*, edited by G. Macmillan *et al.* (New York: Plenum) (in press).

Konz, S., 1983, *Work Design: Industrial Ergonomics* (Columbus, OH: Grid).

Kroemer, K. H. E., Snook, S. H., Meadows, S. K. and Deutsch, S., 1988, *Ergonomic Models of Anthropometry, Human Biomechanics, and Operator–equipment Interfaces* (Washington, DC: National Academy Press).

Laurig, W., 1987, Do you believe in the advantage of expert systems in industrial ergonomics? *International Journal of Industrial Ergonomics*, **2**, 7–13.

Laurig, W., Kühn, F. M. and Schoo, K.-C., 1985, An approach to assessing motor workload in assembly tasks by the use of predetermined-motion-time systems. *Applied Ergonomics*, **16**, 119–125.

Marshall, C., 1989a, Ergonomics is dead: long live ergonomics. Part I. *The Ergonomist – The Newsletter of the Ergonomics Society*, No. **223**, 2–3.

Marshall, C., 1989b, Ergonomics is dead: long live ergonomics. Part II. *The Ergonomist – The Newsletter of the Ergonomics Society*, No. **224**, 1–3.

Mital, A., 1984a, Maximum weights of lift acceptable to male and female industrial workers for extended work shifts. *Ergonomics*, **27**, 1115–1126.

Mital, A., 1984b, Comprehensive maximum acceptable weight of lift database for regular 8-hour work shifts. *Ergonomics*, **27**, 1127–1138.

NIOSH (National Institute for Occupational Safety and Health), 1981, *Work Practice Guide for Manual Lifting* DHSS-NIOSH Publication No. 18–122, (Cincinnati: NIOSH).

Rohmert, W., 1972, Aufgaben und Inhalt der Arbeitswissenschaft. *Die Berufsbildende Schule*, **24**, 3–14.

Rombach, V. and Laurig, W., 1988, ERGON-EXPERT – A knowledge-based approach to the design of workplaces. In: *Trends in Ergonomics/Human Factors V*: edited by F. Aghazadeh (Amsterdam: Elsevier – North Holland), pp. 53–61.

Taylor, N. K. and Corlett, E. N., 1987, ALFIE – auxiliary logistics for industrial engineers. *International Journal of Industrial Ergonomics*, **2**, 15–25.

Troup, J. D. G., 1984, Causes, prediction and prevention of back pain at work. *Scandinavian Journal of Work Environment and Health*, **10**, 419–428.

Chapter 22
The Lifting Stress Calculator: theory and application

F. J. Bejjani, Naiming Xu and Lazaros Pavlidis
Human Performance Analysis Laboratory,
New York University,
35 West 4th Street, Room 876,
New York, NY 10003, U.S.A.

Abstract. The Lifting Stress Calculator (LSC) is a user-friendly device based upon a trigonometric two-dimensional model of static lifting which recommends the optimum method of lifting given the subject's anthropometry and clinical symptoms, and the physical characteristics of the load. The usefulness of the LSC in the field of ergonomics and rehabilitation is demonstrated since that there is no one simple, safe and proper lifting technique and that each lifting task warrants an individualized evaluation. A regression equation summarizing the effects of load height, weight and horizontal distance from the heel on joint forces is obtained.

Keywords: Load characteristics; Lifting; Anthropometry.

Introduction

The impairment of the back is the most frequent cause of activity limitation in persons under 45 years (Kelsey *et al.*, 1979; Andersson, 1981). Even though the etiology of low back pain remains obscure, epidemiological studies implicate such occupational factors as lifting as contributing factors (Andersson, 1981; Frymoyer *et al.*, 1980). Longitudinal studies have shown an almost three-fold increase in low back injury incidence rates in overstressed populations (Chaffin and Park, 1973; Chaffin *et al.*, 1977). NIOSH (1981) stated that over-exertion injuries associated with manual handling accounted for 12 million lost work days. The length of incapacitation is three to four times greater for those engaged in heavy labour (Magora and Taustein, 1969). Perhaps the most significant finding is that 4–6% of the low back pain patients become chronic and this group constitutes 80% of the total cost of low back pain in society (Nachemson, 1984). It has been shown that the sooner the patient resumes activity and returns to work, the better; this helps prevent the psychological consequences of inactivity (Blumer, 1975; Nachemson, 1983). Three approaches have been suggested for prevention of low back pain: (i) screening/selection, (ii) training and (iii) workplace design. Training has been used most extensively and emphasized as an important preventive measure to reduce the incidence of low back pain. Since 1930, it has been recommended to lift with straight back and bent knees. Brown (1972) showed that the use of this method has not produced any great change in the incidence of low back injuries. Many studies have questioned the validity (Chaffin and Park, 1973) and most importantly the practicality of the recommended 'safe' lifting technique (Graveling *et al.*, 1985).

The occurrence of low back pain in industry is highest between 35 and 40 years (Calin *et al.*, 1980; Frymoyer *et al.*, 1980). Knee joint arthritis usually appears later in life, with an age peak above 65 years. At this age, more than 2% of all men and 6·6% of all women in the United States suffer from some degree of knee arthritis (US Department of Health, Education and Welfare, 1971–5). Given that knee lifting is recommended for back pain patients (Miller, 1980; Bendix and Eid, 1983), one would tend to think that ultimately there would be a greater incidence of knee arthritis among these patients.

Therefore the lifting technique is still a major issue in ergonomics, and more so the implementation in the field of all research and laboratory findings on this topic. Thus, in order to convey the message to the worker in his/her work environment, a user-friendly portable computer is the solution of choice. This device would serve as a 'cook-book' of lifting. Besides, blind standardization, using the 'average man' as a model, is one of ergonomics' greatest dangers. Hence, this 'cook-book' should be tailored to the individual worker on a given day. The tool described in this chapter emphasizes these major ergonomical issues: computation and user friendliness; portability; and individualized use.

The Lifting Stress Calculator (LSC) is designed to help reduce the cumulative demands on the musculoskeletal system during lifting. This device has the advantage of being simple, agreeable to use and not constraining to its users. It is ergonomically conceived to educate the workers. It can show a set of ergonomical parameters within seconds. Thus the LSC can be used in an industrial workplace with a large population to help reduce the incidence of occupational diseases related to lifting. The LSC is based on a trigonometric two-dimensional model (Bejjani *et al.*, 1984a, b). This model predicts the knee (JK) and back (JB) joint reaction forces as a function of back, knee and ankle joint angles, and anthropometric measurements for a given physical load condition (weight, height, and its horizontal distance from the heel). Furthermore clinical complaints of the user, with regard to back, knee and ankle symptoms, are also accounted for.

An experimental study was conducted on 53 healthy subjects to determine the effect of load height, weight and distance from the body on both knee and back forces during sagittal bimanual lifting (Bejjani *et al.*, 1984a; Parnianpour *et al.*, 1986a, b). An optimal method of lifting, given the anthropometry, clinical symptoms, and physical characteristics of the load, is also recommended (Parnianpour *et al.*, 1987).

Model

Nomenclature

U	angle between vertical and leg
V	angle of flexion of the back
X	angle of flexion of the knee
Y	angle between the foot and the leg
Z	angle between the thigh and the horizontal
a	distance between T1 and third PIP joint (m)
b	distance between greater trochanter tip and centre of lateral femoral condyle (m)
c	distance between T1 and vertex
e	distance between fingers and ground, standing upright (m)
l	distance between centre of lateral femoral condyle and heel (m)
t	distance between greater trochanter tip and centre of lateral femoral condyle (m)
H	height of the load (m)
D	horizontal distance of the load from the heel (m)
P	weight of the load (N)

W	body weight (N)
Wa	weight of the arms (N)
Wt	weight of the thighs (N)
Wu	weight of the head and torso (N)
Wl	weight of the head, torso and arms (N)
La	lever arm of the arm about G (m)
Le	lever arm of the erector spinae (m)
Lp	lever arm of the load about L5-S1 (m)
L'p	lever arm of the load about G (m)
Lq	lever arm of the quadriceps muscle (m)
Lu	lever arm of the head and torso about G (m)
Lt	lever arm of the thigh about the centre of rotation of the knee, G (m)
Lw	lever arm of W1 about L5-S1 (m)
FE	erector spinae force (N)
FQ	quadriceps muscle force (N)
CB	compression forces on L5-S1 (N)
SB	shear force on L5-S1 (N)
JB	total joint reaction force on L5-S1 (N)
CK	compression force on the tibio-femoral joint (N)
SK	shear force on the tibio-femoral joint (N)
JK	total joint reaction force on the tibio-femoral joint (N)

This is a two-dimensional static model of symmetric sagittal bimanual lifting, based on trigonometry and anthropometry (Figure 22.1). It represents the projection in the sagittal plane of the mechanical axes of the spine and dominant upper and lower limbs of an individual. This position corresponds to the phase when the load is just about to leave the ground. The upper limb is vertical. The elbow is in full extension and the forearm in slight pronation. The proximal interphalangeal joints (PIP) are in contact with the handle. The upper extremity position adopted best minimizes the loads on the rest of the body. The foot is lying flat on the ground. Projection of BG on the horizontal axis gives the following equation:

$$BG = D - l\cos Y \quad (1)$$

where D is the distance between the heel and the load. From triangles FGG' and BCB', we have that:

$$B'G' = CG' - CB' = b\sin V - t\cos Z \quad (2)$$

since $BG = B'G'$, we then have:

$$b\sin V - t\sin Z = D - l\cos Y \quad (3)$$

Projection of the body segments lengths on the vertical axis gives the following equation:

$$a + H = b\cos V + t\sin Z + l\cos Z \quad (4)$$

where H is the height of the load.

Thus, we have a system of two equations with three unknown: Y, Z, V.

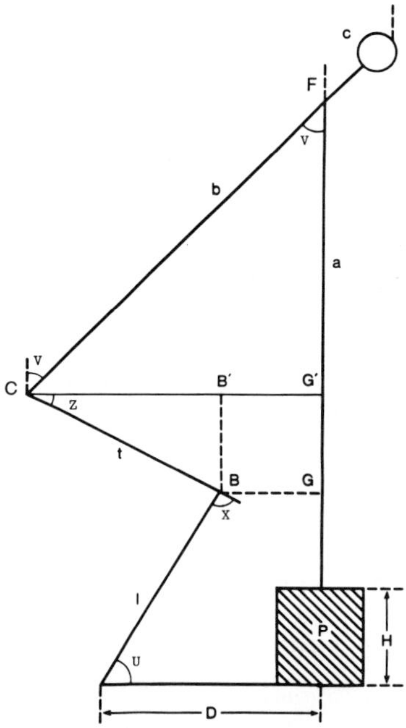

Figure 22.1 Trigonometric and anthropometric model; V = *back flexion angle;* X = *knee flexion angle;* Y = *angle between thigh and horizontal;* Z = *angle between foot and leg;* a = *distance between acromion and third PIP joint;* b = *distance between T1 and third coccygeal vertebra;* c = *distance between T1 and vertex;* t = *distance between greater trochanter and lateral femoral condyle;* l = *distance between lateral femoral condyle and heel;* H = *height of the load;* P = *weight of the load.*

By specifying any of these three angles, the system is reduced to two equations with two unknown, and unique solutions can be found. In the present study, (Y) is allowed to vary from 65 to 90 by increments of 2 and angles V and Z are solved for. The angle of knee flexion (X) is then given by the following equation:

$$X = (90 - Z) + (90 - Y) = 180 - (Z + Y) \qquad (5)$$

A free-body analysis was done on both the back and the knee (Frankel and Nordin, 1980). For the back, we used the free-body above the L4-L5 joint, including the head, torso and arms (Figure 22.2). We obtained the erector spinae force, FE, by applying the moment balance equation (all forces and weights are in Newtons)

$$\begin{aligned} &(Wl * Lw) + (P * Lp) - (FE * Le) = 0 \\ &FE = \{(Wl * Lw) + P * Lp)\}/Le \end{aligned} \qquad (6)$$

where $Wl = Wu$ (weight of head and torso) + Wa (weight of arms).

According to Barter (1957), $Wu = 0{\cdot}47 *$total body weight + 54, $Wa = 0{\cdot}13 *$total body weight + 14, Lw = lever arm of Wl about L5-S1, assumed to be = $b/2\sin W$ (Figure 22.2), P = weight of load, Lp = lever arm of load about L5-S1, assumed to be = $3/4b \sin V$,

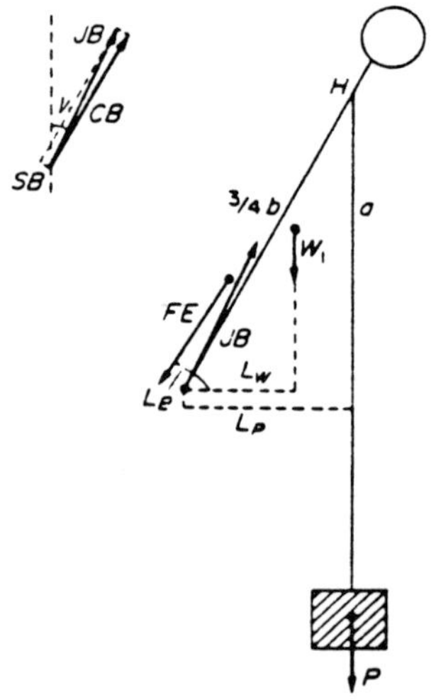

Figure 22.2 Free-body diagram above L5-S1: FE = erector spine; LE = lever arm of the erector spine about L5-S1; Wl = *weight of the head, torso and arms;* Lw = *lever arm of about L5-S1;* P = *weight of the load;* Lp = *lever arm of the load about L5-S1;* JB = *joint reaction force;* CB = *compression force;* SB = *shear force (from Bejjani* et al., *1984a).*

Le = lever arm of erector spinae = 0·05 m. By projecting all the forces on two axes, one along the spine mechanical axis (compression axis) and one perpendicular to it (shear axis), and by applying the force balance equation, we then have (Figure 22.2):

$$CB = Wl * \cos V + FE * \cos V \tag{7}$$

where *CB* is the compression force on the back (L5-S1)

$$SB = Wl * \sin V + P * \sin V \tag{8}$$

where *SB* is the shear force on the back (L5-S1)

$$JB = \sqrt{(CB^2 + SB^2)} \tag{9}$$

where *JB* is the total joint reaction force on the back (L5-S1).

For the knee, we used the free-body above the knee, including the ipsilateral thigh and arm and half the sum of the head and torso. The moment equilibrium equation gives us the value of *FQ*, the quadriceps muscle force (Figure 22.3):

$$FQ = (-P * Lp + Wu * Lu - Wa * La + Wt * Lt)/Lq/2 \tag{10}$$

where *Wa*, *Wu* and *P* have been defined above. *Wt* = weight of the thighs = 0·18 * total body weight + 15 (according to Barter, 1957). *Lt* = lever arm of the thigh about the centre of rotation of the knee, G. Barter's tables show the centre-of-mass of the thigh at 56·7% of the

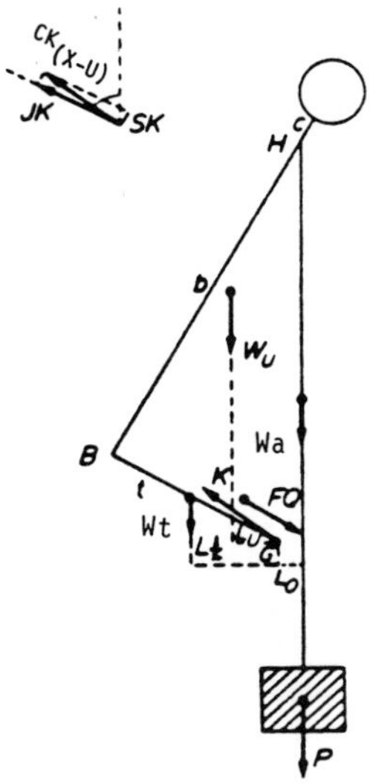

Figure 22.3 Free-body diagram above the knee: Wu *about* G *(centre of rotation of the knee);* Wt = *weight of the thigh;* Lt = *lever arm of* Wt *about* G*;* Wa = *weight of the arm;* Fq = *quadriceps force;* Lq = L′ p = La = *lever arms of* Fq, Pa *and* Wa *about* G*;* Jk = *joint reaction force;* CK = *compression force;* SK = *shear force (from Bejjani* et al.*, 1984a).*

thigh length (t) from G. The value of Lt is therefore: $Lt = 0{\cdot}567 * t * \cos Z$ (Figures 22.1 and 22.3); and if we solve for Z in equation (4), $Lt = 0{\cdot}567 * t * \cos(180 - X - Y)$, Lq = (lever arm of the quadriceps muscle) = Lp' = (lever arm of the load, about G) = La (lever arm of the arm, about G) = 0·05 m, (approximately the distance between the patellar surface and G) (Figure 22.3). Lu = lever arm of the head and torso, about G. Barter's table show the centre-of-mass of the head and torso at 39·6% of b (defined above) + c (distance between $T1$ and the vertex), from the distal end of the spine. This and equation (4) give us (Figures 22.1 and 22.3):

$$Lu = t\cos(180 - X - Y) - 0{\cdot}396\,(b + c)\sin V$$

By projecting all the forces involved on two axes, one along the thigh mechanical axis (compression axis) and one perpendicular to it (shear axis), and by applying the force equilibrium equation, we then have (Figure 22.3).

$$CK = \frac{(Wu + Wa + Wt + P)}{2}\sin(X + Y) + FQ \qquad (11)$$

where CK is the compression force on the knee (tibiofemoral joint). In equation (11), sin $(X - Y) = \cos(X - U)$ because $U = 90 - Y$ (Figure 22.1).

$$SK = \frac{(Wu + Wa + Wt + P)}{2}(-\cos(X + Y)) \qquad (12)$$

where *SK* is the shear force on the knee.

$$JK = \sqrt{(CK^2 + SK^2)} \tag{13}$$

where *JK* is the total joint reaction force on the knee*.

Device

Hardware

The Lifting Stress Calculator is controlled by an Intel 8085 microprocessor unit (MPU) with 32K RAM, a basic interpreter in ROM and a liquid crystal display (LCD), which performs all mathematical and interface functions. The control is accomplished by decoding the most significant address lines on the address bus (Figure 22.4). The calculator memory contains all software pertaining to the algorithms and user interface. Data storage contains anthropometric data and other information on the subjects in question. The information is entered by the user via the keyboard. Data storage has its own, long life, power supply to provide a constant memory, even when the calculator is off. The 32 character liquid crystal display provides sufficient space to ensure simple and precise operation of the calculator.

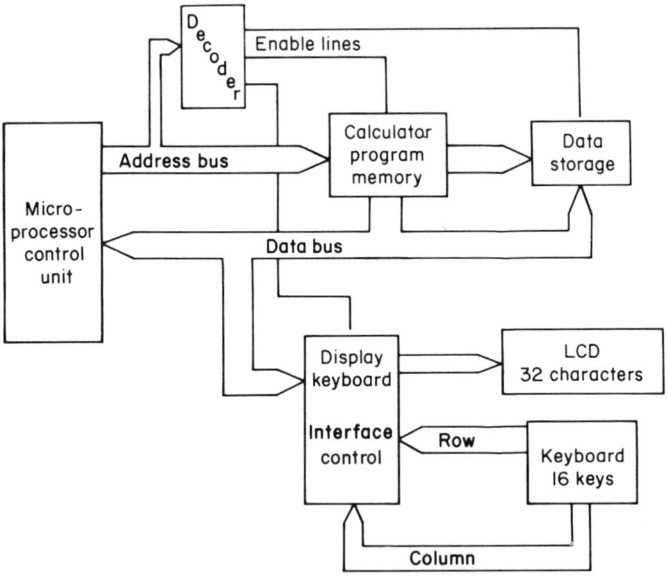

Figure 22.4 Lifting Stress Calculator system diagram (from Bejjani et al., *1984b).*

*The patello-femoral joint reaction force bears a determinate relationship to geometry and extension force, and has not been addressed in this study.

Software

A program was written in 8085 assembler language, to solve the trigonometric and biomechanical equations, incorporating the data entered. All possible combinations of back, knee and ankle angles, within the restrictions of the model (Table 22.1), and then considered and the corresponding forces computed and normalized over body weight (*W*). S1, S2, S3 and S4 are respectively defined as scaling factors for the erector spinae compression force (*FE*), the back shear force (*SB*), the quadriceps compression force (*FQ*) and the knee shear force (*SK*). Their values depend directly on the severity of the clinical syndromes (Table 22.2). The following equation is then computed, for every combination of angles:

$$Z = S1\frac{(FE)}{W} + S2\frac{(SB)}{W} + S3\frac{(FQ)}{W} + S4\frac{(SK)}{W}$$

The program scans for Z minima, and displays the corresponding angles, representing the ideal lifting position. When comparing several users, the name of the one with the smallest Z minima is displayed, as best fit for that particular lifting task.

Table 22.1 Angular measurements (from Bejjani et al., *1984b).*

	Angular measurements	Normal range (degrees)
Back	Angle of flexion of the spine, measured between the vertical and the position of the torso	0–60 (excluding pelvic tilt)
Knee	Angle of flexion, measured between the anterior aspect of the thigh and the leg	0–140
Ankle	Angle between the foot and the leg = 90° – clinical angle of dorsiflexion	65–90

Table 22.2 Clinical data assessment (from Bejjani et al., *1984b).*

Severity of symptoms	0	1	2	3
		Subjective symptoms. No clinical or X-ray signs	Moderate clinical signs (sciatica)	Severe clinical signs (spinal sthenosis)
Back	No symptoms	Mild clinical signs (lumbo-sciatalgia)	1–2 years post discal surgery	Confirmed herniated disc (pre-op or 1 year post-op)
		2–3 years post discal surgery		Need of lumbar brace
Tibio-femoral joint	No symptoms	Mild to moderate pain and/or limitation of range of motion		
		unilaterally and bilaterally		Severe pain and/or disability in one or both knees (confirmed by X-rays)
Patello-femoral joint	No symptoms	Mild to moderate pain going up and down the stairs		
Ankle	Normal range of motion (Table 22.1)	Limited motion due to trauma, pain, etc.		

Method of operation

The Lifting Stress Calculator is designed for one or more users. Each user will have to enter the following information, step by step, answering the questions displayed on the LCD (Figure 22.5):

A. Personal data

Number, sex, age, weight and height.

B. Anthropometric data

(1) Distance between the vertex and T1 vertebra.
(2) Distance between T1 and the third coccygeal vertebra.
(3) Distance between tip of greater trochanter and centre of lateral femoral condyle (approximate distance between the centres of rotation of the hip and the knee).
(4) Distance between the centre of lateral femoral condyle and the heel.
(5) Distance between acromion and third PIP joint, the upper extremity being in full extension.

All data should be measured on the dominant side, using an anthropometric tape, by a colleague or better by a physician. The shortest distance (mechanical axis) should be retained, for all measurements.

Subject: J. R. Age = 28 Sex = M
Dominant side = Right Height = 1.68 m Weight = 83 kg
Distance between T1 and 3rd Coccygeal Vertebra = 54 cm
Distance between T1 and Top of the Head = 26 cm
Distance between 3rd Coccygeal Vertrebra and Top of the Head = 80 cm
Distance between Acromion and 3rd PIP joint = 69 cm
Distance between Greater Trochanter and Femoral Condyle = 38 cm
Distance between Heel and Lateral Femoral Condyle = 48 cm
Distance between Greater Trochanter and Heel = 86 cm
Box Height = 20 cm Box Weight = 10 kg

Severity of Back Symptoms: 1
Severity of Tibio-Femoral Symptoms: 1
Severity of Patello-Femoral Symptoms: 1
Severity of Ankle Symptoms: 0

S1 = 1, S2 = 2, S3 = 2, S4 = 2

	Back	Knee	Ankle	Fe	FQ	SB	SK
Z maxima = 7.790	54°	80°	85°	3085	701	531	393
Z minima = 6.393	19°	110°	85°	1242	1372	213	406

Figure 22.5 Partial output of the Lifting Stress Calculator. FE = erector spine compression force, FQ = quadriceps compression force, SB, SK = shear forces on the back and knee. S1, S2, S3 and S4 = clinical scaling factors, Z = coefficient descriptive of the total force on the body. The angles printed next to Z minima correspond to the ideal lifting position for the subject, under the mentioned load characteristics. All forces are in Newtons (from Bejjani et al., *1984b).*

C. Characteristics of the load

(1) Height = distance between the ground and the handle or the position of the hand on the load.
(2) Weight.

D. Clinical data

(1) Low back symptoms (0–3).
(2) Tibiofemoral joints symptoms (0–3).
(3) Patellofemoral joints symptoms (0–3).
(4) Ankle symptoms (0–1).

The scales for assessing this set of data is explained in Table 22.2 and will be best controlled by each user's physician.

Materials and methods

The anthropometric data were collected, using an anthropometric tape, from 53 subjects (Table 22.3).

A computer program was written to solve the equations previously derived. The program was run several times for each subject, varying one load parameter and keeping the other two constant. The values used for these parameters were ratios of the relevant anthropometric measurements of the subject, in an attempt to individualize the results. The horizontal distance, D, was varied from 1/60 to 1/3 of the trunk length, b, by increments of $b/20$, while the load height H was kept constant at 1/3 of e, a fraction of the subjects's height (e = subject's height $- a - c$); and the load weight P was kept constant at 1/5 of the subject's weight, W. Then H was varied from 1/57 to 1/3 of e by increments of $e/19$, while D was kept constant at $3/b$ and P at $W/5$. P was then varied from 1/90 to 1/5 of W by increments of $W/18$, while D was kept constant at $b/3$ and H at $e/3$. For each of these load conditions, all possible JB and JK were computed. In the meantime, Z, a coefficient of total joint forces, is also calculated by the following equation:

$$Z = (S1 * JK + S2 * JB)/W$$

where W is the weight of the subjects; JK and JB are the joint reaction forces of knee and back, respectively; and $S1$ and $S2$ are the two scaling factors directly determined from the clinical assessment of severity of symptoms of the back (B), the tibiofemoral (TF), patellofemoral (PT) and ankle joints (A) in a scale from 0 (no symptoms) to 3 (severe pain and functional impairment).

Table 22.3 Demographic distribution of the sample with average (± standard deviation) age, weight and height (from Parnianpour et al., *1987).*

	Number	Age (years)	Weight (kg)	Height (m)
Females	26	21·9 (8·9)	57·3 (7·9)	1·60 (0·06)
Males	27	25·1 (8·26)	79·6 (11·9)	1·77 (0·08)
Total	53	23·6 (8·7)	68·6 (15·3)	1·69 (0·11)

The program calculates *JK*, *JB* and *Z* for all the possible combinations, incorporating the entered personal anthropometric and clinical data, and load characteristics (height, weight and horizontal distance from the heel) and recommends a lifting technique which corresponds to minimum *Z*. The optional lifting technique and its associated joint reaction forces are shown numerically and graphically on the display.

Minimum *JB* and *JK*, with their corresponding back and knee angles, were then obtained for each of the load conditions, and normalized over body weight for each subject: JB_{min}/W and JK_{min}/W. Averages of these ratios among all subjects were then tabulated and plotted graphically as a function of each of the three load parameters.

Then, for subject J.R. (28-year-old male; height, 1·68 m; weight, 83 kg), 100 randomly selected load conditions were considered and *JB* and *JK* were computed for each condition. To illustrate the effects of the severity of symptoms on minimum *Z* and recommended angles, a few possible permutations of joint symptoms were considered. A regression analysis of *JK* and *JB* as functions of *D*, *H* and *P* was then performed using the GLM procedure of the SAS statistical package.

Results

As seen in Figures 22.6 and 22.7, increase in load height did not much affect the minimum knee force (JK_{min}/W), while the minimum back force (JB_{min}/W) decreased slowly in the beginning then rapidly. Both knee and back angles, corresponding to JK_{min}/W and JB_{min}/W, respectively, decreased with increasing load height.

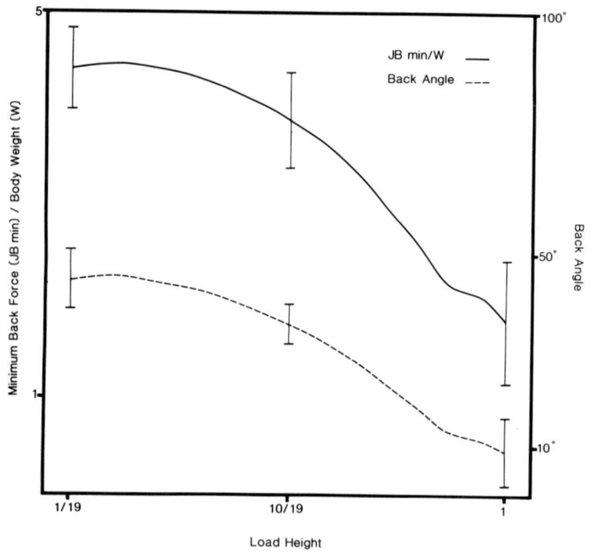

Figure 22.6 Average (± standard deviation) normalized minimum back (L5-S1) joint reaction force as a function of load height, varying from 1/19 to 1 of e/3, in 53 subjects (from Parnianpour et al.*, 1986a).*

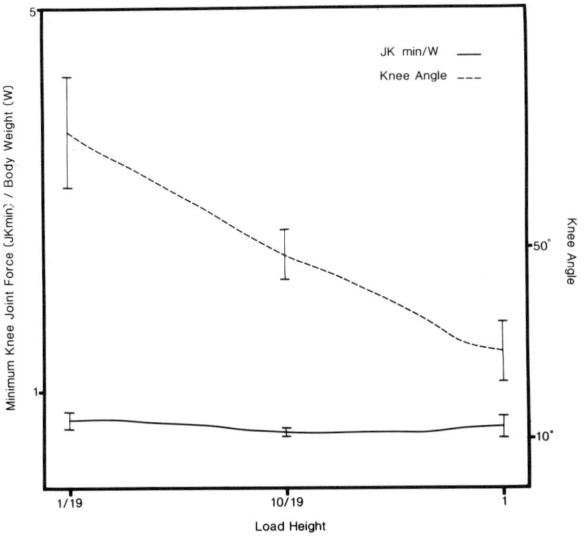

Figure 22.7 Average (± standard deviation) normalized minimum knee (tibio-femoral) joint reaction force as a function of load height, varying from 1/19 to 1 of e/3, *in 53 subjects (from Parnianpour* et al., *1986a).*

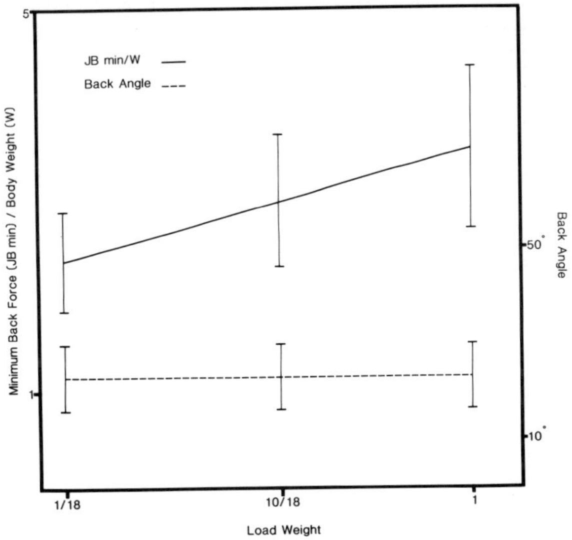

Figure 22.8 Average (± standard deviation) normalized minimum back (L5-S1) joint reaction forces as a function of load weight, varying from 1/18 to 1 of W/5, *in 53 subjects (from Parnianpour* et al., *1986a).*

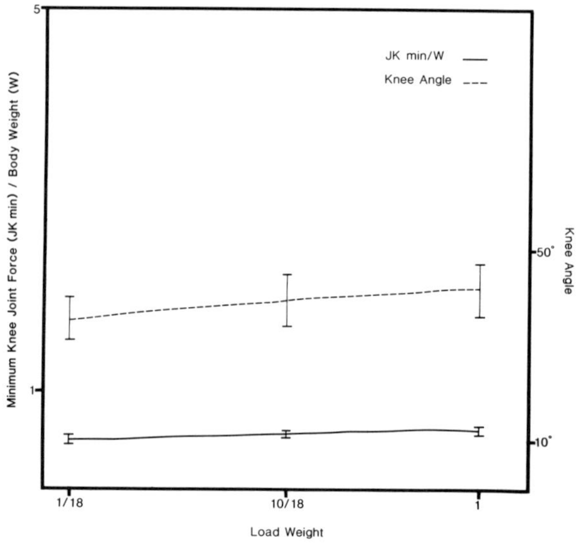

Figure 22.9 Average (± standard deviation) normalized minimum knee (tibio-femoral) joint reaction forces as a function of load weight, varying from 1/18 to W*/5, in 53 subjects (from Parnianpour* et al.*, 1986a).*

Increase in load weight, as seen in Figures 22.8 and 22.9, was followed by a rather rapid increase in the minimum back force and a very slow increase in the minimum knee force. The back angle, corresponding to this minimum back force, remained unchanged; while the knee angle, corresponding to the minimum knee force, increased slowly.

When the horizontal load distance increased, both the minimum back force and its corresponding back angle increased (Figure 22.10). However, minimum knee force first decreased, then reached a plateau at about $b/6$, then rapidly increased, while its corresponding knee angle first decreased till about $b/13$, then increased till $b/6$, then decreased (Figure 22.11).

No simple correlation was found between *JB* and *JK* for the typical subject J.R. The results of the multiple regression analysis showed that the relationship between knee and back force is highly dependent on the load parameters *D*, *H*, and *P*. Therefore, a regression equation was derived for both *JB* and *JK* as a function of *D*, *H* and *P*:

$$JB = 5297*D - 5724*D^2 + 1331*H - 4639*H^2 + 6{\cdot}6*P + 1553 \quad (14)$$

$$JK = 8765*D + 18697*D^2 + 1377*H - 1613*H^2 + 1{\cdot}5*P + 1217 \quad (15)$$

The coefficients of these equations were all highly significant statistically (Table 22.4). The R^2 for equation (14) was 0·92. When D^2 and H^2 are excluded from the equation, the R^2 drops to 0·87. For equation (15) the R^2 was 0·76. When the non-linear elements D^2 and H^2 are excluded from the equation, the R^2 drops to 0·22.

A larger standard deviation was found for back joint reaction forces and back angles which means that anthropometry affects the back more than the knees.

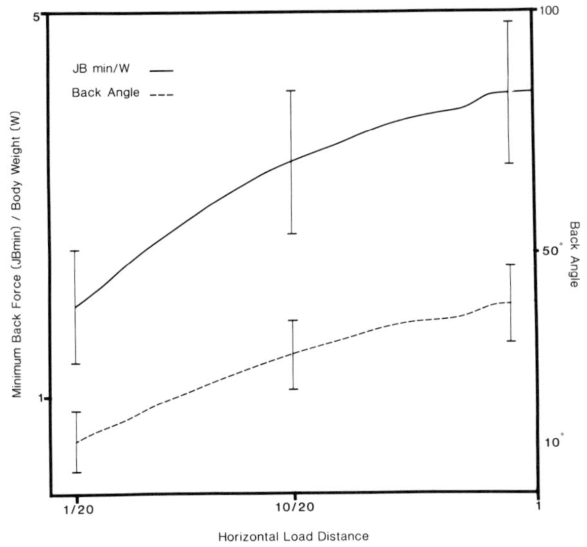

Figure 22.10 Average (± standard deviation) normalized back (L5-S1) joint reaction forces as a function of load horizontal distance, varying from 1/20 to 1 of b/3, in 53 subjects (from Parnianpour et al., *1986a).*

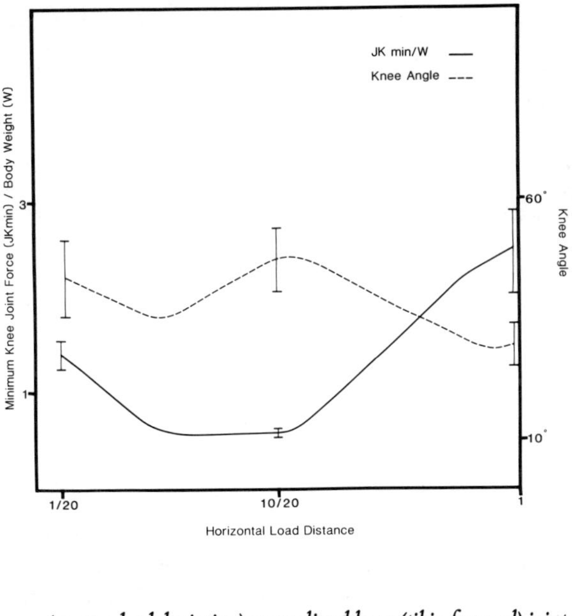

Figure 22.11 Average (± standard deviation) normalized knee (tibio-femoral) joint reaction forces as a function of load horizontal distance, varying from 1/20 to 1 of b/3, in 53 subjects (from Parnianpour et al., *1986a).*

Table 22.4 Estimate (± standard error) and F*-ratios of the regressor variables* D, D^2, H, H^2 *and* P *of* JB [equation (14)] and JK [*equation (15)*]. *All* F*-ratios are significant at* $p<0\cdot0001$.

	JB		*JK*	
	Estimate (S.E.)	*F*-ratio	Estimate (S.E.)	*F*-ratio
D	5297 (148·8)	1267·7	– 8765 (215·5)	1653·9
D^2	– 5724 (275·9)	430·6	18697 (399·6)	2188·9
H	1331 (183·3)	52·7	1377 (265·5)	26·9
H^2	– 4639 (260·3)	317·5	– 1613 (377·1)	18·3
P	6·6 (0·16)	1785	1·5 (0·22)	42
Intercept	1553 (37·7)		1217 (54·6)	
MODEL		2448·8		625·2

Table 22.5 Permutations of the joint symptoms with the equivalent normalized forces, Z values and recommended angles (from Parnianpour et al., *1987*).

Severity of symptoms				Recommended Angles°			Normalized Joint Forces		
B	TF	PF	A	Back	Knee	Ankle	*JB/W*	*JK/W*	*Z*
0	0	0	0	56	61	83	4·93	0·61	5·53
1	1	2	0	59	55	86	5·01	0·60	28·07
2	0	0	0	51	73	77	4·70	0·90	16·80
3	0	0	0	48	78	74	4·27	1·04	22·46

The response and sensitivity of back and knee joint reaction forces were not always in accordance with each other. The minimum knee joint reaction forces and knee angles showed the most non-linear response to the changes of horizontal load distance.

The model recommends different lifting techniques for different joint problems. The joint with more severe symptoms is extended more to reduce its forces and the other joint sustains proportionally more forces (Table 22.5).

The physical characteristics of load are a more significant predictor of joint forces than the lifting technique.

Discussion

As expected, the overall magnitude of minimum forces was larger on the back (2–4 *W*) than on the knee (0·5–2·5 *W*). These results corroborate the findings of Grandjean (1988) and Frankel and Nordin (1980). The larger standard deviations of the minimum back forces suggest a more significant role of individual anthropometry in minimizing back forces, as compared with knee forces.

The joint angles did not always follow the same trends as their corresponding minimum forces, especially for the back when load weight was varied (Figure 22.8) and for the knee when load height (Figure 22.7) or load horizontal distance (Figure 22.11) were varied. In the latter case, the trends for minimum knee force and its corresponding knee flexion angle were practically opposite. The discrepancy between the minimum forces and the joint angles trends suggests that basic biomechanical theory, such as free-body analysis, is not sufficient to predict the optimal lifting posture for a given person. Empirical data, such as individual anthropometry and load parameters, are necessary complementary elements for this prediction. This is in agreement with Grandjean (1988) who discussed the problems of body size in detail and concluded that the workplace should be suited to the body size of the operator.

As seen in Table 22.4, the standard errors of all the estimates were relatively small; therefore allowing the extrapolation of the findings of this study to the general population.

According to the model, the load parameters effect is more linear on the back than on the knee since, by dropping the non-linear components of equations (14) and 15), *JB* is still explained at 87% whereas *JK* is only at 22%. Horizontal load distance, *D*, has a definite positive linear effect and load height, *H*, a definite negative linear effect on the back force (Table 22.4). The coefficient of load weight, *P*, is about four times greater for *JB* than for *JK*, suggesting its more significant impact on back forces, then on knee forces. Horizontal load distance, *D*, seems to be the most important load parameter acting on the knee force, in a non-linear fashion, as seen graphically in Figure 22.11 and statistically in equation (15) and Table 22.4. This finding verifies the conclusion of Grandjean (1988), Pheasant (1986), and Frankel and Nordin (1980), that the distance between the spine and the hands holding the load plays an important role, i.e. the mechanical stress of lumbar spine is greatly decreased if the load is moved closer to the torso.

Conclusions

Optimization of knee and back joint forces

This study stresses the importance of minimizing both back and knee forces during sagittal lifting. This can only be achieved by taking into consideration both individual anthropometry and load parameters. The former seems to have a more prominent role in minimizing forces on the spine. The most significant parameter acting on knee forces seems to be the distance of the load from the body. This parameter acts in a non-linear fashion and is responsible for minimal knee forces up to 2·5 *W*.

This study also shows that no simple generalized relationship can be established between knee and back forces during lifting, because of the uncontrollable variance produced by the anthropometry and to a greater extent by the load parameters.

The efficacy of a single, correct lifting technique

The model shows variability of joint angles for different loading conditions. This is in agreement with Park (1973) who reported that subjects simulating a lifting task assumed a 'free position' resembling neither the vertical back/bent knees nor the stooped back/straight knees method. The concept of distributing the load between the knee and back joint according to a patient's symptoms is attractive but the results point to a crucial problem. The different postures that one can take to lift an object is relatively limited once physical characteristics of the load are determined. The differences in loads are not physiologically significant. This perhaps explains the lack of success in the training of proper lifting technique and the greater success in the re-design of the task and/or workplace for the prevention of injuries. This project aimed at an optimization of the lifting task on an individual basis and our findings show no single 'safe' method of lifting.

References

Andersson, G. B. J., 1981, Epidemiological aspects of low back pain in industry. *Spine*, **6**, 53–60.

Bendix, T. and Eid, S. E., 1983, The distance between the load and body with three bi-manual lifting techniques. *Applied Ergonomics*, **14**, 185–192.

Barter, J. T., 1957, *Estimation of Mass of Body of Segments*. WADC Technical Reports, 57–260 (Wright-Patterson, OH).

Bejjani, F. J., Gross, C. M. and Pugh, J. W., 1983, Stress relationship between the knee and the back. *Proceedings of the International Society for the Lumbar Spine*, Cambridge, UK.

Bejjani, F. J., Gross, C. M. and Pugh, J. W., 1984a, Model for static lifting: relationship of loads on the spine and the knee. *Journal of Biomechanics*, **17**, 281–286.

Bejjani, F. J., Josephson, K. A. and Thornton, R. B., Jr., 1984b, The lifting stress calculator. *Journal of Biomedical Engineering*, **6**, 219–222.

Blumer, D., 1975, Psychological considerations in pain. In: *The Spine*, edited by R. H. Rothman and J. A. Simone (Philadelphia: Saunders).

Brown, J. R., 1972, *Lifting as an Industrial Hazard* (Ontario: Labour Safety Council of Ontario, Ontario Department of Labour).

Calin, A., Kaye, B., Sternberg, M., Antell, B. and Chan, M., 1980, The prevalence and nature of back pain industrial complex. *Spine*, **5**, 201–205.

Chaffin, D. B. and Park, K. S., 1973, A longitudinal study of low back pain as associated with occupational weight lifting factors. *American Industrial Hygiene Association Journal*, **34**, 513–525.

Chaffin, D. B., Herrin, G. D., Keysersling, W. M. and Foulke, A. J., 1977, *Pre-employment Strength Testing*. NIOSH Technical Report, NIOSH Physiology and Ergonomics branch (Cincinnati: NIOSH).

Frankel, V. H. and Nordin, M., 1980, *Basic Biomechanics of the Skeletal System* (Philadelphia: Lea & Febiger), pp. 113, 255.

Frymoyer, J. W., Costanza, M. C., Rosen, J. C., Goggin, J. E. and Wilder, D. G., 1980, Epidemiological studies of low back pain. *Spine*, **5**, 419–423.

Grandjean, E., 1988, *Fitting the Task to the Man* (London: Taylor & Francis), pp. 28, 104.

Graveling, R. A., Simpson, G. C. and Sims, M. T., 1985, Lift with your legs, not with your back: a realistic directive? In: *Ergonomics International 85, Proceedings of the Ninth Congress of the International Ergonomics Association*, 2–6 September 1985, Bournemouth, England, edited by I. D. Brown, R. Goldsmith, K. Coombers and M. A. Sinclair (London: Taylor & Francis), pp. 910–912.

Kelsey, J. L., White, A. A., Pastides, H. and Bishee, G. E., 1979, The impact of musculoskeletal disorders on the population of the United States. *Journal of Bone and Joints Surgery*, **51A**, 959–964.

Magora, A. and Taustein, I., 1969, An investigation of sick leave in the patient suffering from low back pain. *Industrial Medicine*, **38**, 398–408.

Miller, R. L., 1980, When you lift, bend your knees. *Occupational Health Safety*, **45**, 46–47.

Mital, A., Garg, A., Asfour, S. S. and Ayoub, M. M., 1980, Modeling of isometric strength and lifting capacity. *Human Factors*, **22**, 285–290.

Nachemson, A. L., 1984, Prevention of chronic pain: The orthopedic challenge of the 80's, *Bulletin of Hospital for Joint Diseases Orthopedic Institute*, New York, 1–15.

Nachemson, A. L., 1983, Work for all: For those with low back pain as well, *Clinical Orthopedics and Related Research*, **179**, 70–85.

NIOSH (National Institute of Occupational Safety and Health), 1981, *Work Practices Guide for Manual Lifting*. Department of Health and Human Services (NIOSH) Publication No. 81–122, Cincinnati, OH.

Park, K. S., 1973, A computerized simulation model of postures during manual materials handling. PhD dissertation, University of Michigan, Ann Arbor, MI.

Parnianpour, M., Pavlidis, L., Bejjani, F. J. and Nordin, M., 1986a, Effect of load height, weight and distance from the body on knee and lumbar spine forces during lifting. *Proceedings of the North American Congress on Biomechanics*, Montreal, Canada.

Parnianpour, M., Bejjani, F. J., Pavlidis, L., and Bejjani, N., 1986b, The lifting stress calculator: Study of various lifting parameters in 53 subjects. In: *Trends in Ergonomics/Human Factors*, Vol. III, edited by W. Karwowski (Elsevier – Amsterdam: North Holland).

Parnianpour, M., Bejjani, F. J. and Pavlidis, L., 1987, Worker training: the fallacy of a single, correct lifting technique. *Ergonomics*, **30**, 331–334.

Pheasant, S., 1986, *Bodyspace – Anthropometry, Ergonomics and Design*. (London: Taylor & Francis), pp. 206.

US Department of Health, Education and Welfare, 1971–5, *Vital and Health Statistics: Basic Data on Arthritis Knee*. Series 11, No. 213.

PART VI
ARTIFICIAL INTELLIGENCE IN SAFETY ANALYSIS

Chapter 23
Strategy for developing an expert system for safety

KATHRIN KJOS
1294 North First Ave.,
Upland, CA 91786, U.S.A.

Abstract. System safety expert systems can be useful tools in integrating safety in the design process. The need for an expert system for hazard analysis is discussed. An expert system for safety analysis is described along with a framework for development of the system.

Keywords: Hazard analysis; System safety; Expert system.

Introduction

The goal of product/system safety analysis is to identify potential hazards early in the design of a system. Early identification of hazards allows them to be eliminated or controlled without costly design changes and prior to catastrophic accidents. Timely identification of potential hazards also leads to control by design; as opposed to labelling and procedural cautions.

To identify and control hazards, in the design phase of a system, system safety engineers perform the following tasks: (i) identification of potential hazards; (ii) determination of the worst possible effects of the potential hazard; (iii) recommendation of controls to reduce risk; and (iv) identification of design criteria required by company practice, federal standards, contract, etc. The key to controlling a hazard is identifying it. This can often prove to be the most difficult part of safety analysis.

Deficiencies in current approaches

The field of product/system safety combines many disciplines including electrical, chemical and mechanical engineering along with ergonomics, toxicology, etc. As a result, safety engineers are required to act as extensive data bases in their analysis of systems. The problem facing these engineers is the difficulty of being knowledgeable in all the fields to make an adequate risk assessment of the system. Failure to identify a hazard, and thus control it, occurs when the engineer lacks knowledge in an area.

As an example, an engineer may not know that optics are commonly coated with radioactive materials and therefore may not investigate this hazard. The safety engineer could be greatly aided if he had a tool to help point out areas of potential concern (Kjos, 1987).

Checklists are commonly used to aid in this process, yet they have inherent faults. They seem to be best suited for checking compliance with specific standards and regulations.

However, compliance with standards does not necessarily mean the system is safe or easy to use. Another weakness of checklists is that they are usually completed with or by the design engineer who has a bias to say the 'correct' answer automatically without thinking it through or knowing all the information. A checklist question may ask 'Are suspected carcinogens used?' Most design engineers will answer 'no'. Yet, they may not even be aware what materials are carcinogenic; nor should they be expected to. Checklists could be further refined to ask more rigorous questions. In the case of suspected carcinogens one could ask 'Are beryllium compounds used?'; 'Is dimethylamine used?', etc. A problem still exists with this approach as one needs to have a general idea of where beryllium compounds might be used to know if they are being used. A better question might be 'What compounds are the bases of the circuit boards made of?'. Then the safety engineer can determine if this compound is a suspected carcinogen, is toxic, or has compatibility problems.

The role of the safety engineer is to ask the right questions. These questions are based on past experience, military standards, material/chemical properties, previous incidents, etc. The better the data base available to the safety engineer the more complete their analysis will be. Not only must one have this information available, there must also be an efficient way to retrieve it in reference to a particular system. If complete checklists were used they would be hundreds of pages long and probably not be received too well by the design engineering community. To improve the efficiency of safety engineers they need to be able to ask more specific questions. To improve accuracy and completeness they must be aware of what questions to ask.

A second problem is that safety is often not involved early in the design of a system. This may be due to a number of issues such as lack of budget, personnel or politics. This often leads to safety problems that are inherent in the design and therefore costly to change. A method is needed to involve safety analysis as an integrated part of the design process. This integration of safety analysis with the design process would allow design engineers to evaluate the safety of their design in an interactive, real-time fashion.

An expert system approach

Expert system technology provides a great opportunity for tackling these problems. An expert system tries to capture, in software, a data base of knowledge from 'experts' so it can give near human advice on the subject. A system safety expert system would capture knowledge of system safety experts along with experts in various disciplines (e.g. electrical, toxicology, human errors etc). Along with safety knowledge, the system would also include knowledge on systems (such as computers, lasers and missiles).

A system safety expert system allows non-safety experts to perform safety analysis. The system queries the user for information and provides data on potential hazards, its effects and controls. Eventually, this analysis could be integrated into CAD workstation software. This would integrate routine safety analysis into the design. For instance, if an electrical engineer is designing a circuit with 500 V, the expert system would place a tag on the circuit drawing stating that a barrier is needed when the circuit is integrated into the design.

The expert system for safety analysis, under development, contains a knowledge base of relevant standards, historical lessons learned, applicable research data, and approaches to identifying and solving system safety problems. The expert system assists safety engineers in step by step analyses of a particular system. This includes the expert system asking specific questions about the system and predicting possible effects. The expert system also explains how it predicted these hazard effects and recommends solutions that have been used to solve similar problems.

The goal of this program is to be a tool to help safety engineers identify concerns and requirements on a specific system. The program helps the engineer determine what information they may need and will make assumptions if no information is available. These default values can be determined by past experience on similar systems.

It is interesting to note that criteria for determining defaults for safety analyses is fairly simple; choose the worst case. For instance, if we are analyzing a circuit board and do not know what materials are on the board, there are several possibilities, and most of the materials are safe. Yet, past experience also indicates that boards can be made of beryllium substrates, a material that can pose a health hazard. Our default value for the material of the circuit boards would then be beryllium substrates.

Even if this assumption is incorrect it is an important tool to obtain the correct information. Design engineers often do not have time to provide data to system safety engineers. Yet, throw out a safety problem and everyone will go out of their way to prove your assumptions wrong. This is when the system safety engineer can get the actual information if the assumption is wrong. It also ensures that the system safety engineer evaluates the circuit board for toxic materials and knowing that beryllium substrates are hazardous, keys them to check other materials.

Ideally this program will eventually tie into established data bases of information on toxic material, carcinogens, anthropometric data, and human factors/safety requirements. The program would also provide a structured approach to gathering this information for people who do not have a background in system safety. It would also help experienced safety engineers remember the many areas that must be analyzed in a system.

Framework for an expert system for product safety

An expert system is under development to perform two main tasks; (i) aid system safety engineers in identifying, quantifying and recommending control for hazards; and (ii) teach new system safety engineers what information is necessary to identify and quantify a hazard. The expert system attempts to incorporate two different approaches to system safety analysis: energy/hazard based and system/heuristic based. These approaches can be described as:

1. Energy/hazard based. The system safety engineer checks to see if they have energy or situations in the system that can cause a hazard (i.e. they find out what voltages are in the system and analyze whether they pose a hazard or find out what chemicals are used and their effects).
2. System based. Based on past analyses, accident data, and/or experience the system safety engineer knows to analyze specific hazards related to the type of system (i.e. if the system is a laser they know to analyze for: eye and skin damage from the beam, electrical shock hazards from the high voltage power supply, possible rupture of pressurized vessels such as a heat exchanger, harmful gases used in cryogenic systems).

System safety engineers unfamiliar with the type of system they are analyzing rely heavily on the energy/hazard based approach. This approach can be thought of as a hazard checklist. Going through each hazard and determining whether it exists can be a time consuming process. Yet, the approach can result in a very thorough analysis.

System safety engineers who have performed analyses on or worked with the type of system they are analyzing tend to use the, more intuitive, system based approach. This approach will quickly focus on hazards/problems experienced on past systems. Heuristics are often used to pinpoint hazards and recommend controls. This approach can result in quick analysis yet may not analyze hazards associated with new technology.

In reality the two paths or approaches to hazard analysis are rarely distinct and separate processes; both approaches are used simultaneously by system safety engineers. A complete expert system for hazard analysis should attempt to address both paths. The user will be able to use the system through either path.

A typical user scenario based on the energy/hazard flow would go as follows:

1. The user would choose, from a menu (see Figure 23.1), whether to analyze electrical, chemical, mechanical, software, or human error hazards. Let us say electrical was chosen.
2. The user would then decide whether to analyze arcing/sparking, shock, loss of function, temperature, or radiation. Let us say shock was chosen from the menu (see Figure 23.2).

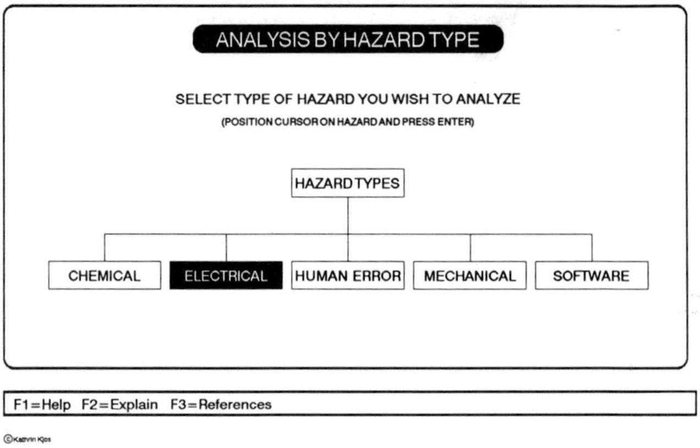

Figure 23.1 Menu for analysis by hazard type.

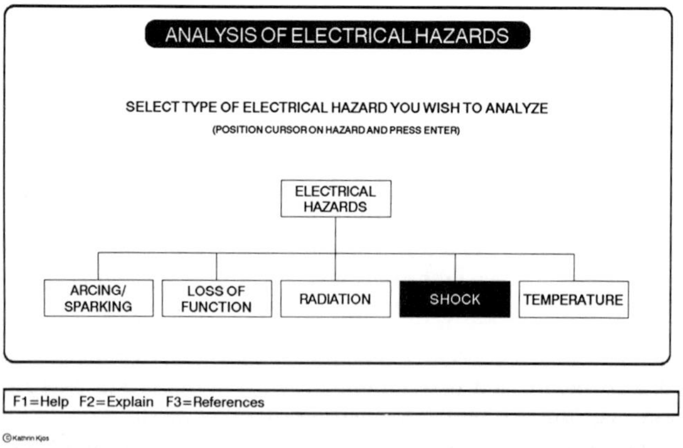

Figure 23.2 Selection of hazard type for analysis.

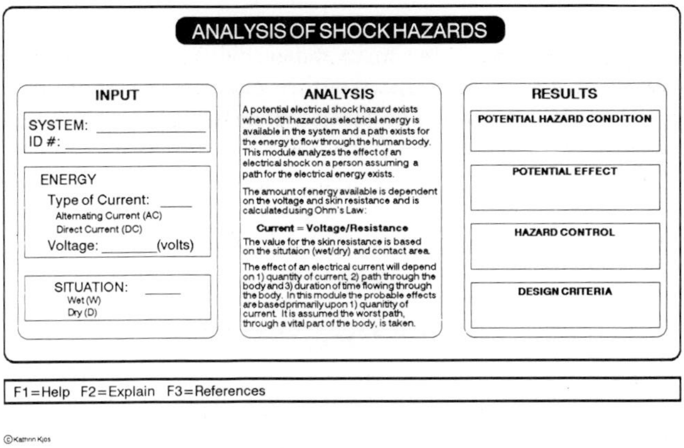

a

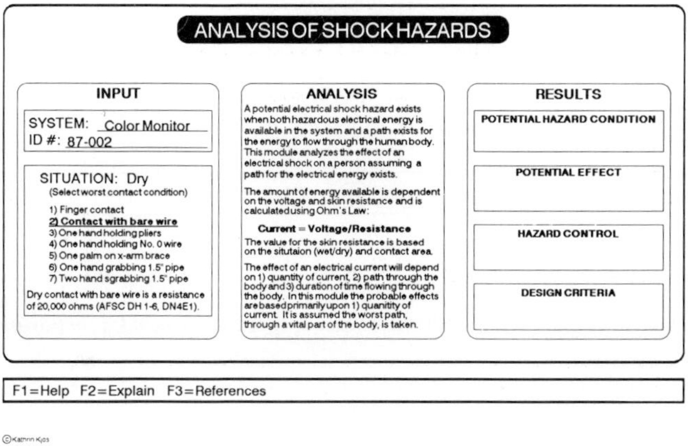

b

Figure 23.3 (a) Module showing request for information from user. (b) Selected input example.

3. The shock module would then ask the user for information regarding the system under analysis (i.e. voltage, type of current, situation, etc.) as shown in Figure 23.3.
4. Using the information supplied by the user and data from tables, calculations and standards the module would output the potential hazard, effect, recommended controls, and applicable design criteria as shown in Figure 23.4.
5. The user could then go on and analyze other shock hazards that may exist in the system (due to different voltages, current). The process would be complete when all five hazard areas (electrical, chemical, etc.) were completely analyzed.

A typical user scenario based on the system based flow would go as follows:

1. The user would select which type of system they are performing a hazard analysis on. These choices could be radar, laser, missile, sensor, thermal imaging system, display console, or test equipment. Let us say thermal imaging system was chosen.
2. The expert system would then 'look up' what it knows about typical hazards found in thermal imaging systems and query the user for confirmation of this information. For

instance, the optics used in many thermal imaging systems are coated with thorium (a potentially toxic compound due to its radioactivity). Therefore, the expert system would ask 'What are the optics coated with?'. If the system safety engineer knows, he would input that information. The expert system would then analyze that compound for chemical hazards. However, if the system safety engineer did not know what coating was used, the expert system would suggest that thorium may be the coating and provide effects, controls, etc. for the potential hazard due to thorium coatings.

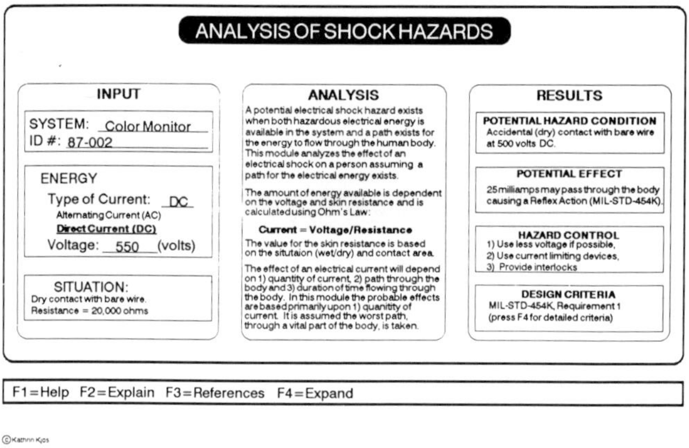

Figure 23.4 System output showing potential hazard, effect, recommended control and applicable design criteria.

Structuring and partitioning the problem

The hazard analysis process has defined inputs and outputs as shown generically in Figure 23.5. The system's energy source will determine the potential hazardous condition. The effect will be determined by the energy source and the situational factors. The control is dependent on the potential effect of the hazard and the level of the energy source.

The development of a hazard analysis expert system can be easily scaled and partitioned. There are two major types of partitioning used for the hazard analysis system: (i) by type of hazard, and (ii) information on particular systems (i.e. lasers, radars, optics). Figure 23.6 shows the partitioning of the system by the type of hazard. This partitioning supports the energy based approach to hazard analysis.

The expert system would also contain information on systems. An example for a laser system is shown in Figure 23.7 (Sliney and Wolbarsht, 1980). This information is basically a model of the systems that would be analysed. It consists of a hierarchy and potential features or characteristics the system might have.

The expert system ties together knowledge from both areas of information (hazards and system). In the analysis of a laser, the expert system asks the user what type of lasing medium is being used. It would provide the user with a list of choices such as neodymiun-yag, ruby, etc., from previously stored knowledge. If the user chose neodymiun-yag, the

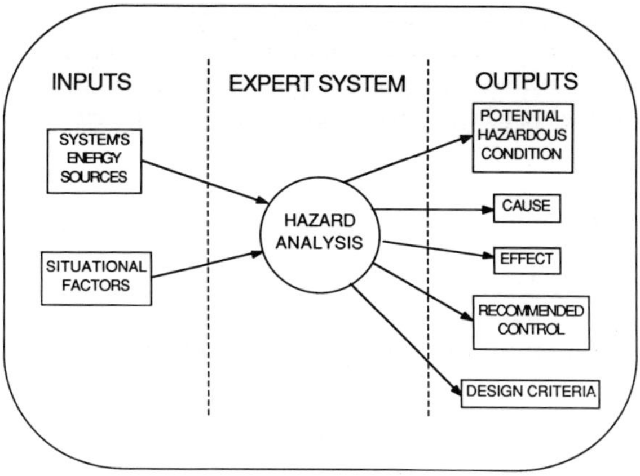

Figure 23.5 *Generic inputs and outputs of expert system for safety.*

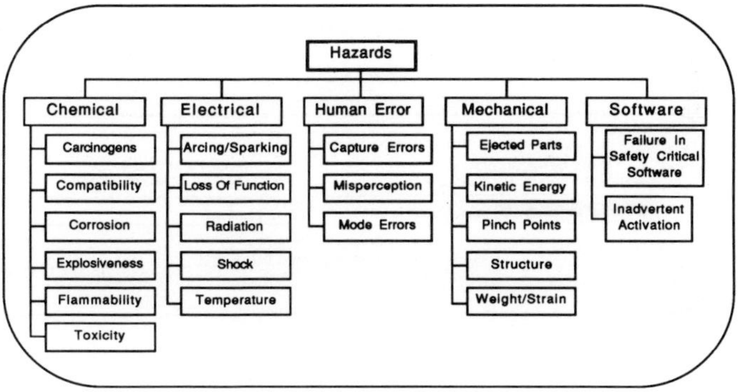

Figure 23.6 *Partitioning of the expert system for safety.*

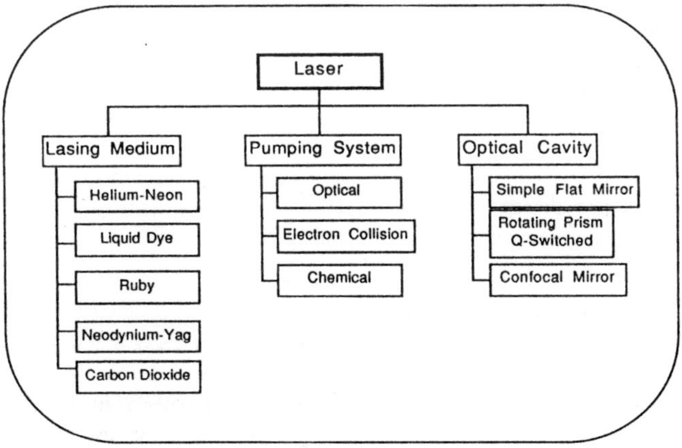

Figure 23.7 *Partitioning of the laser system.*

system would then look up the information it has stored on that lasing medium. It would prompt the user to get the required information for calculating the safe viewing distance and other safety concerns related to lasers and specific to neodymiun-yag lasers. In a similar fashion, hazards would also be postulated on the basis of the type of pumping system.

Development to date

The partitioning allows the development of individual modules one at a time. These modules can then be used while the rest of the system is under development. Certain modules are easier to develop than others since methods for analyzing these types of hazards are more defined.

The expert system for safety analysis is being developed in increments instead of the entire hazard analysis at once. This incremental process will allow the task to be broken down into small portions that are easily defined and programmed. It will also allow portions of the system to be in use while the entire system is under development.

The expert system has been developed in phases as shown in Figure 23.8. This allows the system to be developed starting from a narrow scope, and problems that are easier to address, and then expanded. It also maps out a plan for future development of the system and its more difficult aspects.

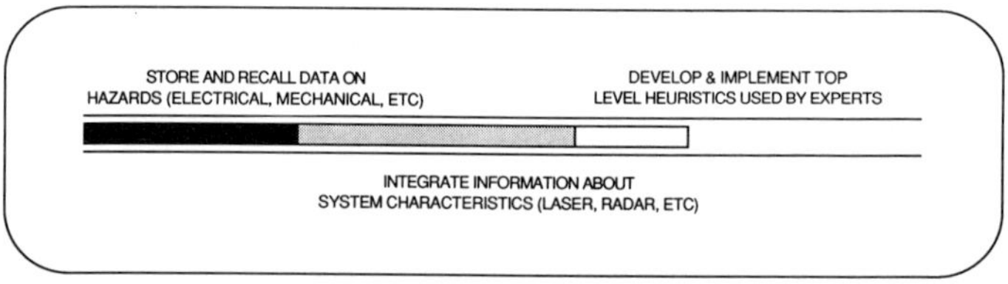

Figure 23.8 Phase of development.

Conclusion

Expert system technology can be used to create valuable tools for safety analysis. The expert system does not replace the need for safety engineers but provides assistance in performing routine hazard analysis. This allows safety personnel to become more involved in the non-routine areas of safety design and analysis. It also enables designers to perform their own safety checks and provides a vehicle for educating them in safety aspects of their design.

References

Kjos, K., 1987, Expert systems; a tool for system safety engineers. *Hazard Prevention*, **23**, 6–10.

Sliney, D. and Wolbarsht, M., 1980, *Safety with Lasers and Other Optical Sources*, (New York: Plenum Press), pp. 39–45.

Chapter 24

Knowledge-based safety analysis in the CAD environment

JOUKO SUOKAS, PERTTU HEINO and IRIS KARVONEN

Technical Research Centre of Finland,
P.O. Box 656,
SF–33101 Tampere, Finland

Abstract. The design of safe and reliable process systems requires safety knowledge and effective methods of using that knowledge. The use of CAD provides new possibilities for taking safety aspects into account more comprehensively by using knowledge-based software. HAZOPEX is an expert system for the HAZOP analysis of process systems. Experience from the prototype version of HAZOPEX indicate that the most effective way of achieving a high level of safety in new process installations is the development of computer tools for CAD environment.

Keywords: Safety; Process; Knowledge; CAD; Analysis.

Introduction

Computer-aided design (CAD) is increasingly used in process and machine design. In addition to the basic CAD software, the CAD environment could contain supportive software. The designer would have easy access to software packages for the analysis of a design, verification of the design requirements, and decision support for the decisions to be made on potential changes.

While the earlier applications of CAD mainly considered drawing tasks, the current and future applications include knowledge on the system behaviour. Among other purposes, this information can be used as necessary input for the knowledge-based safety analysis of the system. Safety and reliability are important factors affecting the life cycle costs and economy of a process system. The design of a new system is of ultimate importance in achieving high level safety and reliability targets with reasonable costs.

The development of complex process systems has made it difficult for a designer to consider carefully even the most important hazards. On the other hand, the increased size of plants has made it possible for accidents with far reaching consequences to occur. Flixborough (1974), Seveso (1976), Bhopal (1984), and Mexico (1984) are just a few examples. Safety analysis is increasingly used as a tool for managing the safety critical features of the designed process system. The major problems of manual safety analysis methods are the resources, competence, and calendar time required.

HAZOPEX is an expert system supporting the safety analysis of process systems. It is being developed in the Technical Research Centre of Finland. HAZOPEX is based on hazard and operability study (HAZOP) which is widely used in the European process industry.

The main objective of this chapter is the development of knowledge-based safety analysis tools for the CAD environment. The role of safety analysis in the design of process systems is discussed in the next section and HAZOP methodology is described and an overview of computer support for safety analysis is presented. The HAZOPEX expert system is then described and results from the evaluation of the prototype version of HAZOPEX are presented. The last section ends the text with concluding remarks.

Safety analysis in process design

The role of safety in process CAD

Common methods of assuring the safety of the designed process installation are (Haastrup, 1984): (i) the use of standards and internal design instructions, (ii) quality control procedures, (iii) risk and safety analyses, (iv) safety checks and inspections, (v) start-up inspections, (vi) skilled designers and systematical design procedures, and (vii) cooperation of different branches of design and operation and maintenance.

There is a logical connection between graphics and data bases in modern CAD systems. Each piece of information is input to the CAD system only once and stored in the data base for common use. This minimizes the amount of required data input and makes large amounts of data available to the designer.

The use of CAD provides new possibilities for taking safety aspects into account more comprehensively. The design can be analyzed automatically or interactively from the safety viewpoint, the design requirements can be verified automatically, and decisions on safety critical characteristics of the system can be supported by presenting relevant information to the designer. The effort put into such supportive tasks is likely to be small compared with the profits gained in the later phases of the system life cycle by diminished modification and maintenance costs.

Hazard and operability study

Systematic methods have been developed for the identification of hazards and for modelling the event sequences which have the capacity to lead to accidents. Hazard and operability study (HAZOP) is probably the most common method used for hazard identification in process systems. HAZOP is based on flow sheets or pipe and instrumentation diagrams (PID) of a process system.

The purpose of HAZOP is to identify all possible deviations from the way the design is expected to operate and all hazards associated with these deviations. In the search for deviations, a list of guide words is applied to the process units, such as pipelines, tanks and reactors. The aim is to identify potential deviations in the parameters, such as flow, temperature, pressure and viscosity having both causes that are conceivable and consequences that are potentially hazardous. For example, Anon (1977), Kletz (1983) and Wells (1980) give lists of deviations and their contributors to be considered in a HAZOP study.

The results of a HAZOP study are usually documented on forms similar to those shown in Figure 24.1. The documentation is mainly done manually but a few computer programs have been developed for this purpose in recent years.

The benefits of HAZOP are based on the early identification of design failures. The use of HAZOP has reduced the need for modifications during the commissioning and start-up of new process systems. It also has reduced the amount of process disturbances and accidents. Despite these advantages, HAZOP still has certain problems. One is the calendar time and

OCCUPATIONAL SAFETY ENGINEERING LABORATORY HAZARD AND OPERABILITY STUDY	SYSTEM: BLACK LIQUOR RECOVERY BOILER PLANT SUBSYSTEM: EVAPORATION PLANT 1 UNIT: EVAPORATION UNIT 1 – STRONG LIQUOR TANK		PAGE: 8 DATE: 18.2.1982 DRAFTED BY: HAZOP TEAM
DEVIATION	**POTENTIAL CAUSES**	**CONSEQUENCES**	**ACTION REQUIRED**
NO FLOW	LINE BLOCKAGE, VALVE ERRONEOUSLY CLOSED, PUMP FAILURE, FAILURE IN ELECTRICITY SUPPLY	15. OVERFILLING OF THE EVAPORATION UNIT 1	
HIGH FLOW	LEAKAGE OF THE PUMP WASH WATER LINE VALVE OR THE VALVE IS ERRONEOUSLY OPEN	16. DILUTION OF THE STRONG LIQUOR – ENTRY OF WATER INTO THE STRONG LIQUOR LINE	USING AS-NEEDED MOUNTED HOSES INSTEAD OF FIXED WATER PIPE LINES
	VALVE V1 ERRONEOUSLY OPEN	17. DILUTION OF THE STRONG LIQUOR – ENTRY OF WEAK LIQUOR INTO THE STRONG LIQUOR LINE	REMOVING THE VALVE V1 AND INSERTING BLANKS
	LEAKAGE OF THE VALVES HV1 AND HV3 LEAKAGE OF THE VALVE HV4 WHEN WASHING THE STRONG LIQUOR LINE FALSE OPERATION OF THE INTER-DEPENDENT AUTOMATIC VALVES HV1/HV4 AND HV3/HV4 WHEN WASHING THE STRONG LIQUOR LINE	18. DILUTION OF THE STRONG LIQUOR – ENTRY OF INTERMEDIATE OR STRONG INTERMEDIATE LIQUOR INTO THE STRONG LIQUOR LINE	CHANGING THE VALVE TYPE OF HV1, HV3 AND HV4 INTO BALL VALVES ALARM ON THE FALSE OPERATION OF THE VALVES

Figure 24.1 *An example of the documentation in HAZOP analysis (Reunanen and Suokas, 1986).*

effort needed in the analysis. When the analysis is carried out using team work the timing with a tight project schedule may also cause problems.

HAZOP analysis is normally carried out when the design is almost completed. However, the benefits from HAZOP analyses in the earlier phases of design seem remarkable. Computer support for HAZOP in the CAD environment could pre-date the use of HAZOP in the design process and lead to better quality design.

Computer-aided safety analysis

Computers have aided safety and reliability analysis in several ways. Figure 24.2 (Suokas *et al.*, 1987) illustrates the development of computer aids.

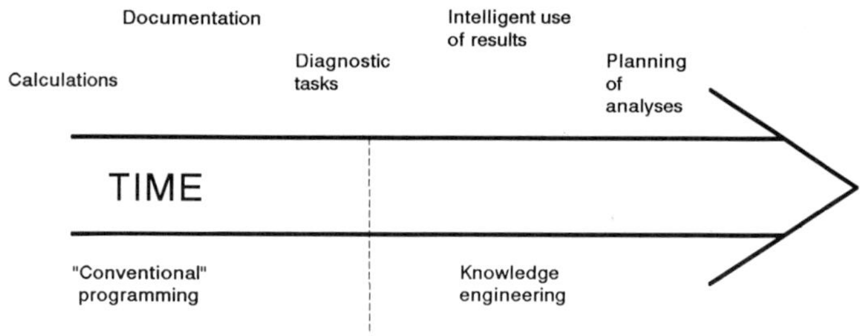

Figure 24.2 The development of computer aids to safety and reliability analysis (Suokas et al., *1987).*

The first applications concerned mainly routine calculations of reliability, availability, and accident frequency based on fault trees, and calculation of gas releases and dispersions based on physical models. Examples are RELVEC (Anon, 1986a, b) for aiding fault tree calculation, and CRUNCH (Jagger, 1983), DENZ (Fryer and Kaiser, 1979) and DEGADIS (Havens and Spicer, 1985) for gas dispersion calculations.

The next phase was to get assistance in the documentation of an analysis, for example drawing a fault tree or documenting the results of HAZOP.

The third phase in computer-aided safety analysis was the development of programs aiding the diagnostic tasks in an analysis. This meant the automation of several earlier manual tasks. Examples of this type of program are RIKKE (Haastrup *et al.*, 1985) and CAFTS (Poucet, 1983) developed for fault tree construction.

These programs were first made using conventional programming techniques and languages. The fourth phase – now under progress – is to use knowledge engineering. The promising views of knowledge-based systems and the wide positive experiences in HAZOP encouraged us to start the development of an expert system for HAZOP analyses.

Knowledge-based applications

The use of knowledge engineering in safety analysis and management has just been started. However, the preliminary results seem so promising that several new projects have been planned in the field. In the following, a few current projects in Europe are shortly reviewed.

Fault trees are often used for modelling accidents. For the construction of fault trees from a system description, for example from PIDs, two extensive projects have been started. The project called STARS (Software Tool for Advanced Safety and Reliability) includes the qualitative modelling of a process plant on a general level and the construction of more detailed fault trees from PIDs (Poucet and Petersen, 1988). The other project called FIABEX concerns the construction of fault trees and the application of fault trees in the diagnosis of disturbances (Barbet *et al.*, 1987). In both projects knowledge-based expert systems are to be developed for achieving the goals. In the STARS project, the initial cooperation between the Joint Research Centre (Italy) and Risø (Denmark) has been extended to cover also ENEA (Italy) and VTT (Finland).

In addition to the analysis of system design, other possible areas for the use of knowledge-based systems in the CAD environment are the automatic verification of design requirements and decision support. The knowledge obtained during the design might also be useful in the later phases of the process system life cycle. If possible malfunctions of the process have been analyzed during the design, the diagnosis of a process disturbance can be made more reliable by employing this information. The knowledge could also be used in operator training in a similar manner.

HAZOPEX is an expert system supprting the safety analysis of process systems. It is being developed in the Technical Research Centre of Finland. HAZOPEX is based on hazard and operability study (HAZOP) which is widely used in the European process industry. HAZOP is a systematical investigation on a process system intending to identify hazardous deviations from the normal state of process parameters. Traditionally, HAZOP analysis is carried out in teams but HAZOPEX enables the designer to perform the analysis alone. The current HAZOPEX prototype contains 400 rules and a large set of search functions for analytical reasoning.

Knowledge-based HAZOP

Overview of HAZOPEX

The aim of HAZOPEX is to support the HAZOP analysis of process systems. Relevant deviations from the normal process state are analyzed and their causes and consequences are identified and remedies planned. In addition to aiding the interactive analysis of deviations HAZOPEX also assists in focusing the analysis and provides information about safety analysis and HAZOP.

HAZOPEX is designed for the use of a process designer. When installed in a CAD workstation used in the process design a HAZOP analysis can fairly easily be made. One alternative is to carry out separate HAZOP analyses whenever the design of a subprocess has been completed. Another choice is to have HAZOPEX active in the background all the time and shift to it whenever something has changed that might cause safety problems.

The intention of HAZOPEX is to pre-date the use of HAZOP analyses to the earlier phases of process design. This means earlier identification of defects and errors in the design and hence better and more economic possibilities for achieving a safe and reliable process system. By using HAZOPEX, it is possible to find relevant deviations and their causes in different parts of the process and propagate the effects of these deviations. This reduces the effort required for a systematic HAZOP study to a lower level and guarantees that the results are of good quality.

HAZOPEX provides expertise of safety analysis and typical faults and hazards in a process environment and a mechanism for logical inference based on this expertise. The structure of

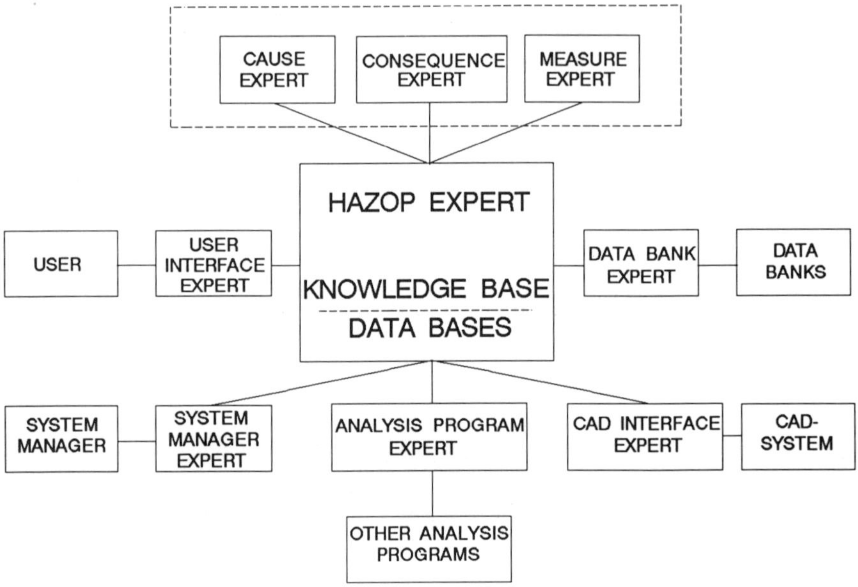

Figure 24.3 The structure of HAZOPEX.

HAZOPEX is presented in Figure 24.3. The knowledge base includes knowledge about (i) the HAZOP method, (ii) potential causes and consequences of deviations, (iii) suggestions for measures to be taken to prevent the deviation or its consequences, and (iv) heuristics for steering the analysis.

The main task of HAZOPEX is to find valid potential causes for each deviation in each part of the process. This has been implemented in a rule-based manner. There are two kinds of rules for tracing the causes in HAZOPEX: general level rules, and rules for detailed search. The general level rules are classified according to deviation type. They are of the form:

if (deviation)
(conditions combined with the process)
then (cause)

The rules for detailed search have a similar form but are common to all deviations.

Use of HAZOPEX

The pipe and instrumentation diagram (PID) and complementary information on the process and its components are required as input. In the prototype version of HAZOPEX the process is modelled using a graphical model editor which allows the user to present the process in the form of a PID. A type hierarchy consisting of the descriptions of certain typical process components forms the basis for the process description. Each component in the process becomes an instance of the type hierarchy. In the current prototype version the type hierarchy is incomplete, for example boilers and reactors are missing.

In the production version the input of the PID is to be automated. Effective use of HAZOPEX requires that the data is transferred automatically from the CAD system used in

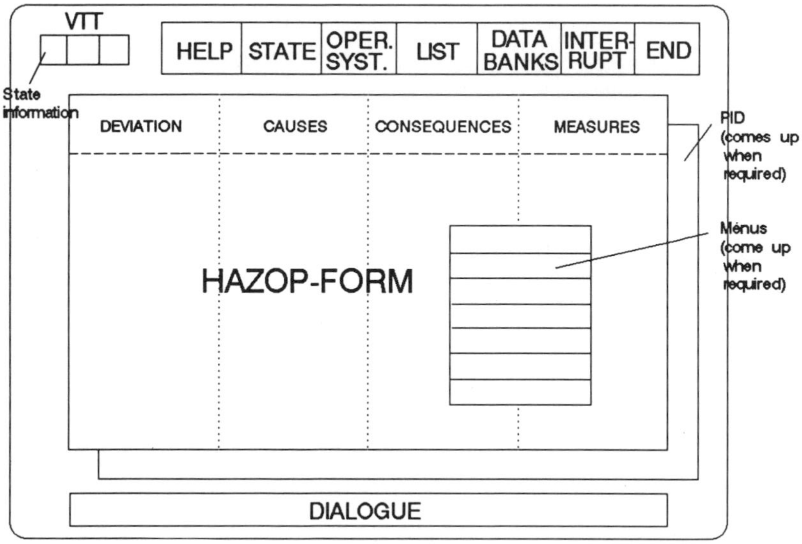

Figure 24.4 The user interface of HAZOPEX.

the design of the process system. There is also a need for automatic input of PIDs being available only as paper prints. This can be done by scanning the drawings into digital form with a scanner and giving manually the necessary complementary information on flows, normal states of components, etc.

There are several optional choices that the user can make in order to affect the course of the analysis. For example, it is possible to declare some of the deviations to be analyzed fully automatically in the background. The results of the analyses of these deviations will appear in the complete listings but the analyses will require no interaction from the user unless dangerous consequences are found. Other possible choices are language and default order of deviations.

During the analysis there is a HAZOP form on the display where the results of the interactive analysis are presented. The PID of the process system is on the screen beneath the HAZOP form and can be accessed easily. The user controls the analysis through an interface based on menus. The user interface of HAZOPEX is presented in Figure 24.4.

Two basic types of HAZOP documentation are produced during the analysis: a complete and a short document. The short HAZOP document contains only those results that were produced interactively whereas the complete document contains also the results of the automatic background analysis and notes about the decisions made during the analysis sessions. Another basic type of documentation is the deviation diagram where the relations between the causes and consequences of a deviation are presented in a graphical form. Also several other output operations have been planned, for example fault trees and listings based on consequence classification. The course of HAZOP analysis and the allocation of tasks to HAZOPEX and the user are presented in Figure 24.5.

HAZOPEX goes systematically through all possible deviations in different parts of the process. It generates analysis results and makes suggestions which the user can accept, reject or complement. However, the user can choose the analysis object or the deviation of most interest if the systematic procedure seems unsuitable.

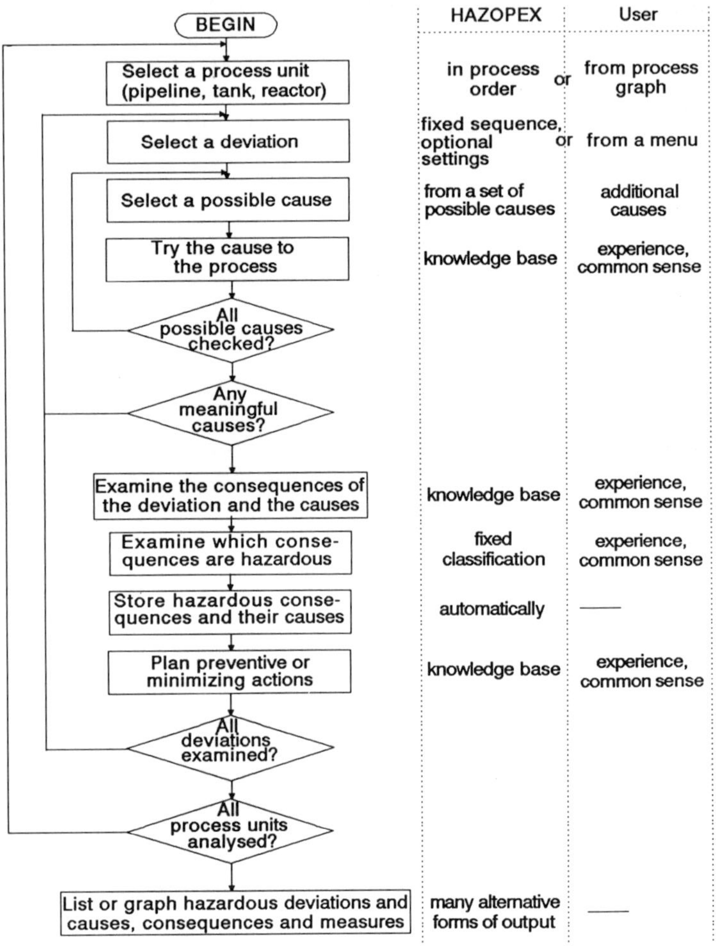

Figure 24.5 The course of HAZOP analysis and the tasks of HAZOPEX and the user.

Example

In this example, the deviation HIGH TEMPERATURE is analyzed in a selected part of an example process system consisting of tanks, valves, pumps and a heat exchanger. The PID of the example process is presented in Figure 24.6.

A complete HAZOP analysis of the example system would consist of analyses of all possible deviations in each functional part of the system. The pipeline considered in this example is indicated with a broken line in Figure 24.6. Figure 24.7 shows the situation after HAZOPEX has generated initial suggestions for causes, consequences and measures.

After the initial suggestions have been generated the user probably wants to examine some of the causes in further detail. In this example the causes 1.7, 1.7.1, 1.7.1.2 and 1.7.1.2.1

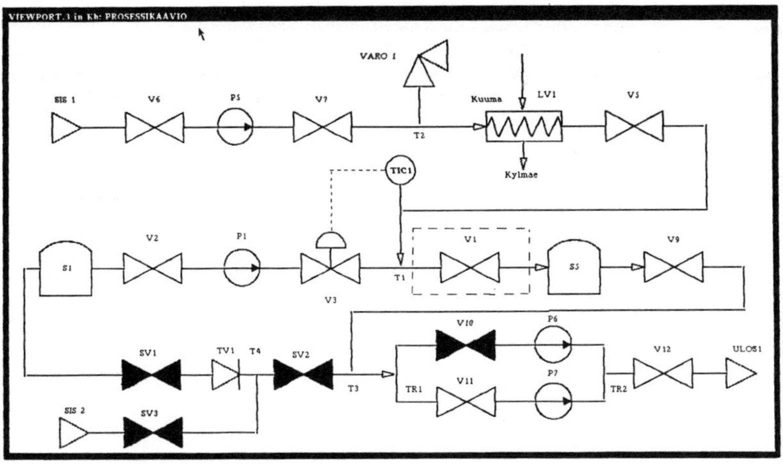

Figure 24.6 The PID of the example process.

HAZOP Form

Deviation	Causes	Consequences	Measures
1 HIGH.TEMPERATURE		1.0.1 EXTRA CHEMICAL REACTION	
		1.0.2 DISTURBANCES IN.PRODUCT.QUALITY	
		1.0.3 PRESSURE.INCREASES	
	1.1 HIGH.PRESSURE CAUSES.INCREASE OF.TEMPERATURE		
	1.2 IN.TANK HIGH.TEMPERATURE UPSTREAM		
	1.3 TOO.HIGH.WARMING		
	1.4 INTO.SUBPROCESS HIGH.TEMPERATURE UPSTREAM		
	1.5 OUT.OF.NORM.CLOSED BR. EXTRA.FLOW UPSTREAM CAUSES.INCR.OF TEMP.		
	1.6 EMISSION.OF.HEAT FROM.ENVIRONMENT		
	1.7 IN.T-BRANCH.INPUTS RELATION.OF.FLOWS CHANGED UPSTREAM		1.7.1 CONTROL RELATION.OF.FLOWS IN.T-BRANCH
	1.8 IN.HEAT.EXCHANGER FLOW.DOES.NOT.COOL ENOUGH UPSTREAM		1.8.1 CONTROL FLOW.OF LINE AND FLOW.AND.TEMP OF.COLD.LINE
	1.9 EXTRA.EXOTHERMIC REACTION CAUSES.INCREASE OF.TEMPERATURE		
	1.10 FAILED.ELECTRIC.DEVICE HEATS.UP		

Figure 24.7 The causes, consequences and measures generated by HAZOPEX for the deviation HIGH TEMPERATURE in an example process system.

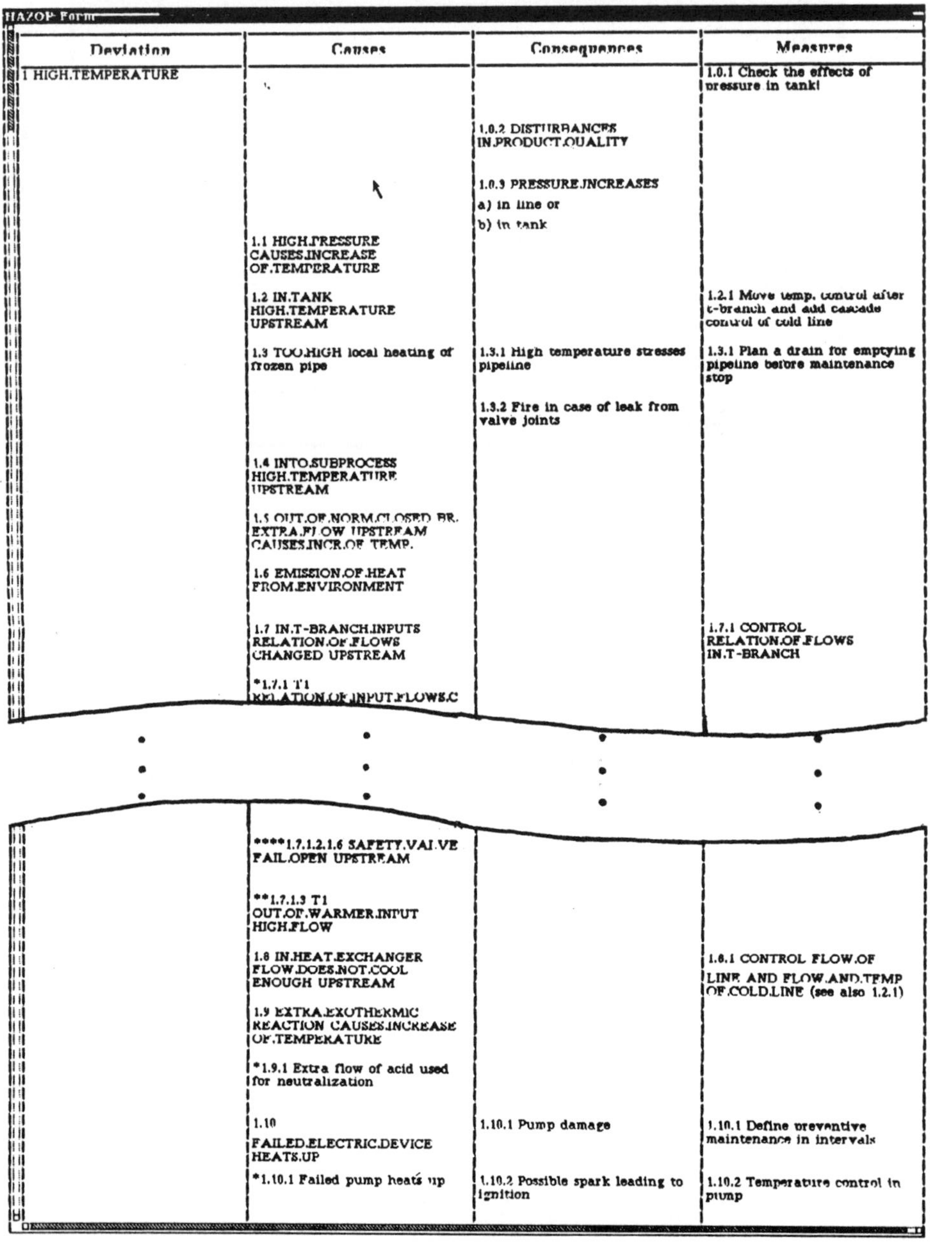

HAZOP Form

Deviation	Causes	Consequences	Measures
1 HIGH.TEMPERATURE			1.0.1 Check the effects of pressure in tank!
		1.0.2 DISTURBANCES IN.PRODUCT.QUALITY	
		1.0.3 PRESSURE.INCREASES a) in line or b) in tank	
	1.1 HIGH.PRESSURE CAUSES.INCREASE OF.TEMPERATURE		
	1.2 IN.TANK HIGH.TEMPERATURE UPSTREAM		1.2.1 Move temp. control after t-branch and add cascade control of cold line
	1.3 TOO.HIGH local heating of frozen pipe	1.3.1 High temperature stresses pipeline	1.3.1 Plan a drain for emptying pipeline before maintenance stop
		1.3.2 Fire in case of leak from valve joints	
	1.4 INTO.SUBPROCESS HIGH.TEMPERATURE UPSTREAM		
	1.5 OUT.OF.NORM.CLOSED BR. EXTRA.FLOW UPSTREAM CAUSES.INCR.OF TEMP.		
	1.6 EMISSION.OF.HEAT FROM.ENVIRONMENT		
	1.7 IN.T-BRANCH.INPUTS RELATION.OF.FLOWS CHANGED UPSTREAM		1.7.1 CONTROL RELATION.OF.FLOWS IN.T-BRANCH
	*1.7.1 T1 RELATION.OF.INPUT.FLOWS.C		
⋮	⋮	⋮	⋮
	****1.7.1.2.1.6 SAFETY.VALVE FAIL.OPEN UPSTREAM		
	**1.7.1.3 T1 OUT.OF.WARMER.INPUT HIGH.FLOW		
	1.8 IN.HEAT.EXCHANGER FLOW.DOES.NOT.COOL ENOUGH UPSTREAM		1.8.1 CONTROL FLOW.OF LINE AND FLOW.AND.TEMP OF.COLD.LINE (see also 1.2.1)
	1.9 EXTRA.EXOTHERMIC REACTION CAUSES.INCREASE OF.TEMPERATURE		
	*1.9.1 Extra flow of acid used for neutralization		
	1.10 FAILED.ELECTRIC.DEVICE HEATS.UP	1.10.1 Pump damage	1.10.1 Define preventive maintenance in intervals
	*1.10.1 Failed pump heats up	1.10.2 Possible spark leading to ignition	1.10.2 Temperature control in pump

Figure 24.8 The HAZOP form after the initial suggestions have been completed and modified by the user.

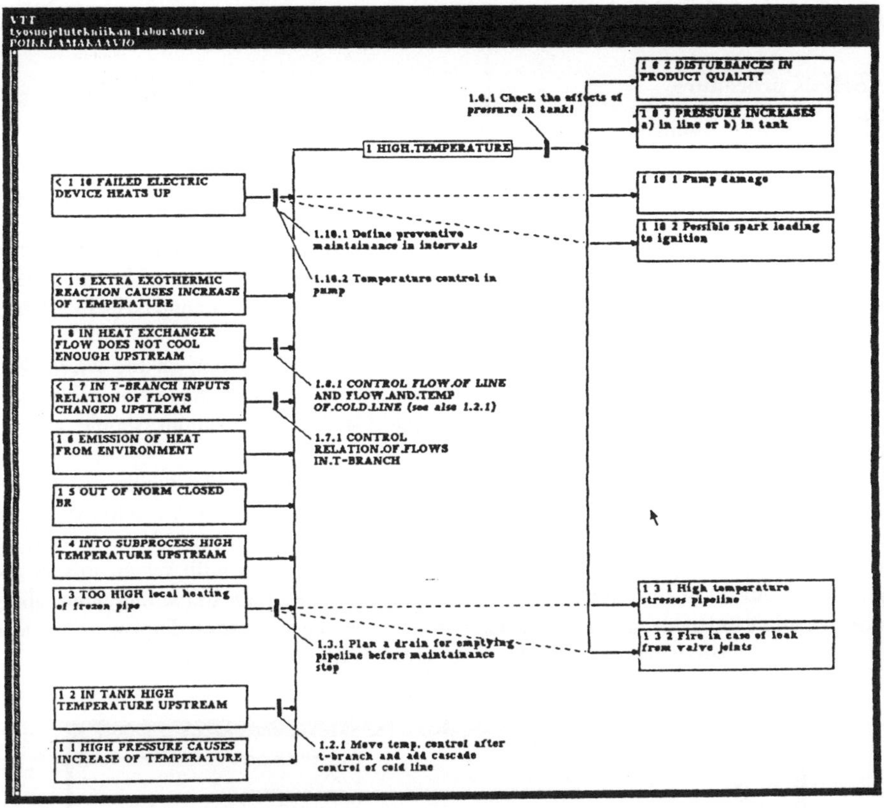

Figure 24.9 The deviation graph of the final results in the example case.

have been examined. As a result six causes related to the deviation LOW FLOW have been obtained on the fifth level of detail. The level of detail of each cause is indicated by a variable number of asterisks preceding the cause number.

The following modifications have been made to the suggested results by the user:

Cause 1.3 has been edited.
Detailed causes 1.9.1 and 1.10.1 have been added.
Consequence 1.0.1. has been removed.
Consequence 1.0.3 has been edited.
Consequences 1.3.1, 1.3.2, 1.10.1 and 1.10.2 have been added.
Measure 1.8.1 has been edited.
Measures 1.0.1, 1.2.1, 1.3.1, 1.10.1 and 1.10.2 have been added.

As a result of the modifying actions the contents of the HAZOP form on the display have changed. The resulting HAZOP form is presented in Figure 24.8. A graphical representation of the HAZOP information, the deviation graph, is presented in Figure 24.9.

Evaluation of the HAZOPEX prototype

Analysis procedure

The results of preliminary evaluation indicate that the basic principles of HAZOPEX have been successfully chosen. However, a faster and more flexible version of HAZOPEX is needed to meet the requirements of cost-effective production use. Integration to CAD is also essential.

The principles of the model editor for process graph input and the interactive implementation of the HAZOP form seem appropriate. Mouse input should be extended to all tasks. Possibilities for reversing actions and terminating execution should be included in all menus.

The long number chains associated with each piece of information on the screen were found confusing. A better way of representing the location of a certain piece of information in the information hierarchy should be found. Conjunctive relations should also be represented more clearly.

A real ammonia storage system was used in the evaluation. The whole system was analyzed manually in another research project, and a small part of it was selected for the evaluation of HAZOPEX. The selected part consisted of two parallel condensers, an accumulator tank, a gas evacuator unit, and connecting pipelines with valves, mixers and dividers. The PID of the selected part is presented in Figure 24.10. In the following, also a storage tank, a separator, and a group of compressors will be mentioned. They were included in the manual analysis but left outside of the HAZOPEX-aided analysis.

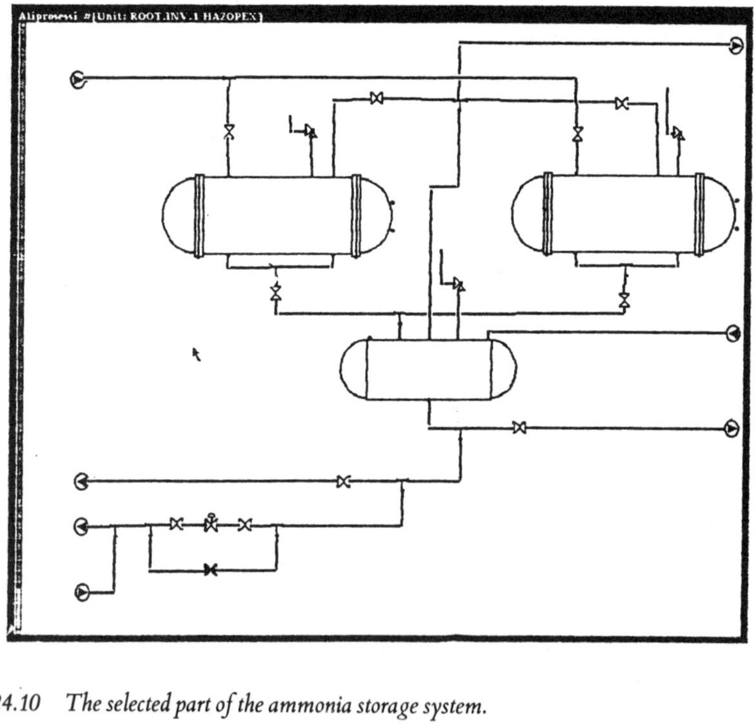

Figure 24.10 The selected part of the ammonia storage system.

Table 24.1. The coverage and acceptability of the initial general level causes generated by HAZOPEX.

No.	Deviation	Causes	No Changes	Modified	Rejected	New	Coverage (%)	Acceptability (%)
1	No flow	5	4	0	1	0	100	80
2	High flow	7	5	2	0	5	58	100
3	High level	3	0	3	0	3	50	100
4	No flow	6	4	0	2	0	100	67
5	No flow	8	5	2	1	2	78	88
6	High flow	7	1	0	6	2	33	14
7	Extra flow	0	0	0	0	1	0	–
8	High temperature	8	1	0	7	0	100	13
9	High pressure	7	0	4	3	2	67	57
Total		51	20	11	20	15	67	61

In the first phase of the evaluation, nine deviations were analyzed. HAZOPEX generated 51 general level causes for these deviations. Approximately 65% of these initially suggested general level causes were accepted by the user (a HAZOP specialist) either as suggested or in a modified form. These causes formed approximately 60% of the final general level causes. The coverage and acceptability of the initial general level causes generated by HAZOPEX is presented in Table 24.1.

Comparative evaluation of the results

The results of the HAZOPEX-aided analysis were compared with the results of a manual HAZOP analysis of the same process system. The comparison proved to be difficult because of the different natures of the analyses. The analysis with HAZOPEX was detailed and covered only a small part of the whole process system, whereas the manual HAZOP study covered the whole installation and its results were of a more general level. Another difficulty was that the manual HAZOP team had chosen to analyze only flow deviations. The differences may indicate that HAZOPEX is better suited for the automation of routine analysis tasks integrated with the design than for the automation of traditional HAZOP.

In the following, the results of the analysis of the deviation NO FLOW are discussed. It was not possible to go through the whole process system with HAZOPEX because of the detailed nature of its analysis procedure. The following causes found in the manual HAZOP can be thought to correspond to the cause 'No input flow to the subprocess' in the HAZOPEX-aided analysis because their location is outside the part of the process which was analyzed with HAZOPEX:

Input line valve erroneously closed.
Separator input valve erroneously closed.
Input valve of the compressor in use erroneously closed.
Output valve of the compressor in use erroneously closed.
Compressor does not start or stops.
Compressor does not create pressure.

The causes found in the manual HAZOP which are located inside the part of the process which was analyzed with HAZOPEX are presented in Table 24.2. The cause 'No flow of cooling water in the condensers' was the only significant cause missing from the HAZOPEX-generated results. Finding that cause with HAZOPEX would have required

Table 24.2. Comparison of the results of manual HAZOP and HAZOPEX-aided analysis.

HAZOP team	HAZOPEX
The input valves of both condensers erroneously closed	Valve V.3 erroneously closed and valve V.4 erroneously closed
The output valves of both condensers erroneously closed	____ (would have been found in covering analysis)
The valve in the end of the main line erroneously closed	No flow out of the subprocess
No flow of cooling water in the condensers	____

complicated reasoning since the direct cause of no flow in that situation was too high a temperature which suppressed the correct function of the condenser and led to no flow. However, in the HAZOPEX-aided analysis of the deviation HIGH TEMPERATURE which was not included in the manual HAZOP at all, this cause would have been found. Other examples were also detected where certain causes appear in the analyses of different deviations in manual HAZOP and HAZOPEX-aided analysis.

There were also some additional causes found in the HAZOPEX-aided analysis. In the following, these causes are listed together with the comments of the members of the manual HAZOP team:

1. Storage tank full
   Comment: 'Logically impossible since the function of the analyzed subprocess is to circulate the ammonia in the storage tank'.
2. Storage tank empty
   Comment: 'Does not seem meaningful but reminds of the potential problems of filling in the tank'.
3. Blockage in line
   Comment: 'Was intentionally left out of the manual analysis'.
4. No flow in the gas evacuator unit valve group and no flow in the passing gas evacuator unit line.
   Comment: 'A limiting case because two simultaneous deviations are required. The event was detected in the analysis of the deviation LOW FLOW in the manual HAZOP; however, the consequences are different'.

Needs for further development

During the evaluation sessions several development needs were detected. The concepts used in the analysis should be clarified, a method for making references should be built, and more flexibility should be introduced. In some phases of the analysis there are redundant tasks that should be eliminated, and certain tasks that occur repeatedly in a similar manner should be automated.

There is a potential danger in the use of knowledge-based computer tools in safety analysis that the amount of creative thinking decreases compared with manual work. Features that support creative thinking should be added to HAZOPEX in further development, and the user should be kept aware of the limits of knowledge-based analysis.

There are several ways to introduce more flexibility in the analysis procedure. A facility for reversing false actions should be provided, and the user should be able to terminate the analysis easily at every stage. The phases where step by step procedure is required should be modified to allow shortcuts. The problem of repeating similar tasks could be solved by implementing a facility that monitors the user's actions and learns to perform similar tasks automatically.

The conditions which determine whether, for example, a cause is a valid cause for a deviation are made as complete as possible in HAZOPEX. Still, it must be realized that it is impossible to create a set of rules that would eliminate the need for user modification of the results completely. It would make the interaction much simpler if all the suggested analysis results were first automatically checked on the basis of the former course of the analysis. That would prevent the system from making the same invalid or incomplete suggestions all over again.

Speeding up the system and improving the quality of results by detecting similarities can also make the system more user-friendly. The similarity check could be made based on the structure of the analyzed part of the process and the values of other variables that have an influence on the analysis. The rate of similarity could be evaluated using an algorithm so that the existing results could also be used in nearly similar situations.

The foundation for the automation of redundant tasks is a set of models where all the significant characteristics of the results and the state of the analysis are described. The significant characteristics are represented with attributes, and the values of these attributes in individual cases can be compared in order to detect such similarities that indicate a possibility for automation. The most simple way of implementing a learning procedure is to associate a weight with each attribute. The weight is then adjusted according to example cases. More sophisticated methods are techniques for learning a set of classification rules (Cohen and Feigenbaum, 1982), techniques of fuzzy reasoning (Zadeh, 1988) and connectionistic networks (Dreyfus, 1987a, b). Experiments with different techniques are performed in order to find a suitable way of implementing the learning facility in HAZOPEX.

Conclusions

The development of an expert system to support the safety analysis of process systems has raised great international interest. It seems possible that the use of HAZOPEX could increase and pre-date the use of hazard and operability studies as a design tool.

The next phase of the development work is to transfer HAZOPEX to a production environment. In the selection of tools for the implementation of the production version portability and price have to be considered together with properties. Advanced knowledge engineering tools being available, there are remarkable possibilities for the development of software which is effective, easy to maintain and easy to use. However, the software development project required for a production version of HAZOPEX is large and has to be planned and organized properly in order to succeed.

A production version of the HAZOPEX expert system could be developed by extending the knowledge base of the current prototype version and improving its flexibility. The production version must also have an effective integration with CAD. The results obtained in the development of the HAZOPEX prototype form a good basis for the specification of the production version and for starting a product development project.

Besides the functionality of the software there are many other things to be considered when taking a new computer-aided tool into production use. The present safety analysis procedure is brought closer to design. Creative synthesis and presently separate creative analysis are joined into a new incremental safety analysis methodology. Introduction of a new methodology requires organizational adaptation.

The following potential profits motivate the development of a new computer-aided safety analysis method and the associated change of the design process:

1. The safety of alternative designs is assessed earlier which leads to smaller and better controlled effects of changes,
2. More detailed specifications to steer the system design are produced by the early assessment of alternatives,
3. Certain quality level of safety analyses is guaranteed by the division of safety analysis into a routine part integrated to design and a creative manual part to be carried out in the later stages of the design,
4. The results of computer-aided safety analysis are represented in a more comprehensive and flexible manner than the results of manual analysis,
5. It is easy to update and extend the results when more experience is gained or changes are made to the process.

There are several interesting areas of further research in the field of safety. The planning of safety analyses and the evaluation of the quality of an analysis are interesting problems. Structuring the expertise of experienced analysts would help the development of computer tools for such tasks. The definition of an appropriate content of an analysis in different design situations and the evaluation of the quality of an analysis presented to an authority for obtaining an operating licence are typical situations where expert systems could give support.

The examples and plans presented in this paper show a considerable interest and growing effort in using knowledge engineering, mainly expert systems, in the field of safety analysis and safety management. This trend will continue in increasing international cooperation for developing systems and tools corresponding to the needs in several countries. The most effective way of increasing the use of safety analyses seems to be the development of computer tools for the CAD environment. The goal is to provide the designer with a set of tools with a high practical value.

References

Anon, 1977, *A Guide to Hazard and Operability Studies* (Tonbridge: Chemical Industries Association Ltd and Tonbridge Printers Ltd).

Anon, 1986a, *RELVEC Manual*, Vol. 1, *Modelling* (Technical Research Centre of Finland).

Anon, 1986b, *RELVEC Manual*, Vol. 2, *User's Guide* (Technical Research Centre of Finland).

Barbet, J. P., Melis, M. and Oliviero, M., 1987, Computerized safety and reliability assessment: From computer codes development to expert system techniques application. In: *International Symposium on Risk Analysis in Environmental Impact Assessment*, Milan, 26–27 November 1987, pp. III/53–III/62.

Cohen, P. R. and Feigenbaum, E. A., 1982, *The Handbook of Artificial Intelligence*, Vol. 3 (London: Pitman).

Dreyfus, S. E., 1987a, Neural nets: an alternative approach to AI. *Applied Artificial Intelligence Reporter*, September, 6–7.

Dreyfus, S. E., 1987b, Multi-layer neural nets recognize regularities. *Applied Artificial Intelligence Reporter*, October, 7, 19.

Fryer, L. S. and Kaiser, G. D., 1979, DENZ – *a Computer Program for the Calculation of the Dispersion of Dense Toxic or Explosive Gases in the Atmosphere*. Safety and Reliability Directorate, SRD R 152 (Warrington, UK: UKAEA).

Haastrup, P., 1984, *Design Error in the Chemical Industry*. Risø–R–500. (Roskilde, Denmark: Risø National Laboratory).

Haastrup, P., *et al.*, 1985, *RIKKE Users Manual*. Risø–M–2480. (Roskilde, Denmark: Risø National Laboratory).

Havens, J. and Spicer, T., 1985, *Development of an Atmospheric Dispersion Model for Heavier-than-air Gas Mixtures*. Report No. CG–D–22–85, CG–D–23–85, CG–D–24–85, Vols I–II. (US Department of Transportation, United States Coast Guard).

Jagger, S. F., 1983, *Development of CRUNCH: A Dispersion Model for Continuous Releases of a Denser-than-air Vapor into the Atmosphere*. Safety and Reliability Directorate, SRD R 229. (Warrington, UK: UKAEA).

Kletz, T. A., 1983, HAZOP and HAZAN, *Notes on the Identification and Assessment of Hazards* (Rugby: The Institution of Chemical Engineers, Hazard Workshop Modules).

Poucet, A., 1983, *Computer Aided Fault Tree Synthesis*. Report EUR 8707 EN. (Ispra: Joint Research Centre).

Poucet, A. and Petersen, K. E., 1988, Knowledge based safety and reliability analysis tools and their application in process control design and operation. In: *Reliability, Availability and Maintainability of Industrial Process Control Systems*, Brugges, 23–30 September 1988.

Reunanen, M. and Suokas, J., 1986, *Safety Analysis of a Black Liquor Recovery Boiler Plant*. Research Notes 551. (Espoo: Technical Research Centre of Finland).

Suokas, J., Heino, P. and Karvonen, I., 1987, Some experiences and developments in computer aided safety analysis. *SINTOM Seminar on Datorhjälpmedel för Tillförlitlighets-analys*, Visby, 27–29 April 1987.

Wells, G. L., 1980, *Safety in Process Plant Design* (London: George Goodwin).

Zadeh, L. A., 1988, Fuzzy logic. *Computer*, April 1988, 83–93.

Chapter 25
Application of artificial intelligence in safety training

CHIN H. LEE and JOHN E. BIEGEL
Department of Industrial Engineering and Management Systems,
University of Central Florida,
Orlando, Florida 32816, U.S.A.

Abstract. This paper describes an automated safety training system which integrates the latest technologies in the fields of artificial intelligence and simulation training. The system can deliver safety training at a location and schedule convenient to the user because a human instructor is not required in the active training loop. The system will reduce the cost of training and increase the effectiveness of training.

Keywords: Safety training; Artificial intelligence; Simulation training; Intelligent tutoring.

Introduction

Safety training is required for all employees. This includes top managers as well as hourly labourers. The workers at the lower level must be trained in order to perform their jobs safely. This is particularly true when hazardous conditions exist. Supervisors require safety knowledge to insure that workers perform their jobs safely as well as to monitor and control hazardous conditions around the workplace. The managers who establish policies need to include job safety as a major consideration since occupational injuries and illness may incur large business costs. Therefore, appropriate safety training must be provided to all personnel at all levels in order to be effective. Safety training should include safe job procedures, methods for conforming to safety precautions and the identification of occupational hazards.

Many modes can be adopted to train employees effectively. Different modes utilize different instructors, materials, equipment and facilities. The instructor must be able to assess the trainee's current knowledge and determine specific areas where the trainee lacks knowledge. The instructor makes training/retraining decisions based upon this assessment. Therefore, the instructor is the most important element for a successful and effective safety training programme.

Although safety training is an essential part of industrial operations, not many companies can afford to hire a full-time safety training instructor who is an expert in all safety areas. Instead they obtain the necessary training by utilizing special education and training programmes offered by governmental agencies, professional societies, universities and private consulting companies. Such specialized training programmes can be quite costly due to the additional requirements for travel to the training site, room and board during training, registration fees, etc. Available safety videos can be licensed for either long-term or short-term use. These videos usually require a leader among the trainees to guide the other trainees for better organized and more effective training.

The effectiveness of a safety training programme depends on the ability to train efficient and strategic responses to a given condition. This requires the instructor's awareness of the trainee's current knowledge, awareness of specific areas where the trainee lacks knowledge, and prefered methods of training. This paper describes a conceptual design of an automated intelligent safety training system which will remove the human instructor from the active training loop. Appropriate safety training can be conducted at a convenient site, time and schedule for the trainee.

Recent developments in the field of artificial intelligence make it possible to remove the human instructor from the active training loop. Without the need for an active human instructor the safety training programme can be conducted at the company's convenient location and at the trainee's convenient time. The trainee can also receive the training at an acceptable pace. This will result in reduced safety training costs and more effective training.

Background

Artificial intelligence and intelligent tutoring system

Artificial intelligence (AI) has been defined in several ways. Gevarter (1985) said that AI is concerned with devising computer programs to make computers smarter. Winston (1984) defined AI as the study of those ideas that enable computers to be intelligent. The most widely accepted definition is given by Feigenbaum and Feldman (1963). They defined AI as a branch of computer science devoted to programming computers to carry out tasks that if carried by human beings would require intelligence. The central goal of AI has been to make computers more useful and understand principles that make intelligence possible (Winston, 1984).

A major part of the early research in AI was directed toward language understanding and translation. Game and problem-solving are other major areas of research that had some early success in producing chess and checkers programs. However, improved computation facilities and new techniques of the early 1970s made it possible to develop intelligent tutoring systems (ITSs).

ITSs have utilized various artificial intelligence techniques such as knowledge representation techniques which were generally the first to be applied and remain foundational to most intelligent tutoring systems. Representation of knowledge in AI involves the storing of information in knowledge structures and the development of interpretive procedures and inferencing schemes to manipulate these knowledge structures intelligently. The objective is to utilize this formalized knowledge to simulate or model intelligent, knowledgeable behaviour. Knowledge representation schemes were also used as the basis for student modelling and for encoding various teaching strategies.

Components of an ITS

ITS embodies the knowledge that allows an expert tutor to create a tutoring sequence. Tutoring sequences are dynamically created based on decisions made in reference to the knowledge represented. Hartley and Sleeman (1973) listed three general kinds of knowledge represented in an ITS. They are: (1) the domain knowledge, (2) the student model, and (3) the pedagogical knowledge. In addition to the above list, an interface is required for communication although it is not knowledge itself.

The domain knowledge which includes the content to be taught is explicitly represented in an ITS. It also includes knowledge on how to use the content to solve related problems. The student's understanding of the subject matter being taught is represented in the student model. The student model contains information about the student. Wenger (1985) divided the student modelling module into two sub-modules: (1) a student model (the knowledge structure which contains student information, and (2) a diagnostic process, which collects this information from available sources. The pedagogical knowledge contains knowledge in the form of teaching strategies that decide teaching strategies which integrate domain information, knowledge of the student, and teaching methods. The interface provides the mechanisms for information exchange between the components of ITS. It also handles the communications between the student and ITS.

Simulation

A simulation is the imitation of the operation of a real-world process or system over time (Banks and Carson, 1984). Simulation involves the generation of an artificial history of a system, and the observation of that artificial history to draw inferences concerning the operating characteristics of the real system. Simulation enables the study of, and experimentation with, the internal interaction of a complex system. Informational and environmental changes can be simulated and the effect of these alterations can be observed. By changing simulation inputs and observing the resulting outputs, valuable insight may be obtained into which variables are more important and how variables interact.

Simulation training systems are considered to be excellent training systems. Their ability to represent elements of a task in which the trainee must be proficient and their ability to control additional parameters which support, facilitate, enhance, and ensure effective training are major advantages. Simulation has been applied in a wide variety of contexts for both commercial and military purposes. Applications range from teaching maintenance principles to tactics training for sophisticated aircraft.

Opportunities for applying AI techniques to simulation-based training certainly exist in the area of safety training as a new safety training method.

Intelligent simulation training system

The University of Central Florida (UCF) in Orlando, Florida, General Electric's Simulation and Control Systems Department (GE) and Embry–Riddle Aeronautical University (ERAU), the latter two in Daytona Beach, Florida are collaborating on the design and development of an Intelligent Simulation Training System (ISTS). An ISTS is a program that uses a real-time (or pseudo-real-time) computer simulation of an adaptive training scenario as a basis for providing personalized interactive instruction to a trainee (Draman, 1989). The objective is to develop a generic intelligent simulation training system which will remove the human instructor from the active training loop.

M. Draman (personal communication) listed four types of knowledge which must be encompassed in an ISTS. They are:

1. Knowledge about the simulated application domain,
2. Knowledge about the specific ways of teaching the handling of situations in the simulated application domain,
3. Knowledge about the trainee, and
4. Knowledge about teaching with simulations in general.

In addition to encompassing the above knowledge, an ISTS must be able to monitor the trainee continuously and handle the communications between the trainee and the system. An

ISTS also requires a simulation support and a control mechanism to monitor data flows and pass the control to relevant system modules.

An ISTS system is designed on a modular basis which includes the above listed system requirements. The modules are categorized by the functions they serve. There are six major groups of modules in the system. Each group contains several modules as follows:

1. Interface Group

The Interface Group provides the mechanisms for those modules, external to the Simulation/Tutoring process, to provide information for the Simulation/Tutoring process. This group consists of the Knowledge Acquisition module and the Discourse module. The Knowledge Acquisition module is to provide the training system with the information about a domain necessary for it to perform tutoring of this domain. The Discourse module is to handle the communications between the users and the system.

2. Input Group

The Input Group provides the mechanisms by which the trainee's response to the simulation or the trainee's desire to modify the simulation are entered into the system. This group contains Translator, Input Filter, and Simulation. The Translator parses the trainee's input, which is directly related to the simulation, into a form the system can interpret.

The Input Filter catches the trainee's inputs that are inconsistent with global simulation data and constraints. The Simulation is to provide facts and objects and carry out those instructions accepted by the system.

3. Domain Expertise Group

The Domain Expertise Group provides the knowledge base necessary for intelligent functioning within the domain. This group includes the Intelligent-Pre-Processor, the Domain Expert, and the Domain Expert Instructor. The Intelligent-Pre-Processor transforms raw simulation data into a high level representation (event-list). The Domain Expert is to generate a set of solutions that an expert would give based on an extracted set of events from the simulation. The Domain Expert Instructor gives expert instruction with lesson plans, guidelines and all other expert information on what to teach in a particular domain.

4. Instruction Group

The Instruction Group determines and establishes the cause of the reason within the instructor's guidelines. It also maintains and updates trainee status on a continuing basis of how well the trainee has performed. This group consists of Evaluator, Student Model, and Tutor. The Evaluator evaluates trainee performance. The Student Model serves the individual needs of the trainee. The Tutor conducts a lesson for a student in a manner that best suits that student's needs and preferred mode of instruction. This includes the ability for the Tutor to modify tutoring behaviour depending on the abilities of the trainee and the current mode of the system.

5. Control Group

The Control Group establishes the results of trainee's inputs and controls the flow of the simulation and training. This group includes Interpreter, Control, and Inference Engine. The Interpreter is to establish the changes that are caused by the trainee's input in the event list of the intelligent pre-processor. It also checks continuously the criticality tags of the intelligent pre-processor events. The Control passes control to the proper modules at run time and assures complete communication. The Inference Engine activates and fires the rules in the modules and enters the rules into the system.

6. Simulator

The Simulator contains the simulation facts and objects and carries out those instructions accepted by the system. It illustrates graphically the application domain with interactive objects and to provide for intelligent objects.

Table 25.1. Diagram input/output descriptions

1.	Expert response to scenario
2.	Raw score data
3.	Refined score/status (performance rating)
4.	Remediation level and topics (to be stored)
5.	Skills/objectives to be taught
6.	Scenario modification information
7.	Topic, level of remediation, and final score
8.	Parsed student commands and syntax errors (if any)
9.	All student input other than commands to simulation
10.	Instructor inputs and requests
11.	Domain teaching methods, lesson plans, scenarios, etc.
12.	Student response to scenario (commands)
13.	Parsed and synthetically correct student input
14.	Simulation Data
15.	Lesson plan, weighting factors
16.	Domain expert knowledge base
17.	List of significant events (or possible events) extracted from the simulation
18.	Domain instruction knowledge base
19.	System's communication to student
20.	Results of logical analysis of student's parsed input
21.	Student/instructor request
22.	Communication from system to instructor
23.	Error messages (syntax errors and uninterpretable input)
24.	Inputs refused by simulation objects
25.	List of significant events (or possible events) extracted from the simulation
26.	Changes in events list
27.	Changes in events list
28.	Data read from simulation objects in order to generate expert solutions to events
29.	Logical errors and specific information on student performance within the domain
30.	Keyword and attributes for the domain
31.	'Canned' message for communication purposes
32.	Student/Instructor comments
33.	Access to student records
34.	Why and how explanations to Domain Expert solutions
35.	Executable message to a simulation object
36.	Error message sent to student (input refused by Input Filter)
37.	Off-line input from the system analyst
38.	Tutor requests to activate other modules individually

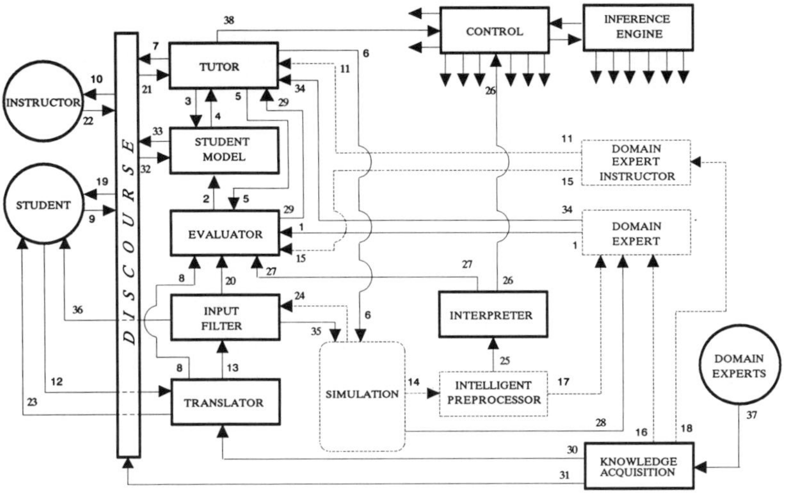

Figure 25.1 ISTS data flow diagram.
—, ISTS modules and data flow; – – –, interfacing modules and data flow.

The detailed data flow between the above-mentioned modules is illustrated in Figure 25.1 and Table 25.1. The ISTS is designed to be a generic system so that various domain areas can be taught with the system.

Application of ISTS in safety training

One of many practical applications of the ISTS is in the occupational safety training area. The Domain Expert module and the Simulation can be developed to contain expert knowledge associated with occupational safety and health. Some examples of such knowledge are: (i) accident investigation, (ii) manual materials handling and back injury prevention, (iii) confined space entry, (iv) container labelling, (v) forklift safety, (vi) hand injury prevention, (vii) hazard communication, (viii) noise, (ix) heat stress, (x) ladder safety, (xi) lockout/tagout, (xii) personal protection, (xiii) respirator safety, (xiv) asbestos, and (xv) eye protection.

The Domain Expert module will include a basic understanding of the effects of occupational hazards, recognition of hazardous conditions, measurements of hazard level, recommended standards, and methods to control occupational hazards. This knowledge can be presented to the trainee in a form of a lecture before the simulation training starts so that the trainee will have basic knowledge of the safety area with which he/she is concerned. The Domain Expert module will also generate expert solutions during the simulation training so that the trainee's solution can be compared with the expert solution.

The Simulation will illustrate graphically the applications of occupational hazard control principles and techniques. It will be a pseudo-real-time process running in parallel with the training system. The Simulation will simulate hazardous working conditions. The trainee will be asked to identify the hazardous condition, level of hazard, and make recommendations to improve the working conditions. The trainee's response will be compared with the expert solution generated by the Domain Expert module in order to evaluate the trainee's performance. The Simulation will present various scenarios to simulate different hazardous conditions. This automated training will continue until the trainee's performance reaches a satisfactory level.

Conclusions

The primary goal of AI is to make computers smarter. Smarter computers can adapt themselves to their human users, instead of the other way around. In a complex world with complex problems, the problem-solving ability of a smarter computer can be a great help. The occupational safety training on various safety areas is one of many representative complex problems. Unlimited opportunities for applying the latest AI and simulation techniques in the area of occupational safety and health training certainly exist.

References

Banks, J. and Carson, J.S., 1984, *Discrete Event System Simulation* (Englewood Cliffs, NJ: Prentice–Hall).

Gevarter, W.B., 1985, *Intelligent Machines* (Englewood Cliffs, NJ: Prentice–Hall).

Feigenbaum, E.A. and Feldman, J., 1963, *Computers and Thought* (New York: McGraw–Hill).

Hartley, J.R. and Sleeman, D.H., 1973, Toward more intelligent teaching systems. *International Journal of Man–Machine Studies*, **5**.

Wenger, E., 1985, AI and the communication of knowledge: an overview of intelligent teaching systems. Unpublished manuscript, University of California, Information and Computer Science, Irvine, CA.

Winston, P.H., 1984, *Artificial Intelligence*, 2nd edn (Reading, MA: Addison–Wesley).

Chapter 26

Systems and events representation in SAFE-EXP using GRAFCET and logical trees

CHRISTIAN FORTIN, DINKAR MUKHEDKAR[1], GEORGES-EMILE APRIL[1] and ROBERT GILBERT[2]

Department of Kinanthropology,
University of Quebec at Montreal

[1]*Department of Electrical Engineering,*
Polytechnic School of Montreal

[2]*Department of Industrial Engineering,*
Polytechnic School of Montreal

Abstract. We are now developing a prototype of an expert system for work saftey. The system will focus on ergonomics, system representation, event and process analyses, human error mechanisms and qualitative fault trees. It will have the capacity to process and merge qualitative fault trees and to help the user to generate scenarios of accidents or other events. The field of the prototype will be manual handling. In artificial intelligence, the main challenge in knowledge-rich domains is to find ways of representing accurately: (i) system decomposition and behaviour; (ii) task, process, and event decomposition and interaction; (iii) spatial and temporal logics, structure and resolution, along with causal interactions. We contend that these problems are not only related, but are in fact similar. The components of a system are themselves systems, as concrete as physical objects, or as abstract as structures, functions, tasks, events or processes. An event on a given time-scale is in fact a complicated process composed of subevents on another scale. The changes of time-scale may be modelled by inertial delays as used in logical analysis of systems. This method can be adapted to the changing of other scales, such as the spatial scales or hierarchical decomposition structures. Finally, processes, events, actions and tasks may be modelled by GRAFCET, a standard form of Petri nets.

Keywords: Knowledge-based expert systems; Ergonomics; Industrial safety; Petri nets; Inertial delays.

Introduction

Over the last 15 years we have developed a multidisciplinary approach to the logistics of accident causes (Gilbert *et al.*, 1975, 1976, 1982; Fortin, 1983; Fortin *et al.*, 1983a, b, c, d, 1987; Fortin and Gilbert, 1986). We gradually established a causal network with about 3000 different variables which is organized using different structures, many categories and abstraction levels. This network, at the start, was to be used mainly to help in the different steps involved in analysis of events by qualitative fault trees, and to serve in the creation of a posteriori scenarios for accidents and a priori scenarios for risks (Fortin, 1983; Fortin *et al.*,

1983a, b, c, d). Others before us had tried such an approach but abandoned their efforts because of its apparent inapplicability (M. deMontmollin, personal communication). They thought in terms of transfer functions, rather than in terms of facts, rules, units of articulated behaviour and the ways human error occurs. In the last 4 years, we began developing an expert system in work safety called SAFE-EXP.

Structure of the Paper

Because of the multidisciplinary nature of this book and of this work, some people may not be completely familiar with some of the concepts used in this paper. It seems appropriate, then, to recall briefly, but systematically, important definitions and interrelationships between the principal concepts. This is done at the start of each respective section where these concepts are introduced.

In the second section, expert systems are defined, the specific development procedure is described and the actual state of development of our prototype is indicated.

In the third section, the very close bonds between safety, ergonomics and reliability are shown. Then, the general goals and functioning of our prototype and the ways by which it may be used to help the user are exposed. The general structure of the causal tree of an accident, as we define it, is also indicated.

In the fourth part, various and important problems of representation of systems and events, particularly in knowledge-rich domains, are discussed. We conclude, as others have, that these problems are considerably alleviated when the state of the world at any point in time is represented by the set of sequences of events that occurred up to that moment. We contend also that these problems are in fact similar. The concepts of structure, frontier and grain-size are essential for an adequate approach of the modelling of systems and for representing in particular the limitations of propagation of a causal effect. Processes, events, actions, tasks and various mechanisms of human error may be easily modelled by GRAFCET, which is a standard form of Petri nets. Several aspects of GRAFCET relate directly to the design of intrinsically safe systems. From the point of view of safety analysis, GRAFCET has the remarkable property that the same external behaviour of a system, may correspond to several possible internal behaviours having various degrees of parallelism.

Expert systems

Definition

Expert systems are essentially computerized systems used for symbolic representation and processing of knowledge. Their goal is to reach, in specific fields and on problems of intermediate difficulty, performances close to those of the best specialists, in areas such as diagnosis, decision making and problem solving, in a world with partial, uncertain, vague and often conflicting information (Farreny, 1985; Laurière, 1986). By intermediate difficulty is generally meant a problem that an expert can solve in less than 1–3 h but not in less than 10–20 min. This goal is actually reached by very few systems. Expert system performance is essentially related to the use of the facts, rules, strategies and heuristics that human experts would use (Bonnet, 1984; Biswas *et al.*, 1985; Hoag, 1985; Lehto and Miller, 1985). All that practical know-how is obtained from experts following an ergonomic analysis of their work (Kuipers and Kassirer, 1983; Feigenbaum and McCorduck, 1984; Bonnet *et al.*, 1986; Hoag, 1985). The term 'ergonomic' is not used by these authors, but ergonomists will easily recognize the familiar, though independently developed methodology (Fortin and Gilbert,

1986). It is interesting that the experts generally have considerable difficulty in verbalizing their actual thought processes. It seems they have developed most of the rules and heuristics unconsciously. They frequently use models that do not match the theoretical models they say they use. All of this brings us back to what we already know about the differences, often important, for the worker or experienced labourer, between formal behaviour and actual behaviour (Ombredane and Faverge, 1955; Cazamian, 1974; Montmollin, 1974). Therefore, it is necessary to observe the expert working on particular and concrete problems (Kuipers and Kassirer, 1983), and, by a normal iterative process in ergonomic analysis of work, expose the concepts, facts, models and heuristics that he/she really uses (Feigenbaum and McCorduck, 1984).

The essential characteristic of an expert system is the control structure, which is distinct from the knowledge, so that rules and facts can be added without any order, and without having to judge in advance how they will be used.

Choice of a subfield

Based on our previous research and the needs of companies interested in the project, we have chosen the field of manual handling. In the first version, we focus on the analysis of causes and leave the synthesis of possible suggestions to the analyst for a later stage.

Specific development procedure

The following steps have been carried out:

1. Construction of the fault trees for at least 60 typical accidents in manual handling.
2. Drafting of the rules for the representation and processing of the trees.
3. Hand-merging of these trees and drafting of the corresponding rules for the necessary operations, to obtain a more general tree.
4. Drafting the rules for representing a given accident, like a subtree of an existing fault tree.
5. Drafting the rules for generating typical accident scenarios from the likely paths in the general tree for a given level of gravity.
6. Testing of several expert system shells.

State of development

All the important structures, concepts, sub-concepts and relations between concepts have been defined, programmed and commented in SAFE-EXP using the taxonomic shell of the Arity expert system shell. This part of the program (280 kbytes) has passed the tests of the various classifiers and analyzers in the Arity system.

Rules and control options are currently being designed, using the rules shell of the Arity expert system shell. Control rules are necessary so that the user can be asked by SAFE-EXP questions in a language understood by the user, and only when it is necessary or appropriate.

Expert systems and work safety

Safety, ergonomics and reliability

Work-related accidents and illnesses can be defined, in a general way, as a result of a breakdown in relations in a production system between the persons, their work and the tools used (Rasmussen *et al.*, 1981; Rasmussen and Lind, 1982; Fortin, 1983).

An expert system in occupational health and safety will have to encompass the available knowledge about the mechanisms of human error and be able to evaluate the effect of the errors on the behaviour of the system under consideration. Human error is an enormous factor in accidents (Fitts and Jones, 1947, 1949; Leplat and Rasmussen, 1984; Leplat and Chenais, 1985). Yet human error can and must be taken, not as a principal cause of accidents, but mainly as a by-product of inadaptations between the person, his work, his tools, his work environment and his work methods. In prevention, the main question is how to correct or create systems which minimize human errors and are immune to residual errors.

In other aspects, a system's reliability is usually defined as the probability of good working order under specified conditions and circumstances during a given time and based on a given set of criteria. Consequently, any undesirable event, such as a work-related illness or accident, presence of a suspected unacceptable level of a particular risk, or disturbance in the system, may be considered a reliability problem. The use of reliability methods (functional diagrams, logical analysis and synthesis of systems, Petri networks, and fault-tree analyses) are then warranted, as shown by Faverge (1967).

Fault-tree analysis in prevention

The analysis by fault trees is certainly the most popular method. It has been adopted, simplified or adapted in some way by numerous researchers and organizations in the prevention field. The fault tree has two excellent qualities: (i) it is easy to understand, (ii) it can be easily formulated and processed in languages such as LISP (Lehto *et al.*, 1984) or PROLOG (Fortin and Gilbert, 1986), and one big drawback: how to take human reliability into account in the analysis.

The problems associated with this last question arise from the difficulties in evaluating the probabilities of human error in a given environment; to this must be added the variability in actual procedures and the instability in the sharing of subtasks by operator and in what is perceived by an observer; the phenomenon of risk homeostasis must also be included. It is possible that the only solution to this problem is to simulate the task using modified Petri networks (Madni and Chu, 1985). In an expert system, such simulation could be triggered by a rule that would itself trigger the execution of the desired procedure.

In prevention, fault trees can help in two ways:

1. Starting with apparent or identified risks, we can find their source, identify their operation modes and try to evaluate their probability of occurrence as a function of the importance of their consequences.
2. Starting with system malfunctions, at the level of components or relations between components, determine the real or potential risks to which they are related.

Either way, it is necessary to represent the composition of the system and of its behaviour in terms of subsystems, tasks, functions, processes and events, as well as in terms of the composition of these elements of the system. This is achieved through various means, particularly functional diagrams, logical analysis and synthesis, qualitative calculus, graph theory, operations research models and Petri networks. In the context of reliability analysis, these methods are oriented, in complex systems, to find the minimal cut sets or the minimal path sets. Then, numerous textbook algorithms may be used to generate the corresponding trees even in cases with multiple feedforward and feedback loops (examples are given in Henley and Kumamoto, 1981, 1985). These methods are applied after at least two preliminary studies: the preliminary hazard listing and analysis study and the failure modes and effects analysis study. These studies are often accompanied by an analysis of work and potential energy exchanges study (taking into account the characteristics of the transfer surfaces) and a criticality analysis study.

Fault-tree Analysis and Development of SAFE-EXP

The development of SAFE-EXP was guided by the next scenario. First, the user describes a foreseeable accident. When needed, he/she answers the system's questions on the sequence of actions and their logical relations. If it is a new accident, the system draws its causal tree. This is a decision tree of the states and events of the accident (an event being here a change of state). In some situations, certain events cannot occur if certain states are not present. Starting with the decision tree, the fault tree is created in an effort to explain each event or state by the variables in the general scheme, or by the specific model of the system in which the accident occurs and by the causal relations between the variables.

Then, the analysis starts. It aims at seven main goals:

1. Identify all the significant causes that can explain the observed facts.
2. Merge the tree of events with those of other versions of the same accident by other witnesses or investigators.
3. Synthesize the corresponding fault tree of the causes of the accident.
4. Merge the resulting fault tree with those of earlier similar accidents. Thus obtain specific models of accident causes and scenarios for specific workplaces, professions, tasks and equipment.
5. Generalize a fault tree, in ascending order, by replacing causes and variables by more general variables or classes of causes, in the general model or in the previously merged fault trees in the data base.
6. Generalize a fault tree, in descending order, by replacing more general variables or classes of causes by the related sets of pertinent and more specific variables or causes, using chains of logical operators OR and AND.
7. Generate, as proposed by Bruun *et al.* (1979), typical accident scenarios, for example of a machine or workstation, by observing the number of historical paths of a given level of gravity in the aggregated fault trees, as well as in the general fault trees.

In merging event trees or even fault trees, conflicts must be resolved if possible with supplementary information or the best hypotheses. For prevention, however, it is generally of no importance to know which particular scenario took place when all the scenarios are likely possibilities. In any case, it is always useful to generate several scenarios of the same accident with different sequences of events or subevents, different circumstances or states, and various levels of gravity, if the goal is to minimize the possibility of similar events in the future. Incompatible scenarios may be represented in the same event or fault tree by using either the conditional operator or the house-event operator, or the exclusive-or operator.

The fault trees of different accidents can in general be merged based on certain characteristics or variables. These are identified first with reference to the general classes of causes that, in the general model or in previously merged fault trees, match the observed environment. The observed environment is evaluated with variables such as the place where the accident occurred, the person involved, the relevant tasks and equipment. The general model or previously merged fault trees may then be used to help the user to:

1. Define or revise the fault tree of a specific accident, actual or potential, by suggesting relevant questions to the user;
2. Generalize a particular fault tree, and its restructuration, as to allow or facilitate its merging with other fault trees at a given level of abstraction;
3. Generate scenarios of accidents but also of preventive measures or solutions.

Causal tree of an accident in SAFE-EXP

In SAFE-EXP, the fault tree of an accident is defined as a general structure having four parts:

(i) the trunk, where the chosen event or sequence of events is represented;
(ii) the backward tree of actual, or potential or likely causes;
(iii) the forward tree of actual, or potential or likely consequences;
(iv) the signature tree, where essential characteristics of the accident are summarized and information for management or statistics is entered.

Examples

The examples available are used as specifications for development of the prototype. Two examples have been developed in great detail. The first was a truck accident. The second was a manual handling accident. An expanded tree of this accident may be obtained in French or English from the authors. It is apparently a very simple accident. A worker carrying a crate trips on a steel rod, even though he had seen it. In his fall, he fractures the back of his neck on the corner of a worktable, sprains his left shoulder, bruises his lower back, and the crate bruises both his knees. He had seen the steel rod in time but thought that he could push it out of his way with his foot, but he erred in his estimation of distance or in his space–time sequence of moves. The analysis, however, relates the accident directly to poor design in an evacuation chute in a functionally unrelated workstation and to the layout of this workstation; they also reveal many interesting contributing factors.

Representation of systems and events

In artificial intelligence, the main challenge in knowledge-rich domains is to find ways of accurately representing: (i) system decomposition and behaviour; (ii) task, process, event decomposition and interaction; and (iii) spatial and temporal logics, structure and resolution, along with causal interactions.

The goal is to be able to derive (or explain or predict) the behaviour of a system from its structure. The representation used must also be logical, coherent, robust and as natural as possible. The difficulties in attaining these objectives arise primarily from the problems of semantics and logics related to the various changes in resolution necessary during the reasoning process: (i) changes from a systemic level to a subsystemic, or vice versa, (ii) changes from the behaviour of a component or a system to the behaviour of its subcomponents or subsystems, (iii) changes in the spatial resolution, and (iv) changes in the temporal resolution. With each change, some details will appear and others disappear, and thus, can change the semantics of the variables used in the reasoning, as well as some or all the previously established causal relationships. In addition, if the position in space of an element of a system or the position in time of an event in a system is changed, or if a relation between two components is added or deleted in a system, any previously established fact or relationship may also undergo changes. These problems are aspects or subproblems of a bigger problem known in artificial intelligence literature as the Frame problem, which is considered an extremely difficult problem in the present state of the art. But, in fact, for the most part, it is a false problem (Georgeff, 1987a, b; Lansky, 1987; McDermott, 1987; Stuart, 1987).

First, most artificial intelligence representations have relied on state-based models in which the world is viewed as a series of states or 'snapshots' that are altered by events. Events are modelled solely and often implicitly by their state-changing function. Now, it is only in this approach that the frame problem seems so difficult. The dual-alternative event based approach represents the world in terms of a set of interrelated events. In this context, the state of the world at any particular point in time is represented in terms of the sets of sequences of events that occurred up to that moment. This approach is very useful for representing multi-agent domains, in which parallel activity is inherent and vital (Lansky, 1987). It reifies events and explicitly represents their causal and temporal interrelationships. In this approach, the frame problem is much easier to circumscribe because it can be posed in terms of:

1. How to design efficient representations of the localization of a causal effect and then of the limitations on its propagation in the semi-orders of systems, space and time;
2. Which layered sets of assumptions to adopt in the reasoning strategy for dealing with this representation and, in particular, the representation of paths of causal interactions, adjacency concepts, the frontier concept, and the ways of manipulating multiple representations or structures of a system, process or event;
3. How to design an overall strategy for reasoning based on progressive relaxation of the assumptions underlying these representations when they are false or seem so because they lead to contradictory or incoherent conclusions;
4. How to manage the search through alternative sets of plans to regulate a given problem, which are the most efficient means of storage and accessing for a large space-time or system-time map, how to determine the proper level of analysis for a given system and draw conclusions from qualitative descriptions of systems and of behaviours of systems;
5. How to design logics of space–time with well understood semantics and allowing for discussion about particular event instances and their various temporal, causal, and simultaneity interrelationships, as well as the way in which particular events constitute a specific nonatomic event and the best means of representing complex structural relationships among all types of events.

Second, logic has in fact severe limitations as a theory of inference, in spite of its tremendous appeal as a unifying theory of knowledge representation, because in practice almost all inference is non-deductive (McDermott, 1987). So, from the start we have to create non-monotonic logics, metalevel logics, logic programming and other bookkeeping tricks to make what we are doing continue to look like deduction (McDermott, 1987). In practice, we generally avoid the assumption that an inference is completely unwarranted unless it can be shown to be deductively valid in some Tarskian logic and in some special sense. This assumption in fact makes the unreasonable demand that we change the concept of deductive validity (McDermott, 1987). Human experts do not work this way. Then why, a priori, would expert systems do otherwise? Abstract reasoning is perforated: it appears seamless only because people are good at giving *post hoc* rationalizations for their decisions or actions, but those rationalizations are not necessarily causally connected to the activity they purport to explain (Chapmann and Agre, 1987). Techniques are often abstract by virtue of the form only, not the content, of the concrete problems they apply to. These considerations explain why expert system performances not only are, but must be, essentially related to the use of facts, rules, strategies and heuristics that human experts would use. Formal logic or models of abstract reasoning may be used as a tool, but only in combination with others. What is needed is not abstract reasoning, but practical reasoning that is as rigorous as possible. In that, practical knowledge is what counts most.

Similarities

Hence, the principal problems in representing systems, events, processes, tasks, space and time are not only related, but are in fact similar because they all stem from the various changes in the focus of analysis and in the power of resolution of the structures, models or scales used in the reasoning. Furthermore, all these concepts may be represented and put in relation to each other by similar abstract structures, such as graphs, nets and semi-orders.

Human beings are so accustomed to fuzzy semantics and logics that, generally, they do not even notice the operations of specialization and generalization of meanings, concepts and operators they continuously engage in almost every time they change models, levels of analysis, or semantic context in reasoning. Moreover, their use of fuzzy semantics and logics actually minimize these operations, mainly to spot-solving of incoherence or ambiguities and to the choice between alternative paths of reasoning. In these operations, information may always be lost or distorted. It is really hard to design a good fuzzy semantics and logic system and then to design ways for scoping it according to the context. Human beings do not often have problems with lost or distorted information, because they may always start their reasoning over again, adapt it or retain along with it the necessary information. Much more frequently, they will simply restore the information by exploiting the redundancies in their knowledge of the problem domain and of the world, and also their common sense.

Structure and causal analysis

Components of a system are themselves systems, either concrete as a physical object, or abstract as a structure, function, behaviour, task, process or event. In turn, an event on a given time-scale is in fact a complicated process composed of subevents on a finer scale. An event is thus organized and defined as a process, and a process as a system. A structure is a representation of a type of relation between components or events of a system, such as the various hierarchical, functional, causal and adjacency structures. The term 'structure' may then refer to a formal representation or to a graph of an aspect of a representation, image, model, or a part of a system. It may, for example, be a partial graph, a subgraph, a tree of hierarchical or functional decomposition, a tree of partial graphs (such as a layered set of levels of structural detail) or a tree of subgraphs (such as a regional decomposition of multiple levels). Structures are essential for representing the limitations of propagation of a causal effect. A system or a subsystem, seen from the exterior, may be defined as an atomic element, a black box, with several characteristics. Similarly, seen from the exterior, an event is not a process, but an atomic event. However, each one must have an identity and be localizable on the various hierarchical, functional, physical, spatial, temporal and other structures. Thus, it becomes possible to use ports through which causal effect may probe outward and inward, as well as various leaking processes (Lansky, 1987). It also becomes possible to assign to each agent or component its functions, tasks, processes and localizations for each of its activities at the appropriate periods and, above all, its local clocks, perhaps one for each significant activity or parameter. This is essential for causal analysis, because a component in a system can only act on or be acted on by its immediate neighbours in the structures (command, control, communications, adjacency, etc.) in which the component is a part.

In domains in which the possibility of representing parallel activity by multiple agents or components is essential, events in space and time will be structured and marked by a set of semi-orders interrelated by the constraints of precedence, adjacence, simultaneity and bounds on intervals or regions on the events. Hence, concrete or abstract spaces and time are modelled using a set of qualitative or quantitative scales having naturally a given power of resolution. In addition, hierarchical structures will also be represented in general by similar sets.

Grain size and inertial delays

The overall representation of a system contains all the structures of the system, plus descriptions of the properties of the system, and the relations between the structures. A frontier always has a certain thickness even if it is represented by a mere line. For example, at a particular grain size, a road in a field may be shown as adjacent, but finer regions would show that they are separated by the sidewalk (Davis, 1986). Any structure may be more or less detailed, according to the level of detail we are interested in and to the precision of the information we have. It is only when we need to look inside a system or a structure that we generally find it useful to add or build more detailed structures, and to resolve some components and relations into more precise ones. Similar problems exist naturally with the structural decomposition of behaviours, processes and events. Moreover, the depth of levels of detail is always arbitrary, because it depends only on the goals of the human user and the meaning of the elements being considered. In other respects, we are not interested in always dragging the entire depth of the structure when we work at a particular level of detail. It is too inefficient and artificial. We must somehow be able to erase or forget that which is not in the focal range, and to restore or rebuild more detailed structures when dropping down a level of structural detail. This is analogous to grain size operators in mathematical morphology.

Inertial delays

The concepts of delays and inertial delays are a good illustration of the effects of grain size in the time domain. In logical analysis of systems, values of variables and of time are discretized at a given finite grain size. We may choose a particular grain size for convenience or because it corresponds to the limit of accessibility to the causality in the case in which we are interested. To every process or event, we may associate a set of time diagrams. The set forms a time map relating to the time map of events. Each time diagram is called a chronogram and represents the variations of a variable or of a parameter. How are the chronograms discretized? How are finer representations replaced by coarser ones on the same scale, when the grain size is larger? Suppose we have two input parameters $x_1(t)$ and $x_2(t)$ and one output parameter $x_3(t)$ as illustrated in Figure 26.1. We want to describe the behaviour of the system by establishing a relation between the variations in the variables. It would be much easier if $x_3(t)$ is as discrete as the other two. To achieve this, we will proceed in three steps, using the following rules. First, we define a new variable $x_4(t)$:

$$x_4(t) = 0 \text{ if } x_3(t) \leq s_0$$
$$x_4(t) = 1 \text{ if } x_3(t) \geq s_1$$
$$x_4(t) = 2 \text{ if } s_0 \leq x_3(t) \leq s_1$$

As we see it, $x_4(t)$ is a quantification of $x_3(t)$. We may associate to each value of x_4 (t) a particular state of x_3 (t). Second, the state where $x_4(t) = 2$ is, in the present case, transitory. If the variations of the inputs are much slower than the duration of this state, and if we do not have a run between two delays, then we may ignore the representation of this state completely, by defining a new variable $x_5(t)$ having only two values or states, as illustrated in Figure 26.1. That representation conserves the idea of delays, but has forgotten their form. In Figure 26.1 delays δ_{01} and δ_{10} are not the same. Third, if δ_{01} and δ_{10} are assumed to be equal to the same constant δ, independent of time, then $x_5(t)$ may be replaced by another new variable $x_6(t)$. Thus, we have obtained the asynchronous logical model of systems. We may note here that if δ is considered negligible, then, in making $\delta = 0$, we obtain $x_7(t)$, which describes the combinatorial logical model of systems (Mange, 1981). The final representation then conserves only important states and the order of magnitude of the delays. We have here an analogy with the concept of structure, in that a structure at a given level of detail refers only to the essential traits of the system or the event at this level.

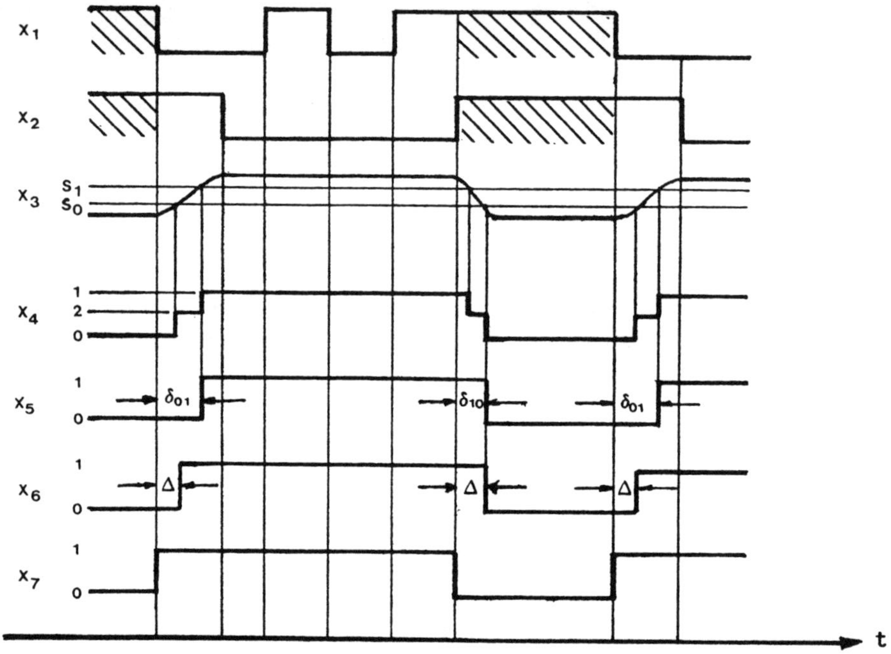

Figure 26.1 Chronograms and process of discretization.

Delays are either pure or inertial. An element of pure delay is such that the output state y_p is the same as the input state x retarded by a constant value δ_p, characteristic of the element. An element of inertial delay is more complex. It acts as a pure delay (a delay function) for slow variations of inputs, but cuts out fast variations (a filter function).

The properties of inertial delays are useful in logical analysis of systems, especially when it comes: (i) to defining transitory and permanent behaviours and (ii) to studying runs between delays, and, consequently, to generating and studying alternative behaviours of a given system systematically. We suggest that similar techniques may be developed to generate and study alternative structures of a system or of an event. The descriptions of the behaviours of a system must be derivable from the structures of the system but we are in fact primarily interested in reasoning from misbehaviour to structural defects: given symptoms of misbehaviour, we want to determine the structural aberrations responsible for the symptoms (Davis, 1983).

Using GRAFCET

Processes, events, actions and tasks are easily modelled by GRAFCET, which is a standard form of Petri nets. Petri nets are particularly useful where the focus is on modelling parallel events (Davis, 1983). GRAFCET was specifically developed for modelling the functioning of complex and parallel systems. It is a synchronous logical model of systems in two ways:

1. Time is discretized. The step or grain size may be as little as desired, but it must be finite.
2. Evolutions are synchronous. If we go from a situation at time t to another situation at time $t+1$, then all valid and receptive transitions take place simultaneously. Each situation lasts at least a time δ, not null.

Those are the fundamental differences from the usual Petri nets, where time is continuous and where a transition is gone through at once when it becomes receptive and valid independently of the other transitions (Blanchard, 1979).

Fundamental concepts of GRAFCET

The behaviour of a system may be represented in GRAFCET by a set of stages to which actions are associated, of transitions to which receptivities are associated, and of oriented connections, or arcs. Each connection binds a stage to a transition, or a transition to a stage. Thus, components and connections between components are clearly separate. This is the same idea as in qualitative physics, where components and conduits, or connections, between components are modelled separately as different types of components and then connected by abstract ties (De Kleer and Brown, 1985; Forbus, 1985; Kuipers, 1985). Similarly, here, a transition may not directly connect with another transition but only with a stage, and a stage may not connect directly with another stage but only with a transition.

In a GRAFCET, each transition may have its local clock or chronometer, and all these clocks are closely interrelated by the general structure of the GRAFCET. It is important to note that stages and actions may be represented by other GRAFCETs, often called subgrafcets. It appears, then, that stages and actions have internal local clocks synchronized in some sense with the beginning of the stage or the action. We see also that the beginning and the end of a stage are given by an external clock. Moreover, GRAFCETs can communicate with each other by subgrafcets and by receptivities. Also, GRAFCET may be autosynchronized.

The fact that each transition has several input stages and several output stages and is passed through as soon as it is validated and receptive, thus simultaneously activating all its output stages, implies that a GRAFCET may have several different valid paths of realization. So, the order and the exact timing of the stages are not necessarily fixed relative to the initial situation and to the beginning time of the GRAFCET, but may vary greatly from time to time. It is like the multiple paths in a precedence net of subtasks. The difference is that the choice of path is determined by the timing of the passing through of the transitions and that an undesirable or dangerous path may be forbidden by an appropriate receptivity on a transition.

That is a very remarkable property. It means that the same behaviour, as defined or seen from the exterior of a system, possibly corresponds to several possible internal behaviours or scenarios having various degrees of parallelism. It also means that some possible internal behaviours in given circumstances and conditions may correspond to an external misbehaviour. We know that in reality alternative behaviours, in one sense or another, are always possible. We know that complex systems never repeat in exactly the same way twice in succession a sequence of situations. It is very practical, then, to have a tool with which we may represent and easily generate alternative operating scenarios.

Several other aspects of GRAFCET are very interesting because they relate directly to the design of intrinsically safe systems. We may cite as examples the properties of being safe or not, living or not, pseudo living or not, and proper or not, as in Petri nets. Other examples are the various methods that exist for marking the redundant stages, transitions and receptivities in a GRAFCET and for then simplifying the GRAFCET. There are also efficient algorithms for finding loops and stressing simultaneously incompatible actions in a GRAFCET. Loops must be analyzed in a GRAFCET, since they lead to uncertainty concerning the future evolutions of the GRAFCET. Simultaneously, incompatible actions must be eliminated or regulated properly for causal coherence and for safety.

Connection between GRAFCET and SAFE-EXP

The connection between GRAFCET and SAFE-EXP is relatively straightforward. Each component of a system has its formal and informal functions, tasks, processes and significant events; it also has its structures, especially its hierarchical, functional, spatial and temporal structures. In particular, each has its local clocks relating directly or not to a system's clock.

Then, events are defined as components of a system having several characteristics. The main ones are:

(i) a distinct identity and a source for the identity;
(ii) a punctual or distributed localization, in the physical and functional structure of the system at a given period;
(iii) one or more informal descriptions, compatible or not, eventually known;
(iv) one or more trees of causes and consequences, eventually known;
(v) one or more formal descriptions, compatible or not, eventually known, and which are GRAFCETs.

Depending on the information available, the causal trees and GRAFCETs will be more or less developed, but we can suppose that at least one exists even if none are known at a given time.

Tasks, processes and events are defined essentially in the same manner. At any time, a situation in a GRAFCET is defined entirely by the set of active stages. On a given scale, an atomic event is a stage or an action with no visible GRAFCET on this scale. More generally, an event is a GRAFCET with no direct loop to its initial situations and receptivities. A process, on the contrary, has at least one return loop and may repeat itself; moreover, a given process may take place by various scenarios, each one corresponding to a different path between the possible situations in the GRAFCET. Finally, the various mechanisms of human error may be represented as specific subgrafcets involving or creating special connections between stages in a GRAFCET.

Conclusion

Before drawing causal trees of systemic behaviour and discussing safety, we must understand the structure and the behaviour of the system considered. In the case of a complex system, this requires a taxonomy, supplied with a set of rules for describing structures and behaviours and a set of rules for analyzing systems, in order to suggest structural defects from given symptoms of misbehaviour or likely misbehaviours from a given structure. We have developed such a taxonomy and such sets of rules.

Even a simple man–machine system with a simple task is in fact a very complex system. If it appears otherwise, the cause can certainly be attributed to the experience of the operator and, more fundamentally, to the adaptability and complexity of human behaviour. There is really no place for simplification when it comes to the identification and analysis of risks in complex systems. The idea behind SAFE-EXP is to put in the machine powerful and complex methods and then to make them as transparent as the user wishes.

Acknowledgements

The authors thank the National Research Council of Canada, the Institute of Research in Occupational Health and Safety in Quebec and the Ergonomic Engineering Department of Genicom Consultants Inc. for their support. They thank Dr Ashraf M. Genaidy for his excellent comments. They thank also Brad McFadyen and Jessica Poitiers for the revision of the translation. Arity is a trademark of Arity Corporation.

References

Biswas, G., Oliff, M. and Sen, A., 1985, Design of an expert system in operations analysis. In: *IEEE 1985, Proceedings of the International Conference on Cybernetics and Society*, pp. 121–125.

Blanchard, M., 1979, *Comprendre, Maîtriser et Appliquer le GRAFCET*. Cepadues-éditions, Toulouse. Collection Automatisation et Production. 174 pp.

Bonnet, A., 1984, *L'Intelligence Artificielle. Promesses et Réalités* (Paris: Inter Editions), 272 pp.

Bonnet, A., Haton, J. P. and Truong–Ngoc, J. M., 1986, *Systèmes-experts. Vers la maîtrise technique* (Paris: Inter Editions), 288 pp.

Bruun, O., Rasmussen, A., Taylor, J. R., 1979, *Cause Consequence Reporting for Accident Reduction. The Accident Anatomy Method*, (Denmark: Riso National Laboratory), Technical Report Riso-M-2206.

Cazamian, P., 1974, *Leçons d'Ergonomie Industrielle, Une Approche Globale*. (Paris: Cujas), 158 pp.

Chapman, D. and Agre, P. E., 1987, *Abstract reasoning as emergent from concrete activity. In: Reasoning about Actions and Plans*, edited by M. T. Georgeff and A. L. Lansky, pp. 411–424.

Davis, E., 1986. *Representing and Acquiring Geographic Knowledge*. (London: Pitman), (Los Altos, CA: Morgan Kaufmann), 224 pp.

Davis, R., 1983, Diagnostic reasoning based on structure and behavior. In: *Qualitative Reasoning about Physical Systems*, (Cambridge, MA: MIT Press), pp. 347–410.

De Kleer, J. and Brown, J. S., 1985, A qualitative physics based on confluences. In: *Qualitative Reasoning about Physical Systems*, (Cambridge, MA: MIT Press), pp. 7–84.

Farreny, H., 1985, *Les Systèmes Experts Principes et Exemples*, (Toulouse, France: Cepadues Editions).

Faverge, J. M., 1967, *Psychosociologie des Accidents du Travail*. Editions des Presses Universitaires de France, Collection SUP, 160 pp.

Feigenbaum, E. and McCorduch, P., 1984, *La Cinquième Génération* (Paris: Inter Editions).

Fitts, P. M. and Jones, R. E., 1947, *Psychological Aspects of Instrument Display I: Analysing and Interpreting Aircraft Instruments*. Memorandum Report ISEAA-694-12A (Wright-Patterson Air Force Base, Dayton, OH: Aero Medical Laboratory, Air Material Command).

Fitts, P. M. and Jones, R. E., 1949, *Analysis of Factors contributing to 460 "Pilot Error" Experiences in Operation Aircraft Controls*. Memorandum Report ISEAA-694-12A (Wright-Patterson Air Force Base, Dayton, OH: Aero Medical Laboratory, Air Material Command).

Forbus, K. D., 1985, Qualitative Process Theory. In: *Qualitative Reasoning about Physical Systems*, (Cambridge, MA: MIT Press), pp. 85–168.

Fortin, C., 1983, *La causalité des accidents*. Mémoire de maîtrise, Département de génie industriel, Ecole Polytechnique de Montréal, 547 pp.

Fortin, C. and Gilbert, R., 1986, *Intérêts et Applications Potentielles des Systèmes Experts au Domaine de la Santé et de la Sécurité du Travail*. Rapport technique EP-86R-40 (Département de génie industriel, Ecole Polytechnique de Montréal).

Fortin, C., Gilbert, R., April, G. E., Schiettekatte, J., Guimont, S. and Mukhedkar, D., 1987, Prototyping an expert system for accident prevention and risk analysis in the work place. In: *Trends in Ergonomics/Human Factors IV*, edited by S. S. Asfour (Amsterdam: Elsevier–North Holland).

Fortin, C., Gilbert, R. and Warmoes, J. C., 1983a, *Typologie des Causes d'Accidents. (Listes et Interrelations)*. Rapport technique EP83-R.27 (Ecole Polytechnique de Montréal), 483 pp.

Fortin, C., Gilbert, R. and Warmoes, J. C., 1983b, *Revue des Modèles Concernant les Systèmes H-Ms et la Sécurité*. Rapport technique numéro EP83-R.28 (Ecole Polytechnique de Montréal), 286 pp.

Fortin, C., Gilbert, R. and Warmoes, J. C., 1983c, *Schéma des Interrelations entre les Diverses Variables Impliquées dans les Accidents*. Rapport technique numéro EP83-R.29 (Ecole Polytechnique de Montréal), 18 plans.

Fortin, C., Warmoes, J. C. and Gilbert, R., 1983d, *Mathématiques Applicables à l'Analyse de la Causalité des Accidents*. Rapport technique numéro EP83-R.26 (Ecole Polytechnique de Montréal), 150 pp.

Georgeff, M. P., 1987a, Many agents are better than one. In: *The Frame Problem in Artificial Intelligence*, edited by F. M. Brown, (Los Altos, CA: Morgan Kaufmann), pp. 59–76.

Georgeff, M. P., 1987b, Actions process and causality. In: *Reasoning about Actions and Plans*, edited by Michael P. Georgeff and Amy L. Lansky (Los Altos, CA: Morgan Kaufmann), pp. 99–122.

Gilbert, R., Mukhedkar, D., Warmoes, J. C., Fortin, C. and Toulouse, G., 1982, The state of the art and future developments in accident research and risk evaluation. In: *Human Factors and New Technology, 15th Annual HFAC Conference*, Toronto, Ontario, Canada.

Gilbert, R., Vo-Ngoc, B. and Rohan, P., 1976, *Les accidents vus du dedans*. Rapport pour l'O.C.Q., Université du Québec (IRNS), 26 pp.

Gilbert, R., Vo-Ngoc, B. and Rohan, P., 1975, *Une Nouvelle Approche en Recherche sur les Accidents*. Rapport Université du Québec (IRNS), 46 pp.

Henley, E. J. and Kumamoto, H., 1981, *Reliability Engineering and Risk Assessment* (Englewood Cliffs, NJ: Prentice-Hall).

Henley, E. J. and Kumamoto, H., 1985 *Designing for Reliability and Safety Control* (Englewood Cliffs, NJ: Prentice-Hall).

Hoag, J. C., 1985, An Expert System for Finding the Causes of Train Derailments. In: *IEEE 1985 Proceedings of the International Conference on Cybernetics and Society*, pp. 112–115.

Kuipers, B., 1985, Commonsense reasoning about causality: deriving behavior from structure. In: *Qualitative Reasoning about Physical Systems* (Cambridge, MA: MIT Press), pp. 169–204.

Kuipers, B. and Kassirer, J. P., 1983, How to discover a knowledge representation for causal reasoning by studying an expert physician. In: *The 8th International Conference on Artificial Intelligence*, pp. 49–56.

Lansky, A. L., 1987, A Representation of paralleled activity based on events, structure and causality. In: *Reasoning about Actions and Plans*, edited by Michael P. Georgeff and Amy L. Lansky (Los Altos, CA: Morgan Kaufmann), pp. 123–160.

Laurière, J. L., 1986, *Intelligence Artificielle, Résolution de Problème par l'Homme et la Machine*. (Paris: Eyrolles), 473 pp.

Lehto, M. R. and Miller, J. M., 1985, An experimental application of knowledge engineering: a generic safety analysis system. In: *IEEE 1985 Proceedings of the International Conference on Cybernetics and Society*, pp. 131–135.

Lehto, M. R., Clark, D. R. and Miller, J. M., 1984, An isomorphism between fault tree analysis and lisp data structures. In: *Human–Computer Interaction*, edited by G. Salvendy, (Amsterdam: Elsevier Science), pp. 297–300.

Leplat, J. and Chenais, I., 1985, *Erreur Humaine, Fiabilité Humaine dans le Travail* (Paris: Armand Collin), 200 pp.

Leplat, J. and Rasmussen, J., 1984, Analysis of human errors in industrial incidents and accidents for improvement of work safety. *Accident Analysis and Prevention*, **12** (2), 77–88.

Madni, A. M. and Chu, Y. Y., 1985, Network model guided expert knowledge elicitation. In: *IEEE 1985 Proceedings of the International Conference on Cybernetics and Society*, pp. 1–13.

Mange, D., 1981, *Analyse et Synthèse des Systèmes Logiques* (Dunod), 394 pp.

McDermott, O., 1987, AI, logic and the frame problem. In: *The Frame Problem in Artificial Intelligence*, edited by Frank M. Brown (Los Altos, CA: Morgan Kaufmann), pp. 105–118.

Montmollin, M. de, 1974, *L'Analyse du Travail Préalable à la Formation* (Paris: Armand Colin).

Ombredane, A. and Faverge, J. M., 1955, *L'Analyse du Travail. Facteur d'Économie Humaine et de Productivité* (Paris: PUF).

Rasmussen, J. and Lind, M., 1982, A Model of Human Decision Making in Complex Systems and its use for Design of System Control Strategies. In: *Proceedings of the American Control Conference*, pp. 270–276.

Rasmussen, J., Pedersen, D. M., Mancini, G., Carnino, A., Griffon, M. and Gagrolet, P., 1981, *Classification System for Reporting Events Involving Human Malfunctions*. Technical report Number Riso-M-2240 (Denmark: Riso National Laboratory).

Stuart, C. J., 1987, Branching regular expressions and multi-agent plans. In: *Reasoning about Actions and Plans*, edited by Michael P. Georgeff and Amy L. Lansky (Los Altos, CA: Morgan Kaufmann), pp. 161–188.

PART VII
PERFORMANCE EVALUATION AND DECISION SUPPORT

Chapter 27

SPES: assessing the effects of the work environment on man with computerized performance testing

FRANCESCO GAMBERALE, ANDERS IREGREN and ANDERS KJELLBERG

Division of Psychophysiology,
National Institute of Occupational Health,
S-171 84 Solna, Sweden

Abstract: This chapter presents a computerized performance evaluation system in use at the Swedish National Institute of Occupational Health. The background and the research experience leading to the system's development and empirical data demonstrating its applicability to different research objectives are described. The role of performance testing in the assessment of unfavourable work environmental conditions is discussed.

Keywords: Performance testing; Computerized; Work environment; Toxicology.

Introduction

Several types of methods may be of value in the early detection of neurotoxic effects or the effects on the nervous system of exposure to other unfavourable physical environmental agents. Traditionally, neurophysiological and neurological testing methods have been used together with subjective reports of symptoms. More recently, performance testing has proved to be an important method of assessment in these contexts. This chapter discusses the role of performance testing in environmental ergonomics and particularly the possibilities offered by computers for such measurements. A computerized system, Swedish Performance Evaluation System (SPES), is presented in some detail and examples are given of studies in which it has been used. The use of computers for collecting subjective reports of mood and other symptoms will also be mentioned since SPES also contains such self-report scales.

The growing interest in the measurement of performance is most probably due to the sensitivity shown by these methods in unveiling changes in the human organism which would otherwise not be detected. By now, the evidence that these changes are some of the earliest indicators of the occurrence of health effects has become unequivocal. In consequence, the measurement of performance has come to be regarded by many as a device of major importance for the monitoring of hazards to health and safety in the work environment.

At our institute, the measurement of performance has undoubtedly constituted the main method of studying the effects on the central nervous system of low dose exposure to the chemical substances which are frequently found in the work environment (Gamberale, 1985; Iregren, 1986a). Thus, we have applied performance tests of various kinds in experimental laboratory studies as well as in field studies and in cross-sectional epidemiological investigations. Through the years, we have made use of performance tests in experimental inhalation studies of different industrial solvents alone and in combination (Gamberale and Hultengren, 1972, 1973, 1974; Gamberale *et al.*, 1975a,b, 1976, 1978; Anshelm Olson *et al.*, 1985; Iregren, 1986b; Iregren *et al.*, 1986; Wigaeus-Hjelm *et al.*, 1990). Psychometric tests have been applied directly at the work site in four investigations of solvent exposed workers (Gamberale *et al.*, 1976; Kjellberg *et al.*, 1979; Anshelm Olson *et al.*, 1981; Anshelm Olson, 1982) and in two investigations of nurses exposed to anaesthetic gases (Gamberale and Svensson, 1974; Kjellberg and Strandberg, 1979). Furthermore, we have used comprehensive batteries of behavioural tests to investigate possible long-term effects of chronic exposure to organic solvents (Knave *et al.*, 1978; Elofsson *et al.*, 1980; Iregren, 1982). A group of foundry workers exposed to manganese were recently studied with a comprehensive battery of computerized tests (Iregren, 1989).

Unfavourable effects on performance of other work environmental factors such as noise, vibration, cold, heat, electric and magnetic fields and physical workload, for example, have been investigated in laboratory experiments as well as in field studies. However, our own experience with the use of psychometric tests to study the effects of non-chemical agents in the physical work environment is relatively modest. Still, behavioural tests have been successfully applied at our laboratory in experiments on the effect on performance of exposure to noise (Kjellberg and Wide, 1988) and to moderate cold and heat (Enander, 1987; Gamberale *et al.*, 1988). Finally, we have used psychometric tests to investigate the possible effects on workers from acute (Gamberale *et al.*, 1989a) and chronic exposure (Knave *et al.*, 1979) to electric and magnetic fields.

Performance as an indicator of environmental load

Because of unsolved theoretical and methodological problems, the choice of relevant performance variables and their measurement poses many difficulties. Although psychologists have studied performance under numerous task and environmental conditions, there is still a considerable lack of agreement among researchers concerning the identification, definition and measurement of performance functions which might be relevant to the study of the effects of different environmental conditions. Some of these difficulties are due to the fact that psychometric tests have been developed and applied within different fields of psychological research and show fundamental differences due to their scope and origins. There are also considerable differences with respect to the extent to which they satisfy psychometric requirements such as *reliability*, *sensitivity*, *specificity*, *theoretical interpretability*, *face validity*, *difficulty level*, *duration*, and *repeatability*.

The relative importance of these requirements also varies considerably when testing is used to access the work environment. High reliability, sensitivity and repeatability would be of crucial importance in performance tests to be used in experimental laboratory investigations. In field studies, face validity is a further important aspect of the tests, which also necessarily must be of short duration.

Undoubtedly the performance tests which have given the most consistent results in experimental exposure studies or in quasi-experimental field studies have been those based on the experiences made in the field of experimental psychology. Usually, the performance functions assessed by these performance tests are better defined and the measurements are more reliable than for the performance tests developed in other fields of psychological research, for example, for clinical purposes.

Since the mechanisms of action are seldom well understood, it is advisable to base the choice of behavioural tests on strictly empirical considerations. From that point of view there is, for example, no doubt that the likelihood of detecting the adverse effect of exposure to industrial solvents will be enhanced if the performance tests used involve sensory-motor, vigilance, and perceptual functions, rather than higher cognitive functions.

Another factor to take into consideration when making the choice of a performance test is what effect the factor under study might have on the arousal level of the person. Although justified criticism has been directed towards the arousal concept, a rough grouping of environmental conditions into arousing and dearousing ones is fruitful when the choice of performance tests is considered. Factors such as monotonous tasks, sleep deprivation, exposure to solvents, prolonged exposure to monotonous noise, vibration, and high room temperature normally lower the person's arousal level, and may lead to an impaired ability to attend to the critical stimuli. In performance this manifests itself as more missed stimuli and as slower responses. The tasks most sensitive to such dearousing factors are generally simple, monotonous tasks which make high demands on sustained attention and which give the subject poor feedback. Typical examples are prolonged simple vigilance and reaction time tasks (Hockey, 1986). Thus, in the experience of our laboratory, a test of simple reaction time (SPES 1) has proved to be most sensitive test in almost all studies of industrial solvents.

Other work loads, such as high work demands and intense and variable noise or vibration, may lead to the opposite effect, i.e. an increased arousal above the optimal level. The effects of such loads on performance are less clear cut, but the most sensitive tasks usually are those which make high demands on information processing and memory. The performance decrement is typically revealed as an increased error rate rather than as a reduced response speed. Among the studies using SPES this was the case in a study of cold (Enander, 1987). The effects often seem to be a consequence of an overly selective attention (Hockey, 1986).

Many of the tests in the SPES battery were originally intended for use in research on the effects of industrial solvents. Thus, they were constructed to be sensitive to the dearousing effects of these agents. The more recent additions to the battery have extended it to comprehend more complex tasks which should be more sensitive to arousing factors.

The SPES battery also contains a scale for ratings of subjective effort or performance. The reason for this is that an effect of an environmental load may sometimes be detected by subjective measures and not by performance measures. In a non-optimal environment, a normal level of performance may still be maintained, at least for a limited time, by an increase in the level of effort. Figure 27.1 illustrates how an effect may manifest itself in performance and

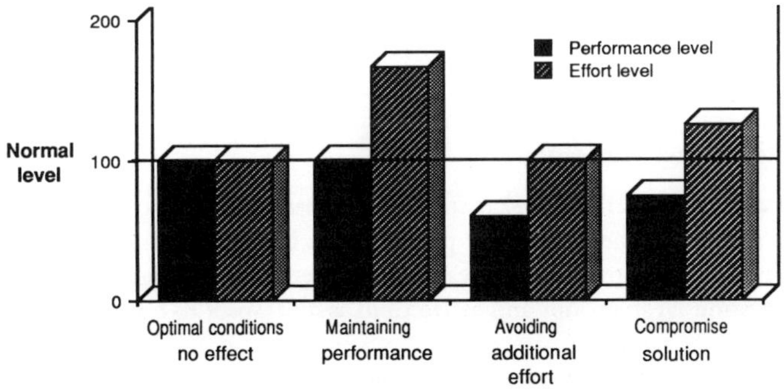

Figure 27.1 Three ways of coping with non-optimal work conditions leading to different effects on performance and effort.

subjective effort given that the person chooses different ways of coping with the non-optimal work conditions. A fourth possibility to take into consideration, which is not illustrated in Figure 27.1, is that the primary action of the environment is on the person's motivation. This would manifest itself as a lowered level of both performance and effort.

In addition to the performance ratings SPES permits the collection of two other types of self-report data: symptoms of both acute and long-term exposure and mood ratings. These are aimed at the detection and description of possible environmentally induced changes in the subject's perception of his physical psychological state.

The benefits and drawbacks of automated testing

Automated testing, whether computerized or not, implies some advantages over traditional paper and pencil testing, such as:

1. The automation of tests gives an excellent opportunity for strict standardization of test procedures, e.g. instructions, test protocols and evaluation variables, thus increasing the possibilities for comparisons across studies.
2. The possibility to perform detailed measurement and analysis of single responses or response components. This type of microanalysis has greatly enhanced the sensitivity of the performance tests.
3. Increased precision in the measurement procedure. By reducing the influence of the investigator the reliability and the validity of the results is increased.
4. Since an automated test could be administered by a technician or a nurse there is a possibility to reduce the workload of psychologists, who therefore could make better use of their skills.
5. It has also been suggested that automation of the testing procedure would render the testing situation less threatening.

Fully computerized testing procedures provide some additional possibilities as well:

6. Computers are flexible, and one system can be used to administer a variety of tests, as well as to perform other routine tasks in the laboratory or clinic.
7. Computers facilitate prompt scoring and evaluation of even very complex tests and questionnaires.
8. Computers make it possible to adapt the choice of items according to the performance capacity of the individual.
9. Computers offer communication possibilities making transfer of data for statistical computations or other purposes very convenient.
10. Modern computers are transportable and in the extreme case even portable, thus making field testing feasible.

Some critical comments regarding computerized testing have also been mentioned. Two critical points have been the supposedly poor rapport between the subject and the machine and the static form of computerized testing. Furthermore, computerized testing implies difficulties in testing large groups and restrictions with respect to both stimuli and response media.

As regards the rapport established, investigations pertaining to this problem generally indicate no difficulties. See, for example, French and Beaumont (1987), or Lukin *et al.* (1985). The 'user-friendliness' of a system is not dependent upon whether it is computerized or not, but on the careful design of the system. As pointed out by Beaumont (1982), the single most

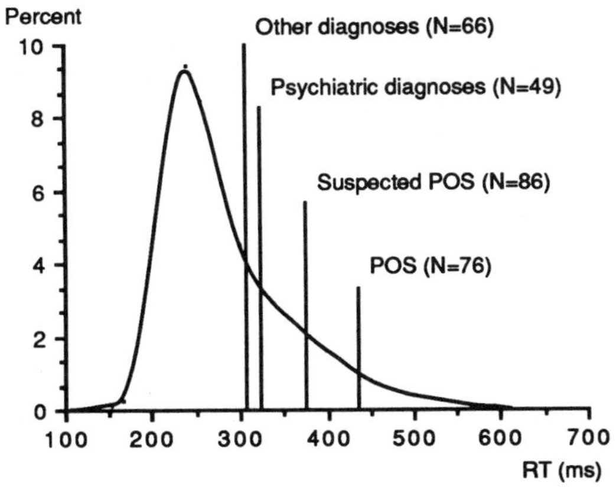

Figure 27.2 Mean reaction times of four groups of workers examined in occupational health clinics. The means are indicated in the frequency distribution of 67840 reaction times obtained from 424 industrial workers. POS indicates 'psycho-organic syndrome'.

important requirement for successful design is the predictability of the system. Thus, no action on the part of the subject should result in accidental termination of the test.

Due to the still relatively high purchase costs even for a microcomputer system, it is difficult to test large groups of people simultaneously. On the other hand, computerized tests often provide much more information than traditional tests within a specified time period. Thus, the possibility to perform simultaneous testing of large groups is somewhat less important with computerized tests.

The objection concerning the static form of computerized tests is by now invalid. There have been successful attempts at making 'tailored', i.e. adaptive, tests, where the test items administered are contingent upon the performance of the subject, see, for example, Weiss and Vale (1987). One of the memory tests available with our system, a version of the digit span test, functions in an adaptive way.

Until today, most efforts at the implementation of performance tests on computers have been using visual stimuli, since administration of auditory or tactile stimuli require the use of rather complicated (probably custom made) external equipment in addition to the computer. Such additions would naturally make it very difficult to standardize the tests to allow for widespread use.

The few response media available present a similar problem with respect to the development of tests. Since input to a computer is normally made via the keyboard, most testing systems presently use this medium. Attempts to use other means of input usually also imply manual performance, as for example with joysticks or touch sensitive screens. At present, the choice of response mode is probably the most limiting factor in the development of computerized tests, since the response medium may easily affect the test results in unintended ways. A fast and reliable system for processing speech input would in many instances provide the only acceptable solution.

The restrictions regarding stimulus presentation and response input are of course steadily diminishing, since the continuous technical development make computers increasingly competent. However, standardized tests using new technical possibilities, such as speech input, are still in the early stages since the development of well standardized tests is a laborious long-term project.

In spite of the restrictions mentioned, efforts at developing computerized tests and testing systems are steadily increasing, and this type of assessment is currently in use in a wide variety of settings. Computerization has made complicated psychometric techniques available even to persons lacking training as professional psychologists. Therefore, this trend accentuates the ethical demands for control of the construction, distribution, and use of these methods (Matarazzo, 1983). The American Psychological Association (1986) and the British Psychological Association (Bartram *et al.*, 1987) have therefore published guidelines for this purpose.

The experience gained at our laboratory using computerized tests has generally been positive, although there are of course difficulties. One significant problem, which applies equally to traditional tests, relates to the time and effort needed for successful test development. Furthermore, computers are still technically complicated machines, thus requiring special skills of psychologists and technicians engaged in this development.

Several recent reviews have treated various aspects of computerized testing, and the reader is recommended those by Bartram and Bayliss (1984), McArthur and Choppin (1984), Space (1981), and Thompson and Wilson (1982).

Description of the SPES

The Swedish Performance Evaluation System consists of a number of semi-automated computerized performance tests and scales for the subjective evaluation of performance and of effort spent on the tests, of mood, and of different kinds of symptoms (Table 27.1). The

Table 27.1 List of the tests and scales in the SPES, presenting the number of items (including practice), the parameters normally used for evaluation of the results and the approximate testing time.

SPES code	Performance tests	No. of items (+practice)	Parameters extracted	Time (min) (approx)
1	Simple reaction time	80 + 16	Mean, SD decrement	6
2	Choice reaction time	112 + 32	Mean, No. of errors	9
3:1	Colour word vigilance	192 + 16	Mean RT, No. of commissions, No. of omissions	8
3:2	Colour word stress	192 + 16	Mean RT, No. of commissions, No. of omissions	7
4	Search and memory	3 × (10 + 1)	Mean RT/search level	10
5	Symbol digit	6 + 4	Mean RT, No. of errors	6
6	Digit span	varies	Length of memory span	8
7	Additions	40 + 3	Mean RT, No. of errors	6
8	Digit classification	240	Mean RT, No. of errors, No. of lags	4
9	Digit addition	120	Mean RT, No. of errors, No. of lags	7
10	Verbal reasoning	64	Mean RT, No. of errors	8
11	Vocabulary	45	No. of correct answers	8
12:1	Finger tapping speed	2 × (3 + 1)	Mean No. of tap/hand	5
12:2	Finger tapping endurance	1	Changes of movement time and resting time over test-time	2
30	Performance	1/test	Percent of max. performance	1
31	Mood	20	Energy score and stress score	3
32	Acute symptoms	17	No. of symptoms reported	4
33	Long-term symptoms	38	No. of symptoms reported	6

system is designed to be dynamic and flexible. Thus, it allows the researcher or the practitioner to choose among the tests and the scales, adapting the battery to the specific purpose of the evaluation at hand. The system is also intended to undergo gradual improvement based on analyses of the results of on-going empirical studies and on future experience with the use of the system.

With few exceptions, the performance tests are non-verbal, i.e. they can be used to assess performance independent of the language of the subjects. Some of the tests, e.g. the Colour word vigilance (SPES3:1), the Colour word stress (SPES3:2) and the Verbal Reasoning test (SPES10) can be easily adapted to other languages and will only require translation of the text files used by the programs. These text files, which are easily edited, contain all the verbal communication with the subject, i.e. the instructions on how to perform the test as well as the test items. The only test which will require a complete new construction and standardization if used with non-Swedish speaking subjects is the Vocabulary test (SPES11). The Mood rating scale (SPES31) will also require a new standardization for each translation into a new language.

Hardware

Any IBM or IBM-compatible PC, XT or AT can be used, equipped with an external clock card, an overlay to the keyboard, an EGA graphics card, a colour monitor, an optional printer, and an external key for Finger tapping endurance (SPES12:2). A hard disk is not necessary to run the test battery. However, since the full system is too large to be contained within a single 360 Kbyte diskette, a hard disk is recommended. Furthermore, a hardware lock, which is provided at cost by the system developers, is needed to run the system.

Software

The programs, which are available in compiled form on diskette, are written in TURBO Pascal. The system includes a master program referring to a number of different test programs, which can be combined to any preferred sequence. This sequence may in turn be repeated any number of times, according to the design of the investigation at hand.

At present, the system consists of the following 14 performance tests, in addition to scales for the measurement of reported mood, symptoms (two scales), and self-rated effort. The tests are simple, choice and complex reaction time (four tests), symbol digit, search and memory test, digit span, logical reasoning, additions, finger tapping (two tests), vocabulary, digit classification and digit addition. A short description of each test is given below, and a more detailed description of the SPES system and a manual for its use is given by Gamberale *et al.* (1989b).

Description of the performance tests

Simple reaction time SPES1

A 6-min sustained attention task measuring response speed to an easily discriminated but temporally uncertain visual signal. The task is to press a key on the keyboard as quickly as possible when a red square is presented on the display.

Choice reaction time SPES2

A four-choice RT task similar to SPES1 with the addition of response selection requirements. The stimulus is a cross, one arm of which is always shorter. The task is to indicate on one of four keys, placed in analogy to the arms of the cross, which arm is the shorter.

Colour word vigilance SPES3:1

A task of vigilance type in the sense that a response is required only to a minority of the signals. The Swedish word for 'red', 'yellow', 'white', or 'blue' (all three-letter words) is presented on the screen. The text can be written in any one of the colours. The task is to press a key as rapidly as possible when there is congruency between the meaning of the word and the colour of the text.

Colour word stress SPES3:2

A version of SPES3 which is constructed to provoke false alarms, and thus primarily measures the ability to inhibit such responses. The stimuli are the same, but the interval between subsequent stimuli is decreased and the proportion of critical stimuli is increased.

Search and memory SPES4

Measures the speed of comparing stimuli shown on the screen with a set of stimuli retained in memory. One, two or three letters are presented on the screen. A row of 30 letters is then presented and the task is to search this row as quickly as possible for the occurrence of any of the critical letters, and each appearance is indicated by the pressing of a key.

Symbol digit SPES5

A revised version of a traditional test of perceptual speed. In one row a key to this coding task is given by the pairing of symbols with the randomly arranged digits 1 to 9, and in a second row the subject is required to key in the digit corresponding to each symbol.

Digit span SPES6

A traditional test of short-term memory capacity. Series of digits are presented on the screen and the subject's task is to reproduce them on the keyboard. Depending on the correctness of the answer, the length of the following series is either increased or decreased.

Additions SPES7

Measures the speed of simple mental arithmetic operations. The task is to add three digits as quickly as possible and to indicate the sum on the keyboard.

Digit classification SPES8

A continuous choice reaction time task. Digits ranging from 1 to 8 are presented one at a time on the screen. The task is to determine whether the digit presented is odd or even.

Digit addition SPES9

A version of SPES8 requiring more complex processing of the signals. The task is to add the digit currently presented to the previous digit and to determine whether this sum is odd or even.

Verbal reasoning SPES10

Measures the speed and accuracy of verbal reasoning. Sentences of varying syntactic complexity are presented on the screen. Each sentence describes an order of presentation of the letters A and B, and it is followed by a combination of these letters. The task is to indicate with one of two keys whether the sentence gives a correct description of the relation between the letters A and B in the example following each sentence at hand.

Vocabulary SPES11

A traditional test of verbal understanding. The task is to indicate among five alternatives which is the synonym of a key word.

Finger tapping speed SPES12:1

Measures the maximum rate of repetitive movement. The task is to tap as rapidly as possible on a key at the keyboard with the index finger.

Finger tapping endurance SPES12:2

A version of SPES12 in which the change in tapping rate over time is assessed. The task is to tap as rapidly as possible with the index finger on an external key during 1 min.

Description of the self-rating scales

Self-rating of effort SPES30

Within the system it is possible to let the subject rate his effort or performance directly after each test. In the standard version, the subject is asked to rate his actual effort in percent of his maximum effort. The question could, however, easily be rephrased.

Self-rating of mood SPES31

The scale consists of 12 mood descriptive adjectives coupled to a six-category response scale. The response categories have verbal labels ranging from 'not at all' to 'very much'. A score is computed in two mood dimensions identified by factor analysis, an Arousal and a Stress dimension.

Acute symptoms SPES32

This program provides a general routine for presentation of a maximum of 50 questionnaire items. There is also a possibility to define the number of steps in the rating scale, as well as the verbal labels for each step. The version of the scale used in toxicological research contains 17 items regarding symptoms of local irritation as well as symptoms from the CNS, coupled to a six-category response scale, where the intensity of each symptom is rated.

Long-term symptoms SPES33

The questionnaire program provides a general routine for presentation of a maximum of 50 items, and also provides a possibility to define the number of steps in the rating scale, as well as the verbal labels for each step. The version of the scale used in toxicological research contains 38 items regarding a wide variety of symptoms, such as vegetative symptoms, concentration deficits, fatigue, tiredness, dizziness and symptoms of peripheral neuropathy. The frequency of occurrence of each symptom during the past 6 months is rated on a four-point scale.

Empirical data and applications

Table 27.2 lists the investigations using SPES tests that have been performed so far. These include experimental studies in the laboratory, and occupational studies of effects from acute or long-term exposures to various agents. In addition to these some studies have directly aimed at a methodological evaluation of the tests (Söderman *et al.*, 1982; Kjellberg and Visung, 1987; Kjellberg and Iwanowski, 1989). In the study by Kjellberg and Visung (1987), the standardization study, a group of 100 subjects went through SPES 1, 2, 3:1, 4, 5, 6, 7, 10 and 12:1. For 59 subjects the testing was repeated 4–5 months later. In another study data from previous field and epidemiologic studies using the simple RT task were re-analyzed for standardization purposes (Söderman *et al.*, 1982). The sample consisted of 730 industrial workers, 306 of whom were exposed to solvents in their work. For 83 of the non-exposed and 149 of the exposed workers measurements had been made twice, before as well as after a workshift. Finally, the development of a Mood Adjective Checklist is reported by Kjellberg and Iwanowski (1989).

Table 27.2 List of the studies where at least one of the SPES tests or scales have been used.

Reference	Environmental agent	Reference	Environmental agent
EXPERIMENTAL STUDIES		OCCUPATIONAL STUDIES	
Lisper and Kjellberg, 1972	Sleep deprivation	Gamberale and Svensson, 1974	Anaesthethic gases
Gamberale and Hultengren, 1972	Organic solvents	Kjellberg and Strandberg, 1979	Anaesthethic gases
Gamberale and Hultengren, 1973	Organic solvents	Gamberale *et al.*, 1976	Organic solvents
Gamberale and Hultengren, 1974	Organic solvents	Kjellberg *et al.*, 1979	Organic solvents
Gamberale *et al.*, 1975a	Organic solvents	Anshelm Olson *et al.*, 1981	Organic solvents
Gamberale *et al.*, 1975b	Organic solvents	Anshelm Olson, 1982	Organic solvents
Gamberale *et al.*, 1976	Organic solvents	Gamberale *et al.*, 1989a	Electric & magnetic fields
Gamberale *et al.*, 1978	Organic solvents		
Anshelm Olson *et al.*, 1985	Organic solvents	Knave *et al.*, 1978	Organic solvents
Iregren, *et al.*, 1986	Organic solvents	Knave *et al.*, 1979	Electric & magnetic fields
Iregren 1986b	Organic solvents		
Enander, 1987 (exp 1)	Cold exposure	Elofsson *et al.*, 1980	Organic solvents
Enander, 1987 (exp 2)	Cold exposure	Iregren, 1982	Organic solvents
Kjellberg and Wide, 1988	Noise exposure	Iregren, 1989	Manganese
Gamberale *et al.*, 1988	Heat & cold		
Wigaeus-Hjelm *et al.*, 1989	Organic solvents		

Table 27.3 Means and standard deviations for the tests in the two test sessions (n = 59). Test–retest and α coefficients (KR20 for dichotomized variables) (n = 59 and 100, respectively). Data from Kjellberg and Visung (1987).

	Session 1		**Session 2**					
Test	Mean	S.D.	Mean	S.D.	*t*	*p*	r_{tt}	*alpha*
1. Simple RT								
Mean (ms)	257·8	42·0	258·8	37·3	0·34	>0·10	0·85	0·99
S.D.	54·4	19·7	54·6	18·9	0·90	>0·10	0·63	
Regression	0·22	0·21	0·15	0·19	2·45	0·012	0·44	
2. Choice RT								
RT mean (ms)	644·2	109·9	635·6	116·8	1·17	>0·10	0·88	0·99
Errors	3·2	5·7	3·1	4·5	0·14	>0·10	−0·02	0·87
3. Colour word vigilance								
RT hits (ms)	535·2	53·4	539·8	57·6	1·39	>0·10	0·90	0·96
No. Correct	41·1	8·0	41·8	8·6	0·91	>0·10	0·74	0·91
4. Search and memory								
RT 1 letter (s)	9·2	2·6	9·0	2·4	1·13	>0·10	0·75	0·94
RT 2 letter (s)	14·6	3·9	13·3	3·5	3·40	0·001	0·71	0·95
5. Symbol digit								
RT 5–10 (s)	25·3	6·5	28·2	8·8	3·74	<0·001	0·74	0·95
No. Errors	0·7	1·2	0·5	1·1	1·15	>0·10	0·15	0·58
6. Digit span								
Median span	7·1	1·3	7·4	1·5	1·81	0·076	0·70	
7. Reasoning								
RT mean (s)	4·6	1·4	4·6	1·5	0·61	>0·10	0·76	0·96
No. Errors	13·5	8·5	11·0	8·5	2·97	0·004	0·70	0·91
8. Additions								
RT mean (s)	2·0	0·9	2·1	0·9	1·66	>0·10	0·94	0·98
No Errors	3·4	2·6	3·2	2·8	0·64	>0·10	0·60	0·87
9. Finger tapping								
Frequency Dominant	66·5	8·8	64·9	10·6	2·05	0·045	0·82	0·91
Frequency Non-dominant	58·5	9·0	57·7	10·7	0·90	>0·10	0·74	0·85

The symptoms rating scales, SPES32 and SPES33, were used in a large study of spray-painters (Iregren, 1986b) in which paper and pencil versions of the scales were distributed to more than 200 painters. A factor analysis of the responses to this questionnaire yielded three factors, two of which were related to solvent exposure. These scales have also been used in a study of manganese exposed foundry workers (Iregren, 1989).

Results

Table 27.3 shows means and standard deviations over repeated measurements of the different performance indices for each of the nine tests included in the standardization study.

Test–retest reliability coefficients and α coefficients for the tests are also given in Table 27.3. Mean response times in Choice RT, Colour word vigilance, Additions and Simple RT all proved to be highly reliable. The test–retest coefficients of Search and memory, Symbol digit, Digit span and reasoning were all below 0·80. As expected, response times were found to be more reliable than error rates.

Individual differences

The only significant difference found between males and females in the standardization study was that women performed somewhat slower in the Tapping test with the dominant hand.

Longer response times have been obtained among older subjects (45–65 years) for Simple RT, Choice RT and Symbol digit, Colour word vigilance and Tapping (dominant hand). Since the error rates were very similar in the different age groups the prolonged response times probably did not just reflect a more cautious strategy in the older group.

No differences between educational groups were obtained in Simple RT, Colour word vigilance, Choice RT to Tapping. In the other tests response times were shortened and/or errors rates lowered in the higher education groups.

No performance differences were obtained between groups who were more or less experienced with work on computers or between subjects who had or had not participated in experiments using computerized tests.

Fatigue and learning effects

Training effects and other effects of increased experience with the tests have been analyzed both within the tests and between the repeated testings.

In most tests no significant within test changes occur after the prescribed training period. However, in Additions and the Verbal reasoning test the performance normally is slightly better in the second half of the test. The opposite is true of Simple RT, where response times generally are gradually prolonged during the test period.

As shown in Table 27.3, performance improved between the two sessions in Search and memory (two letters), and Reasoning (number of errors). The performance decrement in Simple RT was also less in the second session.

Time of day

In Simple RT afternoon performance generally has been found to be slightly better than morning performance in non-exposed groups, although not always significantly so (Gamberale and Svensson, 1974; Gamberale *et al.*, 1976, 1989a; Anshelm Olson, 1982; Kjellberg and Visung, 1987).

Sensitivity to effects of environmental agents

This list of studies provided in Table 27.2 indicates the environmental agent investigated. In the majority of these studies at least one test proved to be sensitive to effects from the agent under study. Response times in, for example, the simple reaction time task have also been found to be associated with the effects of long term exposure to neurotoxic compounds. For more detailed results the reader is referred to the original publications.

Mood and symptoms

The present version of the mood adjective checklist has only been used in one experimental study which primarily dealt with the effects of temperature on performance (Gamberale *et al.*, 1988), where rated stress increased when the room temperature was 5°C, but was unaffected at higher temperatures. Rated arousal decreased in all conditions, most markedly when the room temperature was 40°C. The symptoms scales have been found useful in experimental as well as epidemiological applications, e.g. in studies of spray painters (Iregren, 1986b) and of manganese-exposed foundry workers (Iregren, 1989).

Possible future development of the SPES

The increasing technical competence of computers will certainly broaden the range of mental abilities that can be tested. However, due to the laborious procedure of test development, well standardized and validated new tests are still a few years off. A test of tracking performance has been developed, but it is presently not included in the battery. The SPES-battery is thus undergoing continuous change, and the intention of the developers is especially to improve the evaluation side of the battery. This is to be done by introducing programs for generating performance reports from the SPES, and hopefully, in the long run, to include a data base containing norm data for the various tests.

However, the immense variety of performance measures in use is probably the single factor which at present has the greatest effect on the rate of growth of knowledge about detrimental effects from workplace exposures. The possibilities for making comparisons across studies performed at separate institutions and in different countries are of major importance, and initiatives to facilitate the standardization of computerized test protocols have been taken within the European Community (EC). One significant problem in this process is the slightly different primary uses of various tests and test systems, since the intended use of a test naturally affects the way in which it is implemented. However, the development of standardized protocols is now in progress, and efforts to accomplish this task have been made at our laboratory as well as elsewhere.

References

American Psychological Association, 1986, *Guide-lines for Computerbased Tests and Interpretations* (Washington, DC: APA).

Anshelm Olson, B., 1982, Effects of organic solvents on behavioral performance of workers in the paint industry. *Neurobehavioural, Toxicology and Teratology*, **4**, 703–708.

Anshelm Olson, B., Gamberale, F. and Grönqvist, B., 1981, Reaction time changes among steel workers exposed to solvent vapor. A longitudinal study. *International Archives of Occupational and Environmental Health*, **48**, 211–218.

Anshelm Olson, B., Gamberale, F. and Iregren, A., 1985, Coexposure to toluene and *p*-xylene in man: central nervous functions. *British Journal of Industrial Medicine*, **42**, 117–122.

Bartram, D. and Bayliss, R., 1984, Automated testing: Past, present and future. *Journal of Occupational Psychology*, **57**, 221–237.

Bartram, D., Beaumont, J. G., Cornford, P., Dann, P. L. and Wilson, S., 1987, Recommendations for the design of software for computer based assessment – Summary statement. *Bulletin for The British Psychological Society*, **40**, 86–87.

Beaumont, J. G., 1982, System requirements for interactive testing. *International Journal of Man–Machine Studies*, **17**, 311–320.

Elofsson, S. A., Gamberale, F., Hindmarsh, T., Iregren, A., Isaksson, A., Johnsson, I., Knave, B., Lydahl, E., Mindus, P., Persson, H. E., Philipson, B., Steby, M., Struwe, G., Söderman, E., Wennberg, A. and Widén, L., 1980, Exposure to organic solvents: A cross-sectional epidemiologic investigation on occupationally exposed car and industrial spray painters with special reference to the nervous system. *Scandinavian Journal of Work Environment and Health*, **6**, 239–273.

Enander, A., 1987, Effects of moderate cold on performance of psycho-motor and cognitive tasks. *Ergonomics*, **30**, 1431–1445.

French, C. C. and Beaumont, J. G., 1987, The reaction of psychiatric patients to computerized assessment. *British Journal of Clinical Psychology*, **26**, 267–278.

Gamberale, F., 1985, The use of behavioral performance tests in the assessment of solvent toxicity. *Scandinavian Journal of Work Environment and Health*, **11**(suppl. 1), 65–74.

Gamberale, F. and Hultengren, M., 1972, Toluene exposure II. Psychophysiological functions. *Work Environment Health*, **9**, 131–139.

Gamberale, F. and Hultengren, M., 1973, Methylchloroform exposure II. Psychophysiological functions. *Work Environment Health*, **10**, 82–92.

Gamberale, F. and Hultengren, M., 1974, Exposure to styrene II. Psychological functions. *Work Environment Health*, **11**, 86–93.

Gamberale, F. and Svensson, G., 1974, The effect of anaesthetic gases on the psychomotor and perceptual functions of anaesthetic nurses. *Work Environment Health*, **11**, 108–111.

Gamberale, F., Annwall, G. and Anshelm Olson, B., 1976, Exposure to trichloro-ethylene III. Psychological functions. *Scandinavian Journal of Work Environment and Health*, **4**, 220–224.

Gamberale, F., Annwall, G. and Hultengren, M., 1975a, Exposure to methylene chloride II. Psychological functions. *Scandinavian Journal of Work Environment and Health*, **2**, 95–103.

Gamberale, F., Annwall, G. and Hultengren, M., 1975b, Exposure to white spirit. II. Psychological functions. *Scandinavian Journal of Work Environment and Health*, **1**, 31–39.

Gamberale, F., Annwall, G. and Hultengren, M., 1978, Exposure to xylene and ethylbenzene III. Effects on central nervous functions. *Scandinavian Journal of Work Environment and Health*, **4**, 204–211.

Gamberale, F., Anshelm Olson, B., Eneroth, P., Lind, T. and Wennberg, A., 1989a, Acute effects of ELF electromagnetic fields. A field study on linemen working at 400 kV. *British Journal of Industrial Medicine*, **46**, 729–737.

Gamberale, F., Iregren, A. and Kjellberg, A., 1989b, SPES: The computerized Swedish Performance Evaluation System. Background, critical issues, empirical data, and a user's manual. *Arbete och Hälsa. National Institute of Occupational Health, Solna, Sweden*, **6**, 1–77.

Gamberale, F., Kjellberg, A. and Razmjou, S., 1988, *Effects of Moderate Cold and Heat on Performance of Sensory-motor and Cognitive Tasks*. Report No. 692 (University of Stockholm: Department of Psychology).

Gamberale, F., Lisper, H. O. and Anshelm Olson, B., 1976, The effect of styrene vapour on the reaction time of workers in the plastic boat industry. In: *Adverse Effects of Environmental Chemicals and Psychotropic Drugs*, edited by M. Horvath (Amsterdam and New York: Elsevier), pp. 135–148.

Hockey, G. R. J., 1986, Changes in operator efficiency as a function of environmental stress, fatigue and circadian rhythms. In: *Handbook of Perception and Human Performance*, Vol. II, edited by K. R. Boff, L. Kaufman and J. P. Thomas (Chichester: John Wiley).

Iregren, A., 1982, Effects on psychological test performance of workers exposed to a single solvent (toluene) – A comparison with effects of exposure to a mixture of organic solvents. *Neurobehavioural, Toxicology and Teratology*, **4**, 695–701.

Iregren, A., 1986a, Effects of industrial solvent interactions. Studies of behavioral effects in man. *Arbete och Hälsa. National Board of Occupational Safety and Health, Solna, Sweden*, **11**, 1–60.

Iregren, A., 1986b, Subjective and objective signs of organic solvent toxicity among occupationally exposed workers. An experimental evaluation. *Scandinavian Journal of Work Environment and Health*, **12**, 469–475.

Iregren, A., 1989, Psychological test performance among foundry workers exposed to low levels of manganese. *Neurotoxicology and Teratology* (in press).

Iregren, A., Åkerstedt, T., Anshelm Olson, B. and Gamberale, F., 1986, Experimental exposure to toluene in combination with ethanol intake. Psychophysiological functions. *Scandinavian Journal of Work Environment and Health*, **12**, 128–136.

Kjellberg, A. and Iwanowski, S., 1989, The Stress/Energy Questionnaire. Development of a scale for rating mood at work. Report 1989:2, National Institute of Occupational Health, (in Swedish).

Kjellberg, A. and Strandberg, M., 1979, *The Effects of Anaesthetic Gases on Reaction Time of Anaesthetic Nurses*. Report No. 11. (Sweden: National Board of Occupational Safety and Health) (in Swedish).

Kjellberg, A. and Visung, H., 1987, *Some Metrical Properties in a Computer Administered Test Battery for Use in Behavioral Toxicology*. (Solna, Sweden: Report No. 1. National Board of Occupational Safety and Health) (in Swedish).

Kjellberg, A. and Wide, P., 1988, Effects of simulated ventilation noise on performance of a grammatical reasoning task. In: *Proceedings of the 5th International Congress on Noise as a Public Health Problem*, Vol. 3, edited by B. Berglund, U. Berglund and T. Lindvall (Stockholm: Swedish Council for Building Research), pp. 31–36.

Kjellberg, A., Wigaeus, E., Engström, J., Åstrand, I. and Ljungquist, E., 1979, Long-term effects of exposure to styrene in a polyester plant. *Arbete och Hälsa, National Board of Occupational Safety and Health, Solna, Sweden*, **18**, 1–25 (in Swedish).

Knave, B., Anshelm Olson, B., Elofsson, S., Gamberale, F., Isaksson, A., Mindus, P., Persson, H. E., Struwe, G., Wennberg, A. and Westerholm, P., 1978, Long term exposure to jet fuel. A cross sectional epidemiologic investigation on occupationally exposed industrial workers with special reference to the nervous system. *Scandinavian Journal of Work Environment and Health*, **4**, 19–45.

Knave, B., Gamberale, F., Bergström, S., Birke, E., Iregren, A., Kolmodin-Hedman, B. and Wennberg, A., 1979, Long-term exposure to electric fields. A cross-sectional epidemiologic investigation of occupationally exposed workers in high-voltage substations. *Scandinavian Journal of Work Environment and Health*, **2**, 115–125.

Lisper, H. O. and Kjellberg, A., 1972, Effects of 24-hour sleep deprivation on rate of decrement in a 10-minute auditory reaction time task. *Journal of Experimental Psychology*, **96**, 287–290.

Lukin, M. E., Dowd, E. T., Plake, B. S. and Kraft, R. G., 1985, Comparing computerized versus traditional psychological assessment. *Computers in Human Behavior*, **1**, 49–58.

Matarazzo, J. D., 1983, Computerized psychological testing. *Science*, **221**, 323.

McArthur, D. L. and Choppin, B. H., 1984, Computerized diagnostic testing. *Journal of Educational Measurement*, **31**, 391–397.

Söderman, E., Kjellberg, A., Anshelm Olson, B. and Iregren, A., 1982, *Standardization of a Simple Reaction Time Test for Use in Behavioral Toxicology*, Report No. 27. (Solna, Sweden, National Board of Occupational Safety and Health) (in Swedish).

Space, L. G., 1981, The computer as psychometrician. *Behavior Research Methods Instrumentation*, **13**, 595–606.

Thompson, J. A. and Wilson, S. L., 1982, Automated psychological testing. *International Journal of Man–Machine Studies*, **17**, 279–289.

Weiss, D. J. and Vale, C. D., 1987, Adaptive testing, *Applied Psychology*, **36**, 249–262.

Wigaeus-Hjelm, E., Hagberg, M., Iregren, A. and Löf, A., 1990, Exposure to methyl isobutyl ketone. Toxicokinetics and occurrence of irritative and CNS symptoms. *International Archives of Occupational and Environmental Health*, **62**, 19–26.

Chapter 28
Computer-aided decision support for the battlefield of the 21st Century

PHILIP H. WARREN, HEATHER L. DRAGSBAEK[1] and LAWRENCE E. SYMINGTON

U.S Army Natick Research,
Development and Engineering Center,
Natick, MA 01760-5020,
U.S.A.

[1]*GEO-Centers, Inc., Newton Center, MA 02159, U.S.A.*

Introduction

Predicting the outcome of a mission in the modern battlefield is complicated by the difficulty in predicting the performance of individual soldiers. There are many models for predicting human performance in limited conditions of temperature, load, terrain, etc. (Aghazadeh and Ayoub, 1985; Giambra and Quilter, 1987; Genaidy *et al.*, 1988). However, there are few mathematical models that provide good predictions of human performance under a wide variety of conditions (Pandolf *et al.*, 1986; Haslam and Parsons, 1987). These models can be used to assist military commanders as they make decisions to establish battlefield objectives and to allocate resources. With the use of computers, and especially the techniques of artificial intelligence (AI), there is now the potential to automate these models and make them available to more users under numerous conditions.

The combination of a broadly based model of human performance and the technology of expert systems can be used to solve important problems which have resisted solution by more traditional means. Such a problem is that of the soldier's load. Historically, infantry soldiers have marched into combat carrying loads that are too heavy (Marshall, 1950). This load, combined with high heat, humidity, or a requirement to move quickly contributes to a reduction in combat effectiveness of the soldier, increases the chances that he will become a casualty, and, ultimately, jeopardizes his mission. Indeed, with the increase in sophistication of man's ability to conduct warfare, he has gradually increased the load a soldier must bear in combat.

Research in human factors engineering, physiology and biomechanics has yielded limits for the maximum load a soldier can carry and the maximum amount of work he can perform. Beyond these limits, the soldier is at risk of becoming a casualty prior to engaging the enemy and, therefore, of failing to complete his mission. As a result, the Army has adopted the doctrine that a soldier's marching/approach load should not exceed 45% of his body weight, and his combat load should not exceed 30% of his body weight (Military Standard 1472C).

For the typical soldier of 160 lbs (72 kg), these percentages are equivalent to 72 lbs (32·4 kg) and 48 lbs (21·6 kg), respectively. Unfortunately, soldiers often carry a great deal more than this; in some cases, weights exceed 150 lbs (67·5 kg). For example, the total load for the assistant Dragon gunner, in a temperate climate, is 167 lbs (75·2 kg). For the assistant machine gunner, it is 152 lbs (68·4 kg). In fact, the total load for each position in a rifle platoon exceeds 110 lbs (49·5 kg) (Sampson, 1988). If loads are tailored, that is, if decisions are made about what to carry and what not to carry, load weights can be reduced significantly, but they still exceed the doctrinal limits by substantial amounts. Therefore, soldiers are still at risk of becoming casualties.

Traditional approach

The traditional approach to this persistent problem has been to develop lighter materials for clothing, ballistic protection, food, weapons, communication equipment, etc. There have been remarkable achievements in this arena, and goals for future weight reductions range from 12 to 50% (Sampson, 1988). While making lighter equipment will remain critical, doing so does not solve the soldier load problem. The reason is that commanders make decisions about what to carry. The result is that for every reduction in weight there is a concomitant increase in capability for fire power, communication, and detection of the enemy, each one of which adds weight to the soldier's load (Nordeen, 1986). Thus, real-life decisions about what to carry involve complex cost–benefit analyses, often based on minimal hard information and when there is little time to consider alternatives and their consequences. Under these conditions, it is easy to understand how the decision-maker would tend to err on the side of carrying too much with him: 'We don't know what's ahead, so let's make sure we are prepared for every contingency' (see Dubik and Fullerton, 1987).

New approach – decision support

What is needed is a new approach to solving the soldier load problem. Decision support technology is such an approach. It provides a means for assisting commanders to make decisions about tailoring soldier loads based on mission information. Decision support technology takes the form of expert systems which are computer programs that process information in ways similar to the ways human experts do. Expert systems have been developed for use in a wide variety of situations, such as diagnosis of medical conditions (Barnett, 1987), location of mineral deposits (Campbell *et al*, 1982), determination of the structure of organic compounds (Buchanan and Feigenbaum, 1978), and others (Gevarter, 1985; Lechner and Barth, 1985; Siegel and Warren, 1987). They use many elements of information to make deductions, inferences and recommendations. A significant advantage of these systems is that deductions and inferences are made without being degraded by most of the kinds of stressors to which humans are susceptible. The deductions are derived from hard rules that constrain the output in ways similar to the way a database program does. In contrast, inferences are made using rules, or heuristics, which are true only some of the time, e.g. boys like to play baseball (that may be true only 85% of the time). On the basis of these deductions and inferences, the expert system makes recommendations which aid the user in making many decisions rapidly.

A soldier Load Expert System (LES) is under development at the US Army Natick Research, Development and Engineering Center (Natick). LES will accept input about the soldier and his mission and provide output in the form of recommendations about what to carry and possible changes to the mission scenario. The program will also provide dialogue and 'what if' facilities. In the dialogue mode, the user can respond to recommendations or questions from the program. This facility might lead the user/decision-maker to change his mind about what to carry or some other aspect of performing his mission. This leads directly to the 'what if' capability. Here the decision-maker can change his input to determine if it improves the assessment of the mission by the expert system.

The initial users of LES are projected to be company and field grade commanders in training at the Infantry School, Fort Benning, GA. In their current training they have a block of instruction on load configuration and the importance of not overloading the soldier. With LES we can go beyond telling the commander, 'The decisions you make about load can mean the success or failure of a mission'. LES will allow the commander to receive feedback on his decisions faster than he would under actual field conditions. He can then modify parts of the input and test the effects. Thus, he can learn by following the consequences of his decisions.

Development of the system

After defining the problem clearly and establishing who the users would be, the next step in the development process was to gather the expertise to be used in LES.

Knowledge engineering

In this phase the knowledge or expertise that can be used to solve the problem must be identified, collected and organized into a form that a programmer can put into code. Two sources of knowledge have been identified for LES. The first is a body of scientific information related to the heat stress model developed at the US Army Research Institute for Environmental Medicine (USARIEM). The details of this model have been published in the scientific literature over the last two decades (Givoni and Goldman, 1972; Berlin *et al.*, 1975; Soule *et al.*, 1978; Pandolf *et al.*, 1986). An important reason for choosing this model is that it was developed and validated largely using soldiers as subjects. The model provides excellent predictions of rectal temperature of a soldier under a wide variety of load, heat, humidity, acclimatization and other mission related conditions. This information can then be used to determine how long a soldier can work at a given rate, compute work/rest cycles, determine water requirements and the chances of the soldier becoming a heat casualty. While other models of predicting casualties are possible, and some are available (Haslam and Parsons, 1987), we believe the USARIEM model provides the most robust and accurate predictions over the widest range of conditions the soldier is likely to encounter.

The second phase of knowledge engineering for this project is capturing the knowledge of commanders who have experience making decisions about load configuration. This process requires a different approach than that used above. In this case commanders are interviewed about how they make load configuration decisions. They are provided with mission scenarios with which to work, and, with their help, test cases are developed to evaluate the expert system during development. One of the most difficult challenges of this phase is converting the expertise of the commanders into clear rules that can be coded into the expert system.

Implementation

Software

The present prototype of the system uses two computer languages; Turbo-C and Turbo-

Prolog (both products of Borland International, Inc.). C is used for the mathematically intense portions of the program in which several equations are used to calculate heat exchange, rectal temperature, water requirements, etc. Prolog is used for all other portions of the system. Prolog is specifically designed for use in artificial intelligence (AI) applications. It is very good for supporting user/system dialogue and making inferences based on a set of rules. Turbo-Prolog is particularly strong in supporting screen control and comes with many useful tools for software development. The program modules developed in these two languages can be compiled thus allowing them to be run on any MS-DOS compatible computer.

Hardware

Our objective is to develop a system that is accessible to the widest possible range of users. Therefore, the minimum hardware requirements to support this system are a PC which runs MS-DOS, has 640 Kbyte of core memory, and a hard disk. Because of the amount of processing that occurs a fast processor is recommended. Similarly, a colour monitor is not required, but is recommended.

How the system works

The present version of LES is a working prototype. It will solve the soldier load problem, but only within a limited domain. LES will accept user input in several categories and provide output in the form of summary information and recommendations to be used by the decision-maker. In order to illustrate the functions of LES we will step through the program using a simple scenario in which a rifleman has a mission of marching 20 miles in 10 h.

Figure 28.1 shows the first screen in LES. The main menu is shown across the top. It provides for entering information about the soldier, the load he will carry and various mission related parameters about which the user has some information and/or control. The Advisor option is used to determine if the mission can be completed. The Reports option will provide for hard copies of the transactions between the user and LES for future reference. The Quit option allows the user to clear all values and automatically reinstate the default values

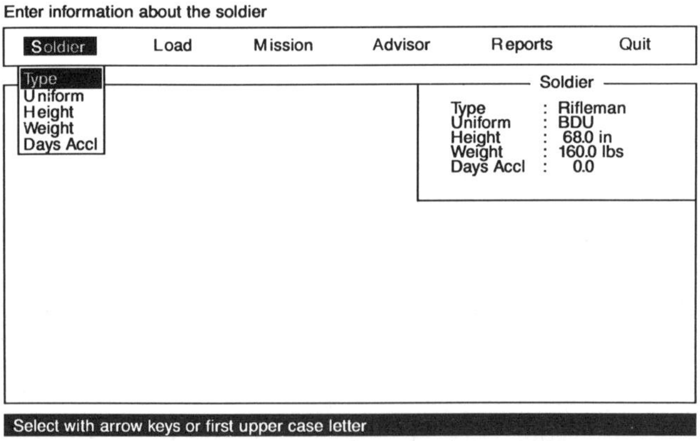

Figure 28.1 The 'Soldier' option and default values on the main menu of the LES program.

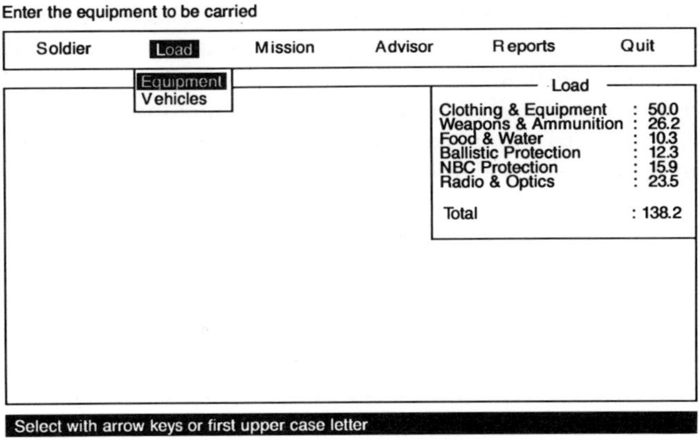

Figure 28.2 The 'Load' option.

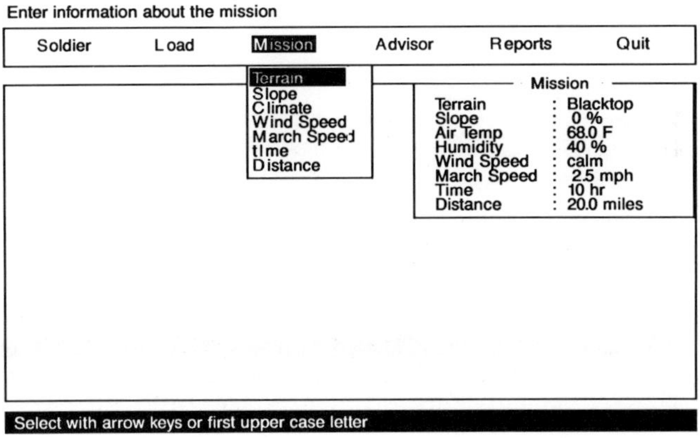

Figure 28.3 The 'Mission' option.

for all parameters available. It also allows the user to exit the program. By pressing RETURN when the Soldier option is highlighted, a popdown menu appears which allows the user to make specific entries for the parameters shown. On the right side of the screen a status window shows the default values for the indicated parameters. The present prototype of LES works only for a rifleman in battle dress uniform (BDU). However, height, weight and days of acclimation to the heat can be changed.

The Load option (Figure 28.2) allows the user to alter information about equipment. By pressing RETURN when the Equipment option is highlighted, a menu appears which allows the user to change how much equipment, from the indicated categories, the soldier will carry with him. The categories of loads are shown on the right with their default subtotal weights. Two statements can be made about the total weight of 138 lbs. First, it is too heavy for optimal performance for the soldier. Second, the weight is realistic. Soldiers often do carry loads to the field that are well above half their bodyweights. When we work through this scenario, we will make several decisions about which particular items to carry and which to leave behind in order to lighten the load. In the completed version of the system the Vehicles option will permit the user to include vehicles as a means for moving some of the soldier's load to his destination.

The Mission option (Figure 28.3) permits the user to change the values of a wide variety of mission and environmental parameters. The default values, shown at the right, define a 20 mile mission to be completed within 10 h under very comfortable environmental and terrain conditions.

When the Advisor option is activated LES determines the likelihood that the soldier will be able to complete his mission using the USARIEM heat stress model along with several constraints and rules built into the system that apply to the particular scenario in question. The result of the processing is presented in Figure 28.4. For the present scenario LES has determined that it is unlikely that the rifleman will be able to complete a march of 20 miles. LES's recommendation is to consult the Advisor for assistance in reconfiguring the rifleman's load and other variables in order to achieve success.

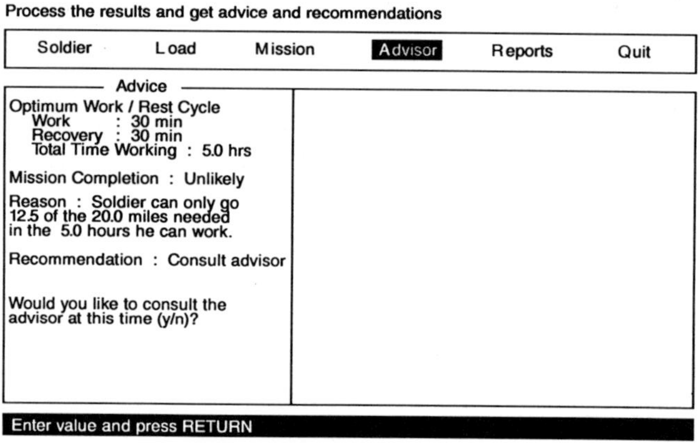

Figure 28.4 The result of activating the 'Advisor' option.

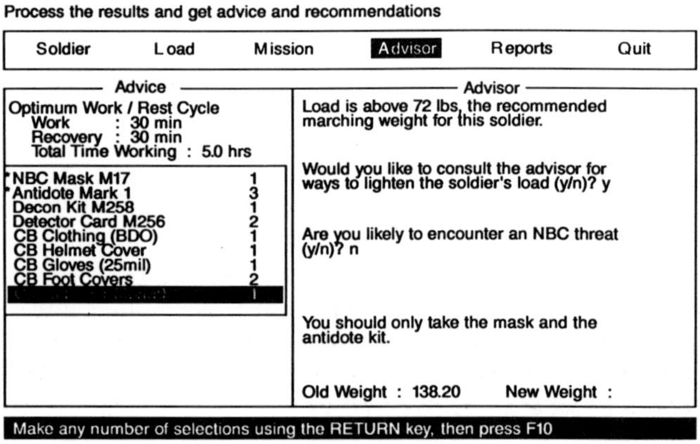

Figure 28.5 The initial dialogue screen for example scenario.

By answering 'y' to the question 'Would you like to consult the advisor at this time (y/n)?' the user is engaged in a dialogue with LES, as shown in Figure 28.5. (In order to save space here multiple questions and answers are presented in a single figure.) LES knows that 72 lbs is 45% of this rifleman's weight and that his load is 138·2 lbs. Therefore, LES finds ways to reduce the rifleman's load. In this scenario, we tell LES that we are not likely to encounter an NBC threat. LES then makes a recommendation and automatically presents the menu which allows the user to carry out the recommendation, which is to delete all but the M17 Chemical Protective Mask and the antidote kit (see Figure 28.5, carried items are asterisked).

Figure 28.6 shows that the result of doing this is to reduce the load to 126·8 lbs; which is still too heavy. LES knows that the load is still too heavy and pursues another route to reduce it. We are automatically put into a menu system which allows us to delete (or add) various items of equipment as appropriate. LES keeps us informed of the running total weight for the load as decisions are made about what the rifleman will carry on this mission.

Once we achieve the desired load, in this case 72 lbs or less, LES proceeds to the next parameter for adjustment. Marching speed is a significant determinant of heat stress. Therefore, LES asks if more time is available for the mission and if it is possible to reduce the marching speed.

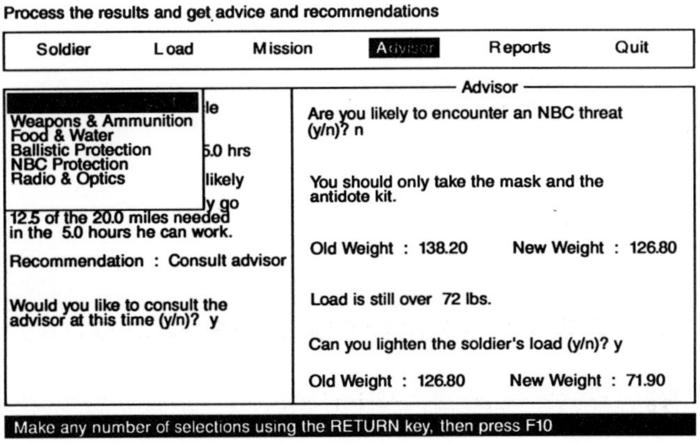

Figure 28.6 The second dialogue screen for example scenario.

In Figure 28.7 we show a change of average marching speed from 2·5 to 2·2 miles per hour. Once that information has been entered LES has enough information to recalculate the likelihood that the rifleman will complete his mission without becoming a heat stress casualty.

Figure 28.8 shows that by reducing the rifleman's load and allowing him to march at a slower pace it is likely that he will be able to reach his destination and complete his mission.

From this point it can be demonstrated that raising the temperature of the environment to 85°F and the humidity to 70%, somewhat more realistic conditions, that the rifleman can no longer complete his mission, even marching at 2·2 mph with a 72 lb load. However, by making the rifleman heat acclimated (see Figure 28.1), it is possible for him once again to complete the mission.

As the system develops, more functions will be provided. For example, future outputs will include recommendations about how much food and water the rifleman should consume in

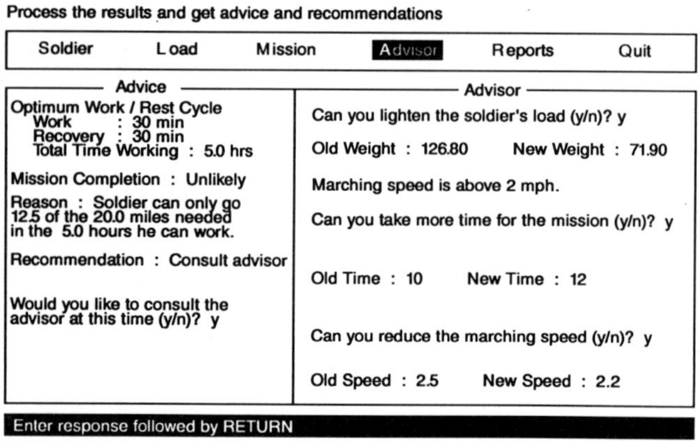

Figure 28.7 The last dialogue screen for example scenario.

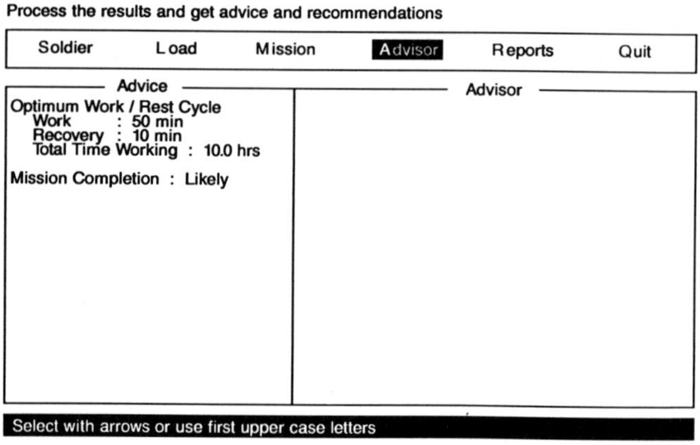

Figure 28.8 Final screen showing mission completion is likely.

order to meet his energy needs and to avoid dehydration. Finally, if the system detects a problem, for example not enough time has been allowed for covering the required distance, LES will automatically initiate a dialogue with the user and recommend that he consult the expert advisor portion of the system. The dialogue will begin with a statement of the problem and will include recommendations for solving it. For example a soldier might be configured with a load that weighs 105 lbs (51·8 kg) and be required to march 15 miles over uneven terrain in 5 h in an environment of high temperature and high humidity. The expert system would state that this mission would result in a significant risk of the rifleman becoming a heat casualty and would recommend either a lighter load (Can other means be found for transporting heavy equipment to the destination?), or a change in the conditions (Can another route be used? or Can the marching begin in the very early morning when it is cooler?).

The user will have available a 'what if' capability. Thus, he will be able to test certain hypotheses on the spot. 'What happens if I leave my night scope (3·5 lbs, 1.6 kg) and sleeping bag (7lbs, 3·2 kg) behind, and bring only my chemical protective mask, leaving behind the rest of my NBC gear (a saving of 9·3 lbs, 4·2 kg). LES would then recompute to determine if those decisions were helpful enough. Other alternatives can be considered until the user finds a set of conditions with which he is satisfied.

The most direct benefit of LES in the classroom is that it will show the commander in training the consequences of his decisions. He can define different mission scenarios and get feedback about whether there is sufficient benefit derived from the costs associated with not having the soldier carry selected items of equipment. He will learn the relative impact of dropping 10 lbs (4·5 kg) of weight versus slowing the rate of marching (if time permits) versus altering the route (this could be used to improve the terrain, but it might increase the required time at a given marching speed). Finally, by becoming better at solving problems related to the soldier's load, the commander will be better able to deal with both logistical and operational problems when LES is not available to assist him.

System evaluation

LES will go through several phases of evaluation. User-friendliness will be evaluated using two methods. The first method involves a new technology: automated analyses of screen layout. These analyses will be most useful in assuring a system that is optimally efficient from the viewpoint of the decision-maker and, we hope, increase the likelihood of acceptance of the system. Secondly, of course, the user must give us his feedback. This system, like all true expert systems, is a dynamic system, and therefore, can grow with the needs of the user. During the next phase of development we will test LES with decision-makers at Fort Benning and use the information we gain to improve the system.

Validation of the computations and recommendations of LES will depend on three processes. First, test cases will be developed which represent diverse but likely mission scenarios. These test cases will be developed with field experts and will be performed every time a change is made in LES. The second procedure involves continuous use of the system by instructors at the Infantry School as a normal part of the training process. This latter procedure will provide important feedback from several students and instructors who will test the system with their own mission scenarios. Finally, when the first version of LES is complete it should be tested in a 'shoot-off' where the success of actual field missions are compared using LES against new commanders. LES will be a success if it helps commanders get more soldiers to their destinations and if those soldiers are more capable of carrying out their missions.

Conclusion

Throughout history the infantry soldier has had to conduct combat missions while carrying extreme loads. Military commanders recognize the significant effects this can have on a soldier's performance in combat. The traditional solution, lighter materials, has not solved the problem. This paper describes a computer-aided decision support system which combines the USARIEM heat stress model, expert knowledge and expert system technology to produce the soldier load expert system (LES). The field commander can use LES to assist him in making decisions about tailoring loads. However, the system goes beyond simply tailoring the load. Mission information such as rate of marching, terrain footing, meteorological information, etc. are put into the system. This additional information allows for a much broader range of decision support. The LES advisor can suggest that specific items be excluded from the load, that alternative routes be considered, or that a march begin at a certain time during the day. The system is interactive and dynamic. That is, the advice given depends on the input provided and will change as the user revises his information base about the mission under consideration.

LES is designed initially for the commander in training. During training the commander will be presented with facts related to a combat mission as a classroom exercise. He will then use LES to define the mission, determine the outcomes of several alternative mission scenarios, and plan operational and logistical support. Through this process, the commander will get immediate feedback about the consequences of his decisions. As the commander works with the system, he becomes more expert at predicting the consequences of his decisions. Ideally, he will progress to the point where he can make expert decisions without the requirement for LES. Eventually, commanders may want to have LES with them in the field to aid them in real-life decision-making. Just as the commander in training will become better at making load configuration and other mission related decisions, LES will also grow. As we acquire more knowledge about factors that impact on the success of missions of the infantry soldier, we can expand the capabilities of LES. This gradual growth of expert systems is a natural process for AI applications. Instead of writing and rewriting new versions of old programs, as occurs with more conventional software like word processors, AI applications, which are heavily rule-based, can be added to in a gradual manner as new information becomes available. Also, individual rules can be changed as part of the refinement process. The other side of this feature of easy gradual improvement is that users get in the habit of making suggestions and expecting those changes to be implemented quickly.

The most recent models show an increasing capability for providing excellent estimates of human performance in an ever widening range of situations. Expert system technology is emerging as the most useful area in the AI arena. By combining the USARIEM heat stress model with the technology of expert systems we have produced a computer-aided decision support system which could be of significant benefit to both footsoldiers and their commanders.

References

Aghazadeh, F. and Ayoub, M.M., 1985, A comparison of dynamic- and static-strength models for prediction of lifting capacity. *Ergonomics*, **28**, 1409–1418.

Barnett, O., 1987, Director of Laboratory of Computer Science, Project DXplain, Massachussetts General Hospital.

Berlin, H.M., Stroschein, L. and Goldman, R.F., 1975, *A Computer Program to Predict Energy Cost, Rectal Temperature, and Heart Rate Response to Work, Clothing, and Environment*. ED-SP 75011 (Edgewood Arsenal, MD: Department of the Army).

Buchanan, B.G. and Feigenbaum, E.A., 1978, Dendral and meta-dendral. *Artificial Intelligence*, **11**, 5–24.

Campbell, A.N., Hollister, V.F., Duda, R.O. and Hart. P., 1982, Recognition of a hidden mineral deposit by an artificial intelligence program. *Science*, **217**, 927–929.

Dubik, J.M. and Fullerton, T.D., 1987, Soldier overloading in Grenada. *Military Review*, January, 38–47.

Genaidy, A.M., Asfour, S.S., Mital, A. and Tritar, M., 1988, Psychophysical capacity modeling in frequent manual materials handling activities. *Human Factors*, **30**, 319–337.

Gevarter, W.B., 1985, Expert systems: limited but powerful. In: *Applications in Artificial Intelligence*, edited by S.J. Andriole (Princeton, NJ: Petrocelli Books), pp. 125–139.
Giambra, L.M. and Quilter, R.E., 1987, A two-term exponential functional description of the time course of sustained attention. *Human Factors*, **29**, 635–643.
Givoni, B. and Goldman, R.F., 1972, Predicting rectal temperature response to work, environment, and clothing. *Journal of Applied Physiology*, **32**, 812–822.
Givoni, B. and Goldman, R.F., 1973, Predicting effects of heat acclimatization on heart rate and rectal temperature. *Journal of Applied Physiology*, **35**, 875–879.
Haslam, R.A. and Parsons, K.C., 1987, A comparison of models for predicting human responses to hot and cold environments. *Ergonomics*, **30**, 1599–1614.
Kennedy, S.J., Goldman, R. and Slauta, J., 1973, *The Carrying of Loads Within an Infantry Company*. Natick Technical Report No. 73–51-CE, C and PLSEL-108 (Natick MA: US Army Natick Research, Development, and Engineering Center).
Lechner, P.E. and Barth, S.W., 1985, Expert systems on microcomputers. In: *Applications in Artificial Intelligence*, edited by S.J. Andriole (Princeton, NJ: Petrocelli Books), pp. 109–124.
Marshall, S.L.A., 1950, *The Soldier's Load and the Mobility of a Nation* (Washington, D.C: The Combat Forces Press).
Military Standard 1472C, 1981, pp. 209; Notice 1, 1983; Notice 2, 1984, Notice 3, 1987.
Nordeen, L.O., 1986, Today's infantryman. *National Defense*, January, 35–40.
Pandolf, K.B., Stroschein, L.A., Drolet, L.L., Gonzalez, R.R. and Sawka, M.N., 1986, Prediction modeling of physiological responses and human performance in the heat. *Computers in Biology and Medicine*, **16**, 319–329.
Sampson, J., 1988, *Technology Demonstrations for Lightening the Soldier's Load*. TR-88/027L. (Natick, MA: US Army Natick Research, Development and Engineering Center).
Siegel, S.F. and Warren, P.H., 1987, Designer compacts: artificial intelligence strategies. *Proceedings of the Research and Development Associates, Natick Research, Development and Engineering Center, Natick, MA*, **39**(1), 183–190.
Soule, R.G., Pandolf, K.B. and Goldman, R.F., 1978, Energy expenditure of heavy load carriage. *Ergonomics*, **21**, 373–381.
Tullis, T.S., 1983, Predicting the Usability of Alphanumeric Displays. Doctoral Thesis, Rice University.

Chapter 29
Can artificial intelligence outperform real people? The potential of computerized decision aids in medical diagnosis

GITTE LINDGAARD and THOMAS J. TRIGGS
Monash University,
Melbourne, Australia

Abstract: Systems designers tend to believe that human judgement and decision making is close to perfect, and that systems should aim to emulate human judges. The psychological literature concerning judgement and decision making suggests that many types of decisions are less than optimal and prone to certain biases and errors with potentially serious consequences. This paper outlines ways in which current technology can be applied constructively to assist human judges making better decisions. If human judgement and decision making is in need of improvement, the conditions under which it falls down must be clearly exposed in order to pinpoint where and how technology can help. In order to achieve this the relevant human judgement literature is reviewed with particular emphasis on studies related to medical diagnosis. A range of artificial intelligence tools and techniques applicable to medical diagnosis are introduced, and the ways in which these could be integrated into a computerized decision aid is discussed. A computer system, developed in three distinct phases and relying on a formal, normative decision model introduced earlier, is outlined, describing the benefits of such a system to the medical diagnostician in each phase in some detail.

Keywords: Human judgement; Artificial intelligence; Expert systems; Heuristics; Medical diagnosis.

Introduction

According to Hammond (1987), both Artificial Intelligence (AI) and judgement/decision making research attempt to capture the cognitive activity of experts; both seek to evaluate and, if necessary, improve it, thus assisting experts as well as novices, and so to advance the study of cognition. Yet, they differ widely in their working hypotheses, assumptions, methods, procedures and premises. As a consequence of this, the conclusions reached also differ within the two fields. Whereas such differences are of importance when defining and sharpening research issues and attracting research dollars, they need not be barriers to the fruitful application of outcomes from both areas.

Gevarter (1984) argues that the present field of AI emerged from a conference convened in 1956 at Dartmouth College by a group of 10 scientists. They predicted that, by 1981, some 25 years later, 'we would all be involved in recreational activities, while computers would be doing all the work' (Gevarter, 1984, p. 8). The assumption was that intelligent behaviour was the application of smart (human) reasoning techniques which could readily be made explicit and incorporated into computer programs. The technology necessary for accommodating such reasoning techniques was evolved rapidly, and it was believed that this development was the crucial factor in producing intelligent computer systems.

Since the release of ELIZA (Weizenbaum, 1966, cited in Gevarter, 1984), a natural language understanding program simulating a non-directive Rogerian psychotherapist at MIT in 1966, the technology has come a long way towards providing an environment suitable for complex problem solving and decision making. Why, then, are computers not doing 'all the work' as was so euphorically predicted in 1956? Apparently, the assumption that human reasoning techniques could readily be extracted from experts, and that such techniques would inevitably lead to optimal judgements and decisions, turned out to be less than accurate: human judgement, problem solving and decision making is fraught with error and bias; the way in which judgements are arrived at is neither fully understood nor completely representable by mathematical models and formulae, although it is well approximated by these. Indeed, some studies suggest that the more human reasoning techniques are uncovered and explained, the less sophisticated, and more suboptimal they appear to be (Nisbett and Ross, 1980; Kahneman *et al.*, 1982).

This chapter seeks to expose some of the shortcomings of human judgement and decision making in order to determine the extent to which computerized decision aids could assist the human judge making better decisions. The notions of AI and expert systems are defined first, and the state of the art in advanced technology assessed in order to explore the nature of those AI tools which could be applied in a system serving as medical decision support. Some of the most important methods applied in the study of human judgement and decision making are then described. A distinction is drawn between formalistic strategies which measure human judgement against predefined formal models and pragmatic strategies which begin with real decision problems for which a structure is then developed to reduce the problem to a manageable size (Triggs, 1975). This is followed by a review of formal decision models in the search for one which may feasibly be applied in medical diagnosis.

Some important formal models in human reasoning, judgement and decision making are described. Ways in which human judgement systematically deviates from such models, principally by applying heuristics, or rules of thumb are outlined. Certain biases to which human judgement is also prone are then discussed. Along with descriptions of these heuristics and biases, the relevant experimental literature is reviewed with specific emphasis on studies involving medical data and diagnosis. Some of these studies demonstrate that a mere reorganization of existing knowledge, provided in a computer system, could assist expert medical diagnosticians in improving their judgement strategies to the benefit of their patients even in the short term.

Since human judgement is shown to be inefficient along a number of dimensions, the issue of the extent to which the relevant research may itself be biased, is addressed. This analysis, together with the studies reviewed, suggests that medical diagnosticians may need help in ways other than with judgement outcomes. The points at which help may be needed and could be provided in a decision support system are made explicit. A system offering support at all levels identified is proposed, the development of which is suggested to take place in three phases. Each of these phases is shown to be useful in formulating medical diagnoses. The description of the system shows also how the AI tools introduced earlier could be applied. The system could lead to the automation of some of the decision processes currently undertaken by human judges, so reasons are given for why people as such should not be replaced by smart computers. Finally, some of the barriers to the introduction and use of AI tools in medical diagnosis are outlined.

AI and expert systems: what are they?

Scown (1985) states that 'AI research is that part of computer science that investigates symbolic, nonalgorithmic reasoning processes and the representation of symbolic knowledge for use in machine intelligence'. It is, she claims further, 'the science of making machines do things that would require intelligence if done by men' . . . Its goals are 'attempts to make computers more useful and to understand the principles that make intelligence possible' (p. 8). Unfortunately, no definition is offered for the term 'intelligence', and nor is it clear what distinguishes 'machine intelligence' from other forms of intelligence. The ways in which computers should be made more useful, and to whom, are not specified either, which renders it difficult clearly to ascertain when a given system meets the criteria for falling into the AI category.

Scown says explicitly that 'AI is a growing set of computer problem-solving techniques that are being developed to imitate human thought, or decision-making processes or to provide the same results as those processes' (p. 9). It is thus presumed that human thought and decision processes can be accurately described, and that such processes result in good judgements. As will be shown later, the human judgement and decision-making literature suggests otherwise; inputs that result in similar, or even identical, outcomes from the computer and the human judge say nothing about the intervening processes (Hoffman, 1960); it is therefore difficult to establish the degree to which a machine imitates human thought. Human judgement is shown to be biased and erroneous in many types of judgement tasks throughout the relevant psychological literature. The assumptions made about people in the AI area thus appear to rest on intuition rather than on a thorough search of the human judgement literature. This inaccuracy is probably the main reason for the difficulties of AI to meet the early expectations. In fact, the enthusiasm for AI seems to have waned over time; articles in the daily press have, for example, reported 'recent severe losses in earnings and numbers of employees' (New York Times, 4 March 1988, cited in Hammond, 1988) in the AI area; one AI entrepreneur is said to have stated that 'the trouble with AI is that "people believed their own hype" ' (Hammond, 1988, p. 3). Even if AI cannot entirely replace human labour, a number of interesting tools have emerged which can be employed usefully in areas such as medical diagnosis.

In describing what AI does rather than what it is, Alty and Coombs (1984), for example, argue that 'AI solves real-world problems by incorporating explicit domain knowledge into problem solving'. The application of such knowledge is called 'knowledge programming' (p. 85). Knowledge programming has proven most successful for problem solving in areas requiring a rich body of personal expertise. Computer programs which mechanize such expertise have become known as expert systems (pp. 85–87). Expert systems have been designed and applied to certain domains within the area of medical diagnosis; MYCIN assists in defining the offending organism in infectious diseases and advises on treatment; PUFF deals with respiratory disorders; GRAVIDA with pregnancy advice; CLOT with blood disorders, and several INTERNIST systems with internal medical disorders (Alty and Coombs, 1984). Despite the success of many such systems, statements such as 'one notable feature of the INTERNIST systems is that they explicitly seek to model human diagnostic thinking (p. 120) and the claim that they actually achieve this (p. 128) suggest that isomorphism is assumed between human thinking and machine processes. Yet, human diagnostic thinking and reasoning is far from perfect; it is suggested here that rather than seeking to imitate, or even determine the extent to which AI can or does imitate, human judgement, attempts should be made to isolate aspects of the diagnostic process in which AI tools can improve human decision strategies.

Which AI tools could be applied in medical decision support systems?

A range of AI tools are currently available which can be incorporated into computer systems. Rather than providing an exhaustive list of these, only some of those which could be of relevance to diagnostic decision support systems are introduced here. Issues such as the construction and coding of these tools as well as hardware requirements are not discussed. Instead, the conceptual usefulness of the techniques is emphasized. The following are some examples of AI tools which could be applied in a medical decision system such as that described later in this paper:

Predicate calculus: this is simply a way to write expressions about the world.

Connectives are used to combine propositions to form longer statements. The most common connectives correspond to the English words *and*, *or*, *not* and the phrase *if* → *then*. Such connectives could be used to reduce the initial problem of formulating plausible hypotheses by excluding diseases that are not suitable candidates for further consideration.

Confidence, or *certainty factors* are used in an expert system to provide an index of how strongly the system's knowledge supports its final, cumulative conclusion.

Semantic networks are generally illustrated by diagrams consisting of arcs and nodes, where nodes represent objects, actions or events. The advantage of semantic networks is seen to be inheritance of meaning across concepts, such as in the following example:

Australian crawl $\overrightarrow{\text{is a}}$ swimming $\overrightarrow{\text{is a}}$ aerobic sport

(Scown, 1985, p. 66)

Whereas it is not clear how this particular aspect of semantic networks apply to medical diagnosis, the notion could be useful in a decision support system by storing keyword lists of synonyms such as

elevated temperature = fever = pyrexia

Such lists could reduce the need for a very restricted vocabulary to be used by the physician-user. However, it must be noted that this seemingly uncomplicated function can generate its own problems. Consider, for example, the following:

nation = state = condition

The conclusion that nation semantically equals condition is obviously wrong. It may be very difficult to specify a large sample of permissible medical terms in a totally unambiguous manner, although methods by which the system can learn cumulatively and iteratively exist and could be used.

Production rules are of the if → then type, applicable to decisions such as "if symptom X is present then diagnosis Y is a candidate for consideration". Chaining along paths to be searched in the firing of production rules can go forward or backward. *Forward chaining*, also called bottom-up processing, proceeds by manipulating data on antecedents in the knowledge base to produce a modified status of the system. This is useful when aiming for synthesis as in the examination of 'what if' questions. In a medical decision support system, this would enable the diagnostician to manipulate hypothetical actions conceptually, for example, by assessing the value of a given test to the diagnostic process before it is carried out. *Backward chaining*, also called top-down processing, breaks down the final judgement, working backwards to see if the necessary and sufficient antecedents that would satisfy the goal are present in the data base. Situations arise in medicine that require instant action to treat a life-threatening condition. Cardiac arrest or heavy bleeding are examples of such conditions. In some instances, the cause of the immediate danger, such as the severe bleeding, may not be clear; the diagnosis may thus need to be verified *post hoc*, that is, after initial treatment has been administered.

Human judgement and decision making

It was stated earlier that two different approaches are taken to research on decision-making; studies of formalistic and pragmatic strategies (Triggs, 1975).

Pragmatic strategies include protocol analysis, a method whereby the decision maker 'thinks aloud' during the decision-making process. Protocol analysis may be applied 'in order to tease out whatever structure is inherent in a situation, or alternatively, to discover what structure can be imposed on the situation without changing its character to the point of discarding the problem of interest' (Triggs, 1975, p. 156). The observation that people have difficulties verbalizing their thoughts is underscored by an example of a study in which a well-known radiologist was questioned about his diagnostic processes with X-ray film. The expert was only able to respond that the diagnosis was accomplished by the way the film looked (Lusted, 1960, cited in Triggs, 1975). Although protocol analysis can be informative for providing issues and potential relations between issues worth researching in more detail, there is a danger of oversimplifying decision-making strategies. In the present context, protocol analyses are of little relevance, as (potential) strategies used by individuals may be idiosyncratic and say little about strategies that are common to groups of people; individual strategies may not help to define properties of a decision support system to be used by many diagnosticians.

Formalistic strategies begin with models against which human performance can then be measured, and which can be applied to uncover strategies common to many people. Two major lines of research have been pursued along formalistic, model-based lines: one is the so-called regression approach, and the other the Bayesian approach (Slovic and Lichtenstein, 1971). Within the regression approach, two paradigms have been applied, namely correlations and Analysis Of Variance (ANOVA). Both these approaches focus on the manner in which items of information influence judgemental outcomes. Research within the Bayesian framework has mainly concentrated on exploring the degree to which human decision processes are consistent with the requirements of Bayes' Theorem.

Regression approach: the ANOVA technique

The ANOVA technique is applied in cases in which judgement cues, or stimuli, are regarded as categorical treatment factors, and judgements made on the basis of the cues are dependent variables. Structurally, the ANOVA model is a specific formulation of a general linear model.

Assuming that the effect of a limited number of discrete, categorical stimuli is being explored within the ANOVA paradigm, one may distinguish between so-called interactions and main effects. A statistically significant main effect for, say stimulus 1 implies that the judge's responses varied systematically with stimulus 1. Similarly, a significant interaction between stimuli 1 and 2 implies that the judge responded to particular patterns of these; that is, the effect of variation of stimulus 1 upon the judgements differed as a function of the level of stimulus 2. In general, the model represents judgements as a weighted additive combination of the factors, both main effects and interactions (Hoffman, 1968). Significance tests can thus provide an indication of the judge's utilization of a particular cue individually or in combination with other stimuli.

Using the ANOVA model, it is possible to calculate an index of the importance of the individual or patterned use of a cue, relative to the importance of others. One such index is based on the degree to which the mean judgement shifts as the levels of a factor are varied (Slovic, 1969). Another is a simple transformation of these mean effects into an estimate of the proportion of the total variation in a person's judgements that can be predicted from knowledge of the particular levels of a given stimulus or pattern of stimuli (Hoffman *et al.*, 1968).

As a research tool, the ANOVA model tends to focus on judges' systematic use of stimuli, thereby stressing the need to control the environmental situation. Stimuli and stimulus levels are varied systematically, often in a factorial manner, in laboratory experiments. A certain degree of realism thus tends to be lost in the process. As a decision aid, the ANOVA model can only help the judge in so far as it can ascertain the degree to which the effect of one cue on the dependent variable differs significantly from, or interacts with, others. In calculating such effects, the computer is employed merely as a sophisticated number cruncher relying on technology that was available in the early 1970s; up-to-date AI techniques are thus not necessary for the performance of ANOVA models.

Correlation

In correlational models, the judge's integration of information is described by means of correlational statistics. The basic approach requires the judge to make quantitative evaluations of a number of stimuli, and to combine these into a single judgement. The approach has developed along two main streams: one focuses on the individual judge by describing his/her method of combining and weighting stimuli and developing mathematical explanations representative of these combinatorial processes (Hoffman, 1960). The other takes into account both the judge and the ecological representatives of experimental research. That is, it considers both the judge and the environment within which the judgement is made, following the philosophy of 'probabilistic functionalism' (Brunswik, 1955). Brunswik was the first to make explicit the fact that modelling tasks is different from modelling people, to search for tools for modelling tasks, and to provide linkages between models of tasks and models of people (Edwards, 1971). In pursuit of this, Brunswik designed the so-called 'lens model' which was later applied in studies including clinical inference as an operational mathematical technique (Hammond *et al.*, 1964). The so-called 'bootstrapping' technique (Dawes, 1971) evolved from Brunswik's lens model as a means for comparing the performance of judges with models based on their own judgements.

Bootstrapping the judge

The intriguing possibility that a model may actually outperform the judge on whose behaviour it was developed, was first entertained by Yntema and Torgerson (1961). They suggested that 'men and computers could cooperate more efficiently . . . if a man could tell the computer how he wanted decisions made and then let the machine make the decision for him' (p. 364). Their hypothesis subsequently proved correct: computers can yield more accurate judgements in a variety of situations than judges upon whose estimates computer judgements are modelled by means of the bootstrapping technique. In this technique, a linear model is constructed which is based upon an expert's judgements and weightings of variables entered into the judgement. A large number of such judgements is required to generate a model of a given judge. When the judge is replaced by his own model, the performance of the model is generally found to be superior to that of the judge himself/herself. The technique has been applied in a range of different areas including forecasting success in graduate school

(Wiggins and Kohen, 1971); in evaluation of applicants for graduate school (Dawes, 1971); in clinical diagnosis, distinguishing between psychosis and neurosis (Goldberg, 1970), and in production scheduling (Bowman, 1963; Kunreuther, 1969). All of these studies have indicated superiority of the bootstrapping model over the judge.

The bootstrapping method appears to provide a means whereby judgements could be taken over by computers, much in line with the predictions made by the early AI fraternity. Unfortunately, there are some problems with this. Goldberg (1970), for example, built a model of each of 29 clinical diagnosticians on data originally provided by Meehl (1959). Using the Minnesota Multiphasic Personality Inventory (MMPI) and profiles of psychiatric patients who had been diagnosed as being either psychotic or neurotic, Goldberg found that the models were at least as valid as the judges for 28 of the 29 individuals. He also calculated values for a 'composite' judge by taking the average judgements of all 29 judges to each MMPI profile. This composite judge was found to perform as well as, and sometimes better than, each individual judge. Goldberg concluded that the composite judge or the model of a given individual judge should be used for formulating future diagnoses. The problem, however, is to locate the most accurate judge; to do so requires analysis of a large number of judges and their models which, for each judge, is based on a substantial number of judgements. These judgements would, of course, all need to be based on a given (large) sample of identical profiles; all in all, a most expensive and time-consuming task!

If the time and expense of establishing a suitable composite model were the only barriers to success, it would still seem appropriate to computerize certain problem solving and judgement tasks. Yet, once the model has been established, it is static; different judgement conditions or the inclusion of even a single new variable to be considered in the judgement would require a totally new set of judgements to update the model. In order to avoid the necessity of such updatings, one would therefore need to assume that the variables included in the original model were both valid and sufficient for calculating the outcomes and continue to be so as time progresses. These criteria are very difficult, if not impossible, to meet in practice.

As is true for the ANOVA model, the actual computation of multiple regressions is relatively unsophisticated, relying primarily on the ability of computers to handle large amounts of data. Although this ability *per se* is very important for researchers, it does not utilize the power of present-day computers to combine data in novel ways or to search for the 'best' of all possible combinations: AI techniques are thus superfluous for the performance of regression models.

Summary of regression approaches

Both regression methods reviewed here have certain merits and disadvantages; the ANOVA paradigm is an excellent tool with which to explore stimulus utilization of judges in a controlled environment over a small number of combined stimuli, but it loses some degree of realism in the process. One form of multiple regression, the bootstrapping technique, is capable of retaining the realism lost by the ANOVA paradigm, but both the selection of variables and the encoding into the computer of a model on which to base future predictions must be done by human judges. It is also inflexible to the extent that the incorporation of new variables requires the model to be re-established from scratch, rather than allowing a more graceful integration. Neither method seems promising in terms of easing the burden for experts making complex judgements by computerizing parts of these judgements other than the sheer arithmetic calculations.

Bayes' Theorem

A different approach to the study of human judgement and decision making that has generated a great deal of research is the application of Bayes' Theorem, which acquired its name after the Reverend Sir Thomas Bayes in 1772. In contrast to the bootstrapping method described above, the outcome of a Bayesian analysis is not a single prediction, or numerical value; rather, it is a distribution of probabilities over a set of hypotheses. It is a normative model in the sense that it specifies certain internally consistent relationships among probabilistic opinions that serve to prescribe how opinions should be revised. In its simplest operational form, Bayes' Theorem states that

$$P(H \mid D) = \frac{P(D \mid H)\, P(H)}{P(D \mid H)\, P(H) + P(D \mid \bar{H})\, P(\bar{H})} \tag{1}$$

where H is the hypothesis being evaluated; D is the probability of the given datum, and $\bar{H}$ is an alternative hypothesis. This alternative hypothesis is included because the rule requires a hypothesis to be contrasted with one or more alternative hypotheses.

The term $P(H \mid D)$ is the posterior probability of hypothesis H, given the datum, D; $(D \mid H)$ is the individuating information, the datum D evaluated in the light of hypothesis H, whereas $(D \mid \bar{H})$ is the same datum viewed in the light of the alternative hypothesis $\bar{H}$. $P(H)$ is the prior probability, or base rate, of hypothesis H, and $P(\bar{H})$ the base rate of hypothesis $\bar{H}$. The base rate summarizes what is known about H prior to it being entered into the equation and being evaluated in the light of the particular datum, D, and similarly for $\bar{H}$.

Stated slightly differently, the posterior probability is the outcome of a revision of the base rates in light of the individuating information and with respect to two competing hypotheses. To illustrate this, assume that the probability of a given patient being psychotic is to be estimated, given that he hallucinates. The prevalence of psychosis is contrasted with that of neurosis in a given hospital patient population of, say, 100 patients. The components of Bayes' Theorem outlined in equation (1) are now the following:

$P(H \mid D)$ = the posterior probability of psychosis, given hallucinations;
$P(D \mid H)$ = the probability of hallucinations, given that the patient is psychotic;
$P(D \mid \bar{H})$ = the probability of hallucinations, given that the patient is neurotic;
$P(H)$ = the prevalence of psychosis in the patient population of 100;
$P(\bar{H})$ = the prevalence of neurosis in the same population.

Replacing the components by numerical values, the posterior probability may then be calculated, where $P(D \mid H) = 0{\cdot}60$; $P(D \mid \bar{H}) = 0{\cdot}15$; $P(H) = 0{\cdot}70$; and $P(\bar{H}) = 0{\cdot}30$.

$$P(H \mid D) \text{ is now} = \frac{(0.60)\,(0.70)}{(0.60)\,(0.70) + (0.15)\,(0.30)} \tag{2}$$

$$= 0.90$$

Hence, the probability that this particular patient is psychotic is 0·90, or 90%.

Note the terms $P(H)$ and $P(\bar{H})$ sum to 1·0. This is because everyone in the entire patient population considered suffers from either psychosis or neurosis; $\bar{H}$ is thus the complement of H in the sense that $1 - P(H) = P(\bar{H})$, and similarly, $1 - P(\bar{H}) = P(H)$. Note also that $P(D \mid H) + P(D \mid \bar{H})$ do not sum to 1. In consideration of the present data, it is thus possible to be either psychotic or neurotic without hallucinating. In cases in which the value of $P(D \mid H)$

is equal to that of $P(D \mid \bar{H})$, the individuating information is non-diagnostic to the extent that it supports hypotheses H and $\bar{H}$ by an equal amount. In such cases $P(H \mid D) = P(H)$, since the datum D should have no effect on the magnitude of the posterior probability whatsoever.

One major advantage of Bayes' Theorem is that it is iterative; it allows for sequential integration of data. A patient may, for example, be assumed to present with more than a single symptom; the overall posterior probability incorporates the effect of every one of these integrated one by one. For each additional datum evaluated, the formula expands by using the entire previous calculation as the new base rate. Although the procedure is conceptually simple, the formula soon becomes unmanageable in practice when calculating the posterior probability manually. For a computer, however, the length of the formula poses no problem; its superior computational power is well exploited when several items of individuating information are to be integrated into the calculation of Bayesian posterior probabilities. In addition to allowing the integration of several items of individuating information, a computer data base is also able to update the base rate values as these change over time and the data base grows. This ability to integrate data sequentially into an existing model and to modify existing data in accordance with new incoming evidence goes beyond mere calculation of outcomes. Bayes' model can thus make better use of the capabilities of modern computers than either of the regression approaches discussed earlier.

Contrary to the bootstrapping method, Bayes' model continues to evolve with the addition of new data; the accuracy of base rates may be assumed to become increasingly representative of actual population values. One limitation of the model is perhaps that a large data base is required before it is of sufficient value to generate realistic possibilities. A basic characteristic of probabilities is that they are more accurately reflected in large than in small samples. The probability of obtaining a 'heads' when tossing an unbiased coin, for example, is known to be 0·50, and the same for 'tails'. However, this may not be confirmed if the coin is tossed only twice. Instead, one could observe h,h, or t,t. Over hundreds, or thousands, of coin tosses, however, the probability of 0·50 for heads will be confirmed. Similarly, the base rate of a disease such as bacterial meningitis generated on the basis of 10 cases, is unlikely to reflect the prevalence of meningitis in a population of thousands of patients presenting with headaches and photophobia. Factors affecting the occurrence of meningitis, or any other disease, other than a small number of cases include seasonal, climatic, social fluctuations as well as potential epidemic outbreaks. Hence, the larger the number of cases observed, and the longer the time over which observed cases are registered in the database, the more accurate the base rates are likely to be.

Another problem associated with using Bayes' Theorem in a data base containing medical data is that the population from which to draw the base rates may be difficult to specify. Imagine diagnosing a case concerning a 63-year-old male truck driver, who is a smoker, is overweight, who does not engage in regular physical exercise, and who shows possible signs of angina pectoris. Should the term $P(H)$ constitute all males over 60 years of age? All truck drivers? All people, male and female, over 60 who smoke? All those with chest pain treated within a particular hospital or region? The size of $P(H)$ has a profound effect on the magnitude of the posterior probability, so the selection of the 'right', most appropriate and representative sample population from which to draw the base rate is very important. The formula itself cannot do that, and nor can the decision be made by the computer, initially at least. Hence, although Bayes' Theorem is one formula that provides flexibility by integrating data sequentially and updating the probabilistic values of the components entering into the calculation of posterior probabilities over time, it does pose certain problems. It is, however, worth noting that these problems are not inherent in the model itself, but rather a function of the way it is applied.

As mentioned earlier, a number of computer systems, some of which use Bayes' Theorem, have been produced to aid the medical diagnostician with impressive results: outcomes, or recommendations, made by such systems compare favourably with those made by physicians (Schwartz and Griffin, 1986). In the present context, however, the intention is not only to compare outcomes, but also to evaluate the processes underlying probabilistic outcomes; the integration of information as such is important for identifying ways to help diagnosticians improve their assessment of medical data.

Do human judgements conform to Bayes' Theorem

Investigations of decision problems within the framework of Bayes' Theorem have assumed a great deal of importance over the past two or three decades. Early studies consistently found that experimental subjects tended to adjust their posterior probabilities somewhat less than required by Bayes' Theorem. That is, subjects' confidence in the apparently correct hypothesis failed to increase as quickly as the accumulating evidence suggested it should. Phillips and Edwards (1966) and Edwards (1982) labelled this tendency 'conservatism'.

The issue of consevatism gave rise to a lively debate in the literature (Slovic and Lichtenstein, 1971). This debate focused mainly on exploring whether conservatism was due to people's failure to appreciate the actual diagnosticity of the evidence, or a failure to combine those diagnosticity assessments according to Bayes' Theorem. Conservatism was seen to be the final problem to be explained concerning the degree to which people adhered to the normative Bayesian principles. Indeed, an extensive review of the early literature led to the optimistic conclusion that 'man is an intuitive Bayesian' (Peterson and Beach, 1967).

A major upheaval caused a revitalization of interest in this line of research when, in 1972, Kahneman and Tversky boldly claimed that 'man is not a conservative Bayesian; he is not a Bayesian at all!' (Kahneman and Tversky, 1972, p. 450). The claim was based on findings from experiments that differed radically from the types of problems presented in earlier studies. The results showed clearly that subjects tended to neglect consideration of the base rates of their Bayesian probability estimates. As a consequence of these findings, Bayesian research took on an entirely new dimension, and a whole family of heuristics that people apparently use was subsequently revealed. Some of these are worth describing here, as they are important for an understanding of certain shortcomings of human judgement.

The use of heuristics

In exposing people's tendency to neglect base rates, Kahneman and Tversky (1972) and Tversky and Kahneman (1973, 1974) proposed that people rely on heuristics rather than following the normative principles of probabilities in assessing the likelihood of uncertain events. Although the term was not explicitly defined, the authors appeared to use it in the same sense as in computer science. In that context, heuristics refer to short cuts, or simplification strategies (Lenat, 1983a,b) that serve to reduce a large 'problem space' (Newell and Simon, 1972) to a manageable size in the search for solutions to problems. While, as Nisbett and Ross (1980) point out, judgemental heuristics are often 'well founded and helpful guides to inference' (p. 271), they can lead to gross and systematic judgemental errors.

The potential negative effect of the use of heuristics on judgemental accuracy is apparently not recognized by some authors. Feigenbaum, for example, a pioneer in expert systems (Feigenbaum, 1982, cited in Gevarter, 1984) asserts that heuristics are 'mostly private, little discussed rules of good judgement rules of plausible reasoning, rules of good guessing that characterize expert-level decision-making in the field' (Gevarter, 1984, p. 71). The reasoning underlying reliance on heuristics may well seem plausible, at least to the judge, but it does not always constitute application of 'good judgement rules'. Feigenbaum further claims that the knowledge necessary to perform at a high level of problem solving, 'plus the inference procedures used, can be thought of as a model of the expertise of the best practitioners in the field' (p. 71). It is thus assumed that expert judgement is very good. Unfortunately, this assumption is not supported in the judgement and decision-making literature; experts are 'not endowed with better memories or better perceptual skills' (Schwartz and Griffin, 1986, p, 214), and nor do they appear to be deeper thinkers than novices. The following discussion will seek to demonstrate that attempts to emulate human behaviour may not lead to the best, most accurate judgements, but rather that the role of the computers applying AI techniques should be to help judges improve their judgements. The focus here is thus on the fallacies of human judgement as a consequence of applying heuristics and using other biases.

Representativeness

The notion of representativeness rests on features of similarity; it is judged by the features that are characteristic for members of a given category rather than by the frequency of members in that relative to other categories. A robin is, for example, judged a more typical bird than a chicken, although robins are less frequent (Rosch, 1978; Kahneman and Tversky, 1982), Similarly, a patient displaying a particular symptom characteristic for appendicitis, Rovsing's sign, would probably be judged more typical of an appendicitis case than a patient with an unspecified pain in the lower right abdominal quadrangle, even though Rovsing's sign occurs rather infrequently (Pschyrembel, 1965; Toohey, 1965; Weinstein *et al.*, 1980). At the same time, Rovsing's sign is highly indicative of appendicitis, whereas unspecified abdominal pain may be indicative of several complaints.

Alternatively, representativeness may be largely controlled by causal beliefs which may or may not be valid. A symptom may, for example, be representative of a given disease because it is frequently associated with that disease. The fact that it may be equally representative of alternative diseases is ignored, perhaps because the diagnostician has less experience with that alternative disease, or the disease under consideration may be more serious or novel than the alternative disease.

Tversky and Kahneman (1974) identified six potential cognitive errors, which may underlie the use of representativeness. These are insensitivity to base rates, to predictability or to sample size, misconceptions about randomness and statistical regressions, and illusions of validity. Each of these errors contravenes some aspect of the normative Bayesian model.

Subjects tend, according to Kahneman and Tversky (1982), to use the representativeness heuristic because it is easily accessible, because it often correlates with the probability of a given event, and because people tend to overestimate this correlation.

Availability

The availability heuristic is said to be employed whenever a person 'estimates frequency or probability by the ease with which instances or associations come to mind' (Tversky and Kahneman, 1973, p. 208). It is thus ease of retrieval, association and construction that provides the estimate of frequency, not the sum total of examples or associations that come to mind (Taylor, 1982).

The availability heuristic is important in the area of medical diagnosis. Physicians asked to rate the mortality of some 42 diseases were found to overestimate the lethality of diseases that receive a great deal of journal coverage in one study (Christensen-Szalanski *et al.*, 1983). The number of journal articles devoted to a disease, rather than their actual content, was a good predictor of physicians' estimates. Christensen-Szalanski *et al.* (1983) noted that the same tendency may influence medical testing to the extent that, a disease judged to be more lethal than it actually is because of a high journal coverage, may give rise to ordering unnecessary screening tests.

Elstein *et al.* (1978) and Schiffman *et al.* (1978) have shown that a symptom characteristic for a disease that is novel or very serious tends to receive more attention than one characteristic for a more common or less serious disease. Similarly, Kleinmuntz (1968) has demonstrated that data that are not related to the physician's mental diagnosis are forgotten. Thus if the most salient datum is not representing the disease from which the patient actually suffers, it may lead the physician astray by causing him/her to pursue an incorrect diagnosis.

The availability heuristic can bias medical judgements by giving easily imagined events more importance than warranted by their objective frequency. Indeed, Leaper (1975) has shown that clinicians' estimates of the diagnostic value of a range of symptoms were less accurate than objective values generated by a computer, and that computer generated diagnoses using real-life data were more accurate than those proposed by the clinicians. He concluded that people do not know actual disease probabilities.

Overconfidence

Because human information processing capacity is limited, it is sometimes counterproductive to gather more information than a sound judgement requires; the excess data may serve to confuse and mislead the diagnostician rather than clarify the issues at hand. An overload of the information processing capacity frequently results in judgemental errors. These are rarely perceived by decision makers who tend to be very confident in their own judgement. In fact, their confidence tends to increase as more data become available.

Overconfidence sometimes arises from distortions to originally available data, due to selective attention and subsequent selective retrieval from memory. Such selective retrieval was demonstrated in a study by Arkes and Harkness (1980) in which subjects were asked to recall symptoms presented earlier in case histories. They found a clear tendency to forget symptoms that were not representative of the final diagnosis but which had been presented, and a tendency incorrectly to recall symptoms that were consistent with the diagnosis but had not been presented. The act of discarding unrelated symptoms and including unpresented ones suggested that subjects were 'cleaning up' the data for cases to resemble textbook examples more closely. This led to greater confidence in the final diagnosis.

The finding that people are very confident in the correctness of their judgements or answers to questions has been reported in numerous studies across a wide range of tasks. In a study of medical diagnosis, Christensen-Szalanski and Bushyhead (1982) required physicians to estimate the probability of pneumonia in patients presenting with a cough. The physicians did not have access to radiology reports of chest X-rays on the basis of which the diagnoses were verified. A total of 1531 such cases were evaluated, and no patient was seen by more than one physician. Although performance varied across subjects, the judged probability of pneumonia was 80% in one subgroup of patients; the actual occurrence of pneumonia was 20% in that group! Hence, medical diagnosis is apparently no exception to the notion that people tend to judge with greater confidence than warranted by the available data.

Confirmation bias

A confirmation bias is described as a tendency for subjects to seek information and confirm an initially favoured hypothesis (Ross *et al.*, 1975; Snyder and Swann, 1978; Nisbett and Ross, 1980). A confirmation bias can be inferred from two different forms of behaviour, namely (i) failure to change one's opinion in the face of nonsupporting or even contradictory evidence (Ross *et al.*, 1975; Wells, 1980; Wells and Lindsay, 1980) and (ii) selection of data favouring one's hypothesis while ignoring data which might be likely to contradict that hypothesis (Doherty *et al.*, 1979; Klayman and Ha, 1987). Both these aspects of confirmation bias have been supported experimentally.

The notion of confirmation bias is of significant relevance to medical diagnosis. Pellegrino (1979), for example, asserts that the problem of diagnosing a disease from a number of observed signs and symptoms must be transformed into a problem of manageable proportions; Elstein (1979) says that to 'allow all disease to enter into consideration each time we are called to a bedside is foolish in the attempt and impossible in performance' (p. 19), and Eddy and Clanton (1982) argue that by selecting a single salient symptom presented by the patient and using this as a 'pivot' around which to gather further information, a diagnostician's memory is 'not stressed by the need to explain all the aggregate findings' (p. 1265). Kruglanski and Ajzen (1983) argue that people have a 'need for structure' (p. 16), and that this need is assumed to exert an inhibiting or braking influence on the hypothesis-gathering process because the generation of alternative hypotheses endangers the existing structure. In essence, these arguments all agree that it is easier to establish one hypothesis and to pursue that than to consider a number of alternatives because of human cognitive limitations. Problems arise, of course, when the hypothesis entertained turns out to be wrong, and the process must be remembered.

Primacy effects

A primacy effect is the tendency for items occurring early in a sequence to receive a disproportionally higher weighting than items occurring later in the sequence. The weighting of a given item thus varies as a function of its serial position within a group of items (Anderson, 1981, Sections 2.5, 3.3; Lopes, 1983). Several theoretical explanations for the occurrence of primacy effects have been proposed, but one of these, in particular, has received a great deal of attention and support in the literature, the attention decrement hypothesis. According to this hypothesis, the weighting of later presented items decreases due to a progressive reduction in attention over a number of items. This is considered to take place irrespective of any relation between the items presented.

In a study of medical diagnosis in which medical students and physicians were free to gather as much information as they wished, Wallsten (1981) found that information gathered early in the diagnostic process was given more weight than later information. In fact, data gathered late in the process was distorted to support evidence for opinions formed early. In a similar fashion, a study carried out within a Bayesian-type framework and analyzed using the ANOVA technique, revealed clear primacy effects even in case histories that were objectively non-diagnostic (Lindgaard, 1985). Nondiagnosticity was achieved by careful selection of symptoms characteristic for different diseases tested. These symptoms were pre-tested for subjectively perceived frequency of occurrence under the testing diseases to generate an index of high and low diagnosticity. In the pre-testing procedure, a highly diagnostic symptom was judged to occur with high frequency under one disease of a pair, and only rarely under the other disease in the pair. A low-diagnostic symptom was perceived to differ less in its frequency of occurrence between the two diseases, and non-diagnostic symptoms occurred with approximately equal frequency under both diseases of a pair.

The probability estimates, $P(H \mid D)$ were found to be significantly higher when the high-diagnostic symptom confirming the disease to be evaluated, was presented first than when the same symptom was presented last in the sequence of three symptoms. Similarly, the probability estimates were very low when the high-diagnostic symptom characteristic for the alternative, disconfirming, disease was presented first than when presented last in the sequence. Subjects, who were highly experienced clinical nurses, were thus very sensitive to the serial position in which symptoms were presented as well as to their diagnosticity. The finding emerged from two different disease pairs tested on different subjects and under various modes of presentation and experimental requirements. The tendency to weight items occurring first in a sequence is apparently pronounced, and the weighting of a given item is likely to change as a function of its serial position.

State of the art: is judgement and decision-making research biased?

The picture of human judgement that emerges from the preceding discussion is less than flattering; people are seemingly quite error prone, and judgements are often biased for a number of reasons, some of which have been highlighted here. The studies presented have, of course, been selected to support a particular line of argument, and do not represent all studies carried out. It must be noted that the quality of judgements can be viewed from different perspectives and by adopting different criteria (Cohen, 1979). Judgements deemed biased within one perspective may thus appear perfectly sound when viewed from a different perspective. Furthermore, the tendency to recall fewer examples from the literature of good than of poor performance has been documented (Christensen-Szalanski and Beach, 1983). Some authors also argue that laboratory research on judgement heuristics is too artificial to be of relevance to the 'real' world (Ebbesen and Konecni, 1980; Edwards, 1983), and that, because these lack 'ecological validity' (Brunswik, 1955), findings from such studies are of no consequence to what people actually 'do' (Gaeth and Shanteau, 1984). It is therefore reasonable to question whether the type of research presented here is representative of the way people make judgements, and whether the studies discussed have been properly conducted and analysed.

While authors such as Christensen-Szalanski (1985) agree that cognitive biases exist, he also argues that they have little effect on the outcome of clinical decisions. In reanalysing data from a series of studies investigating physicians' evaluations of patients potentially suffering from pneumonia, Christensen-Szalanski found most of the physicians' diagnostic decisions to be close to optimum despite the presence of typical judgemental biases. Elimination of all these biases, he calculated, would have helped the average physician in one such study to avoid one incorrect diagnosis per annum. Since these misdiagnosed patients received the same antibiotic treatment as those correctly identified as having pneumonia, the judgemental biases had no practical effect.

This finding may well be clinically encouraging, but, as pointed out by Schwartz and Griffin (1986), research typically focuses on the outcome rather than on the process by which clinical diagnosis are arrived at. It is possible that clinicians 'make the "right" decision for the "wrong" reasons' (Schwartz and Griffin, 1986, p. 81). If the diagnostic process as such is suboptimal when judged in terms of normative principles, this justifies a consideration of the time and costs associated with that process itself. Costs incurred by a patient occupying a hospital bed longer than the disease warrants due to a delay in the diagnostic process, are as readily calculated as those associated with the prescribing of unnecessary tests, treatments and medication. The hidden costs associated with additional pain, suffering and discomfort to patients, however, cannot be estimated exclusively in monetary terms. From a humanitarian or ethical viewpoint, these hidden costs would seem of sufficient importance to seek to reach a better understanding of the cognitive processes associated with the evaluation of medical data.

The removal of an inflamed appendix is, for example, a relatively minor surgical intervention. Yet, the consequences of a perforated appendix can be lethal. Apart from placing the patient in acute life-threatening danger, the recovery time is substantially longer, and the cost of additional treatment of, say diffuse peritonitis, is considerably greater. The time between initial medical assessment of a patient and the proposal of a diagnosis leading to quick action can, in some instances, be as important as, or even more important than, the fact that the diagnosis is correct. This consideration alone would seem to justify an exploration of the extent to which AI techniques may be employed to assist the diagnostician in the act of diagnosing diseases.

When may diagnosticians need help?

It was stated earlier that computer systems already exist, the probabilistic outcomes of which compare favourably with diagnosticians' estimates and recommendations. The above discussion of judgemental heuristics and biases suggests that the outcomes of human judgement may, at least in some instances, not be worthwhile reproducing in a computer system. Focusing on the diagnostic process instead of on outcomes of that process, the following points are worth noting:

1. Because of cognitive limitations, it is not possible for the human judge to entertain a large number of hypotheses simultaneously. While it may be advantageous to limit the search for a particular diagnosis early in the process, it may also be the case that the hypothesis entertained is incorrect. When this hypothesis turns out to be wrong, the time, effort and cost associated with arriving at that conclusion, is largely wasted. Meanwhile, the patient may have become sicker, the road to recovery correspondingly longer, and the medical costs higher. Having arrived at a wrong diagnosis, the diagnostician must review the entire evidence and pursue a new hypothesis, so the diagnostic process must recommence. If a tool could be provided to allude the diagnostician to a range of possible hypotheses based on his/her initial clinical findings, this could save precious time pursuing dead-end lines of investigation. Such a tool would thus assist in the very early stages of the diagnostic process, before a hypothesis is isolated and pursued.
2. The results obtained by Lindgaard (1985) suggest that people may fail to understand the nature of diagnosticity; subjects appeared to evaluate symptoms only in terms of their support for the disease to be evaluated, rather than by a consideration of the term $P(D|\bar{H})$ as well, as required by Bayes' Theorem. Along similar lines, Fischoff and Beyth-Marom (1983) have argued that subjects have a poor understanding of the concept of diagnosticity, and that the relevance of $P(D|\bar{H})$ is not readily comprehended. In order to assess the diagnosticity of a given sign or symptom, it must be evaluated in terms of both competing hypotheses, H and $\bar{H}$. This requirement appears to be violated in some judgements (Beyth-Marom and Fischoff, 1983).

   It was stated earlier that, when applying the representativeness heuristic, subjects ignore the fact that a given symptom may be equally representative of a hypothesis not currently entertained. Hence, even though the diagnostician may 'know' that the symptom occurs under other diseases, (s)he may not consider this because of the early formation of a single hypothesis. Tools providing a set of potential hypotheses as claimed under (1) above, would thus be valuable in terms of helping to overcome potential cognitive limitations.
3. Probability estimates obtained in the Lindgaard (1985) study suggested that subjects relied upon the subjectively perceived frequency of occurrence of symptoms under the

diseases evaluated; Leaper's (1975) findings suggested that people may not know the actual value of symptoms. Frequencies of occurrence of symptoms are not available in current medical textbooks. Traditionally, medicine has been taught in a manner that encourages medical knowledge to be stored in memory according to disease (Eddy and Clanton, 1982). Instead of supplying the frequency of occurrence of individual symptoms, medical textbooks present lists of symptoms that are typical for each disease described (Weinstein *et al.*, 1980; Hickman *et al.*, 1982). Such texts fail to emphasize the 'fact that the same signs and symptoms can be seen in many diseases' (Barrows and Bennett, 1972, pp. 276–7). This argument underscores the point made in (2) above concerning the limitations imposed by early hypothesis formulation. The system suggested here would also provide objective diagnosticity indexes of symptoms, signs, tests and treatments.

4. According to Eddy (1982) who explored physicians' understanding of probabilistic reasoning, the distinction between $P(H|D)$ and $P(D|H)$ is poorly understood. Apart from Eddy's finding that, in an informal sample of some 100 physicians, 95 confused the meaning of the two terms, he also found that this particular error was made explicitly by authors in the medical literature. In one article that sought to rectify this very error, Eddy found that the author in committing it overestimated the probability of malignancy of breast tumours by more than a factor of 20. Similar findings have emerged from other studies (Snyder, 1966; Schwartz *et al.*, 1973; Schwartz and Griffin, 1986). Such studies suggest that surgery could, on occasion, be carried out in cases in which it would not be warranted on the basis of the evidence, but justified instead on misinterpretations of probabilistic data. The empirical findings, taken together, underscore the importance of providing aspiring physicians with formal reasoning skills during their training.

How can AI techniques help?

From the points made above, one may conceive of a computer program that assists in at least three different ways, namely (i) by providing information about diseases under which the symptom configuration observed in any one patient may occur; (ii) by yielding objective estimates to the relative frequency of occurrence of individual symptoms under a range of diseases, and (iii) by calculating posterior probabilities using Bayes' Theorem, based on the information provided by the diagnostician. All of this information could be of value to the expert diagnostician as well as to the medical student wishing to test and further his/her understanding of probabilistic reasoning skills. One may thus propose a computer system that evolves gradually, although three distinct phases are distinguishable. It is important to note that each phase would be of value in its own right, and that, therefore, diagnosticians could quickly benefit from such a system.

Prerequisites for suitable AI tools

Phase 1: a database of medical texts

It was noted earlier that very large databases would need to be established to yield useful posterior probabilities that accurately reflect both the base rates and indexes of diagnosticity of symptoms. The advantage of the present proposal is that databases consisting only of data taken from current medical texts should prove useful very quickly. As was argued earlier, it is important that physicans consider several potential diagnoses before pursuing one particular hypothesis. However, since hypotheses are apparently formed very early in the diagnostic process, and since a single salient symptom may suffice for the formulation of a hypothesis (Eddy and Clanton, 1982), it is also important that the initial data-gathering process be conducted in an unbiased manner.

In addition to the possibility that the most salient symptom may not be indicative of the disease from which the patient is suffering, it has also been established that people find it difficult to attend to the absence of cues (Chrisensen-Szalanski and Bushyhead, 1982). Yet, the absence of a given symptom can be as informative as its presence. Absence of important evidence may, for example, lead to the exclusion of an otherwise attractive hypothesis. Hulse *et al.* (1975) found that the requirement of physicians to fill in a predefined checklist of signs and symptoms when examining patients increased their attention to absent cues. In a similar fashion, deDombal *et al.* (1974) suggested that physicians' performance could be improved by using checklists when examining patients with abdominal pain. Such checklists would thus seem to be a good way to ensure uniform and unbiased data-gathering procedures.

These simple techniques could help diagnosticians overcome the cognitive limitation compelling them to determine and pursue a single hypothesis virtually straight away merely by intervening between the initial gathering of patient data and formulation of an early hypothesis. Although the method cannot entirely preclude the possibility of an incorrect hypothesis being pursued, it can at least make the diagnostician aware of potentially feasible choices, thereby widening his/her initial search for the best candidate.

Applicable AI tools

Predicate calculus and *connectives* could be used to locate the most likely diseases worthy of further consideration by the diagnostician. If exhaustive lists were to be provided to the diagnostician containing all or some of the symptoms present in a given patient and entered from the completed checklist, these lists would, on occasion, include so many diseases that they would be of no value. Similarly, a rule imposed stating that 'diseases containing at least three of the entered symptoms or more, be listed, would not be efficient either. Rovsing's sign was mentioned earlier to be highly diagnostic of appendicitis. Hence, the presence of Rovsing's sign alone would increase the probability of appendicitis considerably and exclude a number of otherwise possible candidates. In other instances, a much longer list of symptoms would have to be present for correct diagnosis.

Where single symptoms, or clusters of symptoms, are known to increase the probability of a given disease, and to exclude other potential diseases, this information could be flagged and stored along with the symptom lists, retrievable by means of connectives. Consider the following example:

Epigastric pain [and] distention [not] cyanosis→
consider indigestion [not] angina pectoris

and reversed:

Epigastric pain [and] cyanosis [not] distention→
consider angina pectoris [not] indigestion

Of course, this example is simplified immensely; in reality more data would be evaluated, but the example shows the way in which connectives and predicate calculus could be applied.

Production rules may be used to rank order the diseases recommended for further consideration. For example,

If {Rovsing's sign} then [appendicitis]
[terminate search here + output to screen]

As primacy effects are to be expected, one may assume that the topmost potential diseases listed where several result from the database search will receive most attention by the clinician. The ability of the system to rank order potential diseases is thus as important as the decision to terminate a search in the presence of certain flagged symptoms.

Phase 2: generating diagnosticity indexes

If the initial data base is to increase in value, it must be expanded. First and foremost, it needs feedback from the diagnostician. Information about tests conducted and their results, about the diagnoses entertained as a result of the initial computer output, the correctness of such diagnoses, recovery patterns, treatments, complications arising, side effects from medication, and so on, would need to be provided continually for the computer gradually to be able to generate diagnosticity indexes.

The establishment and maintenance of such comprehensive databases could be a barrier to their practical application, as this would require continual dialogue between the diagnostician and the computer, even after the patient has been discharged from hospital or has died. While such an achievement may seem to require an insurmountable amount of effort, it should be noted that all this information is stored anyway in current medical records. The difference is that, whereas test results, medication, treatments and treatment effects are currently stored in the records of individual patients, the type of database suggested here would store the same information electronically, although information about individual patients would neither be available nor be of interest in the current context. Yet, diagnostic databases could well be integrated with patient records stored on-line. Technologically, it is quite feasible for a computer program containing AI components to pick out the relevant data from patient records, if such data were to be stored electronically. However, the provision of patient records on-line could prove ethically problematic as it would be both possible and desirable, from the treating physician's point of view, to access individual patient data used to generate diagnosticity indexes. The security in such systems would have to be very tight to guard against unauthorized access; it is doubtful whether current encryption techniques are sufficiently developed to ensure complete privacy. In principle, patient records could form the basis, together with textbook data, for interactive databases.

Given the apparent reliance on frequency of occurrence of symptoms for estimating posterior probabilities as found by Lindgaard (1985), and the lower accuracy of subjective frequencies relative to real-life computer generated frequencies (Leaper, 1975), it would seem feasible to establish, diagnostic aids in the form of data bases. Symptom diagnosticity indexes generated in this fashion would provide valuable information to both experts and trainees; such information is currently not publically available; only experts possess sufficient knowledge to generate subjective indexes and one's experience may or may not be typical or representative of actual occurrence.

Applicable AI tools

Confidence factors: diagnostic values would be stated in terms of actual instances stored in the database, expressed in probabilistic terms as $P(D|H)$ and $P(D|\bar{H})$. *Production rules* would be used to assess the diagnostic value of the symptoms entered:

if [$P(D|H) = P(D|\bar{H})$] then [tell user symptom is nondiagnostic]
if [$P(D|H) > P(D|\bar{H})$ then [favour H]
if [$P(D|H) < P(D|\bar{H})$ then [favour $\bar{H}$]

Phase 3: Generating posterior probabilities

Assume that the fully evolved database is able to generate base rates and diagnosticity indexes from the data contained therein. Such a system should now be in a position to estimate posterior probabilities using Bayes' Theorem. The actual calculation of probabilities only takes a few lines of code. Actual Bayesian calculations could assist the diagnostician in the evaluation of symptoms, signs and test results by generating subsets of the most likely diseases given the configuration of findings entered by the diagnostician.

In addition, it would also give the diagnostician a tool with which to evaluate the effect on the posterior probability of certain tests not yet carried out. The diagnosticity index alone would provide information about the diagnostic value of a given test or procedure in the light of the competing hypotheses under consideration. That is, the diagnostician could interrogate the computer in a 'what if' fashion; in deciding whether to perform a given test (s)he could request the relevant $P(H \mid D)$ assuming that the test outcome were positive (or negative), and in the light of the nearest competing hypothesis, that is $P(D \mid \bar{H})$, given a specified alternative hypothesis. That value can then assist in the decision to conduct or to forego the test. The ability quickly to assess such values should, at least in some instances, enable the diagnostician to avoid tests that are unnecessary because it can be shown a priori that it would provide little or no new information. Similarly, the decision to undertake tests that are very uncomfortable or even dangerous to the patient, could be guided by objective probability estimates of its potential decrease in diagnostic uncertainty. Thus, safer patient care and reduced cost of such care ought to ensue.

Applicable AI tools

Confidence factors and *production rules* would be applied to rank order the values of $P(H \mid D)$ for the diseases tested; the disease with the highest value, that is nearest 1, would be displayed on top of the list. Similarly, a cutoff point could be determined, below which a $P(H \mid D)$ value would not be displayed to the user. *Forward chaining* should be used in the 'what if' situations, and *backward chaining* for verifying diagnoses post hoc, as mentioned earlier.

System capabilities

The type of system described here would need to possess a range of sophisticated capabilities enabling it to conduct selective data searches, to combine data in novel ways, possibly from different data bases, and to make specific recommendations to the system user. The way in which a comprehensive system of this kind is structured and the interactive user–computer dialogue depends largely on the philosophy adopted by the designer. The premises upon which such philosophies rest, and even the philosophies themselves, are not of specific relevance here. Important in the current context is that technologically, the development of such a system is quite feasible, and that it would require the incorporation of up-to-date AI tools. The application of Bayes' Theorem, as well as the fulfilling of prerequisites for its practical usefulness, is thus one area in which AI techniques could be of benefit to medical diagnosticians.

Should the human judge be replaced by smart computers?

Although the computational power and accuracy of the machine will, at some level, surpass that of the human judge, the purpose of allowing Bayes' Theorem to enter into the diagnostic process as a reasoning tool also renders it a valuable decision aid. Indeed, the fact that computers are better and faster at combining evidence and performing calculations should not be seen as a threat; it is still up to the human judge to manipulate the variables and order the computer what to do, despite the fact that, at some stage, the computer would be able to make valid suggestions to the diagnosticians. The fear that computers would replace human judges is not realistic; there will be situations in which the computer need not be consulted, although the relevant data should be entered anyway for storage in the data base as cumulative evidence. Similarly, some data elements cannot be encoded meaningfully because they may be idiosyncratic to a given patient or situation, and diagnosticians often have such knowledge available. The link, for example, between the patient's occupational demands, personal habits, or specific environmental constraints, and the current clinical picture may be perfectly obvious to the physician but not easy to encode into the computer, in particular if the conditions of the patient are inconsistent with other, pre-existing cases.

In addition, the computer can only make use of the knowledge and data contained within its data base. The occurrence of new diseases could not be ascertained from existing data; the Acquired Immunity Deficiency Syndrome (AIDS) would, for example, have been unlikely to be identified in such a system. Compound clinical pictures to which several simultaneously occurring diseases may contribute would also be extremely difficult to isolate. In such instances the human judge is likely to excel and to outperform the computer. Similarly, the diagnostician must be able to check when the computer is likely to have combined data to produce nonsense. Much as schoolchildren still need to learn the small tables despite the availability of inexpensive electronic calculators, so decision aids in the form of databases cannot replace, and does not eliminate the need for human expertise. Hence, the early AI euphoria, which led scientists to predict that computers would do all the work, does not ring true; people may err in their judgements and be prone to be biased, but they are not replaceable!

Barriers to using AI tools

Apart from the obvious high cost of establishing the large data bases necessary to fulfil the prerequisites for phases 2 and 3 described above, acceptance of formal decision tools is a critical factor (Schwartz and Griffin, 1986). Physicians are not generally taught to use formal decision analysis and techniques, and the resistance to using statistics is, according to Eddy (1982), quite remarkable. Statements such as 'the patient is a case of one', 'statistics are for dead men', and 'statistical methods can only be applied to a population of thousands' (Eddy, 1982, p. 259) occur frequently in the medical literature. An increasing number of diagnostic aids and training packages is becoming available, and it is to be hoped that the inclusion of decision analysis in the formal medical curriculum in the future will eventually change these traditional attitudes.

In addition to mistrusting statistical methods and procedures, physicians are not generally well acquainted with computers either. The barriers precluding the introduction, establishment and maintenance of large databases such as those suggested here can therefore be assumed to be enormous, as the validity of the data relies on continued dialogue between physician and computer. In the short term, such databases are unlikely to become a reality, but as young people increasingly are exposed to computers prior to university entry, one may hope that computerized tools and decision aids will gradually find their way into clinical medicine.

Summary

Human judgement and decision making is not as accurate as was assumed within the AI fraternity in the early years. It is prone to a range of biases, and people tend to employ heuristics which, whilst they may not inevitably prevent good decisions from being made, can nevertheless give rise to serious judgemental errors. The consequences of such errors in clinical medicine can lead to prolonged morbidity and an increase in pain and discomfort for the patients, apart from incurring an unnecessarily high cost of patient care. Whereas existing computer systems utilizing formulae such as Bayes' Theorem in medicine tend to focus on the outcome of decisions, a case was made here for the desirability of systems that can be of assistance earlier in the diagnostic process as well; the data-gathering procedure and consideration of alternative hypotheses before a given hypothesis is entertained were argued to be of prime importance. A fully fledged system with a large up-to-date database, it was argued, would also be capable of assisting physicians in making decisions concerning whether or not to perform a given test, take a biopsy or to prescribe a certain type of medication. Even so, the three-phase system described here would be useful diagnostic decision aids at every stage throughout the diagnostic process.

Because computerized procedures could help to reduce the cost of patient care by reducing the number of tests and treatment options, the implementation of such data bases could also contribute directly to safer care of sick people. Despite the fact that such systems would be beneficial even in the short term by providing cross references of symptom configurations, the notion is unlikely to find much support within the medical fraternity very quickly. This is partly due to misconception of statistical and formal decision tools, and partly to a limited exposure to computers by physicians. However, it cannot be denied that the potential for computerized decision aids in medical diagnosis is an attractive reality; it could be a fruitful tool given a chance. The realization that AI techniques are potentially supporting, rather than competing agents against which to measure human performance, is one precondition for such acceptance. Sure, AI can outperform real people, but so what? The fact that aeroplanes can fly and that cars can reach a speed not even attempted by man does not seem to pose the same threat as being beaten intellectually. This is interesting, since the mechanical use of mathematical formulae to large amounts of data or the selection of certain specified items from lists stored in a data base seem to have precious little to do with smart thinking. The most significant barrier deferring the introduction of AI based decision aids in clinical medicine may, in the end, turn out to be sheer human vanity.

References

Alty, J. L. and Coombs, M. J., 1984, *Expert Systems: Concepts and Examples* (Blackburn, Lancashire: NCC Publication).

Anderson, N. H., 1981, *Foundation of Information Integration Theory* (New York: Academic Press).

Arkes, H. R. and Harkness, A. R., 1980, The effect of making a diagnosis on subsequent recognition of symptoms. *Journal of Experimental Psychology: Human Learning and Memory*, **6**, 568–575.

Barrows, H. S. and Bennett, K., 1972, The diagnostic (problem solving) skill of the neurologist: experimental studies and their implications for neurological training. *Archives of Neurology*, **26**, 273–277.

Beyth-Marom, R. and Fischoff, B., 1983, Diagnosticity and pseudodiagnosticity. *Journal of Personality and Social Psychology*, **45**, 1185–1195.

Bowman, E. H., 1963, Consistency and optimality in managerial decision making. *Management Science*, **9**, 310–321.

Brunswik, E., 1955, Representative design and probabilistic theory. *Psychological Review*, **62**, 193–217.

Christensen-Szalanski, J. J. J., 1985, Toward an understanding of human judgment: medical pills for psychological ills. Paper presented at the *10th International Research Conference on Subjective Probability, Utility and Decision Making* (Helsinki, Finland).

Christensen-Szalanski, J. J. J. and Beach, L. R., 1983, The citation bias: fad and fashion in the judgment and decision literature. Paper presented at the *21st Bayesian Research Conference* (Los Angeles, CA).

Christensen-Szalanski, J. J. J. and Bushyhead, J. B., 1982, Physicians' use of probabilistic information in a real clinical setting. *Journal of Experimental Psychology: Human Perception and Performance*, **7**,(4), 928–935.

Christensen-Szalanski, J. J. J., Beck, D. E., Christensen-Szalanski, C. M. and Koepsell, T. D., 1983, The effect of journal coverage on physicians' perception of risk. *Journal of Applied Psychology*, **68**, 278–284.

Cohen, L. J., 1979, On the psychology of prediction: whose is the fallacy. *Cognition*, **7**, 385–407.

Dawes, R. M., 1971, A case study of graduate admission: application of three principles of human decision making. *American Psychologist*, **26**, 180–188.

deDombal, F. T., Leaper, D. J., Horrocks, J. C., Staniland, J. R. and McCann, A. P., 1974, Human and computer-aided diagnosis of abdominal pain: further report with emphasis on performance of clinicians. *British Medical Journal*, **1**, 376–380.

Doherty, M. E., Mynatt, C. R., Tweney, R. D. and Shiavo, M. D., 1979, Pseudodiagnosticity, *Acta Psychologica*, **43**, 111–121.

Ebbesen, E. B. and Konecni, V. J., 1980, On the external validity of decision making research: what do we know about decisions in the real world? In: *Cognitive Processes in Choice and Decision Behavior*, edited by T. S. Wallsten (Hillsdale, NJ: Lawrence Erlbaum).

Eddy, D. M., 1982, Probabilistic reasoning in clinical medicine: problems and opportunities. In: *Judgment Under Uncertainty: Heuristics and Biases*, edited by D. Kahneman, P. Slovic and V. Tversky (Cambridge: Cambridge University Press).

Eddy, D. M. and Clanton, C. H., 1982, The art of diagnosis: solving the clinicopathological exercise. *New England Journal of Medicine*, **306**, 1263–1268.

Edwards, W., 1971, Bayesian and regression models of human information processing – a myopic perspective. *Organizational Behavior and Human Performance*, **6**, 639–648.

Edwards, W., 1982, Conservatism in human information processing. In: *Judgment Under Uncertainty: Heuristics and Biases*, edited by D. Kahneman, P. Slovic and A. Tversky (Cambridge: Cambridge University Press).

Edwards, W., 1983, Human cognitive capabilities, representatives, and ground rules for research. In: *Analysing and Aiding Decision Processes*, edited by P. Humphreys, O. Svenson and A. Vari (Amsterdam, North-Holland: Elsevier Scientific Press).

Elstein, A. S., 1979, Human factors in clinical judgment: discussion of Schriven's "Clinical Judgment". In: *Clinical Judgment: A Critical Appraisal*, edited by H. T. Engelhart, S. F. Spicker and B. Towers (Dordrecht, Holland: Reidel).

Elstein, A. S., Shulman, L. S. and Sprafka, S. A., 1978, *Medical Problem Solving: An Analysis of Clinical Reasoning* (Cambridge, MA: Harvard University Press).

Fischoff, B. and Beyth-Marom, R., 1983, Hypothesis evaluation from a Bayesian perspective. *Psychological Review*, **90**(3), 239–260.

Gaeth, G. J. and Shanteau, J., 1984, Reducing the influence of irrelevant information on experienced decision makers. *Organizational Behavior and Human Performance*, **33**, 263–282.

Gevarter, W. B., 1984, *Artificial Intelligence, Expert Systems, Computer Vision and Natural Language* (Park Ridge, NJ: Noyes Publications).

Goldberg, L. R., 1970, Man versus model of man: a rationale plus some evidence, for a method of improving on clinical inference. *Psychological Bulletin*, **73**(6), 422–432.

Hammond, K. R., 1987, Toward a unified approach to the study of expert judgment. In: *NATO ASI Series F: Computer and Systems Sciences:* Vol. 35, *Expert Judgment and Expert Systems*, edited by J. L. Mumpower, L. D. Phillips, O. Renn and V. R. R. Uppuluri (Berlin: Springer), pp. 1–16.

Hammond, K. R., 1988, Moving results from research to practice. *Judgment/Decision Making*, **7**(2), 3–4.

Hammond, K. R., Hursch, C. J. and Todd, F. J., 1964, Analysing the components of clinical inference. *Psychological Review*, **71**(6), 438–456.

Hickam, D. H., Sox, H. C., Jr, Scheff, K. M. and Chin, D., 1982, A study of the implicit criteria used in diagnosing chest pain. *Medical Decision Making*, **2**(4), 403–414.

Hoffman, P. J., 1960, The paramorphic representation of clinical judgment. *Psychological Bulletin*, **47**, 116–131.

Hoffman, P. J., 1968, Cue-consistency and configurality in human judgment. In: *Formal Representation of Human Judgment*, edited by B. Kleinmuntz (New York: John Wiley).

Hoffman, P. J., Slovic, P. and Rorer, L. G., 1968, An analysis-of-variance model for the assessment of configural cue-utilization in clinical judgment. *Psychological Bulletin*, **69**, 338–349.

Hulse, S. H., Deese, J. and Egeth, H., 1975, *The Psychology of Learning*, 4th edn, (New York: McGraw-Hill).

Kahneman, D. and Tversky, A., 1972, Subjective probability: a judgment of representativeness. *Cognitive Psychology*, **3**, 430–454.

Kahneman, D. and Tversky, A., 1973, On the psychology of prediction. *Psychological Review*, **80**, 237–251.

Kahneman, D. and Tversky, A., 1982, On the study of statistical intuition. In: *Judgment Under Uncertainty: Heuristics and Biases*, edited by D. Kahneman, P. Slovic and A. Tversky (New York: Cambridge University Press).

Kahneman, D., Slovic, P. and Tversky, A., 1982, *Uncertainty: Heuristics and Biases* (New York: Cambridge University Press).

Klayman, J. and Ha, Y., 1987, Confirmation, disconfirmation, and information in hypothesis testing. *Psychological Review*, **94**, 211–228.

Kleinmuntz, B., 1968, The processing of clinical information by man and machine. In: *Formal Representation of Human Judgment*, edited by B. Kleinmuntz (New York: John Wiley).

Kruglanski, A. W. and Ajzen, I., 1983, Bias and error in human judgment. *Euopean Journal of Social Psychology*, **13**, 1–44.

Kunreuther, H., 1969, Extensions of Bowman's theory on managerial decision-making. *Management Science*, **15**(8), 415–439.

Leaper, D. J., 1975, Computer-assisted diagnosis of abdominal pain using estimates data provided by clinicians and survey data obtained from real life. In: *Human Factors in Health Care*, edited by T. J. Triggs and R. M. Pickett (London: Lexington Books).

Lenat, D. B., 1983a, Theory formation by heuristic search: the nature of heuristics II: background and examples. *Artificial Intelligence*, **21**, 31–59.
Lenat, D. B., 1983b, EURISKO: a program that learns new heuristics and domain concepts. *Artificial Intelligence*, **21**, 61–98.
Lindgaard, G., 1985, Weighting of individuating information elements and base rates in a nursing decision-making task involving non-diagnostic case information. Unpublished Masters Thesis, Monash University, Melbourne.
Lopes, L. L., 1983, Toward a procedural theory of judgment. *Psychological Documents*, **13**(2), MS no. 2590.
Meehl, P. E., 1959, A comparison of clinicians with five statistical methods of identifying psychotic MMPI profiles. *Journal of Counseling Psychology*, **6**, 102–109.
Newell, A. and Simon, H. A., 1972, *Human Problem Solving* (Englewood Cliffs, NJ: Prentice-Hall).
Nisbett, R. E. and Ross, L., 1980, *Human Inference: Strategies and Shortcomings of Social Judgment* (Englewood Cliffs: NJ: Prentice-Hall).
Pellegrino, E. D., 1979, The anatomy of clinical judgments: some notes on right reason and right action. In: *Clinical Judgment: A Critical Appraisal*, edited by H. T. Engelhardt, S. F. Spicker and B. Towers (Dordrecht, Holland: Reidel).
Peterson, C. R. and Beach, L. R., 1967, Man as an intuitive statistician. *Psychological Bulletin*, **68**(1), 29–46.
Phillips, L. D. and Edwards, W., 1966, Conservatism in a simple probability inference task. *Journal of Experimental Psychology*, **72**(3), 346–354.
Pschyrembel, W., 1965, *Klinisches Woerterbuch* (Berlin: Walter De Gruyter and Co.).
Rosch, E., 1978, Principles of categorization. In: *Cognition and Categorization*, edited by E. Rosch and B. B. Floyd (Hillsdale, NJ: Lawrence Erlbaum).
Ross, L., Lepper, M. R. and Hubbard, M., 1975, Perseverance in self-perception and social perception: biased attributional processes in the debriefing paradigm. *Journal of Personality and Social Psychology*, **32**, 880–892.
Schiffman, A., Cohen, S. and Novik, R., 1978, Initial diagnostic hypotheses: factors which may distort physicians' judgment. *Organizational Behaviour and Human Performance*, **21** 305–315.
Schwartz, S. and Griffin, T., 1986, *Medical Thinking: The Psychology of Medical Judgement and Decision Making* (New York: Springer).
Schwartz, W. B., Gorry, G. A., Kassirer, J. P. and Essig, A., 1973, Decision analysis and clinical judgment. *The American Journal of Medicine*, **55**, 459–472.
Scown, S. J., 1985, *The Artificial Intelligence Experience: An Introduction* (Maynard, MA: Digital Equipment Corporation).
Slovic, P., 1969, Analyzing the expert judge: a descriptive study of a stockbroker's decision process. *Journal of Applied Psychology*, **53**, 255–263.
Slovic, P. and Lichtenstein, S., 1971, Comparison of Bayesian and regression approaches to the study of information processing in judgment. *Organizational Behavior and Human Performance*, **6**, 649–744.
Snyder, R. E., 1966, Mammography: contributions and limitations in the management of the breast. *Clinical Obstetrics and Gynecology*, **9**, 207–220.
Snyder, R. E. and Swann, W. B., 1978, Hypothesis testing processes in social interaction. *Journal of Personality and Social Psychology*, **36**, 1202–1212.
Taylor, S. E., 1982, The availability bias in social perception and interaction. In: *Judgment Under Uncertainty: Heuristics and Biases*, edited by D. Kahneman, P. Slovic and A. Tversky (New York: Cambridge University Press).

Toohey, M., 1965, *Innere Medizin fuer Krankenschwestern* (Stuttgart: Georg Thieme).
Triggs, T. J., 1975, Decision aids to the individual medical practitioner. In: *Human Factors in Health Care*, edited by T. J. Triggs and R. M. Pickett (London: Lexington Books).
Tversky, A. and Kahneman, D., 1973, A heuristic for judging frequency and probability. *Cognitive Psychology*, **5**, 207–232.
Tversky, A. and Kahneman, D., 1974, Judgment under uncertainty: heuristics and biases. *Science*, **185**, 1124–1131.
Wallsten, T. S., 1981, Physician and medical student bias in evaluating diagnostic information. *Medical Decision Making*, **1**, 145–164.
Weinstein, M. C., Fineberg, H. V., Elstein, A. S., Frazier, H. S., Neuhauser, D., Neutra, R. R. and McNeil, B. J., 1980, *Clinical Decision Analysis* (London: W. B. Saunders).
Wells, G. L., 1980, Lay analyses of causal forces on behavior. In: *Cognition, Social Behavior and the Environment*, edited by J. Harvey (Hillsdale, NJ: Lawrence Erlbaum).
Wells, G. L. and Lindsay, R. C. L., 1980, On estimating the diagnosticity of eyewitness nonidentification. *Psychological Bulletin*, **88**, 776–784.
Wiggins, N. and Kohen, E. G., 1971, Man versus model of man: the forecasting of graduate school success. *Journal of Personality and Social Psychology*, **19**(1), 100–106.
Yntema, D. B. and Torgerson, W. S., 1961, Man–computer cooperation in decisions requiring common sense. In: *Decision Making*, edited by W. Edwards and A. Tversky (Ringowwd, AUS: Penguin Books).

Chapter 30
Reducing human subjectivity in mechanical piece parts classification

ANIL MITAL
Ergonomics Research Laboratory,
Department of Mechanical and Industrial Engineering,
University of Cincinnati,
Cincinatti, Ohio 45221–0072, U.S.A.

Abstract. It is quite common to group mechanical piece parts on the basis of their geometric shape similarity to increase the efficiency of design and manufacturing. Frequently, such coding of parts is done in the absence of part drawings. Since, in these cases visual cues and human subjectivity play a major role in describing the overall geometry of the part and determination of its code, the outcome is an inherently insensitive part coding process. In this paper, a concept to increase the sensitivity of the coding process for mechanical piece parts and a computer-based procedure to utilize the new concept are proposed. The concept requires quantification of subjective descriptors of part geometry in order to determine the part code.

Keywords: Part coding; Subjective descriptors.

Introduction

Grouping and coding of parts on the basis of geometric shape similarity, in order to increase the efficiency of design and manufacturing of mechanical piece parts, is a widely accepted practice. Frequently, parts are grouped together if the processing requirements are similar. This basic concept has been practised for decades. As early as the beginning of this century, Taylor developed a classification and coding system for formation of part families (Hathaway, 1920). In the 1960s, the first classification and coding system was developed and implemented on a computer (Houtzeel, 1985). The recent interest in advanced manufacturing industries to enhance manufacturing productivity has revitalized interest in classification and coding systems for the formulation of machine cells, grouping of tools, parts, etc.

Group technology

Group technology (GT), probably the most common part classification and coding system in use today, is a concept or a manufacturing philosophy which identifies and exploits the sameness of parts and operation processes in design, manufacture, and assembly (Ham *et al.*, 1985). In classification and coding, parts may be grouped together on the basis of geometric shape similarity or similarities of processing requirements (grouping according to part characteristics). The method of grouping parts into a family, or category, involves uniquely identifying attributes of each part by means of a code number which not only indicates the part's similarities with other parts in the family but specific differences as well. The coding system is generally based on the following attributes:

Main shape (rotational, box form, sheet),
Shape elements (such as cones, holes, slots),
Element position,
Dimensions (main, auxiliary, and their ratios),
Accuracy (tolerances),
Materials,
Piece time,
Set-up time,
Quality control information,
Other relevant information (concerning the operation, equipment, assembly, etc.).

The existing part codes, based on GT, reserve the first digit for the overall geometry of the part. Other digits describe such discrete details and characteristics as the nature of machining, presence/absence of auxiliary holds, material, etc.

Of the various parts classification and coding systems, the following three are most widely recognized and used:

1. Opitz system,
2. MICLASS system, and
3. CODE system.

The Opitz system

The basic code consists of nine digits and can be expanded to 13 digits. The first five digits are called the 'form code' and describe the primary design attributes of the part. The next four digits provide attributes useful in manufacturing (e.g. dimensions, work material). Four additional digits may be used to identify production operation type and sequence or some other particular needs.

The MICLASS system

The Metal Institute Classification (MICLASS) system can have from 12 to 30 digits. The first 12 digits are a universal code that can be applied to any part (e.g. main shape, dimensions, tolerance codes, material codes). Up to 18 additional digits can be used to code data that are specific to the particular company or industry.

The CODE system

Developed by Manufacturing Data Systems, Inc., this system was developed for retrieval of part design data. However, it can also be used in inventory control, manufacturing process planning, tool design, and many other applications. The system has eight digits with up to 16 values for each digit. As in other systems, the first digit indicates the part geometry. The interpretation of the remaining seven digits depends on the value of the first digit.

Among the three classification systems described above, the Opitz system is the best known. The general interpretation of the nine digits is shown in Figure 30.1 (Opitz, 1970).

The first digit of the code is very useful in sequencing and scheduling of parts for manufacturing and/or assembly. Frequently, the first digit is also critical in defining the processes involved in its manufacturing.

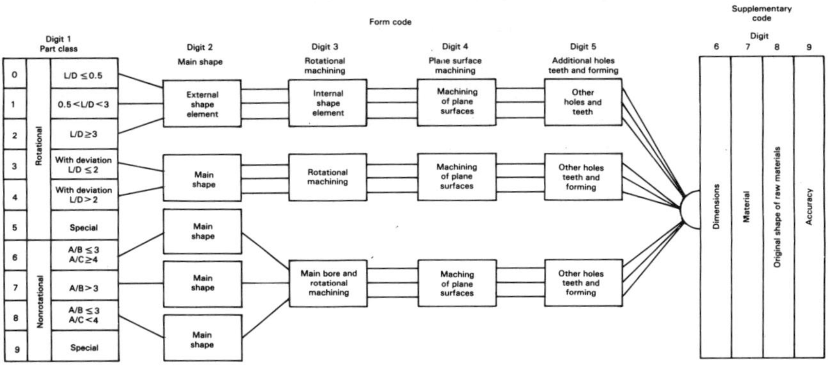

Figure 30.1 Basic structure of the Opitz system, (Opitz, 1970).

Sources of subjectivity

The overall shape of a part is represented by the first digit of the code and is expressed by the ratio, or ratios, of its physical dimensions (Figure 30.1). For rotational parts, the ratio of length (*L*) to diameter (*D*) is used to specify the value of the first digit of the code. Ratios of various sides of a part are used to determine the value of the first digit for flat, long, or cubic components. Contrary to popular belief, in many cases, the determination of these ratios, and hence the first digit, is primarily based on visual cues instead of part drawings (K. Nyamekye, personal communication).

Moreover, even if the various ratios are measured precisely, determination of the first digit is inherently insensitive as parts with different ratios are still assigned to the same category and, therefore, the same value of the first digit. For example, rotational parts are divided into three categories based on their L/D ratios: $L/D \leq 0{\cdot}5$ (category 0), $0{\cdot}5 < L/D < 3$ (category 1), and $L/D \geq 3$ (category 2); the first digit of the code is the category (family) to which a part belongs. The L/D ratio of two parts could be very different, for example 0·4 and 2·6, and yet both would belong to the same category (0). This coupled with the fact that human subjectivity and vagueness often are an integral part of the coding process, since coding in many cases is primarily based on visual cues instead of precise measurements, make the entire coding process quite insensitive.

Problem description and objective

In coding a part and assigning it to a family, operators, in the absence of precise part drawings, rely on visual cues and their experience in judging the geometry of parts. The information on which the decisions are based is qualitative (subjective). The value of the first digit of the part code could be based on the L/D ratio, for rotational parts, or the ratio of the most deviant side to one of the two almost equal sides, for cubic parts, or the ratio of thickness to width, for flat parts. Without precise measurements, determination of all these ratios (critical ratios) requires guesswork. Also, it is easier for individuals to describe the geometry of a part in terms of subjective descriptors of critical ratios. For instance, for rotational parts, individuals, in the absence of exact measurements, could use subjective descriptors such as 'L/D is less than one-half', 'L/D is approximately one-half', 'L/D is more than one-half', to generate a code that is more sensitive than the codes developed by means of the existing GT method (Opitz system). For non-rotational parts, descriptors such as 'the part is almost flat' or 'the part is almost a cube' may be used to describe their shape.

The purpose of this work is to show how subjective descriptors can be used to develop part codes for geometrical shape that are more sensitive than codes provided by the Opitz system. The fuzzy set theory (Zadeh, 1973; Karwowski and Mital, 1984; Karwowski *et al.*, 1986; Karwowski and Mital, 1986a) was used to quantify subjective descriptors of part geometry in order to provide part codes for shape.

Subjective descriptors of past geometry

If the part dimensions are not available during coding, subjective descriptors of part geometry may be used to describe the shape of a part. For rotational parts, the following descriptors may be used:

Part family 1: length is quite small.
Part family 2: length is approximately one-fourth the diameter.
Part family 3: length is approximately one-third the diameter.
Part family 4: length is approximately one-half the diameter.
Part family 5: length is approximately two-thirds the diameter.
Part family 6: length is approximately three-fourths the diameter.
Part family 7: length is approximately equal to the diameter.
Part family 8: length is approximately one and one-half times the diameter.
Part family 9: length is approximately twice the diameter.
Part family 10: length is approximately two and one-half times the diameter.
Part family 11: length is slightly less or nearly equal to three times the diameter.

For flat and cubic parts, the following subjective descriptors may be used:

Part family 12: part is almost flat.
Part family 13: part is more-or-less flat.
Part family 14: part is not very flat.
Part family 15: part is somewhat smaller than a cube.
Part family 16: part is approximately a cube.
Part family 17: part is somewhat bigger than a cube.

Many additional subjective descriptors of the part geometry are possible. The purpose of this work, however, is to demonstrate the concept and procedure rather than to provide an exhaustive list of subjective descriptors.

Quantification of subjective descriptors

Once a part is described by a subjective descriptor, the subjective descriptor must be quantified to generate the part code for shape. The quantification in this case was accomplished with the help of fuzzy set theory. This was not the first time the fuzzy set theory was used to quantify subjective descriptors. Previous examples include quantification of subjective descriptors for mental stress (Mitel and Ulgen, 1982), work performance standards for man–machine systems (Mital and Karwowski, 1986), stresses in manual materials handling (Karwowski and Ayoub, 1984), and risk analysis (Karwowski and Mital, 1986b) among others.

First, the primary term for each subjective descriptor was defined on a universe of finite discourse. These primary terms were then reduced to singletons to yield part codes for geometric shapes.

Intuitive definitions of primary terms

The development of primary terms was subjective and based on intuitive logic – parts described by which critical ratios (*L/D* for rotational parts, thickness over width for flat parts, and the most deviant side over almost equal sides for cubic parts) can be considered as belonging to the family of parts described by a subjective descriptor? The experimental validation of subjective definitions of primary terms is described in the next section.

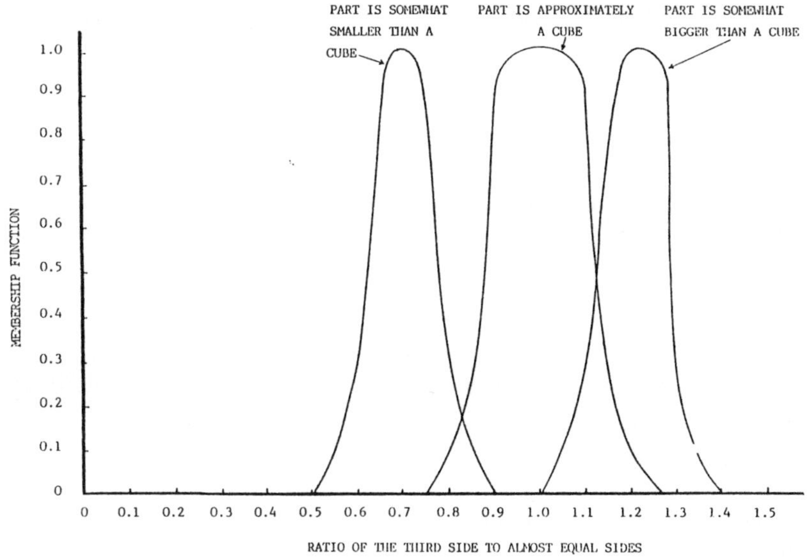

Figure 30.2 Primary terms for rotational parts.

For rotational parts, the primary terms (fuzzy subsets) were defined on a universe of discourse R:

$$R = [(0, 0{\cdot}1, 0{\cdot}2, \ldots, 0{\cdot}9, 1] \ldots \tag{1}$$

covering the membership interval [0, 3] (Figure 30.2 gives examples of these primary terms) as follows:

Length is quite small

$$= [\frac{1}{0}, \frac{1}{0{\cdot}02}, \frac{1}{0{\cdot}04}, \frac{0{\cdot}95}{0{\cdot}06}, \frac{0{\cdot}80}{0{\cdot}08}, \frac{0{\cdot}70}{0{\cdot}10}, \frac{0{\cdot}40}{0{\cdot}12}, \frac{0{\cdot}10}{0{\cdot}14}, \frac{0{\cdot}03}{0{\cdot}16}, \frac{0}{0{\cdot}18}] \tag{2}$$

Length is approximately one-fourth the diameter

$$= \left[\frac{0}{0\cdot 14}, \frac{0\cdot 02}{0\cdot 16}, \frac{0\cdot 20}{0\cdot 18}, \frac{0\cdot 90}{0\cdot 20}, \frac{0\cdot 98}{0\cdot 22}, \frac{1}{0\cdot 24}, \frac{1}{0\cdot 26}, \frac{0\cdot 98}{0\cdot 28}, \frac{0\cdot 90}{0\cdot 30}, \frac{0\cdot 20}{0\cdot 32}, \frac{0\cdot 02}{0\cdot 34}, \frac{0}{0\cdot 36}\right] \quad (3)$$

Length is approximately one-third the diameter

$$= \left[\frac{0}{0\cdot 25}, \frac{0\cdot 02}{0\cdot 26}, \frac{0\cdot 05}{0\cdot 27}, \frac{0\cdot 10}{0\cdot 28}, \frac{0\cdot 70}{0\cdot 29}, \frac{0\cdot 98}{0\cdot 30}, \frac{0\cdot 99}{0\cdot 31}, \frac{1}{0\cdot 32}, \frac{1}{0\cdot 34}, \frac{0\cdot 99}{0\cdot 35}, \frac{0\cdot 98}{0\cdot 36}, \frac{0\cdot 70}{0\cdot 38}, \frac{0\cdot 10}{0\cdot 39}, \frac{0\cdot 05}{0\cdot 40}, \frac{0\cdot 02}{0\cdot 41}, \frac{0}{0\cdot 42}\right] \quad (4)$$

Length is approximately one-half the diameter

$$= \left[\frac{0}{0\cdot 40}, \frac{0\cdot 10}{0\cdot 44}, \frac{0\cdot 20}{0\cdot 45}, \frac{0\cdot 90}{0\cdot 45}, \frac{0\cdot 95}{0\cdot 46}, \frac{0\cdot 99}{0\cdot 47}, \frac{1}{0\cdot 48}, \frac{1}{0\cdot 52}, \frac{0\cdot 99}{0\cdot 53}, \frac{0\cdot 95}{0\cdot 54}, \frac{0\cdot 90}{0\cdot 55}, \frac{0\cdot 20}{0\cdot 55}, \frac{0\cdot 10}{0\cdot 56}, \frac{0}{0\cdot 6}\right] \quad (5)$$

Length is approximately two-thirds the diameter

$$= \left[\frac{0}{0\cdot 58}, \frac{0\cdot 05}{0\cdot 59}, \frac{0\cdot 40}{0\cdot 60}, \frac{0\cdot 60}{0\cdot 61}, \frac{0\cdot 90}{0\cdot 62}, \frac{0\cdot 96}{0\cdot 65}, \frac{1}{0\cdot 66}, \frac{1}{0\cdot 68}, \frac{0\cdot 80}{0\cdot 69}, \frac{0\cdot 40}{0\cdot 70}, \frac{0\cdot 05}{0\cdot 71}, \frac{0}{0\cdot 72}\right] \quad (6)$$

Length is approximately three-fourths the diameter

$$= \left[\frac{0}{0\cdot 68}, \frac{0\cdot 05}{0\cdot 70}, \frac{0\cdot 10}{0\cdot 71}, \frac{0\cdot 90}{0\cdot 71}, \frac{0\cdot 97}{0\cdot 72}, \frac{0\cdot 99}{0\cdot 73}, \frac{1}{0\cdot 74}, \frac{1}{0\cdot 76}, \frac{0\cdot 99}{0\cdot 77}, \frac{0\cdot 97}{0\cdot 78}, \frac{0\cdot 90}{0\cdot 79}, \frac{0\cdot 10}{0\cdot 79}, \frac{0\cdot 05}{0\cdot 80}, \frac{0}{0\cdot 82}\right] \quad (7)$$

Length is approximately equal to the diameter

$$= \left[\frac{0}{0\cdot 9}, \frac{0\cdot 10}{0\cdot 93}, \frac{0\cdot 50}{0\cdot 94}, \frac{0\cdot 90}{0\cdot 95}, \frac{0\cdot 97}{0\cdot 96}, \frac{0\cdot 99}{0\cdot 97}, \frac{1}{0\cdot 98}, \frac{1}{1\cdot 02}, \frac{0\cdot 99}{1\cdot 03}, \frac{0\cdot 97}{1\cdot 04}, \frac{0\cdot 90}{1\cdot 05}, \frac{0\cdot 50}{1\cdot 06}, \frac{0\cdot 10}{1\cdot 07}, \frac{0}{1\cdot 1}\right] \quad (8)$$

Length is approximately one and one-half times the diameter

$$= \left[\frac{0}{1\cdot 4}, \frac{0\cdot 10}{1\cdot 43}, \frac{0\cdot 50}{1\cdot 44}, \frac{0\cdot 90}{1\cdot 45}, \frac{0\cdot 97}{1\cdot 46}, \frac{0\cdot 99}{1\cdot 47}, \frac{1}{1\cdot 48}, \frac{1}{1\cdot 52}, \frac{0\cdot 99}{1\cdot 53}, \frac{0\cdot 97}{1\cdot 54}, \frac{0\cdot 90}{1\cdot 55}, \frac{0\cdot 50}{1\cdot 56}, \frac{0\cdot 10}{1\cdot 57}, \frac{0}{1\cdot 6}\right] \quad (9)$$

Length is approximately twice the diameter

$$= \left[\frac{0}{1\cdot85}, \frac{0\cdot10}{1\cdot89}, \frac{0\cdot50}{1\cdot91}, \frac{0\cdot90}{1\cdot94}, \frac{0\cdot95}{1\cdot95}, \frac{0\cdot99}{1\cdot97}, \frac{1}{1\cdot98}, \frac{1}{2\cdot02}, \frac{0\cdot99}{2\cdot03}, \frac{0\cdot95}{2\cdot05}, \frac{0\cdot90}{2\cdot06}, \frac{0\cdot50}{2\cdot09}, \frac{0\cdot10}{2\cdot11}, \frac{0}{2\cdot15}\right] \quad (10)$$

Length is approximately two and one-half times the diameter

$$= \left[\frac{0}{2\cdot40}, \frac{0\cdot40}{2\cdot43}, \frac{0\cdot60}{2\cdot44}, \frac{0\cdot90}{2\cdot45}, \frac{0\cdot98}{2\cdot47}, \frac{1}{2\cdot49}, \frac{1}{2\cdot51}, \frac{0\cdot90}{2\cdot55}, \frac{0\cdot60}{2\cdot56}, \frac{0\cdot40}{2\cdot57}, \frac{0}{2\cdot6}\right] \quad (11)$$

Length is slightly less or nearly equal to three times the diameter

$$= \left[\frac{0}{0\cdot28}, \frac{0\cdot10}{2\cdot84}, \frac{0\cdot50}{2\cdot9}, \frac{0\cdot80}{2\cdot91}, \frac{0\cdot90}{2\cdot93}, \frac{0\cdot95}{2\cdot94}, \frac{0\cdot98}{2\cdot95}, \frac{1}{2\cdot97}, \frac{1}{3\cdot0}\right] \quad (12)$$

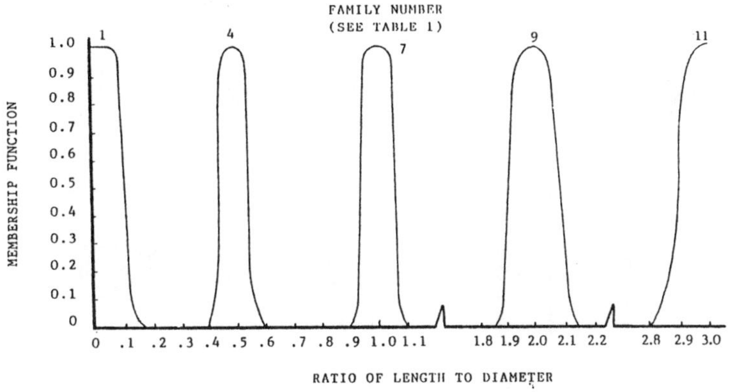

Figure 30.3 Primary terms for flat parts.

In the primary terms described by equations (2)–(12), the numerator is the grade of membership (in the part family) and the denominator is the L/D ratio (from 0 to 3).

For flat parts, the primary terms were again defined on a universe of discourse R (Figure 30.3), covering the membership interval [0, 1]) as follows:

Part is almost flat

$$= \left[\frac{1}{0}, \frac{1}{0\cdot05}, \frac{0\cdot95}{0\cdot07}, \frac{0\cdot90}{0\cdot08}, \frac{0\cdot80}{0\cdot09}, \frac{0\cdot60}{0\cdot1}, \frac{0\cdot50}{0\cdot1}, \frac{0\cdot30}{0\cdot11}, \frac{0\cdot10}{0\cdot12}, \frac{0\cdot05}{0\cdot13}, \frac{0}{0\cdot14}\right] \quad (13)$$

Part is more-or-less flat

$$= \left[\frac{0}{0\cdot 05}, \frac{0\cdot 20}{0\cdot 11}, \frac{0\cdot 60}{0\cdot 12}, \frac{0\cdot 80}{0\cdot 14}, \frac{0\cdot 90}{0\cdot 14}, \frac{0\cdot 95}{0\cdot 15}, \frac{0\cdot 98}{0\cdot 18}, \frac{1}{0\cdot 20}, \frac{1}{0\cdot 30}, \frac{0\cdot 90}{0\cdot 36}, \frac{0\cdot 80}{0\cdot 36}, \frac{0\cdot 60}{0\cdot 38}, \frac{0\cdot 20}{0\cdot 39}, \frac{0\cdot 10}{0\cdot 40}, \frac{0}{0\cdot 45}\right] \qquad (14)$$

Part is not very flat

$$= \left[\frac{0}{0\cdot 3}, \frac{0\cdot 20}{0\cdot 4}, \frac{0\cdot 35}{0\cdot 5}, \frac{0\cdot 86}{0\cdot 6}, \frac{0\cdot 98}{0\cdot 7}, \frac{1}{0\cdot 75}, \frac{1}{0\cdot 8}, \frac{1}{0\cdot 9}, \frac{1}{0\cdot 95}, \frac{1}{1}\right] \qquad (15)$$

In equations (13)–(15), the denominator is the ratio of the thickness to width and the numerator is the grade of membership in the part family.

Similarly, for cubic parts also the primary terms were defined on a universe of discourse R (Figure 30.4) covering the membership interval [0, 1·4]:

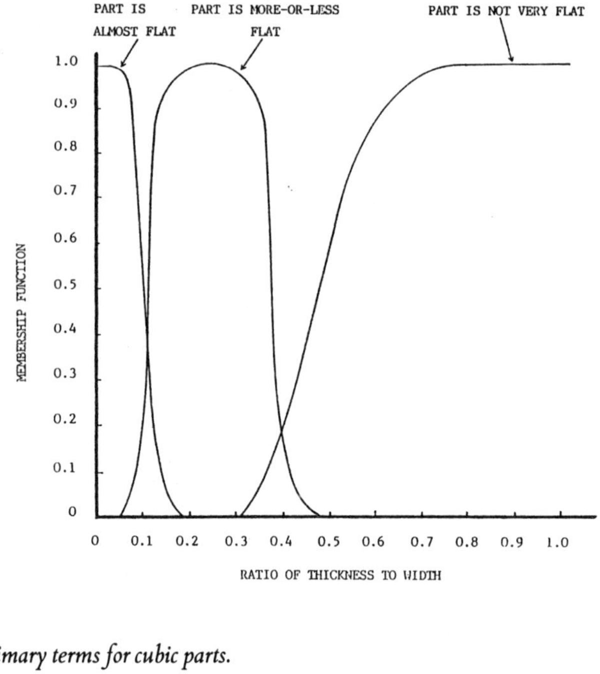

Figure 30.4 Primary terms for cubic parts.

Part is somewhat smaller than a cube

$$= \left[\frac{0}{0\cdot 5}, \frac{0\cdot 30}{0\cdot 6}, \frac{0\cdot 70}{0\cdot 65}, \frac{1}{0\cdot 7}, \frac{0\cdot 7}{0\cdot 75}, \frac{0\cdot 30}{0\cdot 8}, \frac{0}{0\cdot 9}\right] \qquad (16)$$

Part is approximately a cube

$$= \left[\frac{0}{0\cdot75}, \frac{0\cdot10}{0\cdot8}, \frac{0\cdot30}{0\cdot85}, \frac{0\cdot70}{0\cdot9}, \frac{1}{0\cdot95}, \frac{1}{1\cdot0}, \frac{1}{1\cdot05}, \frac{0\cdot70}{1\cdot1}, \frac{0\cdot30}{1\cdot15}, \frac{0\cdot10}{1\cdot2}, \frac{0}{1\cdot25}\right] \qquad (17)$$

Part is somewhat bigger than a cube

$$= \left[\frac{0}{1\cdot0}, \frac{0\cdot20}{1\cdot1}, \frac{0\cdot60}{1\cdot15}, \frac{0\cdot90}{1\cdot18}, \frac{1}{1\cdot2}, \frac{1}{1\cdot25}, \frac{0\cdot80}{1\cdot3}, \frac{0\cdot40}{1\cdot3}, \frac{0\cdot10}{1\cdot32}, \frac{0\cdot10}{1\cdot36}, \frac{0}{1\cdot4}\right] \qquad (18)$$

In equations (16)–(18), the denominator is the ratio of the most deviant side to almost equal sides and the numerator is the grade of membership in the part family.

Validation of primary term definitions

The intuitive definitions of primary terms given by equations (2)–(18) included those part geometry critical ratios which were thought to belong to a particular subjective descriptor. Whether all ratios covered by a subjective descriptor in fact belonged to it or not was determined by carrying out an experimental simulation.

The experiment consisted of projecting two-dimensional cross-sectional views of rotational, flat, and cubic parts on the screen of a personal computer (IBM–XT) in a random order and the worker responding by entering the subjective descriptor, which according to him best described the part geometry, and the numerical value of the part critical ratio. A computer program, written in BASIC, generated the various cross-sections on the screen. Cross-sections of parts representing at least three different critical ratios, but considered belonging to the same family, were projected. In all, 57 different part cross-sections were projected. Table 30.1 shows the family number, subjective descriptor, equation number defining the primary term, and critical ratio for each of the 57 parts projected on the screen. The entire procedure – generation of cross-sections of parts, the order (random) of their presentation on the screen, recording the workers' responses, and determination of membership grade – was computerized.

As each part cross-section was randomly projected on the screen, the subject determined, just by visual cues, its critical ratio and the subjective descriptor (family number) which best described the part geometry. Both responses were automatically recorded on the computer. Each of the 57 parts was apresented to workers experienced in part coding, and their responses (family number and critical ratio) were recorded. A total of 11 workers participated in the simulation. The distribution of subject responses are given in Tables 30.2 and 30.3. Table 30.2 shows, for each of the 57 parts, the actual part critical ratio, the subjective descriptor that was considered to best describe it, and the subjective descriptor each subject thought best described the part. As clearly seen in the table, some subjects thought parts of given critical ratios could be best described by a subjective descriptor other than the one thought to best describe it. In most cases, however, subjects thought the given descriptor best described the part geometry. Nevertheless, some modifications of primary terms based on intuitive logic (equations (2)–(18)) were necessary.

Table 30.3 shows the actual part critical ratio and the mean, standard deviation, and range of critical ratios estimated by workers from the projected part cross-section. As shown in Table 30.3, the average of estimated critical ratio value was quite different from the actual critical ratio value. Both, however, did fall within the same subjective descriptor. This clearly indicates that without measurement, human workers cannot accurately determine the physical size of a part, but with reasonable accuracy, they can determine the subjective descriptor that best describes its shape. This means that while the workers can use visual cues to determine the family of a part, based on geometrical shape, their ability to order parts further within the family is limited.

Table 30.1 Subjective descriptors and critical ratios of the simulated parts used in the validation of primary terms.

Family	Subjective descriptor	Equation of the primary term	Critical ratio	Part no.
			L/D	
1	Length is quite small	(2)	0·03	1
			0·09	2
			0·15	3
2	Length is approximately one-fourth the diameter	(3)	0·17	4
			0·25	5
			0·28	6
3	Length of approximately one-third the diameter	(4)	0·32	7
			0·34	8
			0·40	9
4	Length is approximately half the diameter	(5)	0·42	10
			0·50	11
			0·58	12
5	Length is approximately two-thirds the diameter	(6)	0·60	13
			0·65	14
			0·70	15
6	Length is approximately three-fourths the diameter	(7)	0·72	16
			0·75	17
			0·81	18
7	Length is approximately equal to the diameter	(8)	0·92	19
			1·00	20
			1·08	21
8	Length is approximately one and one-half times the diameter	(9)	1·42	22
			1·50	23
			1·58	24
9	Length is approximately twice the diameter	(10)	1·90	25
			2·00	26
			2·10	27
10	Length is approximately two and one-half times the diameter	(11)	2·42	28
			2·50	29
			2·58	30
11	Length is slightly less or nearly equal to three times the diameter	(12)	2·80	31
			2·90	32
			2·98	33
			Thickness/Width	
12	Part is almost flat	(13)	0·03	34
			0·07	35
			0·10	36
			0·12	37
13	Part is more-or-less flat	(14)	0·15	38
			0·23	39
			0·25	40
			0·42	41
14	Part is not very flat	(15)	0·45	42
			0·55	43
			0·60	44
			0·80	45

Table 30.1 Continued.

Family	Subjective descriptor	Equation of the primary term	Critical ratio	Part no.
			Most deviant side/almost equal side	
15	Part is somewhat smaller than a cube	(16)	0·52	46
			0·68	47
			0·75	48
			0·85	49
16	Part is approximately a cube	(17)	0·92	50
			1·00	51
			1·10	52
			1·15	53
17	Part is somewhat bigger than a cube	(18)	1·18	54
			1·25	55
			1·28	56
			1·38	57

Table 30.2 Subjective descriptor selected by each worker which in his opinion best described the part geometry.

Part number	Subjective descriptor	Critical ratio	Subjective descriptor selected by each subject										
1	1	0·03	1	1	1	1	1	1	1	1	1	1	1
2	1	0·10	1	1	1	1	1	1	1	1	1	1	1
3	1	0·15	1	1	1	1	1	1	1	1	1	1	1
4	2	0·17	2	1	1	1	1	1	2	1	1	2	2
5	2	0·25	2	2	2	2	2	2	2	2	2	2	2
6	2	0·28	2	2	2	2	2	2	2	2	2	2	2
7	3	0·32	3	3	3	3	3	3	3	3	3	3	3
8	3	0·34	3	3	3	3	3	3	3	3	3	3	3
9	3	0·40	4	3	3	3	4	3	4	3	3	4	4
10	4	0·42	4	4	3	3	3	3	3	3	3	3	3
11	4	0·50	4	4	4	4	4	4	4	4	4	4	4
12	4	0·58	5	4	4	4	5	4	5	5	5	5	5
13	5	0·60	5	5	4	4	5	5	5	5	5	5	5
14	5	0·65	5	5	5	5	5	5	5	5	5	5	5
15	5	0·70	6	6	6	5	6	6	5	6	6	5	5
16	6	0·72	6	6	6	6	6	6	6	6	6	6	6
17	6	0·75	6	6	6	6	6	6	6	6	6	6	6
18	6	0·81	7	7	7	7	7	7	7	6	6	7	7
19	7	0·92	7	7	7	7	7	7	7	7	7	7	7
20	7	1·00	7	7	7	7	7	7	7	7	7	7	7
21	7	1·08	7	7	7	7	7	7	7	8	7	7	7
22	8	1·42	8	8	8	8	8	8	8	8	8	8	8
23	8	1·50	8	8	8	8	8	8	8	8	8	8	8
24	8	1·58	8	8	9	8	8	8	8	8	9	8	8
25	9	1·90	9	9	9	9	9	9	9	9	9	9	9
26	9	2·00	9	9	9	9	9	9	9	9	9	9	9
27	9	2·10	9	9	9	9	9	9	9	9	9	9	9

Table 30.2 Continued.

Part number	Subjective descriptor	Critical ratio	Subjective descriptor selected by each subject										
28	10	2·42	10	10	10	10	10	10	10	10	10	10	10
29	10	2·50	10	10	10	10	10	10	10	10	10	10	10
30	10	2·58	10	10	11	10	10	11	10	10	10	10	11
31	11	2·80	11	11	11	11	11	11	11	11	11	11	11
32	11	2·90	11	11	11	11	10	10	11	11	11	11	10
33	11	2·98	11	11	11	11	11	11	10	11	11	11	11
34	12	0·03	12	12	12	13	12	12	12	12	12	12	12
35	12	0·07	12	12	12	12	12	12	12	12	12	12	12
36	12	0·10	12	12	12	12	12	12	12	12	12	12	12
37	12	0·12	13	13	13	13	13	12	12	12	12	13	12
38	13	0·15	13	13	13	13	13	13	13	13	13	13	13
39	13	0·23	13	13	13	13	13	13	13	13	13	13	13
40	13	0·25	13	13	13	13	13	13	13	13	13	13	13
41	13	0·42	14	14	14	14	13	14	14	12	14	13	13
42	14	0·45	14	14	14	14	14	14	14	14	14	14	14
43	14	0·55	14	14	14	14	14	14	14	14	14	14	14
44	14	0·60	14	14	14	14	14	14	14	14	14	14	14
45	14	0·80	14	14	14	14	13	14	14	14	14	14	14
46	15	0·52	15	15	15	15	15	15	15	15	15	16	15
47	15	0·68	15	15	15	15	15	15	15	15	15	15	15
48	15	0·75	15	15	15	15	15	15	15	15	15	15	15
49	15	0·85	15	15	15	15	16	15	15	15	15	16	15
50	16	0·92	16	16	16	16	16	16	16	16	16	16	16
51	16	1·00	16	16	16	16	16	16	16	16	16	16	16
52	16	1·10	16	16	16	16	16	16	16	16	16	16	16
53	16	1·15	16	16	16	16	16	16	16	16	16	16	16
54	17	1·18	17	17	17	17	17	17	17	17	17	17	17
55	17	1·25	17	17	17	17	17	17	17	17	17	17	17
56	17	1·28	17	17	17	17	17	17	17	17	17	17	17
57	17	1·38	17	17	17	17	17	17	17	17	17	17	17

Since in some cases the primary terms needed modifications, as the part was considered belonging to a family described by a subjective descriptor other than the one thought to describe its shape, such modifications were carried out by reassigning some part critical ratios to other families. The modified primary terms are given by equations (19)–(26):

Length is quite small

$$= \left[\frac{1}{0}, \frac{1}{0\cdot 02}, \frac{1}{0\cdot 04}, \frac{0\cdot 95}{0\cdot 06}, \frac{0\cdot 80}{0\cdot 08}, \frac{0\cdot 70}{0\cdot 10}, \frac{0\cdot 40}{0\cdot 12}, \frac{0\cdot 10}{0\cdot 14}, \frac{0\cdot 10}{0\cdot 16}, \frac{0\cdot 10}{0\cdot 17}, \frac{0}{0\cdot 18}, \frac{0}{0\cdot 19}\right] \tag{19}$$

Length is approximately one-fourth the diameter

$$= \left[\frac{0}{0\cdot 16}, \frac{0\cdot 10}{0\cdot 17}, \frac{0\cdot 20}{0\cdot 18}, \frac{0\cdot 90}{0\cdot 20}, \frac{0\cdot 98}{0\cdot 22}, \frac{1}{0\cdot 24}, \frac{1}{0\cdot 26}, \frac{0\cdot 98}{0\cdot 28}, \frac{0\cdot 90}{0\cdot 30}, \frac{0\cdot 20}{0\cdot 32}, \frac{0\cdot 02}{0\cdot 34}, \frac{0}{0\cdot 36}\right] \tag{20}$$

Length is approximately one-third the diameter

$$=\left[\frac{0}{0\cdot25},\frac{0\cdot02}{0\cdot26},\frac{0\cdot05}{0\cdot27},\frac{0\cdot10}{0\cdot28},\frac{0\cdot70}{0\cdot29},\frac{0\cdot98}{0\cdot30},\frac{0\cdot99}{0\cdot31},\frac{1}{0\cdot32},\frac{1}{0\cdot34},\frac{0\cdot99}{0\cdot35},\frac{0\cdot98}{0\cdot36},\frac{0\cdot70}{0\cdot38},\frac{0\cdot10}{0\cdot38},\frac{0\cdot10}{0\cdot42},\frac{0\cdot05}{0\cdot43},\frac{0}{0\cdot60}\right] \quad (21)$$

Length is approximately one-half the diameter

$$=\left[\frac{0}{0\cdot39},\frac{0\cdot10}{0\cdot40},\frac{0\cdot15}{0\cdot44},\frac{0\cdot20}{0\cdot45},\frac{0\cdot90}{0\cdot45},\frac{0\cdot95}{0\cdot46},\frac{0\cdot99}{0\cdot47},\frac{1}{0\cdot48},\frac{1}{0\cdot52},\frac{0\cdot99}{0\cdot53},\frac{0\cdot95}{0\cdot54},\frac{0\cdot90}{0\cdot55},\frac{0\cdot20}{0\cdot55},\frac{0\cdot10}{0\cdot56},\frac{0}{0\cdot60}\right] \quad (22)$$

Length is approximately two-thirds the diameter

$$=\left[\frac{0}{0\cdot58},\frac{0\cdot05}{0\cdot59},\frac{0\cdot40}{0\cdot60},\frac{0\cdot60}{0\cdot61},\frac{0\cdot90}{0\cdot62},\frac{0\cdot96}{0\cdot65},\frac{1}{0\cdot66},\frac{1}{0\cdot68},\frac{0\cdot80}{0\cdot69},\frac{0\cdot40}{0\cdot69},\frac{0\cdot10}{0\cdot70},\frac{0}{0\cdot71}\right] \quad (23)$$

Length is approximately three-fourths the diameter

$$=\left[\frac{0}{0\cdot68},\frac{0\cdot05}{0\cdot70},\frac{0\cdot10}{0\cdot71},\frac{0\cdot90}{0\cdot71},\frac{0\cdot97}{0\cdot72},\frac{0\cdot99}{0\cdot73},\frac{1}{0\cdot74},\frac{1}{0\cdot76},\frac{0\cdot99}{0\cdot77},\frac{0\cdot90}{0\cdot79},\frac{0\cdot05}{0\cdot79},\frac{0}{0\cdot80},\frac{0}{0\cdot81}\right] \quad (24)$$

Part is almost flat

$$=\left[\frac{1}{0},\frac{1}{0\cdot05},\frac{0\cdot95}{0\cdot07},\frac{0\cdot90}{0\cdot08},\frac{0\cdot80}{0\cdot09},\frac{0\cdot60}{0\cdot10},\frac{0\cdot50}{0\cdot10},\frac{0\cdot10}{0\cdot11},\frac{0\cdot05}{0\cdot12},\frac{0}{0\cdot13},\frac{0}{0\cdot14}\right] \quad (25)$$

Part is more-or-less flat

$$=\left[\frac{0}{0\cdot05},\frac{0\cdot20}{0\cdot11},\frac{0\cdot60}{0\cdot12},\frac{0\cdot80}{0\cdot14},\frac{0\cdot90}{0\cdot14},\frac{0\cdot95}{0\cdot15},\frac{0\cdot98}{0\cdot18},\frac{1}{0\cdot20},\frac{1}{0\cdot30},\frac{0\cdot90}{0\cdot32},\frac{0\cdot60}{0\cdot32},\frac{0\cdot20}{0\cdot36},\frac{0\cdot10}{0\cdot39},\frac{0\cdot05}{0\cdot42},\frac{0}{0\cdot43}\right] \quad (26)$$

Table 30.3 Value of critical ratio estimated by each subject for each part.

Part number	Linguistic descriptor	Actual critical ratio	Critical ratio value estimated by each subject											Overall estimated value		
														$\bar{X}$	S	R
1	1	0·03	0·10	0·08	0·06	0·10	0·10	0·08	0·11	0·10	0·01	0·10	0·08	0·083	0·028	0·01–0·11
2	1	0·10	0·15	0·13	0·13	0·20	0·10	0·15	0·13	0·10	0·16	0·16	0·15	0·141	0·028	0·10–0·20
3	1	0·15	0·20	0·15	0·12	0·20	0·20	0·18	0·25	0·20	0·16	0·25	0·20	0·191	0·039	0·12–0·25
4	2	0·17	0·25	0·20	0·15	0·20	0·21	0·17	0·25	0·20	0·15	0·33	0·20	0·210	0·051	0·15–0·33
5	2	0·25	0·25	0·20	0·23	0·25	0·25	0·22	0·25	0·25	0·18	0·25	0·20	0·230	0·026	0·18–0·25
6	2	0·28	0·28	0·20	0·28	0·22	0·25	0·22	0·33	0·20	0·12	0·25	0·33	0·244	0·061	0·32–0·33
7	3	0·32	0·30	0·25	0·25	0·22	0·25	0·22	0·33	0·25	0·20	0·33	0·30	0·264	0·044	0·20–0·33
8	3	0·34	0·33	0·25	0·35	0·30	0·33	0·20	0·33	0·30	0·25	0·33	0·35	0·302	0·047	0·20–0·35
9	3	0·40	0·45	0·33	0·36	0·30	0·50	0·25	0·50	0·30	0·34	0·50	0·50	0·393	0·097	0·25–0·50
10	4	0·42	0·35	0·30	0·48	0·30	0·30	0·30	0·33	0·30	0·30	0·30	0·50	0·342	0·075	0·30–0·50
11	4	0·50	0·48	0·45	0·50	0·50	0·50	0·42	0·50	0·30	0·45	0·25	0·60	0·450	0·098	0·25–0·60
12	4	0·58	0·67	0·50	0·60	0·50	0·75	0·51	0·75	0·60	0·40	0·70	0·67	0·604	0·115	0·40–0·75
13	5	0·60	0·60	0·66	0·60	0·50	0·78	0·60	0·75	0·70	0·66	0·70	0·70	0·660	0·079	0·50–0·78
14	5	0·65	0·66	0·75	0·70	0·67	0·75	0·64	0·75	0·70	0·75	0·75	0·67	0·708	0·043	0·64–0·75
15	5	0·70	0·80	0·85	0·70	0·70	0·90	0·85	0·67	0·80	0·89	0·70	0·70	0·777	0·087	0·66–0·80
16	6	0·72	0·75	0·75	0·76	0·75	0·75	0·80	0·75	0·80	0·75	0·50	0·70	0·732	0·081	0·50–0·80
17	6	0·75	0·75	0·75	0·80	0·67	0·80	0·65	1·00	0·80	0·80	1·00	0·80	0·801	0·111	0·65–1·00
18	6	0·81	0·85	0·99	0·80	1·00	0·90	0·70	1·00	0·80	0·75	1·00	0·95	0·885	0·111	0·70–1·00
19	7	0·92	1·00	1·00	0·90	1·00	1·00	0·90	1·00	1·00	1·00	1·00	1·00	1·00	0·063	0·90–1·10
20	7	1·00	1·00	1·05	1·00	1·00	1·00	1·05	1·00	1·00	1·00	1·00	1·20	1·04	0·063	1·00–1·20
21	7	1·08	1·30	1·20	1·40	1·10	1·00	1·24	1·00	1·00	1·30	1·20	1·30	1·18	0·141	1·00–1·40
22	8	1·42	1·40	1·40	1·70	1·50	1·33	1·38	1·50	1·50	1·60	2·50	1·80	1·60	0·330	1·33–2·50
23	8	1·50	1·40	1·25	2·00	1·50	1·20	1·65	1·50	1·40	1·43	1·50	1·50	1·48	0·211	1·20–2·00
24	8	1·58	1·50	1·30	2·00	1·50	1·40	1·70	1·50	1·50	2·10	2·00	1·70	1·65	0·269	1·30–2·10
25	9	1·90	1·80	1·90	2·20	2·00	1·50	2·10	1·50	2·50	2·20	2·00	2·10	1·98	0·299	1·50–2·50
26	9	2·00	1·90	1·95	2·50	2·00	1·50	2·10	1·50	2·00	2·20	2·00	2·20	1·98	0·291	1·50–2·50
27	9	2·10	2·00	2·25	2·30	2·00	2·10	2·70	1·50	2·50	2·50	2·50	2·20	2·23	0·331	1·50–2·70

Table 30.5 Continued.

Part number	Linguistic descriptor	Actual critical ratio	Critical ratio value estimated by each subject											Overall estimated value		
														$\bar{X}$	S	R
28	10	2·42	2·20	2·60	2·40	3·00	2·40	2·40	2·00	3·00	3·10	2·50	2·40	2·54	0·350	2·00–3·10
29	10	2·50	2·40	2·50	2·40	2·90	2·25	2·40	2·00	2·50	2·34	2·50	2·80	2·45	0·243	2·00–2·90
30	10	2·58	2·40	2·50	3·00	3·00	2·15	3·00	2·50	3·00	3·30	2·50	2·90	2·75	0·352	2·15–3·30
31	11	2·80	2·60	2·75	3·00	3·00	3·00	3·10	3·27	3·00	3·20	3·27	3·00	3·02	0·204	2·60–3·27
32	11	2·90	2·75	2·60	3·40	3·00	2·75	2·75	2·00	3·00	4·00	3·00	2·80	2·91	0·496	2·00–4·00
33	11	2·98	2·90	3·00	3·50	3·50	3·10	3·00	2·50	3·20	3·20	3·00	3·00	3·08	0·278	2·50–3·50
34	12	0·03	0·10	0·10	0·10	0·10	0·10	0·05	0·10	0·10	0·10	0·10	0·05	0·090	0·020	0·05–0·10
35	12	0·07	0·20	0·12	0·12	0·20	0·10	0·15	0·12	0·15	0·14	0·14	0·10	0·140	0·034	0·10–0·20
36	12	0·10	0·20	0·12	0·13	0·20	0·20	0·13	0·25	0·20	0·18	0·20	0·10	0·173	0·046	0·10–0·25
37	12	0·12	0·25	0·14	0·17	0·20	0·42	0·14	0·20	0·13	0·19	0·20	0·13	0·197	0·083	0·12–0·42
38	13	0·15	0·30	0·20	0·16	0·20	0·30	0·21	0·20	0·20	0·22	0·16	0·17	0·211	0·048	0·16–0·30
39	13	0·23	0·35	0·25	0·20	0·25	0·25	0·30	0·33	0·30	0·25	0·25	0·20	0·266	0·049	0·20–0·35
40	13	0·25	0·35	0·20	0·20	0·25	0·25	0·27	0·30	0·25	0·25	0·25	0·20	0·252	0·045	0·20–0·35
41	13	0·42	0·40	0·39	0·33	0·30	0·50	0·50	0·50	0·40	0·40	0·33	0·30	0·395	0·077	0·30–0·50
42	14	0·45	0·40	0·44	0·40	0·30	0·50	0·50	0·50	0·50	0·44	0·50	0·40	0·443	0·065	0·30–0·50
43	14	0·55	0·60	0·59	0·50	0·50	0·80	0·60	0·75	0·50	0·58	0·50	0·60	0·873	0·101	0·50–0·80
44	14	0·60	0·60	0·57	0·50	0·50	0·80	0·45	0·79	0·50	0·57	0·50	0·50	0·570	0·118	0·45–0·80
45	14	0·80	0·75	0·79	0·75	0·75	0·80	0·85	0·90	0·80	0·79	0·80	0·75	0·793	0·047	0·75–0·90
46	15	0·52	0·60	0·50	0·50	0·50	0·80	0·50	0·50	0·50	0·40	0·50	0·55	0·532	0·100	0·40–0·80
47	13	0·68	0·75	0·80	0·60	0·50	0·75	0·50	0·75	0·60	0·75	0·50	0·65	0·650	0·116	0·50–0·80
48	15	0·75	0·75	0·67	0·70	0·70	0·85	0·75	0·75	0·80	0·75	0·80	0·70	0·747	0·053	0·67–0·85
49	15	0·85	0·80	0·80	0·70	0·80	0·90	0·80	0·90	0·80	0·90	0·90	0·80	0·827	0·064	0·70–0·90
50	16	0·92	0·95	0·88	0·80	0·80	0·95	0·85	0·90	0·80	0·90	0·80	0·98	0·873	0·068	0·80–0·98
51	16	1·00	1·00	1·00	1·00	1·00	1·00	1·00	1·00	1·00	1·00	1·00	0·95	0·995	0·015	0·95–1·00
52	16	1·10	1·20	1·10	1·10	1·10	1·00	1·03	1·00	1·10	1·10	1·10	1·20	1·09	0·066	1·00–1·20
53	16	1·15	1·20	1·20	1·20	1·10	1·25	1·10	1·05	1·20	1·20	1·10	1·40	1·18	0·095	1·05–1·40
54	17	1·18	1·30	1·20	1·30	1·24	1·15	1·10	1·10	1·10	1·30	1·20	1·40	1·23	0·100	1·10–1·40
55	17	1·25	1·40	1·20	1·30	1·20	1·35	1·10	1·10	1·20	1·30	1·20	1·60	1·27	0·145	1·10–1·60
56	17	1·28	1·40	1·30	1·30	1·20	1·20	1·15	1·10	1·40	1·20	1·30	1·60	1·28	0·141	1·10–1·60
57	17	1·38	1·50	1·30	1·50	1·30	1·25	1·24	1·15	1·20	1·30	1·30	1·70	1·34	0·161	1·15–1·70

Part classification–application

Once a primary subjective term is used to describe the geometrical shape of a part, the degree of grade of membership can be obtained from equations (8)–(12), (15)–(18), (19)–(26), or similar other equations defining other primary subjective terms which describe part geometry. The grades of membership, which are a fuzzy set, can be reduced to a singleton by selecting the critical ratio(s) corresponding to the highest grade of membership (numerator) in the primary subjective term. In the event there are several grades with the highest value, simple arithmetic average of the corresponding ratios may be taken. For example, if the linguistic descriptor is 'length is approximately three-fourths the diameter', the highest value of the numerator, 1·0, would correspond with ratios of ·74 and ·76. The arithmetical average, or the singleton, would be:

$$(\cdot 74 + \cdot 76)/2 = \cdot 75$$

This would be the part code for shape.

If two rotational parts are described by subjective descriptors – 'length is approximately equal to the diameter' and 'length is approximately twice the diameter' – the Opitz parts classification method (Figure 30.1) would assign the same code, 1 for the first digit, to both of them. The proposed procedure, however, would lead to different codes. For the part which has its length approximately equal to its diameter, the code would be (from equation (8)):

$$= (\cdot 98 + 1\cdot 02)/2 = 1\cdot 00.$$

For the part which has length equal to twice the diameter, the code would be (from equation (10)):

$$= (1\cdot 98 + 2\cdot 02)/2 = 2\cdot 00.$$

The two are quite different from each other and indicate the proposed method is more sensitive. One may also choose to include grades close to the highest grade (1·0) in the averaging process. The contribution of each ratio in the final average should be proportional to its grade. For example, if one wants to include ratios between grades of 0·8 and 1 for the primary subjective term 'part is almost flat', the average part code would be determined as follows:

$$= [1 \times 0 + 1 \times \cdot 05 + \cdot 95 \times \cdot 07 + \cdot 90 \times \cdot 08 + \cdot 80 \times \cdot 09]/[1 + 1 + \cdot 95 + \cdot 9 + \cdot 8]$$
$$= 0\cdot 056$$

The Opitz parts classification system reserves the number '6' for flat parts and the number '8' for cubic components. For the sake of consistency, therefore, a base of 6·0 should be added to part codes, derived on the basis of the proposed method, for flat parts; a base of 8·0 should be added to part codes for cubic parts. Accordingly, if the part code for a flat part is determined to be 0·056, a base of 6·0 would be added to it to determine the final part code for shape:

$$= 6\cdot 056$$

No such bases need to be added to rotational parts. The final part code for shape, thus, may consist of two to four digits instead of one digit yielded by the Opitz system. The various cross-sections of parts can be generated on the computer screen with the help of a graphic software (in this work BASIC was used to generate parts' cross-sections) and presented in either a definite or random order. The operator can then assign a subjective term to describe the part geometrical shape. The operator response can be used to generate the part code as described above. A simple software can automate the entire process with operator's response to part shape being the only input required.

Final remarks

This paper presents a method to increase the sensitivity of the Opitz classification system based on GT. The first digit, which in the Opitz system describes the part geometry, may be replaced by a two to four digit code. This new method is more sensitive than the existing procedure as different codes can be assigned to parts of different dimensions than presently possible.

The experimental simulation of the part coding process, based on subjective descriptors, does indicate that in the absence of precise part dimensions workers can identify parts belonging to different families with the help of subjective descriptors and these descriptors can be reduced to singletons to yield part codes for shape. Seventeen subjective descriptors of part geometry are provided. These descriptors have been tested experimentally and modified where necessary and, therefore, can be readily applied.

The number of subjective descriptors need not be limited; depending upon practical needs, as many such descriptors as needed may be developed. The membership function associated with each primary subjective term, however, should be verified experimentally either by using the experimental simulation described in this paper or by similar other means.

Since the Opitz system is already very popular, methods or procedures which make it more sensitive should be well received.

Acknowledgement

The author acknowledges all individuals who voluntarily participated in the experimental simulation.

References

Ham, I., Hitomi, K. and Yoshida, T., 1985, *Group Technology Application to Production Management* (Boston, MA: Kluwer–Nijhoff).

Hathaway, H. D., 1920, The mnemonic systems of classification as used in the Taylor system of management. *Industrial Management*, **LX**, 173–183.

Houtzeel, A., 1985, Process planning and group technology. In: *CAD/CAM Handbook*, edited by E. Teicholz (New York: McGraw–Hill).

Karwowski, W. and Ayoub, M. M., 1984, Fuzzy modeling of stresses in manual lifting tasks. *Ergonomics*, **27**, 641–649.

Karwowski, W. and Mital, A., 1984, Soft descriptions and classifications in human factors engineering. In: *Trends in Ergonomics/Human Factors I*, edited by A. Mital (Amsterdam: North-Holland), pp. 325–330.

Karwowski, W. and Mital, A. (Eds), 1986a, *Applications of fuzzy set theory in Human Factors* (Amsterdam: Elsevier).

Karwowski, W. and Mital, A., 1986b, Potential applications of fuzzy sets in industrial safety engineering. *Fuzzy Sets and Systems*, **19**, 105–120.

Karwowski, W., Mulholland, N. O., Ward, T. L., Jagannathan, V. and Kirchner, R. L. Jr., 1986, LIFTAN: an experimental expert system for analysis of manual lifting tasks. *Ergonomics*, **29**, 1213–1234.

Mital, A. and Ulgen, O. M., 1982, Mental stress quantification and identification decision modeling. *Proceedings of the Human Factors Society*, Santa Monica, California, pp. 474–478.

Mital, A. and Karwowski, W., 1986, Towards the development of human-work-performance standards in futuristic man–machine systems. *Fuzzy Sets and Systems*, **19**, 133–147.

Opitz, H., 1970, *A Classification System to Describe Workpieces*, Parts I and II (New York: Pergamon).

Zadeh, L. A., 1973, Outline of new approach to the analysis of complex systems and decisions processes. *IEEE Transactions on Systems, Man, and Cybernetics*, **3**, 28–44.

PART VIII
USER–MACHINE INTERFACES

Chapter 31
INTUIT: Intelligent User Interface design Tool

A. J. Russell
System Strategy Centre,
ICL,
Bracknell, U.K.

Abstract. INTUIT has been designed as a demonstrator of a CASE tool in which the process of designing an application is modelled as a step by step process using integrated structured methodologies which are underpinned by a dynamic knowledge-base of appropriate human factors guidance, examples and heuristics. INTUIT is a software development tool which encapsulates expert human factors knowledge to assist a design team in user-centred design. It provides in-line support for the development process and context relevant human factors guidance.

Keywords: CASE tool; Structured design processes; Human factors expertise and software development; Applications design.

Introduction

The European Strategic Programme for Research and Development in Information Technology (ESPRIT) Project 385-HUFIT – Human Factors in Information Technology, aims to provide human factors methods and tools for the design of usable Information Technology (IT) products. The ICL component of the HUFIT project has been to develop an IT-based method of delivering human factors knowledge to the IT product development team. After considerable study of the development processes in its own and other company's environments, it was concluded that the following criteria had to be satisfied.

1. To provide a means by which human factors expertise could be 're-used', due to the endemic shortage of such skills in the design world.
2. To support the whole life cycle of software. Our early work had clearly shown that human factors issues were certainly not confined to the relatively straightforward process of building the interface itself, but instead spread throughout the whole life cycle of a software product.
3. To provide a method(s) by which the user could input sets of constraints which would materially affect the design of the product, such as task and user requirements and general design plans.
4. To provide 'intelligent' advice and guidance about which design plan was applicable to a given situation.
5. To reduce the proportion of empirical design in the interface by providing a structured approach based on human factors advice.
6. To provide a learning environment in which users could explore systematically or by browsing the particular reason why a piece of advice was being offered.

7. To link the advice provided to an appropriate structured design methodology and allow the advice to be delivered either explicitly, in the form of specific advice or examples, (which may or may not be used at the designer's discretion) or implicitly, in the form of *style* in which the designer is provided with human factors advice 'built-in' to the design components, with more limited discretion as to its use.

These criteria had to be satisfied without affecting the innate creativity of the users, i.e. without giving the impression of telling them how to do their job, or de-skilling it. The extent to which these criteria have been realized has still to be validated formally. The following pages outline the general structure of the demonstrator solution.

INTUIT

INTUIT has been designed as a demonstrator of a CASE tool in which the process of designing an application is modelled as a step by step process using integrated structured methodologies which are underpinned by a dynamic knowledge-base of appropriate human factors guidance, examples and heuristics.

The INTUIT design approach was to treat human factors as an integral element of the design process. With this assumption it became clear that the primary task was to construct a method of knowledge delivery which directly mapped onto the current context in which developers are working.

INTUIT, therefore, is a software development tool. Using knowledge engineering techniques, it encapsulates expert human factors knowledge to assist a design team in user-centred design. The driving concern of the tool is in the construction of usable application software, with its main emphasis on the building of user interfaces. However, since the interaction aspects of application design are not completely separable from the rest of the design process, this led to the need to express the whole span of development activity within the process model used by the system. So:

1. INTUIT produces user interface design. Both the display formats and dialogue management components of interface design can be produced with computer assistance from the system.
2. INTUIT may be used as the basis for other aspects of application design. As interface and other design issues are not completely separable, the system will have to cater for design information that is not solely relevant to user interaction. Therefore, there exists an opportunity to perform other activities such as data base design.

The system is intended to provide in-line support for the development process. Human factors expertise should be applied to the development process directly, and a designer should exploit it as a natural part of the job otherwise it is too easily ignored or perceived as mere overhead to the work in hand. INTUIT is intended to operate proactively. Human factors expertise should be used to generate usable systems. A reactive system for evaluation has a role but emphasis on this aspect merely leads to rework and wasted effort in the development process. Ideally, the entire product life cycle needs to be catered for. However, to achieve results in a timely fashion, the requirements for INTUIT have been carefully scoped in terms of user roles, system roles, system tasks and interface styling.

User roles

INTUIT is particularly dedicated to designers. Support for this group of users provides a high degree of leverage in supplying improved usability in the formation of products. This group also presents a reasonably well understood task area to which support tools can be applied. A software designer may use structured methods or prototyping which are both modern and acceptable ways in which a designer may function. Structured methods provide the tools and techniques for transforming a concept into a product. The spirit of such methods is to impose order and mechanizability on to the development process. In so doing, they provide the opportunity to attach human factors expertise directly to the design constructs and transformations used by such methods. Prototyping provides the feedback for evaluation of the designs constructed. In particular, interface evaluation is a popular aspect of prototyping. Prototyping does not presuppose the use of structured methods but is an invaluable adjunct if exploited correctly to evaluate evolving design.

System roles

The system provides human factors support for interface design by a process of stepwise refinement. It provides a window onto the software development process from a user interaction perspective. As indicated below, the initial representation analyses the **task** while the other three loosely correspond to the **semantic, syntactic** and **lexical** levels of interface design. A designer would be expected to transform the initial task analysis into some high level representation of a dialogue which in succeeding steps would be translated into low level screen formats and key strokes. The process adopted is based upon Jackson's system design techniques as embodied in the SSADM methodology for ICL QUICKBUILD implementation (ICL, 1985). INTUIT's support is integrated into this design process.

1. **The task analysis is represented as data flow diagrams (DFD).** These are augmented by the use of entity/relationship models to represent the application object/data of the task domain. One of the first design operations performed upon the task analysis is to identify the user roles and events implied by the data flow diagrams. The use of the word 'implied' is important since we need a representational method or notation which will allow the designer to be more explicit about this critical aspect of design. So the task model represented here is incomplete.
2. **The semantic level of interaction description is represented as data flow structures (DFSC).** This provides an abstract level of dialogue organization in terms of the data flowing between the user and the computer system. The data flow structure can be derived from the data flow diagrams of the task analysis by examination of the data flow associated with the application's events and represented using structure charts.
3. **The deep syntactic level of interaction description is represented as dialogue structures (DSC).** The information in the semantic level data flow structures is regrouped to provide a dialogue structure that meets the requirements of chosen interaction styles and usability. The dialogue structure is primarily represented using structure charts although a corresponding generalized transition network needs to be generated for subsequent refinements.
4. **The lexical level of interaction description is represented as screen formats.** Information about the structure of information represented in preceding stages is translated into choices of interface objects. The surface syntactic level of interaction description is also required, it will constitute an elaboration of the generalized

transition networks derived in the previous phase. The transition networks used to describe details of dialogue structure are based upon the notation developed by Siemens in their work within the HUFIT project on dialogue design (Mittermaier and Haubner, 1987).

For rapid evaluation of evolving designs a prototyping facility is provided. Once a dialogue design has been refined to the lexical or surface syntactic level it can be animated on the INTUIT workbench. These animations make direct use of specific interaction knowledge and encodings of house style. The prototyping activity can thus be supported by a structured methodology in which design information is properly captured and exposed to human factors guidelines.

Interaction styles knowledge will need to be filtered to ensure that it is relevant to the current design task. Interaction styles knowledge will only be relevant to particular types and sets of application objects/data. Certain kinds of user operation also affect interaction relevance. Part of the task of INTUIT is to ensure that only the knowledge relevant to the designer's current problems is provided. INTUIT filters out interaction style information that is inappropriate to the characteristics of the target users and helps focus the interaction styles knowledge base by use of target system information if available.

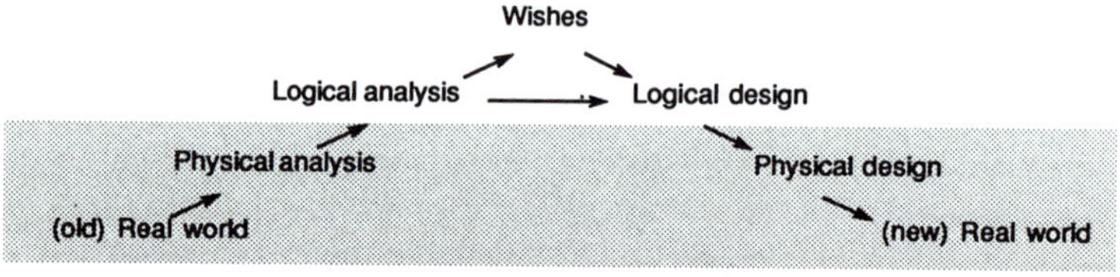

Figure 31.1 The design 'wishbone'.

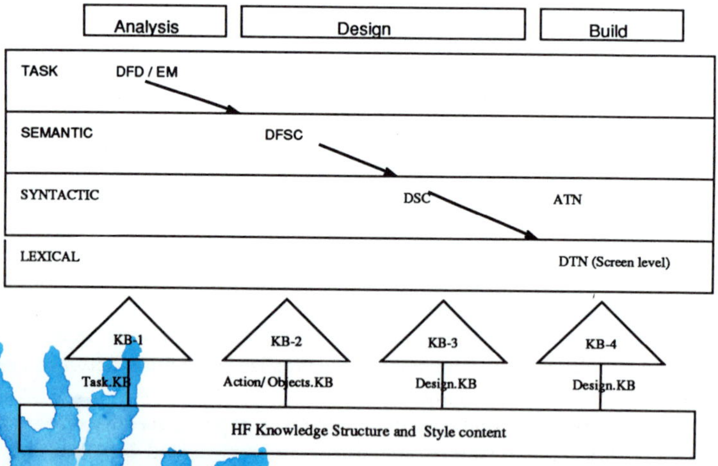

Figure 31.2 The logical processes carried out within INTUIT.

INTUIT will adopt the structured design process as its operating metaphor. Given the complexity of the task, an end-user will need to possess a clear model of how the system operates. This model needs to be based upon the designer's perception of the end-users' existing knowledge (the old problem of matching the designer and end-user models of the task domain). The view is held that the use of a structured methodology represents a communication tool which allows the designer and the end-user to arrive at a commonly understood task model embedded in the system.

Figure 31.1 illustrates the (perhaps) ideal situation in that the logical analysis and design processes naturally relate to user needs (wishes). All too often in fact these logical processes are often omitted from the process and only physical analysis and design is carried out. Figure 31.2 illustrates the INTUIT model of the process.

System tasks

INTUIT supports a range of facilities to enable the designer to develop the interface effectively and with full human factors support, both implicit and explicit. In order to do this it must support the drafting of diagrams. In this respect, it will operate as a computer-aided drafting tool. Only diagrams relevant to INTUIT's interface design process are supported.

The system offers direction to design choices. Where possible, a designer will be able to make design choices by selecting from lists. In the simpler cases this will amount to offering a menu of choices. However, for more complex situations, such as selecting a characterization for a dialogue task, a relatively sophisticated browsing mechanism is supplied.

The system enables interface designs to be prototyped. Automatic screen generation and interface animation allow a designer to evaluate proposed designs and evolve them rapidly.

The system provides guidelines, examples and heuristics. This facility augments the **implicit** provision of human factors expertise offered by the methods listed above. INTUIT will provide this explicit advice to justify its implicit methods and enable its users to learn and improve their own human factors knowledge. This advice will be accessible via the design constructs presented by the drafting facilities. The guidelines and examples presented in any situation will be automatically chosen to be instructive and relevant to that context.

Interface styles

A primary requirement is that the system should be able to store and deliver, in design context, a prescribed interface housestyle. INTUIT contains a generic object set which structures the interface objects and actions such that the interaction style chosen as appropriate to the need can be attached as instances of the generic set.

Architecture

INTUIT consists of four main functional subsystems. These provide a set of design constructs, the human factors knowledge base, the referencer and the user interface. Each of these subsystems is further partitioned and the communication between subsystems is far from simple. However, this section outlines the purpose of each of the components.

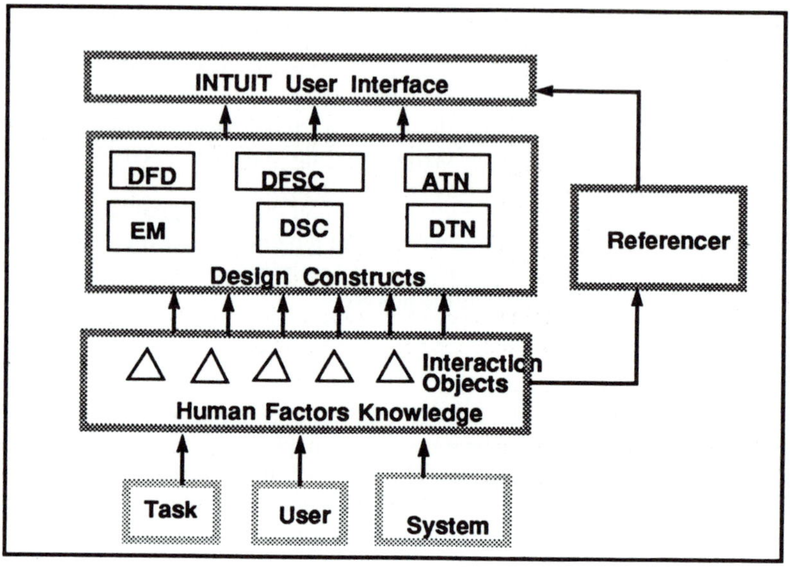

Figure 31.3 The functional subsystems of INTUIT.

User interface subsystem

INTUIT possesses a powerful, graphical user interface. The intended users perform a set of complex tasks requiring high bandwidth communications with the system. Such complex interactions can only be supported in an effective manner using a graphical interface. The interface uses multiple windows in which both text and graphics can be displayed and manipulated using a keyboard and mouse. Application objects are directly manipulable. The basic mode of interaction is by direct manipulation of the application objects presented.

The user interface supports a set of views onto the underlying design constructs and human factors knowledge. The diagrams of a design are displayed in graphical forms in the main work area window. Associated windows are used to access human factors knowledge in such forms as textual guidelines. Note that the designer need not access the taxonomy of human factors knowledge directly, rather this retrieval is performed implicitly by INTUIT via the displayed design constructs. The referencer is that part of the interface that provides the access to guideline and example material either by direct output from a given design context or through the browser mechanism, which also provides hypertext based navigation within the referencer to allow exploration of the advice provided.

Design constructs subsystem (DCS)

The objects in the design constructs subsystem represent the design information required by the development process. These objects are similar to the objects one finds in a data dictionary. They are defined by a rich frame-based schema that describes not only their structure but also the operations that can be applied to them. Inheritance is used extensively to reuse existing definitions and explicitly show similarities. This information consists of the **views** a user can take of the design and the **elements** that represent its details. This subsystem is designed to provide methods of assembling the application objects into a structure representing the design plan. As such it includes SSADM tools for building these plans.

The design constructs subsystem supports transformations. Some of the work a designer would be required to do with INTUIT can be reduced to clerical chores that should be automated. The priority for such automation is in packaging human factors expertise to support designers rather than more general engineering tools for tasks such as data base generation. An example of such automation is in the area of **UIMS** (User Interface Management Systems) where a high level dialogue structure can be used to generate a user interface under the control of a knowledge base encoding a specific interface style.

Figure 31.3 shows that within the design constructs subsystem there are a number of tools required. The views and associated tooling administered by the DCS are such as:

DFD Data Flow Diagrams
EM Entity Models
DFSC Data Flow Structure Chart
ATN Application Transition Network
DSC Dialogue Structure Chart
DTN Dialogue Transition Networks

The DCS also manages the elements contained in the above views. For example, a DFD contains such things as processes, data flows, store, etc. These are all aspects of the structured design methodology known as SSADM (Structured Systems Design Methodology) which was developed by the CCTA (Central Computer and Telecommunications Agency) and approved by the British Government.

One way of defining INTUIT is to regard it as a CASE 'shell' into which knowledge bases have been inserted in order to specialize it to perform a particular function: in this case, design of the interface. The DCS constitutes the kernel of INTUIT. It provides the basic data dictionary and design plan drafting facilities which are effectively what current generations of CASE tools provide.

Human factors knowledge subsystem

The human factors knowledge subsystem is founded on a taxonomy on interaction components. These knowledge bases may be regarded as 'libraries' of components that can be used in the design session.

These components include both the interaction **objects** that a user would manipulate and the interaction **actions** that constitute those manipulations. This taxonomy includes both very generic forms of component such as ones that describe selection actions in general terms and increasingly specific descriptions such as selection from a menu on a particular target system using a particular style. The interaction components taxonomy is partitioned into a collection of knowledge bases which enables its content to be managed in an effective manner as it modularizes the knowledge into distinct chunks.

1. **A generic interaction components** knowledge base lies at the root of the taxonomy. This knowlege base defines both the general structure of the interaction components and how they are organized according to human factors principles. All other interaction component knowledge bases will be based upon this root.
2. **Subsidiary interaction components** knowledge bases will package distinct interface styles, both character and graphic based. The knowledge about each style will be related back to the generic knowledge in the root knowledge base.

3. **Dialects of interface styles** will be accommodated by further knowledge bases. These will allow variants according to target system limitations to be accommodated. It also enables a rich structure of interrelations between interface styles to be represented.

Human factors knowledge and advice is 'attached' to this taxonomy which acts as the framework for delivering this knowledge in the current design context. Rule systems are incorporated at the contextually relevant points in the taxonomy to aid a designer in making choices about how to proceed with the interface design process. Interface design itself can be guided by an orderly migration through the taxonomy. Ideally, design proceeds from general and abstract to more specific. Using the taxonomy, a designer will be directed to make initial, general decisions about dialogues that are to be constructed. Introducing further constraints about the user, the task and the target system will imply movement through the taxonomy of interface components. Eventually, the designer will be located at a set of consistent points in the taxonomy that correspond to a realizable interface design and that lie within the knowledge base for one interface style. Thereby, the design process is directed by the human factors issues appropriate to the choice of task, user and target system.

Design filters

Design filters are used to narrow down both design options and guidance that can be accessed. More specifically, information about **task characteristics, user characteristics** and **target system characteristics** which have been specified early in the development process can be used to limit the range of design options that should be considered at any one stage in the design process, and to ensure that the context-related human factors design guidance and examples are as relevant as possible.

1. Task characteristics will be described as attributes of task components within a task modelling facility. It is assumed that the modelling of components and characteristics would follow similar lines to the modelling of dialogue tasks that is currently supported. As such, a library of task actions and characteristics could be used to assist the modelling process, which would provide the pointers needed for the filtering process. This filtering should help the user make decisions regarding interface styles and dialogue control facilities necessary to support the task.
2. User characteristics are currently represented as a very coarse filter on appropriate interface styles using a simple classification scheme. A system developed by HUSAT known as the Functionality Matrix (described elsewhere in this book) is used to associate functionality with the identified characteristics of a user group. Much research is still going on into methods of defining user models which will provide a much more sophisticated set of filtering techniques (Dunker *et al.*, 1987).
3. Target system characteristics will be described relative to the UIMS that they support. It seems that the most appropriate way of using target system information is as a filter on the interface styles that can be effectively supported. This in turn depends upon the availability of UIM delivery system for those styles on the targets described. Target systems, of course, are the intended delivery vehicle on which the application under design is to be run. The knowledge embedded here will include such material as the nature of the screen, keyboard or other input capabilities and, where intrusive in the dialogue, the characteristics of the operating system.

Application development

Space does not permit a complete description of the process of application development using INTUIT. What follows is a very brief walkthrough of the salient points in the process. (Since INTUIT is built round SSADM [Structured Systems Analysis and Design Method] readers who may be unfamiliar with the syntax will find most of the constructions echoed in other forms of structured methodologies.)

Diagrams are created to represent the real-world application objects ('entities') that are to be modelled in some manner within the system, and their interrelations. Typically they will emerge as data objects stored within the data base of the completed design. The form these diagrams take will have the basic form shown in Figure 31.4.

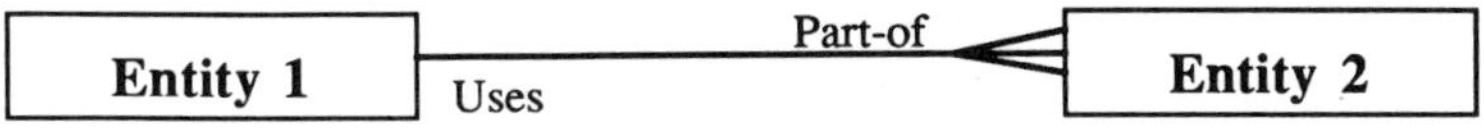

Figure 31.4 Basic form of created diagrams.

As new entries are drawn on-screen, new instances of Entity-type constructs (stored in the Design Constructs subsystem) are created in the Design KB being used to represent the design. As the entities and their relationships are further defined, this information too is stored within the Design KB.

If, for example, we were creating an application for an employment agency, we might see an entity model entry as shown in Figure 31.5, which means that an applicant has a skill or skills which are then further connected to other entities defining the skills requirements for an appointment. Each of these entities or data objects are then further defined within a data dictionary and a form is provided for entering this material for each entity.

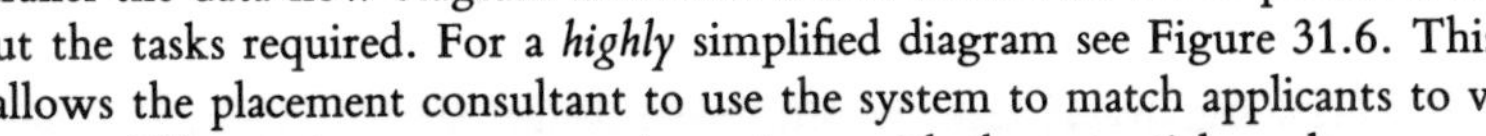

Figure 31.5 Entity model entry for an application for an employment agency.

In parallel the data flow diagram is created which defines the set of processes needed to carry out the tasks required. For a *highly* simplified diagram see Figure 31.6. This simple model allows the placement consultant to use the system to match applicants to vacancies according to skills requirements, set up interviews with the potential employers and record the results in the data store 'Interviews'. What is happening within INTUIT is that all these processes (and subprocesses) and entities are stored as knowledge base frames, or objects, which can be assigned properties and behaviours as task constraints within a task catalogue which will influence – via inheritance and rules systems – the construction of the dialogue.

The next step in the process is to design the overall dialogue steps for each task described in the task catalogues. Construction of Dialogue Structure Charts allows the designer to perform a Jackson-type hierarchical decomposition, gradually breaking down high level tasks into their sub-components, thus determining the dialogue steps the user will have to go

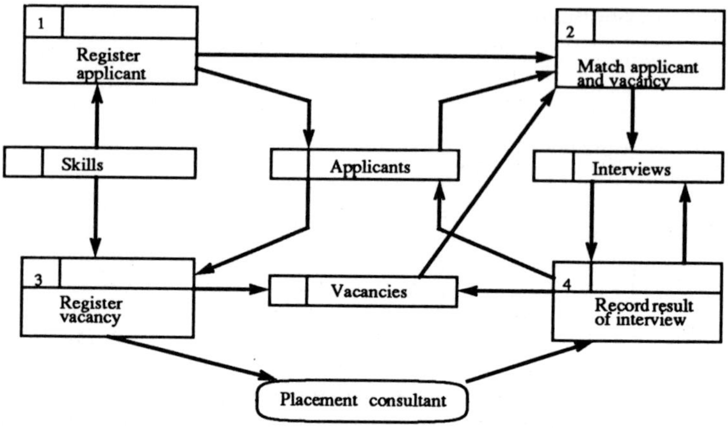

Figure 31.6 Simplified data flow diagram.

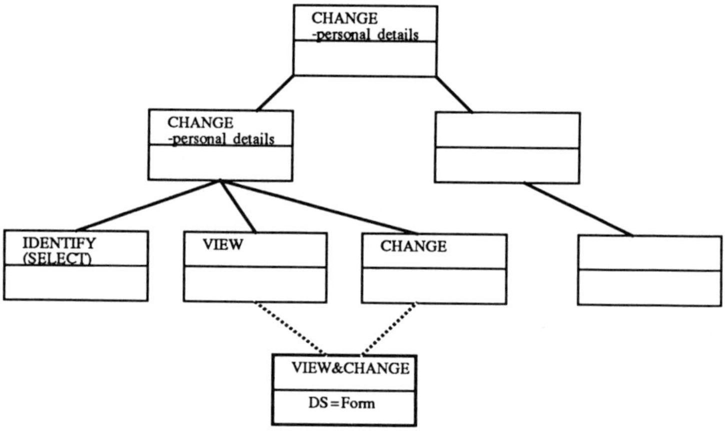

Figure 31.7 Overall dialogue steps.

through to achieve the overall task. In principle a design can be progressed down to the lowest level user interface actions through this process. As shown in Figure 31.7 (again simplified) the overall process involves creating the top-most DSC box to represent the dialogue task under design (in this case Change Applicant Details), then making decisions as to how this overall task will be achieved. In the example shown, the first level of breakdown is into the steps of first changing the details (CHANGE personal details) then changing the applicants local office if a change of address requires it (CHANGE office). The designer has then decided that the first of these tasks (CHANGE personal details) should be broken down into the steps of identifying the applicant being dealt with (IDENTIFY applicant), viewing the current details of the applicant (VIEW personal details) and changing those required

(CHANGE some personal details). He has then decided that all three of these dialogue tasks should be achieved using a forms style interface (DS = FORM), but that while identifying the applicant warrants a screen on its own (SPECIFY applicant id) the tasks of viewing and changing the details should be performed on a single screen. Individual screens are identified by a double border to the box. How each of these tasks is then performed within the screens is not addressed until Screen Design itself.

The bottom leaves of the dialogue structure chart (Figure 31.7) represent screen states, and note that it is within DSCs that the first *explicit* human factors guidance can be accessed (in the current version of INTUIT). When a new DSC box is created on the chart, the designer specifies the type of dialogue action it is to represent – such as CHANGE, CREATE or SELECT. This is done through a menu from the title area of the box. This creates a unit in the underlying Design KB, which is an instance of the appropriate Dialogue Task Action from the library stored in the Human Factors Knowledge subsystem, and also labels the box in the diagram (though the designer may choose a synonym from a submenu if required). The rest of the label for the box may then be specified, which may refer to entities specified earlier in the process – as is the case of 'applicant details' for the DSC box CHANGE Applicant Details. Where this is the case, it will be possible to specify the relevant entities from a list of those modelled, through the DSC box's menu.

Each Dialogue Task Action in the library contains pointers to human factors guidelines that are relevant to that particular type of action; for example, a guideline about data protection may be linked to the CHANGE task action. By specifying components of a DSC to be instances of Dialogue Task Actions, guidance relevant to that type of action (and only that type) can be accessed directly.

This guidance can then be used to assist in the further breakdown and dialogue design for that component. The guidance itself is accessed through a menu from the centre area of a DSC box, and is categorized into guidance on dialogue control, user assistance and guidance on the choice of dialogue style (form, menu, etc.) for that type of action. Equally, examples of good and bad instances are provided. A list of individual guidelines for the category chosen appears in the Guidelines/Examples panel, and on selection an individual guideline will appear in a separate 'referencer' display panel. This panel also contains a facility to access directly other guidelines cross-referred to by the guideline on display as a form of hypertext.

Both selection of a dialogue style and other dialogue design decisions should be based on characteristics of the task the dialogue is aimed to support, and of the users performing the task. To this end, the Task Characteristics information specified earlier can be accessed directly from each box of a DSC. As the characteristics of the overall task are unlikely to be identical to those at lower levels of decomposition (frequency or length, for instance), the characteristics information is inherited down through the decomposition hierarchy but may be altered or redefined at any point.

Dialogue Style Selection of a particular dialogue style for a dialogue step is usually performed at the lower levels of decomposition, and may also be specified through the menu. In addition to placing a label in the box, the effect of this is to change the particular Dialogue Task Action in the library of which the element in the Design KB that this box represents is an instance. Rather than being an instance of the CHANGE task action for example, it is now an instance of the CHANGE.BY.FORM task action, where CHANGE.BY.FORM is actually a subclass of CHANGE in the library's classification structure. Dialogue control actions may be specified from within the DSC at any level, such as Cancel, Backstep, Suspend, etc.

At this point screen design can commence, but with a number of significant differences from normal screen painting. Facilities are provided for animating screen sequences by typing then to a state transition model at either application or dialogue level. Also, where many applications are developed for an organization an interface housestyle may be specified, such that all the organization's applications have a consistent 'look and feel'. Large supplier organizations may also wish to deliver a consistent housestyle and smaller software houses may wish to sell applications that accord to one or more such interface style standards.

To accommodate this, INTUIT allows individual Interface Style knowledge bases to be created. These would be specialized versions of the generic knowledge bases of the Human Factors Knowledge subsystem for Interface Objects and Actions, and would contain specialized versions of individual Objects and Actions from the generic set. As with the generic set, sub-specializations according to dialogue style (Form, Menu, etc.) would exist.

These Interface Style knowledge bases would allow the specification of what standard information areas a screen could contain, what type of interface objects each area should contain (for a particular dialogue style), how these objects should be laid out and displayed, what type of interface actions are allowed (for a particular dialogue style), how they can be invoked and what their effect should be. Again, space limitations do not permit a full description of all the processes of building the interface, including the way in which the interface objects and actions are invoked and inherited, and the manner in which entity requirements are pulled through to the screen design. The following is a listing of some of the facilities available:

Guidelines and dynamic examples at interface object and action level.
Standard housestyled screen structures are provided as templates.
Screens are literal not symbolic, i.e. WYSIWYG.
Menu selection of interface objects within screen components.
Properties such as highlight and protection are embedded in the style.
Default data pulled through from the entity models.
Stepwise refinement of screen functions.
Dialogue transition models can be built for check of consistency/completeness and individual screen designs directly accessible from the state transition nodes.

The next logical step is, of course, to include code generation facilities and integration within a PCTE environment. This is certainly outside the scope of the present HUFIT project but may well be the subject of a separate programme.

Development route

INTUIT was developed using KEE™, a hybrid AI toolkit. It provides frame-based representation facilities together with powerful rule-based programming. Traditional procedural programming can be provided using the underlying COMMON LISP in which KEE is embedded. KEE also offers advanced expert system facilities such as truth maintenance and hypothetical worlds (Dunker *et al.*, 1987).

INTUIT was developed on SUN™ workstations. These combine the ability to host KEE™ with a powerful graphical interface in a UNIX environment. The minimum configuration for development requires a SUN™ 3/60 with 12 Mbytes of memory and 130 Mbytes of filing system.

Conclusion

In order to influence design practice effectively it is essential to provide designers with the information, methods and techniques they require in the environment in which they normally work. INTUIT provides designers with an IT-based method of delivering human factors knowledge. The INTUIT design approach is to treat human factors as an integral part of the design process and deliver it as a part of software development tool.

References

Dannenberg, M. and Ziegler, J., 1987, Dialogue Management Tool for Prototyping and Development of Interactive Systems. HUFIT Working Paper B3.4a, *HUFIT deliverable HUFIT/16-IAO-11/87.*

Dunker, J., Melchior, E-M. and Bosser, T., 1987, *Empirical Methods for the Validation of Models of User Knowledge. HUFIT Working Paper B7.1a/B7.2 HUFIT/3-TOM-12/87.*

ICL, 1985, *ICL:QUICKBUILD User Guide.*

Mittermaier, E. and Haubner, P., 1987, *A Method of Describing Dialogs. HUFIT deliverable 07/-SIE-5/87.*

Chapter 32
Computer-aided design applied to the Flight Telerobotic Servicer user interface

MARK A. STUART and RANDY L. SMITH
Lockheed Engineering and Sciences Company
2400 NASA Road 1
Houston, Texas 77058-3711, U.S.A.

Abstract. The Man–Systems Telerobotics Laboratory (MSTL) of NASA's Johnson Space Center employs computer-aided design (CAD) tools in their design and evaluation of the Flight Telerobotic Servicer (FTS) human/telerobot interface on the Shuttle and on the Space Station. It has been determined by the MSTL that the use of CAD tools can promote more expedient and less costly design endeavours. This chapter describes in detail several specific examples of CAD use applied to the FTS user interface by the MSTL.

Keywords: Computer-aided design; Prototyping; Human factors; Remote manipulators; Telerobotics.

Introduction

Computer-aided design (CAD) techniques, such as software prototype development programs and graphics packages, can serve as an aid in the design, evaluation, and development of user interfaces of many types. These systems design tools can result in the development of ergonomically well-designed workstations in less time with lower costs when compared with the use of other systems design tools.

With the system development process becoming more complex and expensive, more emphasis is being placed on the evaluation of systems during early stages of the development cycle. The design of systems that include human operators is especially complex because determining overall systems performance is dependent upon the interaction of the human operator, hardware components and software components (Chubb *et al.*, 1987). Adequately evaluating the performance of a system during the design cycle is becoming more difficult when using the static evaluation tools traditionally available to the human factors engineer, such as job and task analyses and mock-up development (Geer, 1981). It is becoming more common for systems developers to use CAD and computer simulation as a design tool instead of hardware models (Gawron and Polito, 1985) and for human factors engineers to use CAD and computer simulation to enhance the use of static design tools (Stuart and Smith, 1988).

The Man–Systems Telerobotics Laboratory (MSTL) of NASA Johnson Space Center (JSC) with support from Lockheed has extensively used CAD tools in their design and evaluation of the Flight Telerobotic Servicer (FTS) user interface. It is the goal of the MSTL to use CAD tools to help design, evaluate and develop requirements for the user interface of the FTS. Goddard Space Flight Center is the lead centre in the development of the FTS with other NASA centres and industry playing various roles.

The FTS will be a dual-armed teleoperated robot used to help assemble, service, and maintain NASA's Space Station. There will be an FTS control panel on both the Shuttle and the Space Station. The design of the FTS control panel is especially challenging since it may be physically impossible to have identical control panels on both the Shuttle and the Space Station due to the physical constraints of the Shuttle. The ultimate objective in the design of the FTS control panel is that the human operator's capabilities and limitations have been best accommodated for while ensuring that overall systems goals and requirements are met. The use of CAD will enable NASA to design iteratively a good FTS control panel on the Space Station which does not radically differ from the FTS control panel included on the Shuttle. Radical departures from the control panel used on the Shuttle will increase the likelihood of negative transference or reversal errors. Therefore, design features which take advantage of population expectancies should be a constant feature across both control panels to ensure maximum performance.

This paper will discuss the MSTL's use of CAD tools that have been applied to the design and evaluation of the human–telerobot interface that will be a part of NASA's Shuttle and Space Station. Specific emphasis will be placed upon the use of a prototype development program and on graphics packages. Each example will begin with a statement of the objectives of the task and will then detail the approach taken by the MSTL for that particular application. The discussion of these applications will also include illustrations of the CAD graphics.

Programmable display pushbuttons

The first example given will be an illustration of how CAD was used by the MSTL to establish a set of guidelines concerning the use of programmable display pushbuttons (PDPs) on the Space Station's FTS control panel (Stuart *et al.*, 1988). The CAD tool used during this evaluation was Hypercard. Hypercard is an information management software package which allows the user to organize text, graphics and active 'screen buttons' into cards. The cards can then be linked together in different user-definable stacks. The stacks can then be arranged so that high-fidelity control panel prototypes can be created with relative ease.

This phase of the FTS workstation evaluation covered a preliminary study of PDPs. Since the study of PDPs is now in the early phase of the design cycle, the focus on this evaluation was to use CAD as a means of testing the feasibility of using PDPs on the FTS control panel. The PDP is constructed of a matrix of directly addressable electroluminescent (EL) pixels which can be used to form dot-matrix characters. PDPs can be used to display more than one message and to control more than one function. Since the PDPs have these features, then a single PDP may possibly replace the use of many single-function pushbuttons, rotary switches, and toggle switches, thus using less panel space. Due to space constraints on the Orbiter and the Space Station, an overriding objective of the design of the FTS workstation is that it take up as little panel space as possible. It is of interest to determine if PDPs can be used to perform complex hierarchically structured task sequences adequately.

Other investigators have reported on the feasibility of using PDPs in systems design (Hawkins *et al.*, 1984; Burns and Warren, 1985), but the present endeavour was deemed necessary so that a clearly defined set of guidelines concerning the advantages and disadvantages of PDP use in the FTS workstation could be established. This would ensure that PDP use was optimized in the FTS workstation.

The objective of this investigation, then was to use CAD to compare the performance of experienced and inexperienced manipulator operators while performing a manipulator task on simulated PDP and non-PDP computer prototypes so that guidelines governing the use of programmable display pushbuttons on the FTS workstation could be created. The functionality of the manipulator on the Orbiter was used as a model for this evaluation since the functionality of the FTS at the time of this writing has not been solidified.

CAD in the form of computer prototyping was used as the means of evaluating the two different FTS control panel layouts. Hypercard was used as the prototyping package and it was run on an Apple Macintosh computer. Hypercard was also used as a data acquisition package once testing began. Total task time and the total number of commands activated were recorded. The simulated task consisted of the operations to deploy a satellite on the Space Shuttle. This task required simulated joint manipulations, camera manipulations, as well as other manipulator activities while using the computer prototypes.

The CAD version of the non-PDP control panel is depicted in Figure 32.1. The distinguishing feature of this configuration is that traditional single-function pushbuttons are used in conjunction with a simulated EL panel to activate commands. The EL panel was simulated in this evaluation by displaying single-function commands as they would appear on the EL panel in the upper right-hand corner of the prototyped screen.

The CAD version of the PDP control panel is depicted in Figure 32.2. This control panel utilized simulated PDPs instead of single-function pushbuttons. In Figure 32.2, the PDPs are the 12 pushbuttons located in the lower-middle portion of the display. The portions to the left and top of the display are dynamic status indicators that were used to display various functional states.

When a PDP is selected, the name of that function is then displayed in a small simulated EL display located just above the PDP cluster and the options that follow within that functional category are then displayed by the PDPs. For example, when SINGLE is selected in Figure 32.2, the display changes to that depicted in Figure 32.3. In Figure 32.3, SINGLE is now displayed in the EL display and the PDPs have changed to list the options that follow under

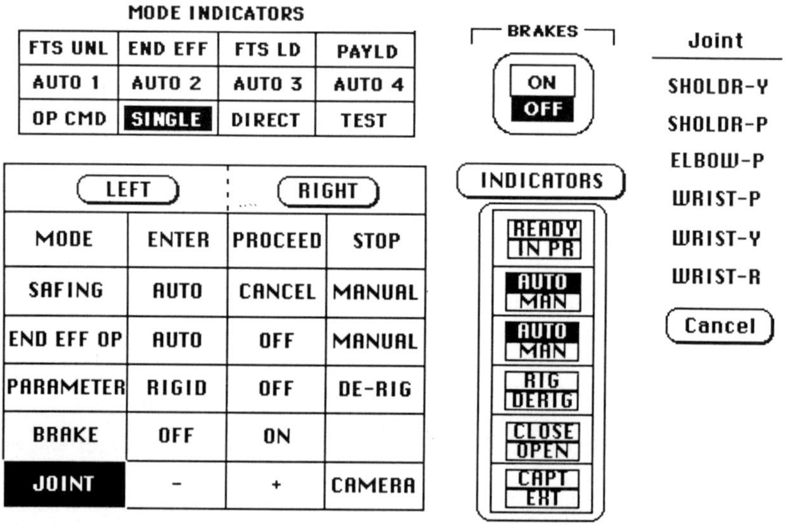

Figure 32.1 Non-PDP control panel prototype.

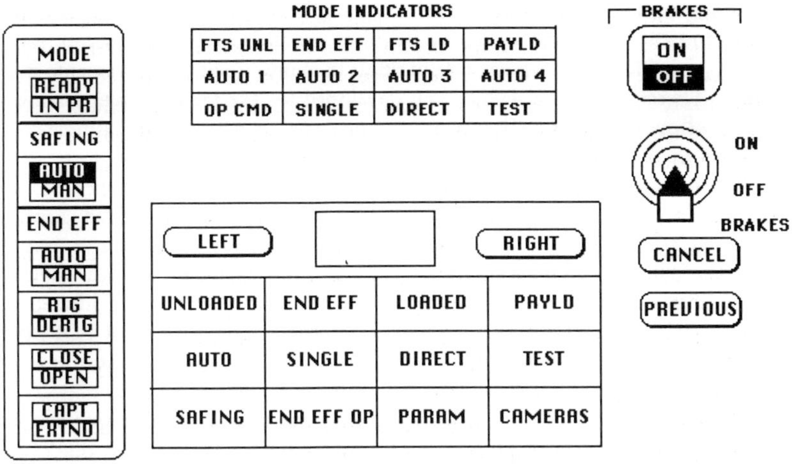

Figure 32.2 PDP control panel prototype.

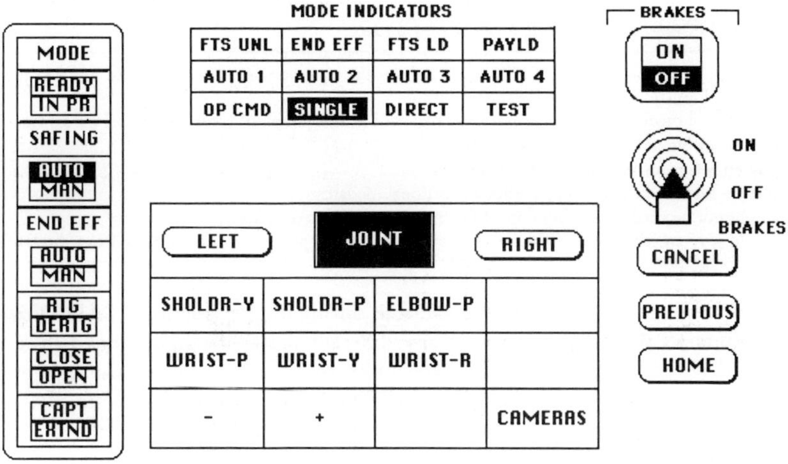

Figure 32.3 PDP control panel prototype with PDP changes and navigational aid.

SINGLE. The small EL display was designed to serve as a navigational aid to help orient operators throughout performance of the hierarchically structured tasks. It was contended that the use of the navigational aid in the PDP hierarchy would be useful since a previous evaluation (Gray, 1986) found that navigational aids were helpful with hierarchical search tasks through menu structures on a computer.

It was determined that there was a problem in manoeuvring across functional modalities during the development of the PDP prototypes because it required many commands. For example, when one was in the joint manipulation mode, it would require several steps, including going back to the 'Home' level of the task hierarchy first, to be able to make camera adjustments. Since the operation requires much manoeuvring across modalities during its use, this PDP arrangement would result in many circuitous movements and much wasted time. Therefore, a special PDP was developed for the present investigation which would readily allow operators to 'jump' across functional modalities with a single command located within the PDP matrix.

After performing the task scenarios on both of the control panel prototypes, each subject was asked to select which of the two control panel prototypes was preferred. Each subject was also asked to complete a questionnaire designed to garner subjective impressions concerning the control panels. Data were collected and analysed with the objective of determining differences in user performance and preference between the two different control panel configurations so that, ultimately, guidelines concerning the use of PDPs could be established. All numeric data were statistically analysed with a repeated measures analysis of variance.

The ultimate objective of this investigation was to use the CAD technique of computer prototyping to establish a set of guidelines concerning the use of PDPs on the FTS workstation. The data collected during this investigation were then used to create this set of guidelines. It is contended that the established set of guidelines will also be generalizable to other workstations as well. These guidelines are listed in Table 32.1. It is contended that the use of this set of guidelines will help to ensure that PDPs will be optimally designed and arranged.

Table 32.1 Guidelines concerning the use of PDPs.

Use PDPs instead of other controls if PDP usage reduces the total number of commands to perform the task and doesn't significantly increase task completion time
A PDP or control capability should be provided that will allow 'jumping' across functional modalities
Navigational aids should be used to help orient users
PDPs may be better for infrequently used sub-tasks
PDPs may be better when working within a functional modality
PDPs should not be used for certain critical functions, such as brake control
PDPs should give an indication of the number of hierarchical steps the operator is away from the 'Home' level

The use of CAD proved to be invaluable during this evaluation. CAD allowed the experimenters to try out iteratively many different design configurations before testing actual, hard-wired PDPs. The MSTL determined that the use of CAD can be a much less expensive systems design and evaluation tool when compared with the use of traditional hardware design and evaluation techniques. For example, it would have been 6·5 times more expensive to conduct the PDP evaluation with the use of hardware control panel configurations instead of CAD control panel configurations. It should be pointed out that this figure may well vary across applications, though. Factors such as the level of complexity of the control panel configurations being evaluated and the learning requirements/ease-of-use of the CAD packages used will have an effect on the cost differential between hardware and CAD design techniques.

Hand controllers and restraint systems

The second example will be a discussion of how CAD was used to evaluate the placement of different types of hand controllers and different types of body restraint systems within various conceptual designs of the FTS workstation on the Shuttle. The CAD tool used during this

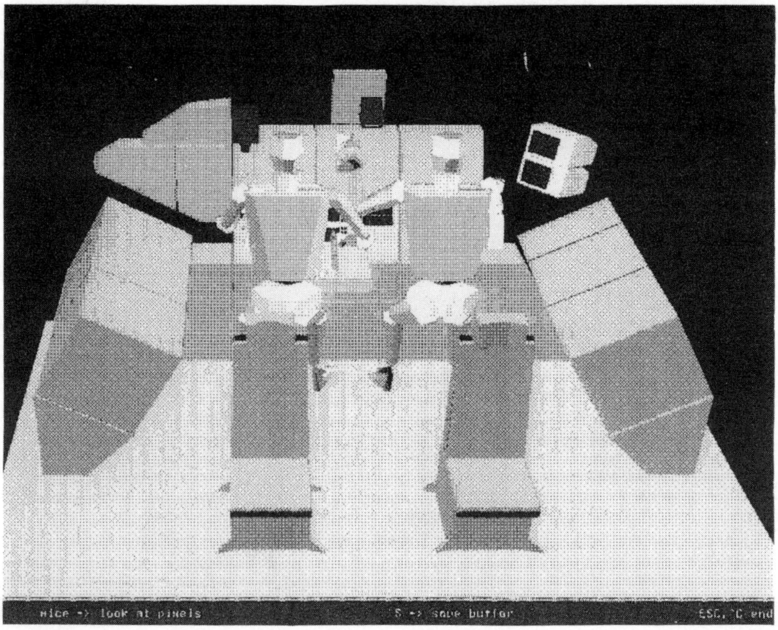

Figure 32.4 PLAID conceptualization of the placement of the FTS workstation in the aft flight deck of the Shuttle.

evaluation was the PLAID graphics package. PLAID is a graphics development package created by the Graphics Analysis Facility of NASA's Johnson Space Center. PLAID enables the creation of three-dimensional graphical images with accompanying animation. The feature of animation enables the MSTL to evaluate different workstation configurations with the interaction of anthropometrically-correct human operators, thereby determining anthropometric reach limits and viewing angles. PLAID also makes it possible to evaluate how well operators of varying physical dimensions can interact with different workstations.

PLAID enabled the MSTL to evaluate FTS workstation layouts interatively within the aft flight deck and the mid-deck of the Shuttle. Figure 32.4 illustrates a conceptual design of the placement of the FTS workstation on the aft flight deck. If the FTS workstation is placed in this location, it will be in close proximity to the Remote Manipulator System (RMS) control panel. Figure 32.4 gives an indication of how two 95th percentile male operators would work together simultaneously. The reader should notice that the PLAID drawing indicates that there will be some shared work space between the two operators. This important finding was made available to the MSTL without the necessity of fabricating full-scale mock-ups. Different sized operators other than the ones examined in this example could also have easily been put into the aft flight deck conceptualizations for evaluation.

Figure 32.5 illustrates how the FTS workstation might be laid out in the mid-deck of the Shuttle and shows a 95th percentile male operator using the workstation located within the bank of lockers in the mid-deck of the Shuttle. Figure 32.6 is a conceptualization of how well a 5th percentile female would be able to reach the controls of the mid-deck FTS workstation. It should be noticed that in Figures 32.5 and 32.6 a restraint system that attaches to the torso of the operators is included for evaluation. This particular restraint system concept was developed by Charles Willits of NASA-Reston.

Figure 32.5 PLAID conceptualization of the placement of the FTS workstation in the mid-deck of the Shuttle with a 95th percentile male operator.

Figure 32.6 PLAID conceptualization of the placement of the FTS workstation in the mid-deck of the Shuttle with a 5th percentile female operator.

Of course, for maximum fidelity and generalizability, the design conceptualizations of the MSTL will be conducted with full-scale mock-ups, but until the design conceptualizations become more definitive, the MSTL will continue to use CAD tools.

Control/display layouts

The third example will be a discussion of the use of CAD in the consideration of the placement of the FTS control panel in the Shuttle. At the time of this writing, it had not been determined where the FTS control panel would be located in the Shuttle. As in the discussion of the use of PLAID in the previous section, two locations were being considered: the aft flight deck and the mid-deck. Many different design features were considered and CAD enabled the MSTL quickly and inexpensively to evaluate the preliminary placement of these features. Some of these features were the size and number of monitors to use, placement of control switches, the types of controls to use, and whether or not a detachable keyboard should be a part of the control panel. The CAD tool used in this example was MacDraw. MacDraw is a graphics development package that is available on Apple Macintosh computer products.

The first examples given will be design considerations made concerning the placement of the FTS control panel in the aft flight deck. Figure 32.7 is a drawing made with MacDraw to illustrate a possible FTS control panel using aft flight deck panel A6-A2. Figure 32.8 is another illustration of panel A6-A2 but with the addition of a detachable keyboard. A slightly different layout is exhibited in Figure 32.9. This is a control panel layout of aft flight deck panel A7-A2 which illustrates the use of two nine-inch monitors.

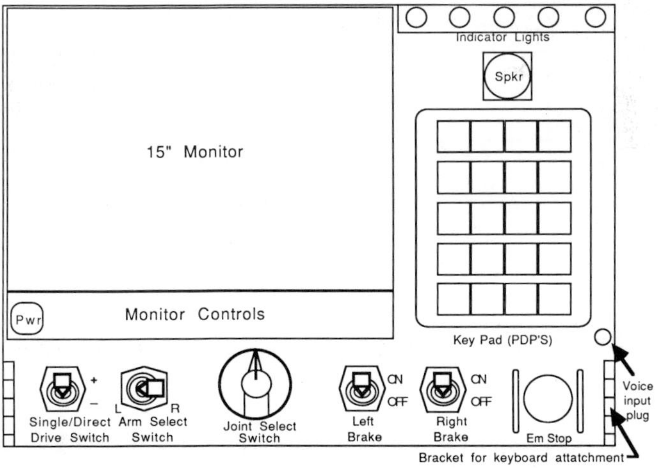

Figure 32.7 FTS control panel in aft flight deck panel A6-A2.

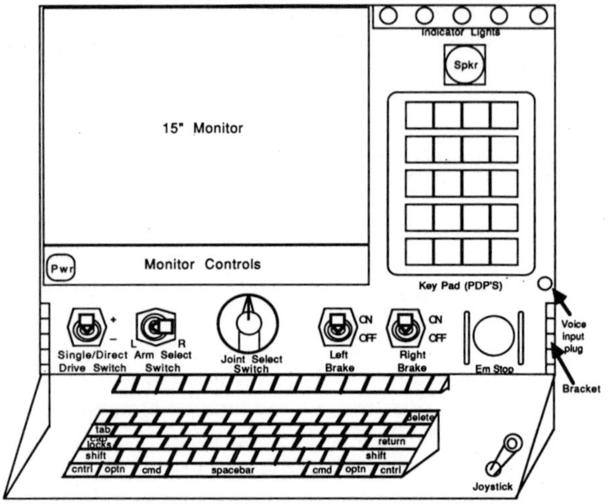

Figure 32.8 FTS control panel in aft flight deck panel A6-A2 with detachable keyboard.

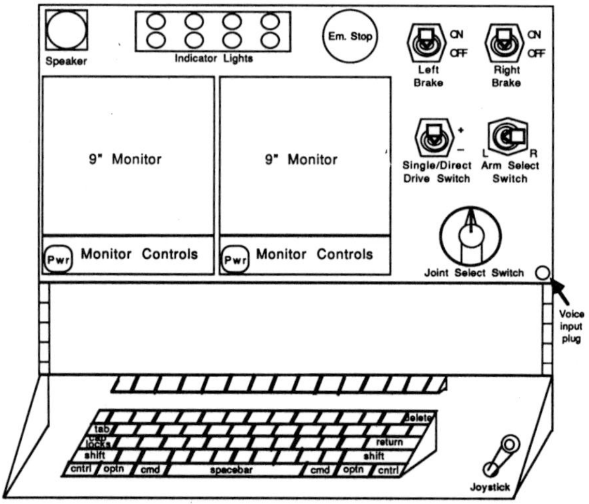

Figure 32.9 FTS control panel in aft flight deck panel with two 9-inch monitors.

The second location within the Orbiter where the placement of the FTS control panel was considered was the mid-deck. There was more space available in the mid-deck for the FTS control panel, so the control panel layouts where slightly different. Figures 32.10 and 32.11 are illustrations of two different panel layouts in the mid-deck. These two control panel configurations also include programmable display pushbutton (PDP) matrices similar to that described in a previous section.

The MSTL has determined that one advantage of the use of CAD is that it will allow a somewhat extensive analysis to take place before any physical mock-ups have been developed. After several CAD design iterations, full-scale mock-ups with varying levels of fidelity can then be constructed.

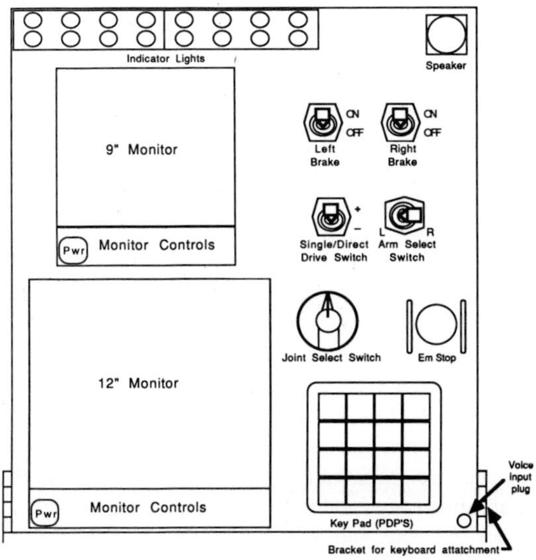

Figure 32.10 FTS control panel in mid-deck with 9-inch monitor and 12-inch monitor.

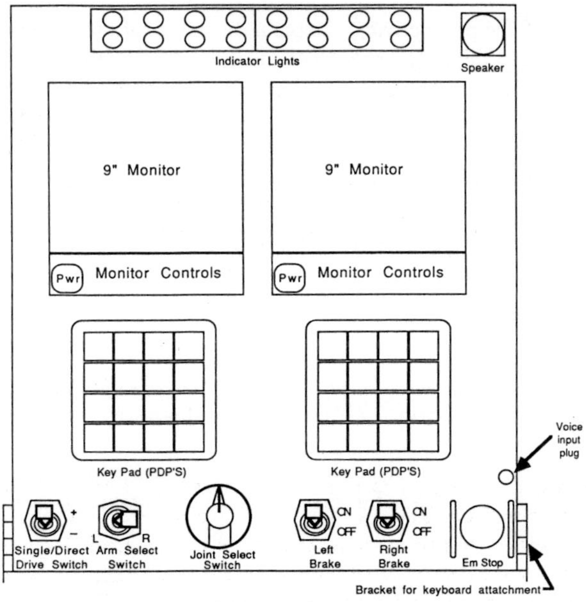

Figure 32.11 FTS control panel in mid-deck with two 9-inch monitors.

Other CAD uses

The MSTL had other proposed uses for CAD at the time of this writing. Since these applications were still in the design stage, the drawings were not available for this publication. None the less, these applications also represent further uses of CAD within the field of human factors. For this reason, then, these projects will be briefly described here.

One project which is currently underway is the use of Hypercard to create 'pulldown' and 'popup' menu-overlays on real-time video images that appear on cathode ray tube (CRT) screens. The video images will be fed from analogue and digital cameras located at remote locations from test subject viewers. The video images will be the subjects' only view of remote work sites of interest. The menu-overlays will enable the MSTL to evaluate the utility of an operator using various input devices to control cameras while performing simulated FTS remote manipulation tasks. CAD will allow the MSTL to reformat computer displays dependent upon the demands of the remote manipulation task as well as on the difficulty of using various camera control devices.

Another project underway at the MSTL was the proposed use of computer-aided measurement tools to monitor and display various indicants of work physiology, especially mental workload. The objective here was to incorporate computer-aided data collection and display technologies so that the MSTL could evaluate the workload trade-offs associated with various workstation components and configurations. In this application of CAD, the work physiology indicants that are collected and monitored throughout the test sessions will be displayed to the test administrators and not the test subjects. The MSTL contends that this is another application of CAD that shows potential for the field of human factors.

Conclusion

The consideration of the productivity, safety, and comfort of the astronaut crewmember is being incorporated into the design process of advanced NASA telerobots through the use of powerful computer-aided systems such as PLAID, Hypercard and MacDraw. The above-mentioned examples serve to illustrate the invaluable role that computer-aided design technologies play in the design and development of the FTS workstation by NASA JSC's MSTL. The MSTL has determined that the use of CAD tools contributes to a more efficient and less costly systems design cycle. CAD tools will continue to be used by the MSTL and should certainly exhibit increased usage throughout the field of human factors.

Acknowledgements

Support for this investigation was provided by the National Aeronautics and Space Administration through Contract NAS9-17900 to Lockheed Engineering and Sciences Company.

Thanks are also extended to Ervette Moore and Terence Fleming of Lockheed and Linda Dumis and Linda Orr of NASA for their invaluable assistance with the CAD packages described in this paper.

References

Burns, M. J. and Warren, D. L., 1985, Applying programmable display pushbuttons to manned space operations. In *Proceedings of the Human Factors Society 29th Annual Meeting* (Santa Monica, CA: Human Factors Society).

Chubb, G. P., Laughery, Jr., K. R. and Prister, A. A. B., 1987, Simulating manned systems. In *Handbook of Human Factors*, edited by G. Salvendy (New York: John Wiley).

Gawron, V. J. and Polito, J., 1985, Human performance simulation: combining the data. In *Simulators*, edited by J. S. Gardenier (La Jolla, CA: Simulation Councils).

Geer, C. W., 1981, *Human Engineering Procedures Guide*. AFAMRL-TR-81-35 (Wright-Patterson Air Force Base, OH: Air Force Aerospace Medical Research Laboratory).

Gray, J., 1986, The role of menu titles as a navigational aid in hierarchical menus. *SIGCHI Bulletin*, 17, 33–40.

Hawkins, J. S., Reising, J. M. and Woodson, B. K., 1984, A study of programmable switch symbology. In *Proceedings of the Human Factors Society 28th Annual Meeting* (Santa Monica, CA: Human Factors Society).

Stuart, M. A. and Smith, R. L., 1988, Simulation of the human–telerobot interface. In *Proceedings of the Second Annual Space Operations Automation and Robotics Conference (SOAR '88)* (Scientific and Technical Informaton Branch, Dayton, OH: National Aeronautics and Space Administration).

Stuart, M. A., Smith, R. L. and Moore, E. P., 1988, Guidelines for the use of programmable display pushbuttons on the space station's telerobot control panel. In *Proceedings of the Human Factors Society 32nd Annual Meeting* (Santa Monica, CA: Human Factors Society).

Chapter 33
A computer model concerning the man–machine interface of mining crewstations

JAMES P. RIDER, AUDREY F. GLOWACKI and RICHARD L. UNGER

U.S. Department of the Interior, Bureau of Mines, Pittsburgh Research Center, PA, U.S.A.

Abstract. The US Bureau of Mines is developing a computer model to analyse the human factors aspect of mining machine operator compartments. The model will be used as a research tool and as a design aid in the development process of machine crewstations. When completed the model will have the capability to perform the following: simulated anthropometric or reach assessment, visibility analysis, illumination analysis, structural analysis of the protective canopy, operator fatigue analysis, and computation of an ingress–egress rating. The model will make extensive use of graphics to simplify data input and output. Two-dimensional orthographic projections of the machine and its operator compartment are digitized and the data rebuilt into a three-dimensional representation of the mining machine. Anthropometric data from either an individual or any size population may be used. The model is intended for use by equipment manufacturers and mining companies during initial design work on new machines. In addition to its use in machine design, the model should prove helpful as an accident investigation tool and for determining the effects of machine modifications made in the field on the critical areas of visibility and control reach ability.

Keywords: Human factors; Operator compartment; Sample population; Visibility; Reach.

Introduction

Because of the unique environment in underground mining, mine workers are exposed to a variety of conditions and stresses that are not common in other industries. Mine equipment is usually massive in size and built to withstand a tremendous workload. Often the design of the mining equipment stresses functionality and production over human operator considerations. This has resulted in mining machines where visibility is at a premium and often the operators must lean outside the safe confines of the operator compartment in order to see and operate the equipment as shown in Figure 33.1. Controls are often unlabelled and placed where they are difficult to reach and distinguish from one another; in panic situations, the wrong control is frequently activated. Because of height restrictions in many underground mines, the operator works in an awkward reclined seating position which leads to fatigue and stress. The Mining Equipment Safety Laboratory of the Mine Safety and Health Administration (MSHA) analysed all fatal accident reports involving underground coal mine mobile equipment for the years 1972–9. During this period 350 fatalities were investigated and 126 of the fatalities were attributed to improper control design, inadequate visibility from the operators' cab, inadequate compartment size, operators leaning outside the cab, and poorly designed seats.

Figure 33.1 Miner working in a cramped, visually obstructed environment.

In summary, approximately 36% of the fatalities involving underground mobile mine equipment for the 7 years in question relate to improperly designed operator compartments (Sanders and Peay, 1988).

In order to address these and other operator design problems the US Bureau of Mines is researching a computer model to assist in the design and analysis, from a human engineering standpoint, of underground mine equipment operator stations. The model known as CAP (Crewstation Analysis Programs) will be used by original equipment manufacturers and mining companies for the preliminary design work on new machines in terms of good ergonomic design principles. With the implementation of the model the designer will be given flexibility to experiment with the development of new equipment which is not practical using conventional techniques.

For the sake of simplicity, CAP can be thought of as being composed of the following sections:

1. Input (machine and sample population),
2. Anthropometric analysis,
3. Visibility analysis,
4. Reach analysis,
5. Illumination analysis,
6. Canopy structural analysis.

This paper briefly describes the input, reach, and visibility sections of the model. Interested readers are invited to contact the Bureau of Mines, Pittsburgh Research Center, for more information on the CAP model.

Machine and crewstation input

Before any analysis can begin, a three-dimensional (3-D), simplified model of the operating station and related mining machine must be entered into the computer. Because of the number and complexity of many CAD packages on the market today, the procedure to input the data often becomes tedious and time consuming, and often the user becomes confused and frustrated. For this reason a method was developed to enter the data as quickly and accurately as possible so the users may concentrate their efforts on the ergonomic analysis of the operator compartment and the related mining machine. The method chosen allows engineering layouts drawn in an orthographic projection format (Figure 33.2) to be input into the computer via a digitizing graphics tablet as a sequence of polygonal geometries. The model performs a spatial 3-D reconstruction using a series of functions which identify specific 3-D geometric primitives in the orthographic projected drawings. The program does not contain any high level sophisticated networks of geometric concepts but uses a concept similar to those of Lafue (1976) and Thornton (1978). By using a function menu to identify 3-D geometric primitives such as polyhedrons, circular cylinders, spheres, and wedges the program reconstructs 3-D objects from two-dimensional (2-D) orthographic projection drawings.

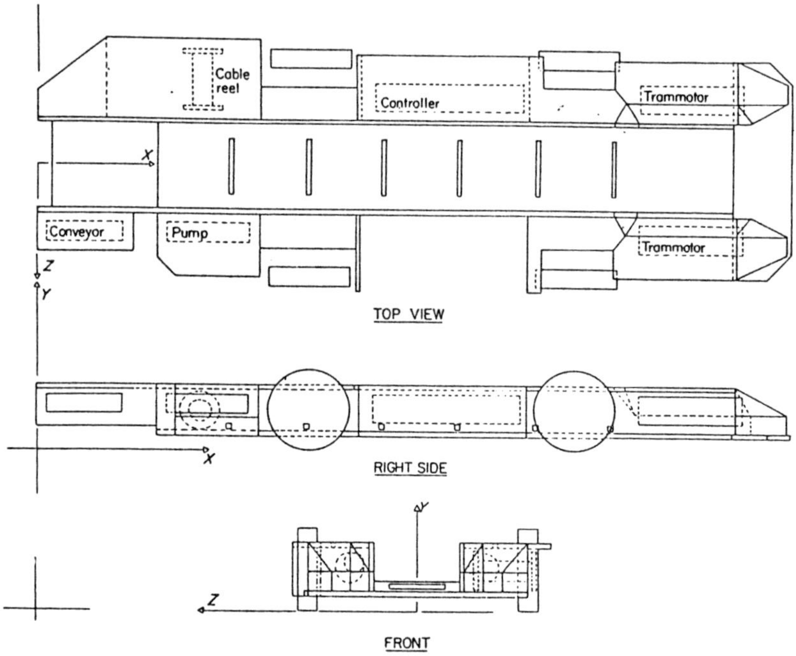

Figure 33.2 Orthographic projection drawing of an underground haulage vehicle (shuttle car) used for digitization.

The model expects a minimum of two views and a maximum of six views. For each view, the user defines a right-handed coordinate system by digitizing a point at the origin as well as points on the positive *x*, *y* and *z* axes. A correction angle accounting for any misalignment between the tablet's coordinate system and the view's coordinate system is calculated and will later be applied to every digitized point in the view. Once the initial step of identifying the necessary drawing set-ups has been completed the spatial reconstruction process begins.

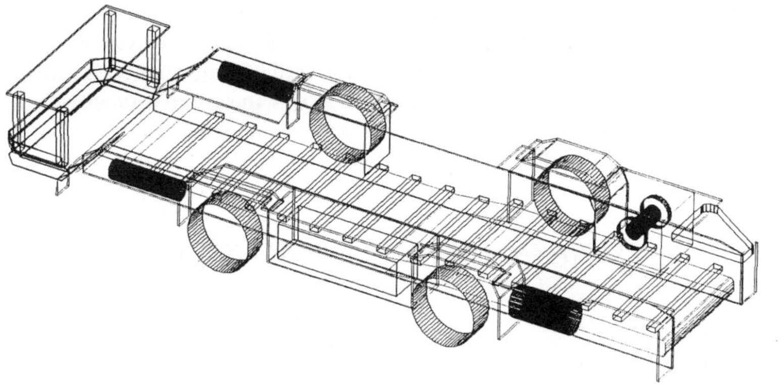

Figure 33.3 Digitized shuttle car with an end mounted compartment.

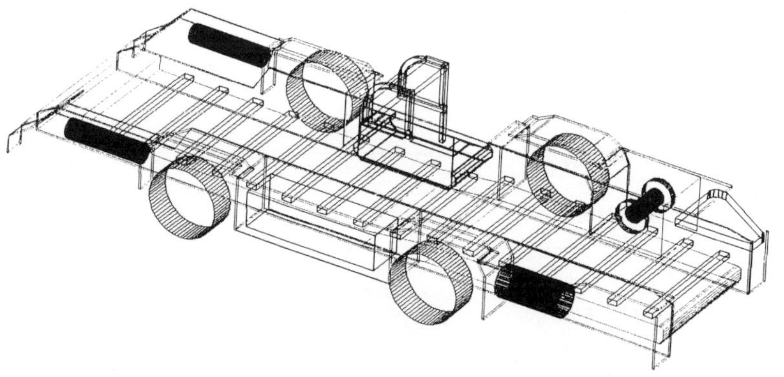

Figure 33.4 Digitized shuttle car with a side mounted compartment.

The program queries the user for a geometric primitive from the function menu. Step by step instructions follow for the digitization of the primitive and the geometric representation of the object is displayed on an attached graphics terminal. Built-in program functions queries the user to identify the depth of the primitive in a corresponding view. Using this information the program spatially reconstructs the digitized primitive and stores the calculated 3-D surfaces and vertices in a drawing file. This procedure is repeated until the operator compartment or mining machine has been completely reconstructed. The model has the capability to merge a series of reconstructed drawing files into a central location which gives the designer the flexibility to experiment with an infinite amount of machine and compartment designs as shown in Figures 33.3 and 33.4.

Sample population

Most of the analysis sections of the CAP model require a sample population for testing. The model allows the user to enter 12 external anthropometric measurements (Figure 33.5) for one or more individuals to build a sample population. Once the 12 measurements have been entered, the model determines the validity of the input and responds according. If the anthropometric input is invalid the model will prompt the user to re-enter the specific measurement in question. If valid data exists the procedure will prompt the user to enter another individual's anthropometric measurements or exit and save the population data base. The 12 external measurements for the sample population are transformed into internal link lengths to create a 3-D link-man. The link-man used in the CAP model is a version of the original link-man in the CAR program (Harris *et al.*, 1980). Each link is a straight line

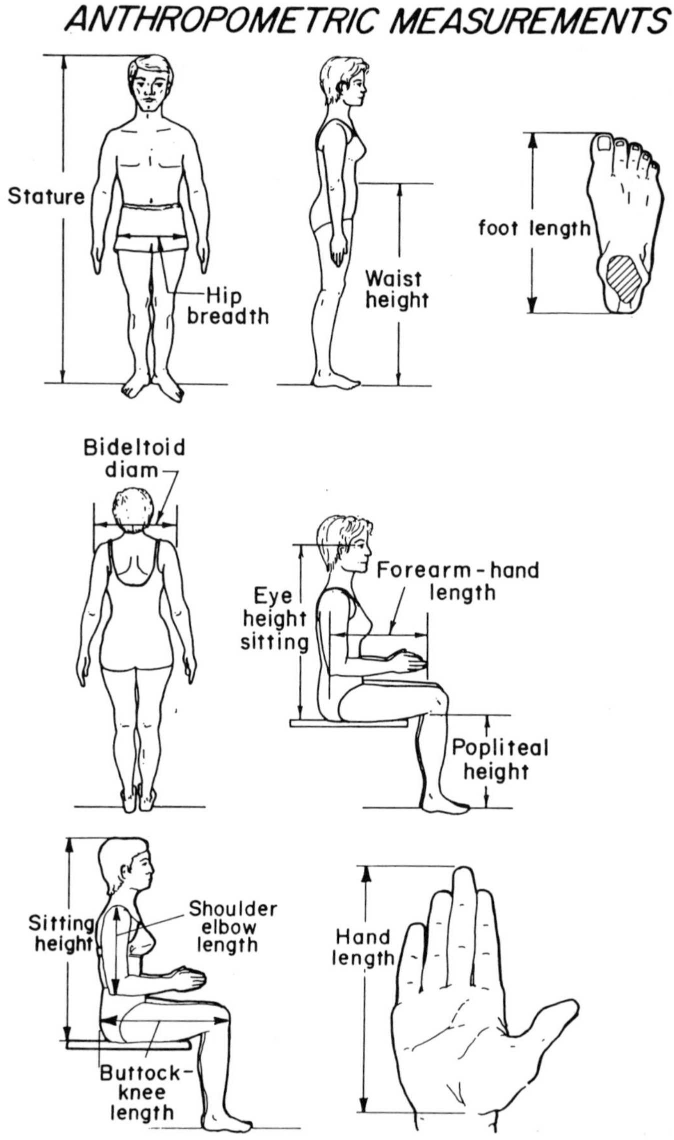

Figure 33.5 External anthropometric measurements used to build the model's sample population.

segment between centres of joint rotation corresponding to human bone structures. The link-man consists of 31 links and is based on work done by Boeing Aerospace Corporation to develop the BOEMAN model (Ryan *et al.*, 1970). CAP computes the link-lengths from external anthropometric measurements through a series of transformations derived primarily from Dempster's (1955) analysis of anthropometric data.

Once the input of the machine, compartment and the identification of the sample population is complete, the user is free to choose any analysis section desired. The flexibility of the model will allow the user to modify either the machine input or population sample based on the analysis results.

Anthropometric/reach analysis

The location of the operator and the placement of the controls is essential in the ergonomic design of operator compartments. The reach analysis sections of the model will address both issues by using an adaptation of the CAR model to position the operator in the compartment and tests if each control is reachable. It consists of a control data input module which defines key parameters needed to perform the reach study, and a link-man model needed to define the anthropometric measurements of a sample population. A tabular printout indicates the percentage of the sample population capable of reaching the defined controls (Table 33.1).

The reach analysis module constructs a link-man to reach for a specific point in space defined by a control location. The link-man is built using a link by link approach. The

Table 33.1 Reach analysis summary of the side mounted compartment shown in Figure 33.4.

Crewstation	Flip-top canopy compartment
Operator sample	50% Male
Head clearance	2·0 inches
Helmet thickness	1·75 inches
Canopy clearance	44·5 inches
Sample size	1
Clothing	Winter

Control summary

Control name	Hand/ foot	Harness lock	Type of grip	Control Location (inches) X	Y	Z	% Not accom.	% Accom.
B. Lift	Both hands	unlocked	clenched fist	12	14	25	0·0	100
Power	Left hand	unlocked	finger tip	2	9	28	0·0	100
Foot Brake	Left foot			21	8	23	0·0	100
Cab Lift	Left hand	unlocked	clenched fist	−4	4	27	100	0·0
		average miss		−0·6	2·9	−3·2		
		worst case miss		−0·6	2·9	−3·2		
Steering	Both hands	unlocked	functional grip	1	28	23	0·0	100

Note: Control location is a function of the seat reference point.

building begins at the lumbar joint which is a function of the seat reference point, the seat pan and seat back angles, and the operator's clothing. Each link is added to the previous link, with the links pointing in the direction of the defined reach point. The model takes into consideration the operator's clothing effect which reduces the angular limits of motion for each link, and alters the link lengths and the position of the joints.

The control analysis identifies the name, reach point location, body part used in the reach assessment, grip associated with the body part, and the harness condition for each of the defined controls. It displays the percentage of the sample population capable of reaching each control. In the case of the sample population not being able to reach the control the program will display the average and worst case distance from the last link to the control's reach point location as shown in Table 33.1

Visibility analysis

Because of the working environment underground, operator visibility is usually at a premium. To address this problem the Bureau sponsored research to analyse the visual requirements of mobile mining equipment and to assess the requirements through a computer model. A task analysis to distinguish the visibility requirements for a mining machine (i.e. what needs to be seen) from the field of visibility (i.e. what can be seen) (Sanders and Kelly, 1981) was conducted. To identify the points in space that need to be seen as opposed to what could be seen, the operator was asked what visual information was required to do the job.

The task analysis concentrated on information that changed as a function of time or location within the mine, such as location of obstacles, position of the machine in the roadway, and the vehicle's speed. A series of visual features were associated with each information requirement and visual requirements were identified. The visual requirements were defined in terms of specific locations known as visual attention locations (VALs). Using this method the VALs are specified with reference to a specific machine point and by identifying the VALs in this fashion, the location of the visual features are independent of the length, width, and height of the machine. Table 33.2 lists the machine and mine variables necessary for the identification of the VALs, Table 33.3 lists some of the recommended VALs associated

Table 33.2 Pertinent machine and mine variables necessary for the identification of the visual attention locations (VALs).

Abbreviations	Description
CH	Canopy height
FE	Front edge
HM	Highest machine point
MCL	Machine centre line
MCH	Maximum cutter height
MMH	Median machine height
NSD	Necessary stopping distance
OCL	Operator centre line
OEH	Operator eye height
OH	Operator's head
RE	Rear end
SH	Seam height
WMPSS	Widest machine point (same side)
WMPOS	Widest machine point (opposite side)

Table 33.3 A list of recommended VALs for a shuttle car.

No.	Fore–aft	Side-to-side	Up–down
1	RE	MCL	OEH
2	RE	MCL	HMP
3	OH	MCL	OEH
4	OH	MCL	HMP
5	OH 10'	WMP (SS) + 3'	OEH
6	FE + 2'	WMP (SS) + 1/2 NSD	Floor
7	FE + 2'	WMP (SS) + 1/2 NSD	HMP
8	FE + 2'	WMP (SS) + 3'	Floor
9	FE + 2'	WMP (SS) + 3'	MMH
10	FE + 2'	WMP (SS) + 3'	OEH
11	FE + 2'	WMP (SS) + 3'	HMP
12	FE	MCL	MMH
13	FE	MCL	OEH
14	FE	MCL	HMP
15	FE	MCL	SH or MHB
16	FE + 2'	WMP (OS) + 2'	Floor
17	FE + 2'	WMP (OS) + 2'	MMH
18	FE + 2'	WMP (OS) + 2'	HMP
19	FE + 2'	WMP (OS) + 1/2 NSD	Floor
20	FE + 2'	WMP (OS) + 1/2 NSD	MMH
21	FE + 2'	WMP (OS) + 1/2 NSD	OEH
22	FE + 2'	WMP (OS) + 1/2 NSD	HMP
23	FE + 1/2 NSD	WMP (OS) + 1/2 NSD	Floor
24	FE + 1/2 NSD	WMP (OS) + 1/2 NSD	OEH
25	FE + 1/2 NSD	WMP (OS) + 1/2 NSD	HMP
26	FE + 1/2 NSD	WMP (OS) + 2'	Floor
27	FE + 1/2 NSD	WMP (OS) + 2'	HMP
28	FE + 1/2 NSD	MCL	MMH
29	FE + 1/2 NSD	MCL	SH
30	FE + 1/2 NSD	OCL	Floor
31	FE + 1/2 NSD	OCL	SH
32	FE + 1/2 NSD	WMP (SS) + 3'	Floor
33	FE + 1/2 NSD	WMP (SS) + 3'	MMH
34	FE + 1/2 NSD	WMP (SS) + 3'	HMP
35	FE + 1/2 NSD	WMP (SS) + 1/2 NSD	Floor
36	FE + 1/2 NSD	WMP (SS) + 1/2 NSD	HMP
37	FE + NSD	WMP (SS) + 1/2 NSD	Floor
38	FE + NSD	WMP (SS) + 1/2 NSD	OEH
39	FE + NSD	WMP (SS) + 1/2 NSD	HMP
40	FE + NSD	WMP (SS) + 2'	Floor
41	FE + NSD	WMP (SS) + 2'	OEH
42	FE + NSD	WMP (SS) + 2'	HMP
43	FE + NSD	OCL	MMH
44	FE + NSD	OCL	HMP
45	FE + NSD	MCL	Floor
46	FE + NSD	MCL	MMH
47	FE + NSD	MCL	OEH
48	FE + NSD	MCL	HMP
49	FE + NSD	WMP (OS)	Floor
50	FE + NSD	WMP (OS)	OEH
51	FE + NSD	WMP (OS)	HMP
52	FE + NSD	WMP (OS) + 1/2 NSD	Floor
53	FE + NSD	WMP (OS) + 1/2 NSD	OEH
54	FE + NSD	WMP (OS) + 1/2 NSD	HMP

Note: For abbreviations see Table 33.2.

with an underground haulage vehicle or shuttle car, and Figure 33.6 displays the shuttle car VALs in the fore–aft and lateral planes.

A large portion of the visual attention locations are calculated as a function of a necessary stopping distance (NSD). This distance can be defined by adding the reaction time distance (the distance the vehicle will move from the time the operator detects a target to the time the brakes are applied) to the maximum braking distance (the distance required to stop the vehicle after the brakes have been applied) (SAE Subcommittee 29, 1979). Table 33.4 identifies the necessary stopping distance for tram speeds from 1 to 10 miles per hour.

Before any analysis can take place the physical configuration of the operator compartment along with the related machine must be entered into the computer. The model will prompt the user to select and enter a digitized crewstation and the related machine data base. The program queries the user for the operator's eye position or designed eye point (DEP). The user has the option to enter a sample population, graphically enter the eye position, or

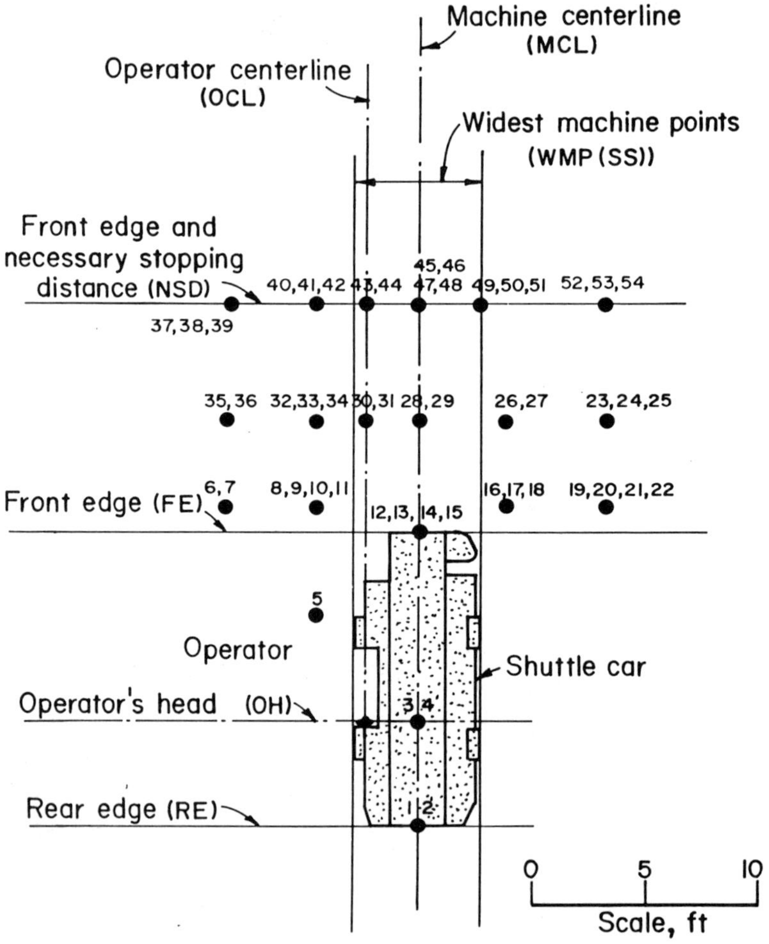

Figure 33.6 Shuttle car VALs in the fore–aft and lateral planes.

manually enter the *x, y, z* coordinates of the eye location. If the sample population option is chosen the model will calculate the eye position by vertically building successive links dependent on the defined anthropometric measurements. If the second or third option is chosen the model will prompt the user for the appropriate information. With an eye position established the model builds the VAL data base to be used in the visual analysis and asks the user to define a focus point. This point is defined to be a point in space which forms a line of sight vector with the design eyepoint. Once again the user may graphically enter the focus point, manually enter the coordinates, or enter a number corresponding to a specific visual

Table 33.4 Necessary stopping distances for tram speeds of 1–10 miles per hour.

	Necessary stopping distances		
Tram speed (miles per hour)	Reaction time distance (ft)	+ Maximum brake distance (ft)	= Necessary stopping distance (ft)
1	3	0·7	3·7
2	6	1·4	7·4
3	9	2	11·2
4	12	3·4	15·4
5	15	4·7	19·7
6	18	6·1	24·1
7	21	7·7	28·7
8	23	9·5	32·5
9	26	11·4	37·4
10	29	13·5	42·5

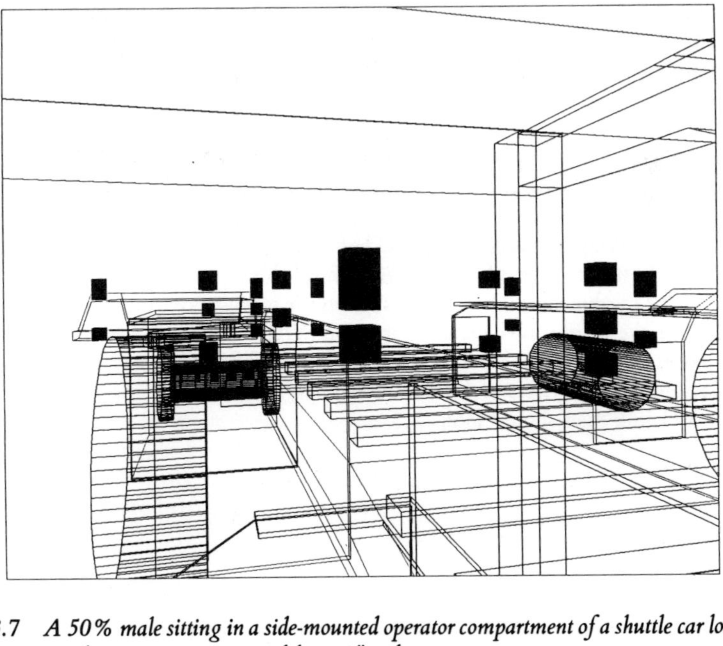

Figure 33.7 A 50% male sitting in a side-mounted operator compartment of a shuttle car looking at VAL # 13. Note the VAL is represented by a 3" cube.

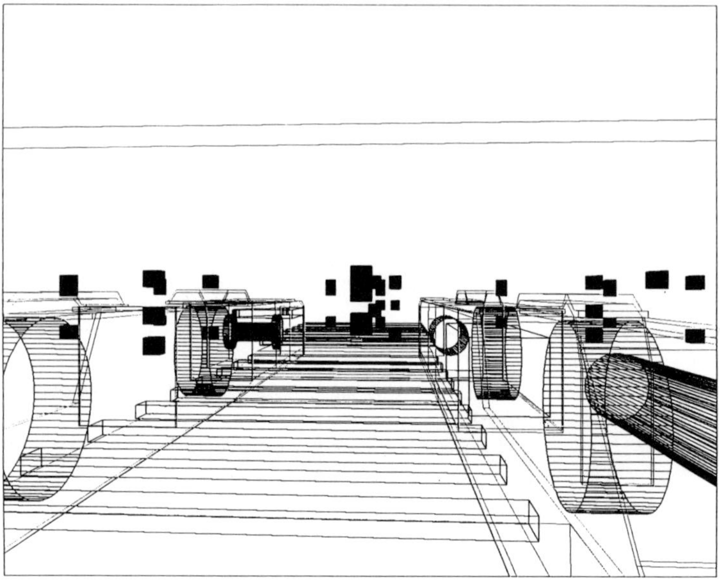

Figure 33.8 A 50% male sitting in an end-mounted operator compartment of a shuttle car looking at VAL # 13.

attention location in which case the model calculates the location of the focus point. The operator's field of visibility is an important parameter in the analysis of the visual requirements of mining equipment. The recommended visual envelope for an operator to view a working display should be within a 30° cone around the principal line of sight (Woodson, 1981).

Once the operator's eye view, focus point and the visual envelope dimension have been identified, the model allows the user to create an operator's eye view of the surrounding environment for an assessment of visibility of the VALs. The view from the operator's station is drawn, taking into account the field of vision, focus point and eye position. Figures 33.7 and 33.8 illustrate views of an average man (50th percentile) looking at a point at the front edge of the machine in the fore–aft plane, on the machine centreline in the vertical place and at operator eye height in the lateral direction (VAL # 13), from the cab of a shuttle car. The user notes which VALs are visible and what modifications may be necessary in the machine design. The analysis is performed repetitively, with the user making judgments as to where the 3-D link-man should look.

Conclusion

Original equipment manufacturers and mine operators will have the capability to use the CAP model as a research tool and as a design aid in the development and modification of new and existing mining machines. The model will have the capability to identify the visual requirements of mining machines, maximize the operator's visibility, and optimize the location of the controls to accommodate an operator sample. This may allow the designers an opportunity to experiment with the development of new equipment and aid designers to better human engineer underground mining equipment.

References

Dempster, W. T., 1955, *Space Requirements of the Seated Operator*. Report WADC-TR-SS-159. (Wright-Paterson Air Force Base, OH: Wright Air Development Center).

Harris, R. M., Bennett, J. and Dow, L., 1980, *CAR II – A Revised Model for Crewstation Assessment of Reach*. Contract N62269-79-C-0235, Boeing Aerospace. Naval Development Center.

Lafue, G., 1976, Recognition of three dimensional objects from orthographic views. *Computer Graphics*, **10**, 103–108.

Ryan, P. W., Sather, H. N. and Bearee, A. W., 1970, *Cockpit Geometry Evaluation Phase II*. Final Report D162-10126-2. (Seattle, WA: Boeing Aerospace).

SAE Subcommittee 29, Working Group 8, 1979, *Recommended Minimum Performance Criteria for Braking Systems for Rubber-tired Self-propelled Underground Mining Machines*. (Warrendale, PA: Society of Automotive Engineers).

Sanders, M S. and Kelly, G., 1981, *Visual Attention Locations For Operating Continuous Miners, Shuttle Cars, and Scoops*. Contract J0387218. BuMines.

Sanders, M. S. and Peay, J. M., 1988, *Human Factors in Mining*, BuMines IC 9182, 153 pp.

Thornton, R. W., 1978, Interactive modeling in three dimensions through two dimensional windows. *CAD*, **78**, 204–211.

Woodson, W. E., 1981, *Human Factors Design Handbook*. (New York: McGraw-Hill). 1049 pp.

Chapter 34
Computer-based analysis of EMG signals: design and use of an RMS-to-DC convertor

J. W. YATES and STANLEY A. HERTEL

Exercise Physiology Laboratory and University Engineering Services,
University of Louisville,
Louisville, KY 40292, U.S.A.

Abstract. Ergonomists frequently use analysis of electromyographic (EMG) signals to assess the involvement of various muscle groups in job performance. Integrated EMG has been the most frequently employed method for quantifying muscle activity. However, the methods used to integrate the EMG vary widely from numerical calculation to the application of various analogue signal processors. Basmajian and De Luca (1985) suggest that calculation of the true root mean square (RMS) is a better method of EMG signal analysis than is integration. While numerical RMS calculations can be performed on digitized data, the high sampling rates required often exceed the capabilities of small systems. This article discusses how RMS-to-DC convertor hardware can improve the analysis of EMG data in a computerized system. In addition it provides schematics for building the convertor and describes calibration procedures and performance characteristics. The convertor can handle continuous, bipolar, zero DC voltage inputs of approximately 0·2–5 V peak-to-peak with a bandwidth of 20–1000 Hz. Differences between the RMS convertor output and numerical calculation of the RMS voltage for identical inputs were less than 4% over the tested amplitudes. Parts cost for a single channel convertor is estimated at less than $100.00.

Keywords: Electromyography; Integration; Ergonomics; Microcomputer.

Introduction

Ergonomists frequently use analysis of electromyographic (EMG) signals to assess the involvement of various muscle groups in job performance. Integration of the EMG (IEMG) signal has been the most frequently reported method of quantifying muscle activity by both ergonomists and physiologists. Electronic integrators have been available from commercial manufacturers for many years. However, a careful reading of the literature shows that many investigators report their data as an IEMG when in fact it may be derived from some other form of signal processing such as smoothing or averaging of a rectified signal (half-wave or full-wave) (Basmajian and De Luca, 1985). True integration of a signal involves calculation of the area under a signal–time curve and thus the magnitude of the IEMG is a function of both muscle output as well as integration time. The units of this parameter are volt second or millivolt millisecond. A raw EMG signal is composed of both positive and negative voltages and has an average value of zero, hence the integrated value will also be zero. Therefore, integration can only be applied to the rectified value (absolute value) of the EMG. Further, IEMG as commonly reported in the literature must be limited to integration of the rectified signal over a fixed time period, since otherwise the area under the curve would increase indefinitely.

Basmajian and De Luca (1985) prefer calculation of the true RMS (root mean square) of the raw EMG over that of EMG integration. The RMS of an EMG signal varies only with changes in input signal magnitude and thus does not require calculation over a fixed time period as does the IEMG. Representation of EMG data as RMS values thus facilitates comparison of data among investigators by eliminating the integration time variable.

Unfortunately, RMS-to-DC convertor units are not readily available from commercial manufacturers. However, computer-based analogue-to-digital conversion systems can be used to acquire time-varying analogue signals such as EMG, and the RMS can be calculated using a mathematical algorithm (Basmajian *et al.*, 1975). Calculation of the RMS is obtained by performing the following operation:

$$\text{RMS}\,[m(t)] = (1/T\int_0^T [m(t)]^2 \mathrm{d}t)^{1/2}$$

The top portion of Figure 34.1 shows a signal processing set-up for collection of EMG data on a microcomputer. This arrangement, which has been recently used in our laboratory, makes use of a telemetered EMG system (Transkinetics) and a FM tape recorder (Teac R-71). The telemetry system uses a small radio transmitter to preamplify the EMG signal and then transmit data to the receiver. The receiver further amplifies the signal which is finally output to the FM recorder. This arrangement facilitates subject mobility and provides protection by electrically isolating subjects from the measurement system. The FM recorder allows up to seven channels to be recorded simultaneously during real-time data acquisition. At some later time the data are digitized by connecting the output of the FM recorder to the analogue-to-digital (A/D) convertor of a microcomputer (Isaac-Apple System). This scheme was used for several projects in our laboratory to compute RMS values or to integrate the raw EMG.

Nyquist's sampling theorem states that when performing A/D conversion, at least two samples per cycle of input data bandwidth are required to reproduce sampled data without loss of information (Ramirez, 1985). Since surface EMG signals often contain information up to 500 Hz (Basmajian and De Luca, 1985), A/D sampling rates must be at least 1000 samples/s. Sampling rates of this magnitude can quickly exceed the capabilities of small systems, especially when lengthy sampling periods are involved or multiple channels are being measured. In addition, since diskette storage of digitized data is frequently required, the time needed to calculate RMS output values for multiple muscles and subjects can become excessive when large numbers of samples are collected.

Projects measuring changes in the EMG during exercise or work usually sample more than one muscle group (often three or four groups). The more muscle groups recorded, the more

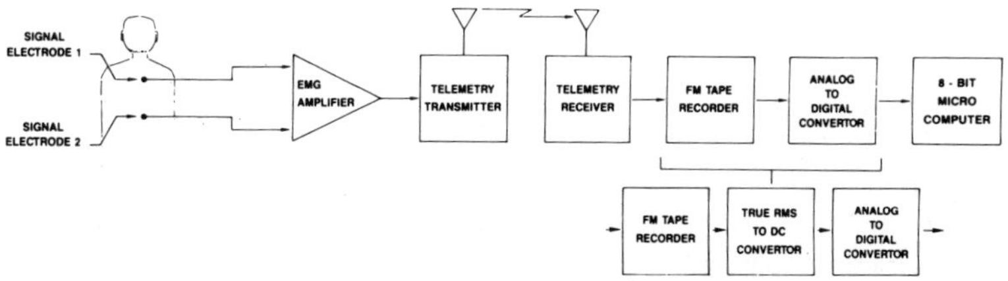

Figure 34.1. Block diagram of EMG signal measurement system. Top portion illustrates data collection system used before inclusion of an RMS-to-DC convertor, whereas the bottom portion shows where the RMS-to-DC convertor fits in the total data collection system.

time required to analyse the data. Because of the high digitization rate required for the analysis of the EMG, multiple runs are required to analyse the data from all muscle groups. In addition, it is not uncommon for 10–15 sampling periods to be acquired from a single experiment. When the number of experiments and subjects are added together, it is easy to see how the volume of data can overwhelm the microcomputer. For example, one recent project in our laboratory required approximately 150 floppy disks to store the digitized data and over 4 weeks of technician time to analyse the EMG data.

When a second project involving the analysis of EMG signals was designed, it was decided that a hardware based RMS-to-DC convertor was needed to make data analysis more efficient. Since RMS-to-DC convertors are not readily available from commercial suppliers, the device described below was designed and tested to fit the needs of rapid RMS-to-DC conversion. The convertor takes an AC signal input and outputs a DC voltage proportional to the true RMS value of that input. If such a processor is placed in the input data stream prior to A/D conversion (bottom portion of Figure 34.1), the required sampling rate becomes dependent upon the maximum RMS amplitude rate of change, rather than on the highest frequency component present in the raw input waveform. Thus, for acquisition of continuous signals whose RMS amplitudes vary relatively slowly, implementation of an RMS convertor greatly reduces computer memory and speed requirements and thus enhances system throughput. In our system, data sampling rates were reduced from 2048 to 250 samples/s, without significant loss of information. Additionally, building four RMS-to-DC convertor channels enabled us to sample and analyse data from four EMG channels simultaneously. The total time required to calculate the RMS values of 3 s of EMG data on four channels was reduced from approximately 20 min to 30 s. The actual time saved for other systems will depend upon the computer system and software used to perform the RMS calculations.

Ideally, a true RMS convertor would produce a DC output equal to the RMS value of its input voltage, independent of amplitude, frequency, or shape of the input waveform. Although a perfect convertor is physically impossible, the circuit illustrated in Figure 34.2 has performed adequately in our electromyographic measurement work. These applications have involved continuous, bipolar, zero DC voltage inputs of approximately 0·2–5 V peak-to-peak and bandwidth of 20–1000 Hz. The circuit has routinely been applied subsequent to EMG preamplification and bandpass filtering, but before A/D conversion.

Since this chapter may be read by individuals with varying levels of technical expertise, it is divided into two sections. The first part of the chapter deals with the methods used to determine the performance of the convertor, while the second section provides schematics and instructions for actually building the device.

Procedures

The performance of the RMS-to-DC convertor was measured by comparing the output of the convertor with numerically calculated RMS values using the equation shown previously. These comparisons were made for input signals produced from both a white-noise generator and from raw EMG signals recorded from surface electrodes on a human biceps muscle.

Signals of various amplitudes were recorded on one channel of a Teac-71 FM recorder (frequency range from DC to 1250 Hz), along with a synchronization signal on another channel. Prior to recording, all data signals were bandpass filtered with cut-offs at 20–1000 Hz. The recorded test signals were then applied either directly to the A/D convertor of an Apple-Isaac data acquisition system or to the input of the RMS convertor. Therefore, in the former case, the raw data were digitized for use in the numerical calculation of the reference RMS values. In the latter case, the raw data signal was input to the RMS convertor whose output was then digitized. The data collection program was written so that, in both cases, sampling was initiated upon receipt of the synchronization signal and continued for 4·39 s. This data collection method ensured that comparisons between RMS convertor outputs and numerically calculated values were based on identical inputs.

Raw input signals were digitized by the 12-bit A/D convertor at a sampling rate of 2048 samples/s and then stored on floppy disk. Later, a single RMS voltage was numerically calculated from these data. Output from the RMS convertor was digitized by the same A/D convertor at a rate of 250 samples/s. These data points were then averaged to yield a single RMS voltage for the sampling time period. Each signal was measured three times and averaged for comparison purposes.

Results and discussion

Table 34.1 compares the numerically determined RMS voltages with those produced by the RMS convertor for white noise input. Inspection of the data shows that the absolute difference decreases from a high of 3·33% to a low of 0·64% as the RMS voltage increases. Closer inspection of the data reveals that a systematic difference occurs in that the RMS convertor voltage is always less than the computed voltage (7–12 mV). When the computed voltages are linearly regressed against the RMS convertor voltages, we get a better view of how well the data agree. The resulting equation (Table 34.1) has a slope nearly equal to one and a *y*-intercept of 9·7. The *y*-intercept emphasizes the systematic difference of approximately 10 mV between the numerically computed and converted RMS voltages. The coefficient of determination shows that 99·999% of the variation in the RMS convertor output is explained by the variation in the computed output of the EMG. In other words, there is less than 0·001% unexplained change between the two methods of determining the RMS voltage. Therefore, while the absolute differences are significant, these appear to be systematic instead of random. Table 34.1 also includes the predicted RMS voltages when the RMS convertor voltages are input into the linear regression equation. When the predicted

Table 34.1 Comparisons between RMS, numerical, and regression predicted voltages (mv) for white-noise data.

RMS convertor	Numerically computed	% Error (RMS vs NUM)	Linear regress predicted	% Error (NUM vs PRED)
261	270	3·33	270	0·0
375	385	2·59	383	0·5
480	489	1·84	489	0·0
591	598	1·17	599	−0·2
721	728	0·96	729	−0·1
807	815	0·98	815	0·0
934	942	0·84	942	0·0
1067	1075	0·74	1074	0·1
1071	1078	0·64	1078	0·0

Predicted voltage = RMS voltage (0·9977) + 9·7
$r^2 = 0·99999$
s.e.e. = 1·07

voltages are compared with the numerically determined voltages, the absolute difference is always less than or equal to 0·5%.

Table 34.2 compares the computed voltage and RMS convertor voltage for actual EMG signals. As was true for the white-noise data, percent absolute difference decreases with increasing signal amplitude. For values less than 200 mV, the absolute differences average approximately 4% and decrease to less than 2% for values exceeding 600 mV. When the data are fit with a linear regression we see that the unexplained variance between the RMS convertor voltage and the computed voltage is very small (less than 0·002%). A comparison between the predicted RMS voltage and the numerically determined voltage reveals differences of less than 2% for all voltages tested.

In order to determine the performance of the RMS convertor, we had to choose a standard. Certainly, calculation of the RMS voltage using numerical methods is the best standard available. However, computation of the numerical RMS voltage required the digitization of the raw EMG data; this step necessitated analogue-to-digital conversion of voltage signals. In our measurements, we have relied upon the manufacturer's calibration of our A/D convertor. Calibration of the RMS convertor, however, was necessarily accomplished in-house with instrumentation unrelated to the A/D system. We believe these separate calibrations largely to account for the systematic differences illustrated by Tables 34.1 and 34.2. In any case, as was shown by the coefficient of determination, the unexplained variance of the RMS convertor was extremely small. Even with these uncontrollable problems, the RMS convertor showed acceptable performance. For all voltages exceeding 200 mV, the absolute difference was always less than 4% and, over the amplitude range germane to our work, averaged 1·4% and 2·44% for the white-noise and EMG data, respectively. This accuracy is within the range typical for commercially available electrophysiological equipment. However, by regressing the RMS voltage against the numerically determined voltage, we were able to remove the systematic difference and show that the RMS convertor performed to within 2% or better when compared with the numerically calculated values over the voltages measured.

The absolute differences revealed here would have little effect on experimental results. The EMG data shown here ranged from low level contractions to maximum contractions of the

Table 34.2 Comparisons between RMS, numerical, and regression predicted voltages (mv) for EMG data.

RMS convertor	Numerically computed	% Error (RMS vs NUM)	Linear regress predicted	% Error (NUM vs PRED)
77	81	4·9	82	−1·2
133	140	5·0	139	0·7
154	159	3·1	160	−0·6
399	405	1·5	407	−0·4
487	500	2·6	496	0·8
518	538	3·7	528	1·9
640	645	0·8	651	−0·8
721	734	1·8	733	0·1
894	903	1·0	908	−0·6
1484	1492	0·5	1504	0·8
1784	1818	1·9	1807	0·6

Predicted voltage = RMS voltage (1·01) + 4·17
$r^2 = 0{\cdot}99985$
s.e.e. = 7·06

elbow flexor muscles. As shown in Table 34.2, the data ranged from less than 100 mV to approximately 1800 mV, a dynamic range of 1700 mV. An absolute error of 10 mV is less than 0·6% of this range. Therefore, even if the worst case difference were to occur, there would be little effect on the experimental data. Therefore, we believe that with careful construction and calibration of this device, the performance attainable is acceptable; hence the additional step performed here of regressing the RMS convertor output against the numerical calculation is optional for other users.

In conclusion, a comparison between the RMS convertor and computed RMS values of white-noise and raw EMG signals, showed absolute differences of less than 4% when used over the suggested voltage range. However, when differences in calibration of instruments were taken into account, the errors were reduced to less than 2%.

General circuit information

Design criteria for the true RMS convertor include adequate performance and reliability, along with ease of construction. Consistent with reasonable simplicity, the circuit is self-protecting to a degree. A measure of immunity is thus provided to occasional electrical misconnections which occur even among experienced researchers in the laboratory. Use of an input buffer amplifier to mechanize input protection affords the added advantage of gain flexibility. Although Figure 34.1 details a convertor whose DC output equals the RMS value of its input voltage, input buffer gain may be conveniently varied to produce different transfer functions.

True RMS-to-DC conversion is done by a Burr Brown Corporation integrated circuit type 4341 (Burr-Brown, 1986). Driven by the input buffer amplifier, the 4341 performs signal squaring, averaging, and square rooting required by the RMS calculation. Use of this analogue signal building block facilitates ease of construction and a level of performance difficult to attain with discrete component approaches.

Integrated circuits IC1 and IC2 are socketed on perforated glass-epoxy board and discrete components are mounted on 'push through' terminals as shown. Integrated circuits IC3 and IC4 are mounted directly to the board. Miniature shielded cable with appropriate connectors is suggested for signal input and output. Experience has shown that board construction time by a competent electronic technician is approximately 16 h and that parts cost for a single channel is approximately $100.

Circuit elements

The true RMS-to-DC convertor system consists of three circuit elements, an input buffer amplifier, IC1, the convertor element itself, IC2, and a power supply containing IC3 and IC4. Although the various circuit elements ultimately function together as a system, they may be described, constructed, and tested separately.

Input Buffer Amplifier

The Input Buffer Amplifier, consisting of IC1 and associated components, is depicted at the left of Figure 34.2. IC1, a National Semiconductor LM308N, is connected as an inverting stage with closed loop gain of -1. Although the input buffer amplifier could indeed be eliminated entirely, connection of input signals directly to IC2 is ill-advised in many laboratory environments. The integrity of IC2, a relatively expensive component, is guaranteed only for inputs less than +15 V and greater than −15 V. These limits might well be exceeded by misconnection of instruments routinely used in the laboratory. The inclusion of a protection stage tolerant of such extremes has proven to enhance reliability and minimize subsequent repair costs. A sinusoidal input of 120 V AC, the largest voltage routinely found in most research laboratories, was tolerated indefinitely without failure by the input buffer amplifier.

True RMS-to-DC convertor stage

IC2 and associated components, shown at the right of Figure 34.2, perform the true RMS-to-DC conversion. The 4341 integrated circuit first squares the signal received from IC1, then averages the square, and finally, extracts the square root of the result. These calculations, all internal to IC2, are continuous analogue operations and produce a DC output voltage proportional to the true RMS value at IC1's input.

The averaging procedure described above requires a specification of averaging time by the selection of capacitor C1. Averaging time ideally should be sufficiently long to produce a constant convertor output at the lowest frequency of operation desired. Since averaging time is that time in which the input signal is 'held' during computation, however, convertor rate of response to input level changes is directly affected by C1 as well. Selection of C1, therefore, becomes a tradeoff between the minimization of convertor output error for low frequency inputs and minimization of response time. The value of C1 specified by Figure 34.2 has been found optimum for our electromyographic application.

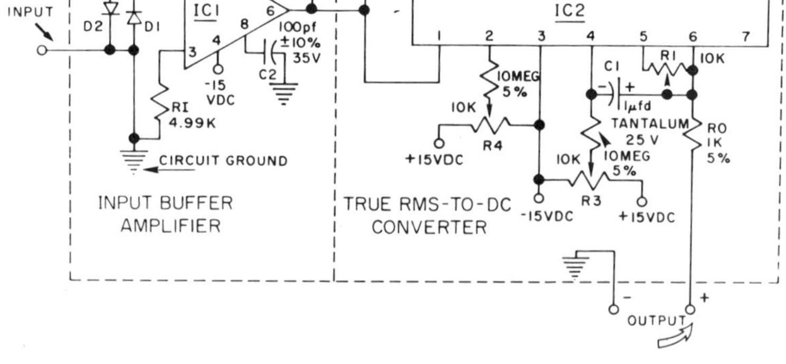

Figure 34.2. RMS-to-DC convertor circuit. All fixed resistors are 1/4 W types, except RIN which is rated at 2 W, and all are 1% tolerance unless otherwise noted. Potentiometers are 25 turn, 1/2 W, screwdriver adjust units.

Power supply

The signal processing elements previously described are powered by the regulated power supplies illustrated in Figure 34.3. IC3, IC4, and associated components develop both a regulated +15 and −15 V DC from a 120 V AC, single phase source. The calibration accuracy of IC2 is a function of both regulated voltages. Regulation provided by the illustrated circuit has ensured the long-term calibration stability required by our application.

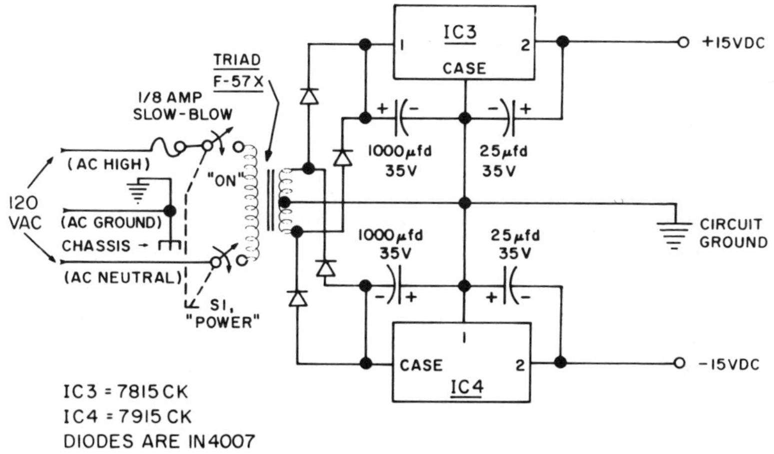

Figure 34.3. Regulated power supply circuit.

Calibration procedure

The following steps are required to calibrate an RMS-to-DC convertor channel. Input voltages must be derived from a regulated DC voltage source of sufficient stability to hold the specified calibration inputs. All input and output voltages are measured with respect to circuit ground using a calibrated DC digital voltmeter capable of resolving to 1 mV while on any range through 10 V DC. Referring to Figure 34.2 to identify input, output, circuit ground, and the relevant potentiometers, calibration proceeds as follows:

1. Alternately apply at the input +10 Volts DC (±0·001 V DC) and −10 V DC (±0·001 V DC). Adjust potentiometer R5 such that the resulting alternate voltages measured at the output are identical to within 0·002 V DC.
2. Apply as input −10 V DC (±0·001 V DC). Adjust potentiometer R1 such that the circuit output is +10 V DC (±0·001 V DC).
3. Alternately apply at the input +0·010 V DC (±0·001 V DC) and −0·010 V DC (±0·001) V DC). Adjust potentiometer R4 such that the resulting alternate voltages measured at the output are identical to within 0·001 V DC.
4. Connect circuit input to circuit ground. Adjust potentiometer R3 such that the output voltage is 0·000 V DC (±0·001 V DC). Repeat Step 3, then proceed to Step 5.
5. Apply as input −0·010 V DC (±0·001 V DC). Adjust potentiometer R2 such that the circuit output is +0·010 V DC (±0·001 V DC).

Warning

EMG measurement work by the authors has involved the acquisition of EMG data from human subjects by wireless (radio) telemetry and its simultaneous recording on an FM tape recorder for later off-line analysis. The RMS-to-DC convertor was in no way connected to this system or to the experimental subjects at any time during EMG data acquisition. The convertor provides no ground fault isolation and only 10 k Ω resistive isolation of the signal input port. THEREFORE, THE CONVERTOR MUST NEVER BE USED IN ANY APPLICATION WHICH ALLOWS THE CONNECTION OF THE CONVERTOR SIGNAL INPUT PORT, POWER SUPPLIES, OR CIRCUIT GROUND BY ANY ELECTRICALLY CONDUCTIVE PATH TO EXPERIMENTAL SUBJECTS.

Acknowledgement

The authors thank Mr Ben Hahn for his excellent construction work in the building of nine RMS-to-DC convertor channels.

References

Basmajian, J. V., Clifford, H. C., McLeod, W. D. and Nunnally, H. N., 1975, *Computers in Electromyography* (London: Butterworth).
Basmajian, J. V. and De Luca, C. J., 1985, *Muscles Alive: Their Function Revealed by Electromyography*, 5th edn. (Baltimore: Williams and Wilkins).
Burr-Brown, 1986, *Integrated Circuits Data Book* (Tucson, Az).
Ramirez, R. W., 1985, *The FFT: Fundamentals and Concepts*, (Englewood Cliffs, NJ: Prentice-Hall).

PART IX
CAD AND THE ERGONOMIC DESIGN PROCESS

Chapter 35
The contribution of CAD to the enhancement of the ergonomist's role in the design process

LOUIS BRENNAN and
ENDA F. FALLON
Department of Industrial Engineering and Information Systems,
Northeastern University,
Boston, MA 02115, U.S.A.

Abstract. This chapter addresses the role that computer-aided design (CAD) can play in enhancing the participation of the ergonomist in the design process. Traditionally, the human factors practitioner/ergonomist has encountered many obstacles in attempting to provide human engineering inputs to the design process. This chapter considers the nature of the design process and the role of computers in design. The necessity for new techniques in ergonomics design is stressed and CAD is seen as a means of implementing these techniques. Some computer-aided ergonomic design systems are also considered.

Keywords: CAD; Design; Ergonomics; Systems.

Introduction

The difficulties encountered by the human factors practitioner/ergonomist in attaining acceptance for human engineering inputs in design can be explained by two main reasons. First, there is the difficulty of specifying personnel performance requirements. Second, there is the inability to demonstrate the advantages of the human engineer's inputs.

New techniques and methodologies are needed to overcome these problems. CAD systems provide a means by which these new methodologies might be implemented. They provide complex data handling facilities which will enable designers to access human factors databases, once they become available, at the point of use. They assist designers in visualizing their creations and they provide facilities for simulation, both graphic and dynamic. CAD systems provide the ergonomist with the means to quantify human inputs in systems design.

This chapter is concerned with the contribution of CAD to the enhancement of the ergonomist role in the design process. It adopts a systems approach to the design activity. The roles of the CAD system and the user are regarded as complementary. The computer excels in analysis and computation while the user's strengths are in pattern recognition and creative ability. The view is put forward that the human is a necessary component in the design process and will remain so in the long term.

The design process

Engineers engage in design to solve problems. Two approaches to the design process can be identified, namely a traditional approach and a systems approach.

Traditional approach

Traditional design might be described as an activity whose purpose is to transform a statement of requirements for a new product into a detailed description of that product. Such a description seems to be constrained in view of recent developments in systems thinking. Shigley (1977) describes the design process (Table 35.1) as an iterative procedure made up of six identifiable stages:

1. Recognition of need: i.e. realization that a problem exists, or of a need for change.
2. Definition of problem: i.e. specification of the need which exists and the possible outcomes. Ackoff (1980) describes a problem as consisting of five types of components (i) the decision makers, (ii) the controllable and (iii) uncontrollable variables, (iv) constraints imposed from within or without on the possible values of the controlled and uncontrolled variables, and (v) the possible outcomes produced jointly by the decision maker's choice and the uncontrolled variables.

   This stage of the design process also involves specification of the constraints of the solution. Such a specification includes physical, functional and operating characteristics and also the cost and quality of the solution.
3. Synthesis: i.e. conceptualization of the solution of the problem.
4. Analysis and optimization: i.e. the concept generated in (3) is subjected to analysis and optimization using appropriate techniques. The process is repeated until the design is optimized. Two main strategies are adopted here. A single (starting) point strategy involves selecting a tentative solution and modifying it by eliminating errors until an acceptable solution emerges. Alternatively, as many tentative solutions as possible are generated and evaluated.
5. Evaluation: i.e. construction of a prototype and evaluation of the design with respect to performance criteria using experimental, mathematical, and simulation techniques. If the design proves inadequate, the whole process is repeated.
6. Presentation: i.e. preparation of final details and data pertaining to the solution.

Table 35.1 Outline of design process (Shighley, 1977).

1. Recognition of need
2. Definition of problem
3. Synthesis
4. Analysis and optimization
5. Evaluation
6. Presentation

The process by which a designer arrives at a solution is not well understood. Methods used in the process include:

(a) A leap of the imagination (creative approach),
(b) Analogy with previously encountered solutions (pattern recognition),

(c) Following a set of rules based on experience (heuristic approach),
(d) Guessing (random search),
(e) Reducing the solution space to a finite number of possibilities and exploring each one in turn,
(f) Transforming the design problem into a mathematical problem and obtaining its solution.

Advances in expert systems and artificial intelligence may allow computer systems to replace the human to some degree in the latter five areas outlined above, but is more difficult to imagine a similar situation occurring with respect to the creative approach.

Systems approach

The various stages in the design process have been outlined above. It has been noted that traditional descriptions of design may be limited in view of developments in systems thinking. It is the authors' view that the design concept should be considered within a systems framework. It is felt that such an approach is necessary for successful participation of the ergonomist in the design process.

What is a systems approach? It is a way of going about tackling problems. It implies not being reductionist, i.e. dividing the problem into separate parts and treating these one by one. A systems approach is a way of associating activities with other activities so that the relationships can be identified and classified appropriately. The systems approach can be divided into three main subsections, (i) systems theory, (ii) systems management, and (iii) systems analysis.

Systems theory is the development of a theoretical framework for explaining general relationships for the empirical world. Systems management emphasis is on the integration of each activity with the overall objective of the organization in mind. Systems analysis is the practical application of systems theory. Its purpose is to provide aid to the decision maker in making decisions about complex issues or systems. The main techniques of system analysis are data gathering, modelling and synthesis.

It is within this systems framework that the overall design process should be seen. The systems approach adopted in this paper views design as extending from the planning stage to the operational stage in systems development.

Systems development

Systems development is a technique or framework for solving problems which arise from the recognition of a need or a want on the part of decision makers. Meister (1982a) has outlined the various stages in system development. They consist of system planning, predesign, detail design, production, test and evaluation and operations. Significant overlaps can occur between some of these stages.

Bailey (1982) has described a six-stage approach to the design process (Table 35.2) In the first stage, the decision maker decides on the objectives of the system and the performance specifications. These usually include a set of requirements and constraints. Requirements help to identify the goals of the system while the constraints identify the boundaries within which the goal must be attained.

At the system definition stage inputs/outputs and functions of the system must be determined and described. Function statements are descriptions of the work which must be accomplished by the system to meet its objectives.

Table 35.2 Systems development (Bailey, 1982).

1. System objectives and performance specification
2. System definition
3. Design
4. Evaluation
5. Production and test
6. Monitoring of system operation

At the design stage the outputs from the previous stages are expanded. A tentative solution is chosen and subjected to the process described in traditional descriptions of design. It must be pointed out that a similar procedure to this is employed at each separate stage although it is more intense here. It should also be pointed out that there is a certain degree of overlap and that no stage functions as an autonomous process.

The interface design stage deals mainly with interaction between men and machines, both subsystems of the overall system. At the facilitator design stage, materials that encourage acceptable human performance are selected.

Once the system has been 'designed' and produced, it must be tested to ensure that it performs as specified. It is usually quite difficult to make major changes at this stage in system development and so the design should be evaluated prior to production. This can be done using various techniques such as mathematical modelling, simulation, testing of prototypes and analysis. Many computer-aided techniques are particularly useful at this stage because of their speed and ability to handle large amounts of data. Finally, once operational the system must be monitored in order to detect and correct initial flaws. This stage is also important in that it provides feedback which enhances the data base and can be used at a later stage in the development of other similar systems.

Computer-aided design

Groover (1984) states that, 'Computer-aided design can be described as any activity that involves the use of the computer to create or modify an engineering design.' This definition of CAD is adopted for the purpose of this paper. He points to a number of reasons for the use of a computer in the design of a product or a system:

1. To increase the productivity of the designer. Computers increase the productivity of designers by reducing or eliminating the time needed at various stages of design.
2. To improve the quality of the design. A CAD system allows the designer to evaluate a number of alternative solutions. Such a system provides greater accuracy leading to reduced error.
3. To improve communications. This improvement might be between the members of the design team or between it and other functions or the client.
4. To create a data base for manufacturing. The design of a system using a computer automatically generates data which can be used in the manufacturing and/or test stages of system development. Thus, CAD provides a vital communication link with other functions.

Role of computers in design

Storage and retrieval of design data

Computers provide a practical and easy method of storing data over the duration of a project. The range of data sources which may be consulted is usually limited by manual storage. The use of manual storage may lead to the neglect of new and effective methods, materials and processes. Computers can help by organizing data and making sources available immediately. It also allows for cross-referencing of data enabling the user to identify common parameters and relationships. CAD systems with a limited or confined expert database may be beneficial in certain applications, e.g. group technology in manufacturing and 'designing for manufacture' in robot or automated installations.

Generation of suboptimal design solutions

Traditionally the design of large scale systems is carried out in a two-stage procedure although for less critical areas this is not necessarily so. Senior design staff create tentative design solutions, often taking the form of rough sketches, drawings, mathematical equations and explanatory notes. These are then transformed into detailed drawings, descriptions and diagrams by the technical staff. The use of computer-aided design systems can offer a number of improvements in this area.

(a) The ability to retrieve data quickly may help to reinforce or stimulate the designers' ideas.
(b) The designer can enter tentative design solutions into the machine and recall them when needed,
(c) CAD systems provide feedback using colour graphics.
(d) The designer can also add verbal comments which can be recalled during the design process.
(e) Editing facilities can be used for initial error elimination.
(f) It is also possible to compare the compatibility and consistency of alternative solutions.

One might ask the question whether computers can help designers to create or invent. While the concepts of creativity and inventiveness are not sufficiently understood, there are two current approaches to the application of CAD which are of benefit in this respect. One is the pattern recognition approach and the other is the heuristic approach. It is possible to create or construct intelligent knowledge-based systems which help designers to use both these approaches for the creation of initial or tentative solutions. Such a system can carry out the following functions.

(a) Store design constraints and requirements.
(b) Elicit from an expert designer a large number of design rules based on his/her experience.
(c) Apply these rules, requirements and constraints
(d) Request additional information from the designer if the knowledge proves to be inadequate.
(e) Explain the procedure it has carried out to the designer.
(f) Learn from its experience and rearrange its own knowledge base accordingly.

There is considerable scope for the use of intelligent knowledge-based systems as components of design systems.

Evaluation

CAD system can be used in a number of ways for evaluating various designs. Some of these are listed below:

(a) Predicting various aspects of product performance by simulation or mathematical analysis.
(b) Comparing predicted and specified performances.
(c) Repeating the whole cycle until all aspects of product performance are acceptable.
(d) Repeating the whole cycle with different initial solutions until the best solution or at least a better solution is found.

These activities can be carried out in one of two ways: (i) interactively, by man and machine, and (ii) automatically, by machine without any involvement of humans.

In the initial stages of development, CAD systems were mainly concerned with checking certain performance measures of proposed design solutions. This process usually involved the submission of data to a central system for processing over a relatively long period of time. However in recent years the nature of the man–machine interaction has changed. Data is now available on demand through the medium of alphanumeric keyboards, graphic displays and light pens. The nature of the functions carried out by the man and machine have also evolved in a similar manner. Brennan (1984) outlines the various methodologies used in the allocation of functions. In particular, he cites as an example a CAD system in which the computer and user are seen as complementing one another. The computer excels in analysis and computation while the user's strengths are in pattern recognition and creative ability. Thus the computer is seen as enhancing the creative abilities of the designer and making his role more effective.

Communication links with other functions

The design function is only one of a number of activities in a complex process which includes planning, marketing, purchasing and manufacturing. CAD systems have two other functions within this system of activities: (i) to provide aids for collecting, storing, processing, retrieving, distributing and utilizing information needed for each activity, and (ii) to provide a means of communication between the various functions.

At an early stage of development CAD/CAM systems were proposed for computer integrated manufacturing systems which provide automation for all aspects of a product life cycle. Current thinking has substituted this idea with the concept of decentralized systems aiding a particular function and connected with other systems through a local area network.

The ergonomist's role in systems development

The design process has previously been described from two different points of view. The first description adopts a traditional approach which limits the overall scope of the process. The second description adopts a systems approach and describes the design activity within the context of systems development. This chapter is concerned with the contribution of CAD to the enhancement of the ergonomist's role in the design process. The approach adopted is one which describes that role within the context of systems development as outlined by Meister (1982a). Traditionally the ergonomist's role in this process was limited to the design of controls and displays (knobs and dials). This has now been extended to include areas such as manpower, training, system information and system maintainability. With this expanded role has come increased responsibility and the need for new techniques and methodologies with which to meet the challenge.

Huchingson (1981) points to three major areas of systems development in which the human factors specialist must take part.

1. Systems analysis to identify and define systems functions and to allocate these to the human element, the equipment element or human–equipment combinations.
2. 'Design' and development of equipment, procedures, work environments and facilities used by the human in performing functions.
3. Test and evaluation of equipment, procedures, work environment, and facilities to verify that human performance and life support requirements are compatible with overall system requirements.

Huchingson divides the life cycle of a system into four phases: (i) the conceptual phase, (ii) definition phase, (iii) acquisition phase, and (iv) operational phase. He equates these with Meister's (Meister, 1982a) phases in the following manner: (i) predesign or initiation phase, (ii) preliminary design phase, (iii) detailed design, production and test phase, and (iv) design verification in the field.

The first phase involves preparing a detailed description of a system that will satisfy the decision maker's requirements. In the second phase the system is described in greater detail. This involves specification of personnel, equipment, facilities and costs. The next phase involves production of the final design. In the final phase the system is tested under near operational conditions.

The initial input of the human factors engineer is in the area of systems analysis. The questions posed at the stage of system definition and system requirement specification are mainly concerned with economics and logistics. There are some behavioural equivalents described by (Holshauser, 1975):

1. What are the changes in the new system that require changes in the number and types of personnel needed to run the system?
2. What changes in personnel selection, training and operations are required due to a change in the nature of the task?
3. How can the impact of the new system on personnel be minimized?

Designers can attempt to answer these questions by extrapolating from older systems which perform a similar function. Unfortunately without a new design these answers may be inaccurate. The allocation of functions is the next important decision to be made. Brennan (1984) outlines the various approaches to the allocation of functions decision. He argues in favour of a scheme which sees the man and the machine complementing one another and acting to enhance the human's ability. Other approaches tend to assign work which machines cannot carry out to human operators. Thus the human is seen as an extension of the machine. At the predesign stage, in which tentative solutions are proposed, a number of questions need to be answered (Meister, 1982a).

1. What is the most effective design from the behavioural performance point of view?
2. Will personnel be able to perform the functions required of them in the allotted time?
3. Will personnel encounter excessive workload?
4. How can one reduce the likelihood of error?

At the detailed design stage the tasks necessary to carry out each function are listed in detail. Task analysis may be used to: (i) determine information and control requirements for equipment selection, (ii) determine skill and knowledge requirements for job specification purposes, (iii) to estimate errors associated with tasks, and (iv) to predict workload. The earlier cited questions are still asked but in greater detail. It should be remembered that as the design process progresses it becomes more difficult to change the basic design.

At the system specification stage two main questions need to be asked: (i) Are personnel performance requirements satisfied? and (ii) Do any design inadequacies exist which need to be updated?

Finally at the operational stage the system needs to be monitored for any behavioural deficiencies and corrected appropriately.

Malone (1984) outlines the essential features of a human factors design process. He points out that it should be: (i) standardized and formalized, (ii) interdisciplinary, (iii) geared to requirements, (iv) emphasize interface, and (v) influence system design. He suggests an interdisciplinary approach, one that ensures human factors specialists are placed in a directive role rather than a supporting role. Unfortunately there are some difficulties associated with this. Meister (1982b, c) has identified the factors which determine whether human engineering inputs to system development will be accepted. They include: (i) the advantages supplied by the human engineering recommendation, (ii) the costs involved in implementing it, (iii) the funding supplied by the customer for the new system, and (iv) the customers oversight of human engineering efforts.

One of the main reasons for the lack of acceptance of human engineering in the systems development process is the human engineer's inability to specify quantitatively human factors inputs. This may account for the customers oversight of human engineering efforts and also the inability of the designer to understand fully the advantages of the human engineering recommendation.

Meister (1982b) outlines a number of behavioural questions which need to be answered in order to overcome these difficulties. He links these to a number of research requirements.

1. A need for techniques which will enable designers to quantify personnel requirements.
2. Techniques to enable designers to infer personnel requirements by extrapolating from systems already in existence.
3. Techniques to translate personnel requirements into usable design procedures.
4. Techniques which will assess the adequacy of a design from the human performance point of view. Such techniques should be able to inform the designer where performance will be less than adequate.
5. Techniques for estimating manpower requirements.
6. Techniques for estimating training requirements.
7. Techniques for predicting personnel capability. These should be able to provide quantitative predictions of the performance of personnel in a proposed design.
8. Techniques for workload assessment.
9. Techniques for evaluating man–machine interface design

Other questions which need to be answered are concerned with the factors contributing to error, and the evaluation of team performance.

Role of CAD systems

It may be possible for human factors researchers to provide new techniques and methodologies to answer these questions. Undoubtedly their implementation will depend almost exclusively on computer-aided systems. Such systems provide the human engineer with the following advantages:

1. Ability to access human factors data bases at the point of use.
2. Ability to demonstrate user–machine space compatibility.
3. Ability to evaluate quantitatively human performance in a proposed design.
4. Ability to quantify advantages of human engineering input at the point of use. This may well enhance the likelihood of acceptance of human factors inputs.
5. The provision of tentative design solutions by expert systems.

At present the human engineering input to systems design must compete with other inputs for acceptance. However, successful implementation of new methodologies and techniques on CAD systems can help alter this and place the human engineering input in a directive role rather than a supporting one.

Fallon and Dillon (1988) note that more complex products are being designed today which rely heavily on a human component interacting with computer systems for their successful operation. Given the complex nature of this interaction, the task of designing these systems cannot be carried out effectively without human factors inputs. To optimize the human factors input to the design process it is argued (Fallon and Dillon, 1988) that there is an increased need for computer-based tools and methods for the following reasons:

1. There exists a diverse and disparate knowledge base within human factors which needs to be organized and coded to make it easily accessible.
2. Given the general shortage of human factors practitioners, the concept of a local human factors expert needs to be supplemented by tools which will allow designers to put human factors into the design process.
3. Because of the multidisciplinary nature of the design activity, there is an increasing need to communicate at an operational level with other disciplines which make extensive use of computers.
4. There is the expectation that the human factors practitioner's inputs will be better received as a consequence of using computer-based systems.

While it has been argued so far in this chapter that CAD systems will enable an enhancement of the human factors input to the design process, CAD technology is also exerting a large influence on the nature of the design process itself. This in turn makes the development and adoption of CAD-based tools imperative for the ergonomist. The design team is now having to concern itself with functions such as machining, process planning, material handling and assembly. This follows from the ongoing trend of pushing activities forward to earlier stages of the design to product cycle. The increasing computer processing power available is making possible the concept of an integrated CAD system or design to product system (Fallon and Dillon, 1988). The human factors practitioner will require appropriate computer-based tools to take advantage of such a system.

While a number of CAD-based ergonomic systems have been developed the system for aiding man–machine interaction evaluation (SAMMIE) is perhaps the best known (Bonney *et al.*, 1974). It can be used at any stage in the design process, but it is felt that it is more effective at the preconcept and prototype stages before design evaluation. The system, designed for usability, allows the designer to remain in control throughout, acting as an aid to his own skills rather than a replacement design methodology. The system provides the following facilities: (i) three-dimensional modelling of equipment and workplaces, (ii) ability to display a model on a variety of computer graphic devices, (iii) the ability to look at perspectives and projections of models, (iv) easy to learn method of communicating with the model, and (v) a man model which can be associated with the workplace model and used for ergonomic evaluations.

Once a tentative design solution has been chosen, SAMMIE can be used to evaluate its appropriateness: (i) tests can be carried out on the models ability to reach items in the workplace; (ii) the model can evaluate the ability of the operator to fit into the work space provided: working postures can be assessed for comfort and feasibility; (iii) the view of the operator in his working environment can be displayed; and (iv) it is possible to simulate the sequence of movements of a work task.

SAMMIE has been used to evaluate the functionality of equipment under design. It allows the ergonomist to have an input at an early stage in design and enables him/her to quantify the effect of different designs on the human body. The ability to evaluate working postures for feasibility and comfort is especially impressive.

Examples of other computer-based systems which have been developed include SPACE (Siegel *et al.*, 1982) and CAPABLE (Bonney and Williams, 1977). Space (Systems Planning for Aiding Configuration Effectiveness) is a computer-aided method which produces near optimum placement of personnel, equipment and materials for workplace layout. CAPABLE (Controls And Panel Arrangement By Logical Evaluation) is a computer-based system to layout controls and panels based on certain ergonomic design criteria.

With the emergence of more powerful computer technologies and artificial intelligence and expert systems, developments in computer-based ergonomic design systems are increasing.

Discussion and conclusions

This chapter is concerned with the contribution of CAD to the enhancement of the ergonomist's role in the design process. A systems view of design is adopted and the design process is seen as extending from the problem definition stage to the implementation stage.

Computer-aided design is described as any activity that involves the use of a computer to create or modify an engineering design. Thus computers used at any stage in the design process are regarded as components of a CAD system. The major roles of computers in design are: (i) storage and retrieval of design data, (ii) the generation of tentative design solutions, (iii) the evaluation of these solutions, and (iv) communications links with other functions.

It is pointed out that the ergonomist is commonly seen as occupying a supporting role in the design process. The argument is put forward that this should be a directive role. A number of reasons are offered for the failure of the ergonomist to occupy such a role. The primary ones are: the lack of specification of personnel performance requirements, the failure to demonstrate the advantages of ergonomic inputs, the failure to quantify human factors inputs.

New techniques and methodologies are required to overcome these problems. CAD is proposed as a means of implementing these methodologies. It offers the following advantages:

1. Ability to access data at the point of use.
2. Ability to visualize human inputs through the medium of graphics.
3. The means by which system evaluation with respect to ergonomic criteria might be carried out (i.e. through simulation and complex mathematical analysis).
4. Ability to carry out sensitivity analysis and demonstrate to the designer the advantages of human inputs.

5. Ability to demonstrate the consequences of omitting human engineering inputs in the system design process.

The authors regard the role of CAD systems in design as a complimentary one. The objective in their implementation is to enhance the abilities of the ergonomist and to place him/her in a more directive role in the system design process.

References

Ackoff, R. L., 1978, *The Art of Problem Solving* (New York: John Wiley).

Bailey, R. W., 1982, *Human Performance Engineering: A Guide for System Designers* (Englewood Cliffs, NJ: Prentice-Hall).

Bonney, M. C., Case, K., Hughes, B. J., Kennedy, D. N. and Williams, R. W., 1974, Using SAMMIE for computer-aided workplace and work task design. *Society of Automotive Engineers paper 740270, SAE Congress*, February 1974.

Bonney, M. C. and Williams, R. W., 1977, CAPABLE – A computer program to layout controls and panels. *Ergonomics*, **20**,(3), 297–316.

Brennan, L., 1984, The influence of new technology on the allocation of functions decision. In: *Proceedings 1st International Conference on Human Factors in Manufacturing* (IFS Publications Ltd.), pp. 79–86.

Fallon, E. F. and Dillon, A., 1988, CAD as an enhancement of the ergonomist's role in the design process. In: *Contemporary Ergonomics '88*, edited by E. Megaw (London: Taylor and Francis), pp. 457–462.

Groover, M. P., 1984, *CAD/CAM Computer-Aided Design and Manufacturing* (Englewood Cliffs, NJ: Prentice-Hall).

Holshauser, E. L., 1975, *Translation of DSARC Milestones into Human Factors Engineering Requirements*. Technical Publication TP75–78, Pacific Missile Test Center, CA.

Huchingson, R. D., 1981, *New Horizons for Human Factors in Design* (New York: McGraw-Hill).

Malone, T. B., 1984, The organization and management of human factors as an influence on systems design. In: *Human Factors in Organizational Design and Management*, edited by H. W. Hendrick and O. Brown (Amsterdam: North-Holland), pp. 545–552.

Meister, D., 1982a, The role of human factors in system development. *Applied Ergonomics*, **13**,(2), 119–124.

Meister, D., 1982b, Human factors problems and solutions. *Applied Ergonomics*, **13**(3), 219–223.

Meister, D., 1982c, The present and future of human factors. *Applied Ergonomics*, **13**(4), 281–287.

Shigley, J. E., 1977, *Mechanical Engineering Design* (New York: McGraw-Hill).

Siegel, A. I., Wolf, J. J. and Pilitsis, J., 1982, A new method for the scientific layout of workplaces. *Applied Ergonomics*, **13**(2), 87–90.

Chapter 36
Designing CAD systems: an open systems approach

P. T. KIDD
Cheshire Henbury Research and Consultancy,
Tamworth House,
P.O. Box 103,
Macclesfield, Cheshire SK11 8UW, U.K.

Abstract: The relationship, and the way design work is shared between the user and a computer-aided design (CAD) system, is often taken for granted or ignored when CAD systems are developed. This paper examines the nature of the design process, and then shows by way of examples, that the design of CAD systems is on the whole leading to the creation of closed systems which are totally inappropriate as a design aid. The concept of an open system is introduced and related to the design of CAD systems. An example describing the way that the open systems philosophy can be used to shape CAD systems for use in applied ergonomics is given, and an alternative and more constructive use for expert systems is discussed.

Keywords: Computer-aided design; Artificial intelligence; Open systems; Decision support systems; IT systems design; Human oriented systems.

Introduction

Computer-aided design (CAD) is a rapidly developing field which has found application in such areas as mechanical design, architecture, industrial design, control systems design and applied ergonomics. Of all these application areas, mechanical design is perhaps the one where CAD has become the most widespread. However, the potential for growth in other areas, such as applied ergonomics, is quite considerable.

CAD is still very much in an early stage of development. In fact, in the case of mechanical design, CAD more correctly stands for computer-aided drafting rather than computer-aided design, as many of the current generation of CAD systems are nothing more than sophisticated computerized drawing boards. Nevertheless, a computerized drawing board can, if used properly, be a very effective design tool and is therefore a significant development in design technology.

The fact that a distinction can be drawn between the somewhat routine job of producing a drawing and the more creative aspects of design is not something that has escaped the attention of researchers and CAD system manufacturers. Although the present generation of CAD systems have had little or no impact on the creative and conceptual aspects of design, attempts are now being made to change this situation through the automation of limited aspects of design and by restricting freedom of action.

While there is nothing new about attempts to automate the design process – various efforts have been made over many years to produce algorithmic solutions to design problems – the emergence of expert systems, knowledge-based systems and relatively cheap but sophisticated and powerful computer workstations, has given the designers of CAD systems more advanced techniques which they can use in their attempts to automate design. Probably for the first time therefore, there is an important and urgent need to address the question: 'What principles should guide the design of CAD systems?'

This paper is primarily concerned with addressing this question and with providing some answers, the objective being to provide the designers of CAD systems for ergonomic analysis and design, with some general guidance which is applicable to a wide range of specific systems.

The paper will first address something that CAD system designers most often overlook, and that is the nature of design itself. The paper will then examine some CAD systems which have been developed for use in other application areas, the aim being to demonstrate the mistakes that are commonly made when designing CAD systems.

Armed with some understanding of the nature of design and some knowledge of the common mistakes which the designers of CAD systems make, the concept of open and closed systems will then be introduced and related to CAD systems. The desired characteristics of a CAD system, in terms of an open system, will then be described, and the role of new techniques such as expert systems will be explored.

The nature of the design

Madni (1988), in a paper describing a CAD system for assisting with allocation of functions, has said that it is important when designing a CAD system to understand the designer's needs. This is true, but the designer of a CAD system also needs to understand the nature of design.

Madni's paper addresses the problem of developing a CAD system to assist human factors and system designers with the modelling and evaluation of function allocation options. In his paper he states that function allocation (a design activity) methods tend to be more of an art than a science. He then goes on to state that recent advances have made function allocation a tractable problem, the implication of this being that it can now be addressed in a more scientific way.

The scientific way is a dominant aspect of Western culture and, as Cooley (1986) has observed, the notion of predictability, repeatability and quantifiability are central to Western scientific methodology. It is not uncommon to find technologists who believe that design should be scientific, that is, should embody the concepts of predictability, repeatability and quantifiability, and if design involves aspects which make it resemble an art, then these aspects should be made, through research and development, more scientific.

Rosenbrock (1977) has specifically stated that engineering design is an art rather than a science, and in saying this he is implying a higher status for engineering design and not a lower one. Design, whether it be in mechanical engineering, architecture, control systems or ergonomics, is an art not a science.

This is not to say however, that scientific knowledge, methods and mathematical analysis do not enter into design in an important way. They do, and will continue to do so. However, as Jones (1981) has observed, design should not be confused with art, with science, or with mathematics. He argues that design is a hybrid activity which depends, for its successful execution, upon a proper blending of all three and is most unlikely to succeed if it is exclusively identified with any one. Furthermore, Rosenbrock (1977) has commented that design also contains elements of experience and judgement. These embody partly knowledge which is not expressible in exact and mathematical form, partly what Polanyi (1962) refers to as tacit knowledge (the things we know but cannot tell), and partly value judgements which are not amenable to the scientific method.

If the nature of design is examined more closely it will be seen that design is inherently subject to a large degree of uncertainty. It is often not possible for a designer to specify in an exact way what is required. There will be constraints on what can be achieved and tradeoffs required between conflicting performance requirements. The solution to a particular design problem has therefore to be a compromise and a great deal needs to be known about the relative advantages and disadvantages of the different design options before a satisfactory choice can be made.

A large number of constraints will also have to be satisfied, and many of these will either be tacit or only vaguely known. Furthermore, some constraints will only become explicit when they have been violated. In addition, the designer's knowledge of the system will always be limited and system models will always be limited representations of reality, because some aspects of the real system will be neglected and simplifying assumptions made in order to make the problem more tractable.

Design, therefore, is a skilled job calling for the exercise of expert knowledge and judgement. It is carried out under conditions of uncertainty, and scientific knowledge and methods are used to support the activity. Design is also a creative activity in which the designer is continually learning new things, addressing new and unsolved problems and extending his skills and knowledge.

Unfortunately the design of many CAD systems often ignores these facts. Various efforts have been made to produce algorithmic solutions to design problems, and the emergence of rule-based techniques such as expert systems has given new impetus to this line of development.

Design is influenced by values and therefore involves making tacit or explicit value judgements. Studies of design (for example see Lawson, 1980) suggest that design is not solely concerned with logical and explicit criteria. Research suggests that tacit criteria and subjective value judgements strongly influence designers.

Value judgements therefore, such as those made by people who believe that design should be more of a science and less of an art, are therefore bound to have an influence on the outcome of the design process. When the design process is concerned with developing a CAD system, the values of the system designers will therefore influence both the form of the final system and the way that the technology will develop in the future.

In essence, if the designers of CAD systems want to treat design as a science and want to automate certain aspects of design, then they will produce a different form of CAD system from those who are willing to accept that design is an art where science plays an important role, but in which the designer's creative role is central.

Given the nature of design as described above, which of the two approaches is more realistic and likely to prove to be the most effective? Before proceeding to answer this question, some examples of CAD systems that have been influenced by the scientific outlook will be examined.

Examples of CAD systems

Cooley (1986) has described a CAD system for architectural design known as the 'Harness

System', where the design of a building is treated in a systematized way. Each building is regarded as a communication route, and stored within the computer are a number of predetermined architectural elements which can be positioned on a computer screen along this communication route in order to produce different building configurations. Only these predetermined elements can be used, and as Cooley (1986) has observed, the architect is reduced to operating a sophisticated 'lego set'.

Most architects would regard this type of CAD system as unacceptable, and the majority of CAD systems in current use are nowhere near as inflexible and restrictive as the Harness System. However, in other application areas there are a number of emerging developments which have similar overtones to that of the Harness System and these are worth examining.

Two applications of CAD will be considered, mechanical CAD and computer-aided control system design (CACSD). Developments that are taking place within these two areas will first be described. Some general observations will then be made before proceeding to discuss how developments in CAD could be shaped by a different paradigm.

Example one - mechanical CAD

Most CAD systems used in mechanical design are computerized drawing boards which offer productivity gains by virtue of the relative ease with which drawings can be modified, geometric shapes replicated, dimensions changed, etc. Most of the productivity benefits come from the information processing and storage capabilities of the computer (a situation which is analogous to that of word processing).

Mechanical CAD systems have evolved from simple two-dimensional drafting systems which simply replicated the traditional office drawing board, into more sophisticated three-dimensional systems that can be used to generate any desired view of a mechanical component. More recently three-dimensional solid and surface modelling facilities have extended the capabilities of the more expensive CAD systems. Within the next few years however, solid modelling will also appear as a standard feature of the less expensive CAD system.

Technological developments, however, are moving beyond these drafting and modelling aids to include data bases containing details of potential solutions to standardized design problems, and details of components and materials which could be used to provide solutions to design problems. Furthermore, techniques which are referred to as parametrics (Owens, 1988), enable standard design solutions to be stored in a data base. The user then merely has to add dimensions to the standard solution selected from the data base. Attempts are also being made to get the computer automatically to dimension parts in some optimum way, and this is seen as a significant step towards automating mechanical design.

In the field of mechanical design it is now widely accepted that design should take account of the methods that will be used to manufacture a product. Because of the way manufacturing companies are organized, design is often carried out in one department, and questions concerning how the product is to be manufactured are addressed in another department. The typical pyramid organizational structure found in many manufacturing companies tends to result in very poor horizontal linking between departments, and as a result, the cross-functional approach which is required for good product design is not very effective. This situation is not helped by the fact that in many countries, designers are not educated or trained in manufacturing techniques and they are therefore unable to adequately deal with the manufacturing aspects of design unaided.

It has been proposed that new information technology techniques must be developed to provide a more formalized influence on the design process, for example by the definition of design rules and the simulation of the downstream production processes at the design stage, the aim being to improve the probability of a 'right first time' design for a new product (CEC, 1987). This concept has been given the name 'design for manufacture' and in the form described above represents a technological fix to a problem that is largely organizational in origin.

Plummer and Hannam (1983) have designed a CAD system (known as the 'Shapes System') that is based upon design for manufacture, and this system illustrates some important points relevant to the discussion on designing CAD systems, because the Shapes System contains a 'built-in' methodology that the designer has to follow (compare with the Harness System).

The Shapes System works on the principle that all components can be made up from a number of standard shapes (a similar idea to parametrics) and all the user has to do is to dimension the selected shapes and position them as required in relation to other selected shapes. In fact, in the Shapes System, the user is not allowed full control over dimensioning as some dimensions are built into the software and reflect existing standards.

The system, however, imposes other constraints on the user. For example, the system will only offer the user shapes that can be manufactured within the company concerned, using raw materials and tooling which are imposed upon the designer, thus forcing the designer to use standardized items and leaving him with no choice in the matter. If the part that the designer attempts to design cannot be manufactured, then the Shapes System rejects the design.

More recently, a fourth generation of CAD system has appeared, which goes by the name of knowledge-aided design. An example of this is the Intelligent CAD (ICAD) system developed by ICAD Inc. in the USA, which makes use of the techniques of Artificial Intelligence (AI) (Hampshire, 1988). The system basically consists of a knowledge- or rule-base structure into which can be entered the sort of rule of thumb or experiential knowledge that design engineers acquire over many years. Thus it is possible to capture some of the knowledge and skill of senior designers, which can, in theory, then be used to enable less skilled people to carry out design work. However, if a less skilled person is undertaking the design work, that person will not have the same freedom as the skilled designer, because he or she will lack the knowledge and experience to act independently of the computer. The less skilled user then tends to become a semi-skilled operator.

Example two – computer aided control system design

Computers have been used as a design tool in control engineering from the time when they first became available. In the early days computers were mainly used by designers for system simulation, for example to study the behaviour of complex dynamical systems, or to examine, using a more complex model, the performance of control systems designed using simple linear models. However, it did not take very long for a different type of approach to emerge that was based upon the application of what are called synthesis procedures.

Synthesis procedures represent the first attempts at automating control system design. They require the designer to specify in mathematical form, and in great detail, what he wants to achieve, so that a unique solution can be generated using an appropriate computer algorithm. Typically, such algorithms are based upon optimization theory and involve the minimization of some functional subject to certain constraints.

On the whole these synthesis procedures have proved to be unsatisfactory, largely because they ignore the uncertain nature of design and the difficulties of formulating the design problem in the detailed mathematical way required. The main contribution of these concepts was not therefore to achieve automated design, but more to show what level of performance was theoretically possible given ideal conditions.

The main and most important developments in CACSD occurred largely in an area which built upon the pencil and paper and slide rule frequency domain-based design techniques that were developed during the 1930s and 1940s (see Nyquist, 1932; Bode, 1940). Using the computer technology that was available in the 1960s it was a relatively straightforward matter to develop computer programs that allowed the designer to use these frequency domain design methods in a highly interactive way, thus enabling these techniques to be used in a faster and more efficient manner than had been possible using pencil and paper and slide rule.

These developments in CACSD have a direct parallel with the emergence of the computerized drawing board in mechanical design. In both cases the existing methods were computerized, but the central creative and judgemental aspects of design remained in the hands of the skilled designer. In CACSD however, the advent of interactive computing also heralded the start of a new era in control system design methods, because it became possible to do things which had hitherto been impracticable.

The pencil and paper and slide rule design techniques of the 1930s and 1940s were only applicable to a certain class of system, those with one input and one output. When applied to systems with several inputs and outputs, where any one input affects more than one output, these early design methods failed to give a satisfactory indication of performance. Attempts to extend these design methods to cater for this class of system were largely unsuccessful. One reason for this was that the structural properties of multiple input, multiple output systems was not well understood. Another reason was that the analysis of these complex systems was too difficult to deal with using pencil and paper and slide rule techniques.

New theoretical insights and the advent of interactive computing changed all this. In the late 1960s and early 1970s new computer-based frequency domain design methods for these more complex systems were developed (Rosenbrock, 1974; Postlethwaite and MacFarlane, 1979). These new design tools were heavily dependent on the power of the computer to perform complex mathematical calculations and to deal with vast and fairly complex amounts of information. In fact, without a computer, the methods are of little use, except perhaps for the most simple and trivial of design problems. However, the new techniques were also highly dependent on the skills of the designer, and were based upon the same theoretical concepts underlying the techniques of the 1930s and 1940s. The new design methods therefore built upon existing skills and allowed these skills to develop.

Over recent years interest in automated control system design has re-emerged and has been centred around the application of rule-based techniques such as expert systems. A number of such systems have been developed (for example, see Taylor and Frederick, 1984; Taylor, 1985; James *et al.*, 1987; Pang, 1987) and with these developments has come the idea of the 'less than expert' user (Taylor and Frederick, 1984).

It is worth noting here that the possibility of using less skilled users was something that Rosenbrock (1981) foresaw when he described two ways that expert systems could be used in medical diagnosis. The emergence of this concept in CACSD also lends support to Cooley's (1986) contention that CAD is being developed to deskill designers and not to liberate them from boring and routine work so that they may concentrate on the more creative aspects of design.

The expert system developed by Taylor and Frederick (1984) sits between the user and the frequency domain design methods that have already been referred to and which require certain skills to use effectively. The user specifies information such as the performance specification, design constraints, etc., and the expert system then uses the rule base and the

CACSD software to produce a controller design satisfying the performance specification and the constraints (James *et al.*, 1987).

Observations

The preceding has attempted to describe some of the developments taking place in two specific CAD application areas. The example of mechanical design shows attempts being made to restrict the designer's freedom of action by building into the CAD system constraints which cannot be violated. Whilst it is clearly necessary for designers to conform to standards and to take into account manufacturing considerations during design, the system described attempts to achieve this objective in a rather authoritarian way.

The system is also inflexible because clearly it is feasible, and sometimes necessary, to subcontract manufacturing work, and although the system's data base can be modified to allow for this, to do this would be inconvenient, time consuming and sometimes difficult especially if the user organization does not have a clear idea who will do the subcontracting work.

Furthermore, by designing a CAD system that restricts what the designer is allowed to do, the designer is prevented from exploring new design options which could lead to an improved product or reduced costs, albeit at the price of acquiring new production machines or changing production methods. By designing CAD systems with the characteristics that have been described, the opportunity to investigate new designs has been foreclosed.

Within the field of CACSD two distinct lines of development can be seen. The first of these has attempted to put the power of the computer at the disposal of the skilled designer and has provided him with more powerful design tools. The second line of development has aimed at automating the design process, first by the use of synthesis procedures, and more recently via the use of rule-based techniques such as expert systems which have been developed with the explicit aim of allowing less than expert users to undertake design.

As with all attempts at automation however, the need for human input cannot be completely eliminated, and thus expert systems used in design are arranged so that design decisions can be overridden by the user. It could be argued that because of this facility, expert systems do not deskill designers since the skilled users can effectively ignore the expert system. This argument, however, ignores a number of points.

First, it is not possible to ignore the expert system if it is the only way one can gain access to the design software. Second, very few organizations spend money on systems that users are not going to use. Pressure, no matter how subtle, will be placed on designers to use the expert systems. Third, if the expert system can in principle be used by a less than expert user, then most companies will organize work in such a way that will allow them to take advantage of this facility. Consequently, design will be done by a large number of less skilled technicians and be supervised by a smaller number of more skilled designers (such a working method forms the basis of the expert system for programming numerically controlled machine tools proposed by Alexander Houtzeel Consultancy, 1987).

Finally, expert systems are clearly incapable of being correct all the time. However, in order to disagree with these systems the user needs not only to be highly skilled, but he also needs continually to practise his design skills. Not surprisingly, those who advocate the use of expert systems in design have not yet made any serious attempt to address this contradiction.

These observations suggest an answer to the question that was posed at the end of the section on the nature of design. Those designers of CAD systems who want to attempt to codify and to automate design are ignoring the realities of design. They will create enormous difficulties for themselves and in the end will still rely upon human skill to overcome the inadequacies of the technology. By rejecting human skill however, they will have made it extremely difficult for people to exercise that skill. The end result therefore will be less effective than a system designed to support and to encourage the development of human skills.

An alternative paradigm

The examples considered have illustrated developments which, if the scientific paradigm is pursued, are also likely to occur in other application areas such as applied ergonomics. The preceding sections have attempted to show that this paradigm is inappropriate. What is required therefore is a new design paradigm, one which accepts that design is an art in which the designer's skills, knowledge, judgement and creativity are central.

Such a paradigm can be found in certain developments already described. The scientific paradigm seeks to subdivide and codify the design process. Attempts are made to incorporate the designer's knowledge into a computer so that design is reduced to a sequence of simple tasks (Rosenbrock, 1977). The alternative is not to use the computer to eliminate skill, but rather to enhance the designer's skill and to make it more productive. Thus a 'skill-based' design paradigm would seek to combine the power, speed and accuracy of the computer, with the skills, knowledge, judgement and creativity of the designer in a way that achieves an effect that cannot be achieved by either human or computer acting predominantly alone.

Unfortunately, this goal is not easily achieved because the concept of a 'skill-based' CAD system is subject to wide interpretation. Whilst clearly there is no single CAD system fulfilling the goals described above, skill-based CAD systems must have some recognizable features which distinguish them from other CAD systems. It is because of the different ways that the idea of a skill-based system can be interpreted, that the concept of open and closed system is useful. Before describing these concepts however, the following example will serve to illustrate the problem.

MacFarlane (1982) has observed that synthesis approaches to design reduce the designer's scope to exercise skill and judgement. He then goes on to say that a CACSD system should provide a skilled designer with a set of manipulative and interpretative tools which will enable him to build up, modify and assess a design put together on the basis of the designer's physical reasoning within guidelines laid down by his engineering experience.

MacFarlane (1982) then goes on to say, quite correctly, that the design of CAD systems needs to consider the way in which the burden of the design work is shared between the designer and computer. The aim, he argues, should be to achieve a situation such that each makes an appropriate contribution to the overall solution, and this involves developing techniques which will lead to a fruitful and effective symbiotic relationship between the computer and designer.

He then observes that the computer is conventionally used for calculation, manipulation and optimization. He then proposes, by way of an example of his thinking, that in any fully developed CAD system, the tuning of controller parameters during simulations is best done by a systematic use of appropriate optimization techniques. Although this sounds like an appropriate sharing of design tasks between the designer and computer, the idea does have a number of problems that are not immediately evident.

At the moment there are two approaches that can be used to undertake the task of controller tuning during simulation studies. The first is to do what MacFarlane (1982) has described and use optimization methods. The second is for the designer to make adjustments himself.

This manual method can result in a very good learning situation. The designer can experiment and observe results, and the process of making these adjustments can give the designer more insight into the system's performance characteristics, and can give him a 'feel' for the system. This approach also requires the use of knowledge and the exercise of skill and judgement. The prime disadvantage of this approach is the time taken to achieve a satisfactory performance (for a complex system this can be several person-months).

The optimization method is, by comparison with the other approach, very fast. However, since the computer is doing all the work there is no learning situation. Furthermore, the designer does not develop any 'feel' for the system, or extend his understanding about the system, and little or no knowledge, skill or judgement is required. Moreover, if the optimization algorithm fails to produce a satisfactory result, then there is no indication from the computer why this has happened, and the designer will have gained no insight which he could make use of to resolve the problem.

Although a superficial examination of the optimization approach suggests that the power, speed, and accuracy of the computer are being used to support the designer, what in actual fact exists is the classic human–machine relationship. The computer is doing all the work, and when it fails, the user is left to cope as best he can.

The problems that arise is this example stem from two sources. First, the use of the optimization method for tuning controllers is based solely on the criterion of the amount of time taken to produce a result. The manual method is slow, and since the computer is better at examining a large number of options and can do so very quickly, it is concluded that controller tuning is best done by the computer. Jordan (1963) has referred to this kind of thinking as 'man–machine comparability'. Unfortunately, man–machine comparability ignores the needs of the designer, which in this case are related to learning and the opportunity to exercise skills and judgement.

Second, the design of the system ignores the fact that the algorithm may fail to achieve a satisfactory result. If any design tool has a 100% success rate or is applicable to all situations then it is worth calling a design aid. If, however, a design tool sometimes fails to produce a satisfactory result, then the user needs to understand under what conditions it may fail and should have access to alternative tools. If there are no alternative tools or there is no understanding of the range of applicability of the tool, then such a design tool has limited value.

The open system philosophy

The example discussed above illustrates a class of system that can be referred to as a closed system. Essentially, a closed system is one in which the actions of the user are determined by the designer of the CAD system to a degree that is more than is actually required, say, by hardware, or software or performance constraints. In other words a closed system is one in which the user–computer relationship is 'over-determined'.

Thus a closed system typically restricts the designer's freedom of action, or forces him to use the CAD system in a particular way that is of the systems designer's choosing and not necessarily of the user's choosing. A closed system also attempts to automate a particular design task, but when the system fails, it leaves the user without any computer-based decision support.

An open system is the opposite of the closed system. An open system does not restrict what the user can do. For example, if the user wants to violate some constraint, then the closed system will not allow this, while an open system will. More fundamentally, however, an open system is not characterized by the situation described earlier in relation to controller tuning. In that situation, controller tuning is either done by the computer acting alone, or by the designer acting alone. Between these two extremes there are an infinite number of possibilities. An open system would allow the designer to make use of any of these possibilities.

In essence therefore, an open system is one in which the relationship between the user and the computer is determined by the user and not the designer of the CAD system. The role of the designer of the CAD system is to create a system that will satisfy the user's personal preferences, and allow users to work in a way that they feel most appropriate given the wide range of problems that the CAD system can be applied to.

At this point it is natural to ask why the open systems philosophy should be used to shape the design of a CAD system? The answer to this question has three parts. First, consider what has already been observed, and that is that design is inherently subject to a large degree of uncertainty. Uncertainty, however, is something that scientists and technologists have for the most part tried to reduce and to eliminate.

The 19th-century mechanical engineer Frederick Taylor, in writing about his system of management, said: 'In our system the workman is told in minute detail what he is to do and how he is to do it, and any improvements which he makes upon the orders given to him is fatal to success' (Taylor, 1907). Taylor's system is a closed system. It leaves no room for initiative or for creativity. Trends in the development of CAD systems, on the whole, appear to be driven, consciously or unconsciously, by this Tayloristic vision. The essence of design, however, is creativity. Creativity does not flourish in closed systems.

Second, consider the wide variety of situations that CAD users have to face. Is it really realistic and feasible for the designers of CAD systems to believe, to think, that they can create one CAD system that would be suitable for all mechanical designers, or for all architects, or for all ergonomists. To admit that this is possible is to suggest that there is only one best way of doing something, only one best solution to any particular design problem. Anyone with any knowledge of design knows that this is not the case.

Third, and perhaps of greatest importance however, the open systems philosophy prevents the designer of a CAD system from imposing on the CAD user, his own views on what the computer should do and what the user should do. Design is influenced by subjective value judgements, and the only way to minimize the effect of the value judgements made by designers of CAD systems is to eliminate as far as possible the opportunity to make these value judgements. The open systems concept is one way of achieving this.

It therefore follows that CAD systems should not be designed as closed systems, but as open systems, and this means avoiding restrictions, using decision support systems in a way that is highly interactive and which allows sharing of the decision-making process in a way that is unique to each user and to each situation, and using expert systems in a far more imaginative and constructive way than at present.

Application in ergonomics

A large part of this chapter has concentrated on applications of CAD in mechanical design and in control system design. Much can be learnt by studying the way other disciplines have developed CAD, especially when these disciplines have a long history of involvement in this area.

Within the field of applied ergonomics there is scope for the use of a wide range of information technology tools such as expert systems, data base management systems, decision support systems, etc., and the question is how can the open systems philosophy be used to shape the development of CAD? The answer to this question can be provided by way of an example.

Consider the problem of designing a workspace where an operator has to carry out the assembly and inspection of some product. The design of the workspace can be improved for the operator by considering the ergonomics of the workspace.

Bullinger *et al.* (1988) has described a computer-based facility for simulating in real time the motion of the human arm, and this can be used to study the performance of the human operator engaged in assembly tasks. By examining such things as the possibility of collisions when performing tasks, the amount of reaching that has to be performed, the effect that workspace design has upon the operator can be examined. A skilled ergonomist can then work through a range of design options or improvements, simulating each one using the computer, until he arrives at a satisfactory solution in terms of the workspace ergonomics.

This example is similar to the use that control engineers made of computers in the very early days of computing, where the designer uses the computer to simulate the performance of the system, but where all design decisions are taken by the designer. Because of the interactive nature of the system, the ergonomist can make as many changes as he likes, the only constraint being the time he has available for the study.

How could this system evolve from this fairly simple design tool into a more sophisticated one? One possibility would be to provide a data base containing information about current design standards. The data base could also contain information about references relevant to the ergonomics of workspace design (Davies, 1987).

The question of using expert systems would seem to be more problematic. In fact it is not. Conventional wisdom would suggest that the rules that the expert uses when considering the ergonomic aspects of workspace design should be elicited from a number of experts and then incorporated into an expert system rule base. The system could then be used to generate an ergonomically satisfactory design.

Ostberg (1988) has described five possible ways that expert systems can be used. The first is the 'transactional mode' where the expert system takes over all responsibilities and performs all the tasks through a human facilitator. The second is an 'interactive mode' where the human performs the task with the help of the expert system and is in charge during the dialogue. The third is the 'commentary mode' where the human performs the task independently and only uses the expert system when in need of a second opinion. The fourth mode is the 'teaching mode' where the expert system teaches the human until he or she has reached task proficiency, or serves as a manual for the infrequent task performer. The fifth and final mode is the 'supervisory mode', where the expert system is a passive supervisor and only calls attention to illegal human actions likely to interfere with developments in other areas of the system of which the task domain is a part.

The conventional approach to expert systems development is almost exclusively concerned with developing systems that operate in a transactional mode. This is the approach of the closed system school. Essentially the system attempts to mimic the expert knowledge of the designer. It tells him what he already knows and does things which he can already do. Thus the ergonomist has to deal with a system that contains expert knowledge from the domain of his own expertise, all he has to do is to specify the problem and then the computer will attempt to solve that problem.

The open systems philosophy however involves a different approach, which is based on the idea of using AI techniques, not to solve problems, but to give advice and to support decision making. An expert system should not be telling people what they already know, or doing things which they themselves can do, but should instead, be making design more productive by bringing to the designer advice about things which he is unsure of, or which he does not know about, and reminding him of the possible consequences of his decisions. This of course is more difficult than just simply attempting to mimic the actions of designers.

Coombs and Alty (1984) have argued that human experts are rarely required to solve well-defined problems, which is what expert systems operating in the transactional mode are designed to do. Instead they work on problems which involve two or more domains of knowledge. What is required therefore, is for the expert in one knowledge domain to receive advice from experts in the other knowledge domains, and this involves using expert systems which contain expertise from these other domains.

The open systems approach uses expert systems to support the designer rather than to replace him, and this requires the designer of the CAD system to establish what extra skills and knowledge the expert user requires in order to do his job more effectively, and to choose an operational mode that leaves the CAD user with freedom to act in the way he wants.

This sort of analysis usually reveals a number of possibilities. For example, many computer-based systems make use of algorithms that are often numerically unreliable, or there may be a number of numerical procedures that could be used for a given task, each having its own range of applicability. The expertise concerning these issues lies in the domain of numerical analysis and not in ergonomics.

Consider another possibility. As the ergonomist works his way towards an ergonomically satisfactory solution, he will make changes to the system. Such changes will have cost implications, or may be impracticable, or mechanically difficult to achieve, or may violate mechanical or system specifications. Again, the expertise that could assist the ergonomist to make more effective use of his time and the CAD system, lies elsewhere outside the area of ergonomics.

Expert systems containing knowledge from a number of knowledge domains could thus assist the ergonomist with his work. These expert systems, however, would not be used in the transactional mode, but instead in the supervisory mode and the commentary mode. Thus, such a system would fulfil the requirements mentioned earlier, because they would be advising the ergonomist about the consequences of his decisions, and telling him about things which he is unsure of or does not know about.

It is in areas such as those described above that the greatest scope for the use of expert systems lie. This is not to say, however, that the proposed use of expert systems would make other experts redundant. The idea is, that by providing expert systems in areas which are peripheral to the main area of interest, the ergonomist is more likely to produce a design that is acceptable to the other professionals involved, because he will have considered aspects that are strictly outside his own domain of expertise. The other professionals involved will still be required, because design is always subject to uncertainty.

Conclusions

CAD has begun to emerge as an indispensable tool for designers in many fields. It has the potential to free the designer from routine and mundane tasks and to provide a vehicle whereby the designer can attempt to do things that would be impossible or extremely difficult without the aid of a computer. However, CAD also has the potential to deskill the designer and to trivialize his work.

Many of the developments that have taken place in CAD have aimed to codify design and to reduce it to a sequence of simple choices. These developments have been influenced by a paradigm that has treated design in a scientific and rather sterile way. This paradign has led to the creation of closed systems.

The design process is subject to a large degree of uncertainty and involves aspects of creativity, skill and judgement that are not amenable to formal scientific treatment. None of these human qualities can thrive in a closed system, and this approach therefore is not the one that should be pursued.

An open systems philosophy for the design of CAD systems has been proposed. This approach will result in CAD systems that allow designers freedom to exercise their creativity, and give them the flexibility that they need. This open systems approach avoids some of the difficulties that arise because the designers of CAD systems make value judgements concerning what the computer should be doing and what the user should be doing. Using the open systems philosophy the designer of a CAD system is designing a system that each individual can choose to use in his own particular way. This is quite a new and challenging thing for the designers of CAD systems to attempt, and will require them to develop new skills and a new outlook (Kidd, 1988).

The question of whether developments in CAD will eventually lead to the complete deskilling and trivialization of design work is a question that no one can answer. However, if people aim, either consciously or unconsciously, to achieve this goal, then they will go a long way towards achieving their aim. The outcome of these developments, judging by previous experience of attempts to automate work processes, will not be very good for the people who have to work with such systems. Of equal importance, however, is the growing belief that such systems will not be very effective.

A better way forward can be followed, and this involves making use of human skill, judgement and creativity, not in a reactionary way that seeks to halt or freeze technological development, but in a constructive and realistic manner which seeks to develop and use technology in a way that supports designers. The open system approach represents one way of creating these human oriented CAD systems.

References

Alexander Houtzeel Consultancy, 1987, *M–TES A Knowledge Based System for Numerical Control: Management Overview* (Alexander Houtzeel Consultancy Ltd., South Park, Hill House Drive, Weybridge, Surrey, KT13 0NQ, UK).

Bode, H. W., 1940, Feedback amplifier design. *Bell System Technical Journal*, **19**, 421–454.

Bullinger, H.-J., Lay, K. and Menges, R., 1988, GRIBS – An approach to a realistic realtime simulation of human arm motion. In: *Ergonomics of Hybrid Automated Systems I*, edited by W. Karwowski, H. R. Parsaei and M. R. Wilhelm (Amsterdam: Elsevier), pp. 599–606.

CEC, 1987, *Draft ESPRIT Workprogramme*. (Brussels, Belgium: Commission of the European Communities).

Cooley, M. J. E., 1986, Problems with automation. In: *Human Factors*, edited by T. Lupton (Kempston: IFS Publications), pp. 3–13.

Coombs, M. and Alty, J., 1984, Expert systems: An alternative paradigm. In: *Developments in Expert Systems*, edited by M. Coombs (London: Academic Press), pp. 135–157.

Davies, D., 1987, HUFIT's role in office systems design. In: *ESPRIT '86: Results and achievements, Proceedings of the third ESPRIT Conference*, edited by the Directorate General XIII, Commission of the European Communities (Amsterdam: Elsevier), pp. 813–824.

Hampshire, N., 1988, The intelligent amendment. *Workstation Magazine*, November, 27–29.

James, J. R., Frederick, D. K. and Taylor, J. H., 1987, Use of expert systems programming techniques for the design of lead-lag compensators. *IEE Proceedings, Part D*, **134**, 137–144.

Jones, J. C., 1981, *Design Methods. Seeds of Human Futures* (Chichester: John Wiley).

Jordan, N., 1963, Allocation of functions between man and machines in automated systems. *Journal of Applied Psychology*, **47**, 161–165.

Kidd, P. T., 1988, The social shaping of technology: the case of a CNC lathe. *Behaviour and Information Technology*, **7**(2), 193–204.

Lawson, B., 1980, *How Designers Think* (London: Architectural Press).

MacFarlane, A. G. J., 1982, Conceptual frameworks for control system CAD. *Proceedings of the 21st IEEE Conference on Decision and Control*, Orlando, FL, pp. 538–539.

Madni, A. M., 1988, HUMANE: A designer's assistant for modelling and evaluating function allocation options. In: *Ergonomics of Hybrid Automated Systems I*, edited by W. Karwowski, H. R. Parsaei and M. R. Wilhelm (Amsterdam: Elsevier), pp. 291–302.

Nyquist, N., 1932, Regeneration theory. *Bell Systems Technical Journal*, **11**, 126–147.

Ostberg, O., 1988, Applying expert systems technology: division of labour and division of knowledge. In: *Knowledge, Skill and Artificial Intelligence*, edited by B. Goranzon and I. Josefson, (Berlin: Springer), pp. 169–183.

Owens, W., 1988, Where CAD Really Pays. *3D*, **1**(6), 22–24.

Pang, G. K. H., 1987, An expert system for CAD of multivariable control system design using a systematic design approach. *Proceedings of the American Control Conference*, American Automatic Control Council, Green Valley, AZ, pp. 555–560.

Plummer, J. C. S. and Hannam, R. G., 1983, Design for manufacture using a CAD/CAM system – a methodology for turned parts. *Proceedings of the Institution of Mechanical Engineers*, **197B**, 187–195.

Polanyi, M., 1962, Tacit knowing: its bearing on some problems of philosophy. *Review of Modern Physics*, **34**, 601–616.

Postlethwaite, I. and MacFarlane, A. G. J., 1979, *A Complex Variable Approach to the Analysis of Linear Multivariable Feedback Systems. Lecture Notes in Control and Information Sciences*, Vol. 12 (Berlin: Springer).

Rosenbrock, H. H., 1974, *Computer Aided Control System Design* (London: Academic Press).

Rosenbrock, H. H., 1977, The future of control. *Automatica*, **13**, 389–392.

Rosenbrock, H. H., 1981, Automation and society. *Systems and Control Letters*, **1**, 2–6.

Taylor, F. W., 1907, On the art of cutting metals. *Transactions of the ASME*, **28**, 31–350.

Taylor, J. H., 1985, An expert system for integrated aircraft/engine controls design. *Proceedings NAECON*, Dayton, OH, pp. 661–669.

Taylor, J. H. and Frederick, D. K., 1984, *Expert Systems Concepts for Computer Aided Control Engineering*, Corporate Research Development Report No. 84CRD127, (Schenectady: General Electric).

Chapter 37

An integrated human factors input to the design of Information Technology products

MARGARET GALER, BRONWEN TAYLOR, MARC DOWD, BERNARD CATTERALL, MARTIN MAGUIRE and GORDON ALLISON

HUSAT Research Institute,
The Elms, Elms Grove,
Loughborough, Leicestershire LE11 3TU, U.K.

Abstract: The ESPRIT project – Human Factors in Information Technology (HUFIT) has developed the HUFIT Toolset. These human factors tools have been produced in order to provide designers in Information Technology (IT) product supplier companies with methods for the user-centred design of their products, in particular the software. The Toolset comprises a number of integrated tools for use by participants in the design process such as in planning, design and quality assurance. They have been designed for use within the time and resource constraints of the design process and by people who do not necessarily have human factors skills and expertise. The Toolset includes the User Requirements Program, Specification and Design Tools and The Documentation Design Guidelines and HyperGuidelines.

The HUFIT Toolset was originally developed as paper-based tools and where appropriate on-line versions have been developed. The user can decide which formats are appropriate to their environment.

Keywords: Life cycle design methods; Human factors tools; Information technology products; User-centred design.

Introduction

As part of the European Strategic Programme for Research and Development in Information Technology (ESPRIT), project 385 HUFIT (Human Factors in Information Technology) has developed The HUFIT Toolset. These human factors tools have been produced to provide designers with integrated methods for the user-centred design of Information Technology (IT) products, in particular the software. The HUFIT project is collaborative involving partners from HUSAT Research Institute (UK), IAO-FhG (FRG), ICL (UK), Olivetti (Italy), Bull (France), Philips (The Netherlands) and Siemens (FRG).

The HUFIT Toolset is one of the major products of the HUFIT project. Other products are INTUIT, and the Computer Human Factors Information Service, an on-line bibliographic database (both described elsewhere in this book); MULTEX, a development environment for integrated multi-media interfaces and DIAMANT, a dialogue management tool for the development of integrated graphical interfaces.

The HUFIT Toolset has been developed by human factors experts in collaboration with designers in IT supplier companies. The users of the Toolset are the participants in the design process in marketing, product planning, development, quality assurance and so on. A survey of design practice in five European IT supplier companies identified the requirements for human factors input to product design and also the constraints within which the human factors tools had to operate. The Toolset has been designed not only to meet the requirements of designers within the constraints of the product design process but also to be usable by people who do not necessarily have human factors/ergonomics expertise. The Toolset is an integrated approach to the design of IT products but the modules can also be used independently. It comprises tools for the planning stage of the design process to help derive user requirements from information about the users and their tasks; tools to facilitate the development of the functional specification; tools to offer human factors guidance to interface designers and to technical authors producing user documentation; and tools to encourage a thorough user evaluation of the product during development and testing. The Toolset is both on-line and paper based and can be used in either mode depending on the circumstances prevailing in the IT supplier company. In this chapter a selection of the Tools are described, namely the User Requirements Program, the Functionality Matrix, the User Computer Interface Design Tool and the Documentation Design Guidelines.

The User Requirements Program

The design of successful IT products relies upon a good understanding of the target users and their tasks. A set of paper-based tools called the User Requirements Toolset has been developed by HUSAT in the HUFIT project to help planners and designers of IT products effectively to use information about users and their tasks. By giving examples which help designers to develop this information the user and task information can be used to identify user requirements. An on-line version of the tools called the User Requirements Program is described.

These on-line tools are intended for use alongside the systems analysis and task analysis methods used by the designers and are complementary to them. The tools are intended for use by planners and designers without any previous human factors training or experience. They present a process to follow, with supporting explanation and examples. The output of the Program is of use to marketeers, designers and evaluators of the product at various stages in the design process.

The need for user requirements tools

The purpose of the HUFIT project is to promote the use of human factors in order to improve the products designed by IT supplier companies. There are two ways to improve products: (i) by applying human factors knowledge and techniques during the design of the product, and (ii) evaluating the product using human factors criteria for user testing.

In order to use human factors guidance to improve design and to carry out human factors evaluation it is essential to understand the characteristics and requirements of the user. A firm basis of user and task information is necessary for successful product design. Figure 37.1 represents these aspects of human factors in design.

In large scale system design, e.g. for bespoke systems, a considerable proportion of the resources in the early stages of design is allocated to systems analysis which may include a method of task analysis. It is increasingly recognized that this must be accompanied by user analysis to gain an equal understanding of who the users are and how they make up the organization which will use the system. In the design of generic products the analysis phase

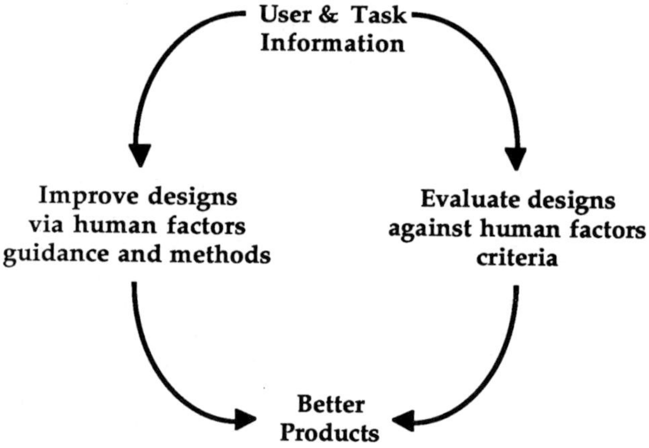

Figure 37.1 Two routes to improving products.

may be shorter but there may already be information held in the company about the users and their tasks. This information is not always accessible to the people who need it.

There is also a problem with many task analysis methods in translating the sometimes voluminous data gathered into product requirements. Tools to help designers use information about users and their tasks were clearly needed. Further the tools should bridge the gap between user/task information and product requirements. For these reasons the HUFIT User Requirements Toolset and Program were developed.

The audience for the tools

In order to provide human factors methods, guidance and tools which will be of use to any European IT supplier the tools must be independent of the design methods of any one company. The tools are aimed at designers of IT products for the office. The process set out in the tools will also be useful for other types of product but the tools concentrate on user and task characteristics and give example solutions which are appropriate to products used in the office rather than elsewhere. They are also aimed at designers of generic products. In bespoke system design a similar but more intensive process of user and task analysis would be required.

Early discussions with designers showed that any methods provided for use in a commercial environment would have to be quick to use as the designers were already under considerable time pressure. Nor should they demand that the users have previous human factors training. The aim in developing this Toolset was to provide quick, easy-to-use methods to help designers take account of user needs without adding significantly to the time-scale or skill requirements of their work, and which will be usable within the constraints of the design process.

Developing the paper-based tools

The development of the tools followed a user-centred design approach. They were developed interactively, involving their intended users and human factors experts. The process began with studies of the actual and prescribed design process in five IT supplier companies. Draft tools were then developed, based on methods used by the Human Sciences and Advanced Technology Research Institute (HUSAT) over a number of years and on ideas developed in work with other organizations, notably development of an extensive methodology called User Skills Task Match (Hutt *et al.*, 1987).

A workshop forum for the use of the tools was adopted so that a small amount of human factors knowledge could be used effectively. In a 'ripple effect' training programme the experience in a company could be spread to other team members during the work processes required to use the tools. The workshop format was also used for evaluation trials of the tools with potential users. The tools have evolved in this way through a series of drafts into a highly structured format. The evaluation trials with designers to try out the tools showed that the tools will be usable by designers with only a short period of training and some support in the early stages. The structure of the tools was also appreciated by people with previous knowledge of the subject matter such as human factors practitioners, as it provides a framework which helps to formulate answers to user needs. The on-line version of the tools retains much of the structure of the paper version, since this was found to be useful to designers.

Developing a PC-based tool

Having developed satisfactory drafts on paper an on-line version was proposed. Some people, especially designers and programmers, are known to prefer to work on a keyboard to paper and pencil. The on-line version will have more appeal for them. There are certain benefits to be gained by putting the tools on-line, in that information input in the early stages of the process can automatically be carried over into the later stages. In addition the explanation and notes can be made more easily accessible on-line.

There may also be disadvantages to the on-line version which will have to be overcome in use. The suggested use of the tools in a workshop forum will be more difficult if the tool is being used on a PC. It is difficult for more than two people to work together around a small screen. One of the major benefits expected from the use of the tools arises out of the workshop process. By bringing together representatives from marketing, planning and development they are able to share their knowledge and discuss the needs of the product user in order to develop appropriate product plans. This discussion process should also benefit the company in the long term by raising the general level of awareness of user and task issues and improving communication between different sections of the company. It may be that the workshop using paper tools will gather information and develop ideas while the on-line tool is used as a convenient way of capturing the results of the discussion for storage and use later in the design process.

The form of the on-line tool

The demonstration version of the tool was developed on a Macintosh™ using Hypercard™. Working versions may be implemented using data base software. It is expected that companies will wish to produce their own versions to be compatible with other on-line design tools.

The Program follows a form filling pattern to retain the benefits of structure perceived in the paper version of the tools. Each title for a field to be filled by the user is supported by HELP which contains an explanation of the term and examples to assist in its completion.

A diagram showing the process of the tool is given in Figure 37.2. The Program has two stages which may be carried out separately. Stage 1 is intended to contribute to the very early

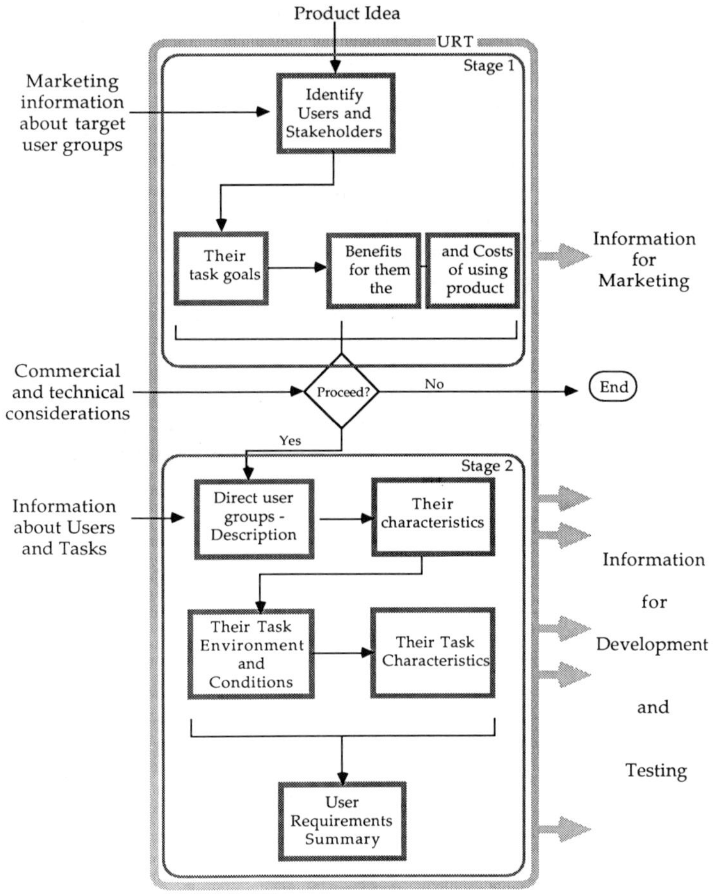

Figure 37.2 User requirements core tools.

development of a product idea in order to decide whether the idea is viable. This period is sometimes referred to as the conception stage of the design process and is generally given too little attention in design methodologies. Stage 2 would be completed for a product idea which is considered viable and is being investigated for development. Both of these stages fall into the planning stage of design, prior to the production of a detailed functional specification.

In Stage 1 the first step is to identify users and stakeholders, i.e. all those who will use the product either directly or indirectly and other people who have a stake in the product such as the customer data processing staff. The job goals for these groups are then identified in order to examine the benefits and costs to them of using the proposed product. When all of this information is available it can be considered along with other commercial and technical information to decide whether the product idea is viable and whether the product should be put forward for analysis and planning.

In Stage 2 the steps all concentrate on the direct user groups which were identified in Stage 1, i.e. those groups who will have hands-on use of the product. The first step in Stage 2 elicits a general description of these user groups. The next step looks at each user group with respect to characteristics of users which have been found to be important for the use of IT products. Product requirements are derived to meet these characteristics. The examples provided in the HELP are intended to support the designer when deriving product requirements. The users' tasks are dealt with similarly in the next steps. A description of the task environment and conditions is elicited and user requirements derived to meet these and the task characteristics for each user group. Finally the product requirements are represented with an editing facility to allow collation and reconciliation to be carried out to produce a summary.

The information from any section of the tool may be printed out and given directly to designers and to system evaluators. Both groups need to know who the users are and what sort of tasks they will perform. The output from Stage 1 is also useful to the marketing department as it highlights the strengths and weaknesses of the product from the users' point of view. The tools provide information required to use other tools being developed in the HUFIT programme: the Specification and Design Tools (described later) and USE, a tool for specifying goals and criteria for human factors evaluation of a product.

Specification and design tools

Having completed the User Requirements Specification phase, the design team must then attempt to amalgamate these requirements with the proposed technical specification and, on approval, move to the design phase. In the design phase they will particularly address the user interface design issues which will have the greatest bearing on the system's usability.

Very often, however, the level of human factors knowledge required of planners and designers is too great for them to be able successfully to assess the degree to which the proposed specification actually meets the underlying user and task requirements. Equally the use of written interface design guidelines, in any practical situation, is often too laborious and the guidelines too intangible to be of any real value. The tools within the Specification and Design Toolset address these issues by providing planners and designers with a tangible yet simple means of linking user, task and technical issues in a way which makes the process both meaningful and purposeful. The Specification and Design Toolset comprises the Functionality Matrix and the User Computer Interface Design Tool.

The Functionality Matrix

The Functionality Matrix supports user and task issues in the development of an enhanced functionality specification by presenting complex human factors, business and technical data in an easily understandable matrix format. The Functionality Matrix, like other tools in the HUFIT Toolset, is both paper based and on-line. A highly simplified schematic view is shown in Figure 37.3.

Used by product planners during the feasibility, planning and specification phase, the matrix requires the cross-referencing of individual items of proposed functionality with each identified user and task characteristic on a simple rating scale. These characteristics are obtained from the User Requirements Program and from the analysis of task content which has been carried out. In addition estimates are made of a number of business factors affecting the specification namely time, cost and development effort. Note is also taken of marketing feedback relating to competition analysis and the views of product purchasers, who often

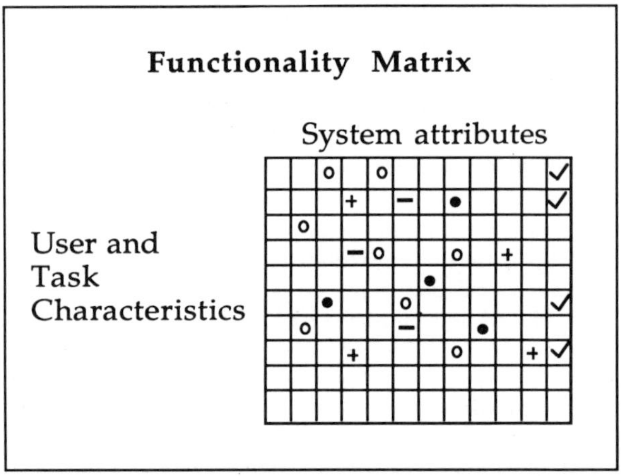

Figure 37.3 Schematic representation of the Functionality Matrix.

have very different requirements from those of product users. Assessments are then made of the degree to which the underlying requirements of a given characteristic are, or are not, being adequately met by the overall specification. Additional functionality required can then be identified.

At this stage the matrix shows the complex interrelationships of the various business constraints affecting the product specification. Each characteristic can be assessed as to whether it is being adequately and demonstrably met by the specification and an assessment can be made of the actual requirement for each functionality item in the context of user and task issues. The trade-off procedure between the various and often conflicting business interests can now be carried out to finalize the specification prior to design. User and task issues are explicitly considered as an important business constraint while representation of the various interests provides the rationale for any given specification decisions. The matrix is completed in a workshop setting with representatives present from planning, design, marketing, and with financial and managerial expertise. This helps to generate a more comprehensive design proposal which is representative to all parties. This design proposal is then captured, recorded and disseminated to the design team.

The paper-based matrix was developed using a standard Macintosh™ drawing package (Macdraw II™) and employed an iconic rating scale for the cross-referencing procedure. This was found to be useful in supporting the assessment procedure by allowing quick visual assessments to be made which efficiently pinpointed extremes. The *process* of completing the paper-based form in a workshop promoted cooperative and effective decision making and was acceptable to most of the designers who were involved in early user trials. A major drawback of its use in this form, however, was the inflexibility of the graphic layout, every revision requiring a major re-draw.

In order to overcome this problem, the matrix has now been adapted for use within a spreadsheet format. Currently Excel™ on the Macintosh has been selected for demonstration purposes. This not only allows for more efficient data entry and retrieval, but allows a more flexible use of layout and storage facilities. For instance, alternative functionality proposals to meet a perceived need can be assessed by opening a number of columns, completing the rating procedure and the least effective solutions can easily be deleted from the spreadsheet. A numeric rating scale has been developed which uses summation to aid the assessment procedure after initial criteria setting.

The Functionality Matrix builds on initial ideas for system functions and helps to produce an enhanced specification by integrating human factors, business and technical considerations. In both paper and spreadsheet forms the design solution proposals which emerge during the *process* of completing the matrix represent a vital input to hardware and interface design.

User-Computer Interface Design Tool

Having identified a range of appropriate product functions, the software designer has to build a usable interface to allow the user to access them. This can be a complex and time consuming process due to the wide variety of interface design styles that can be produced. Much good advice and guidance is available, e.g. the Mitre Guidelines (Smith and Mosier, 1984), but the sheer volume of such information is likely to discourage the busy designer. In addition, most guidelines in the field of ergonomics suffer from being too general and fail to offer a coherent or prescriptive approach. The User–Computer Interface Design Tool attempts to overcome these drawbacks by offering a step by step approach to interface design in a concise format.

The tool comprises five stages:

Stage 1. Choosing relevant styles.
Stage 2. Designing overall appearance.
Stage 3. Designing dialogue structure.
Stage 4. Designing detailed aspects of style.
Stage 5. Providing user support.

The first stage involves the designer deciding whether the user or the computer should principally lead the dialogue. Having done this, the designer then responds to a series of questions about the sorts of activities the user is to perform using the computer. These basic activities include browsing or reading information, entering data, making enquiries, setting operating parameters and so on. For each activity, a number of dialogue styles are recommended based on the likelihood of the users being trained, or experienced in using computers, and whether the designer has opted for user-controlled or computer-guided dialogue. For example, one of the proposed guidelines for the question 'Is the user going to be making enquiries?' suggests that the designer should consider the use of a command language for experienced or trained users interacting with the computer via user-controlled dialogue. The first stage ends with the designer summarizing the user's activities in a table together with the suggested dialogue styles.

Stage two considers the overall appearance of the application. Once again the design progresses in response to a series of questions that cover the use of graphics, windows and screen layout.

A dialogue structure is created in stage three, largely based on the list of functions that have been generated using the Functionality Matrix. This structure enables the designer to think about how the functions link together and how the user is to navigate around them. A diagrammatic notation is suggested in the tool but the exact notation used is not important and the designer may choose to use Jackson's Structured Diagrams (JSD), Data Flow Diagrams (DFD) or other suitable representations. Structure charts should allow the designer to see where shortcuts are needed, whether there are too many states, and how easily a user can move between functions.

Having developed an overall view and structure for the interface, stage four helps to develop the detailed aspects of each basic dialogue style used. The following dialogue styles are considered: structured menus, flexible menus, commands, question and answer, form-filling, natural language, function keys and buttons, icons, and direct manipulation. Advice is also given on how to enhance the dialogue with support features and shortcuts. This normally involves incorporating other style components, thus producing a mixture of styles. Finally, further construction advice is included to cope with special needs such as input feedback, hardware required and so on.

Finally, stage five is to consider user support needs for help, feedback and status information, error handling and control actions. Illustrations from existing systems known to confuse users are presented in order to demonstrate how easy it is to fail to understand users' real needs.

An on-line version of this tool called Dialogue Adviser has been implemented in Hypercard™ on the Macintosh™. This comprises two parts: as interactive checklist where the designer specifies the sorts of activities the user will be carrying out with the system and a detailed reference section where the designer can obtain advice about constructing and implementing the various forms of dialogue style. In hypertext fashion the designer checks boxes and then presses a button to jump directly to the appropriate design advice. A facility is incorporated for compiling the selected sections of advice into a brief report which can then be printed for use off-line. Planned versions will also include animated examples of dialogue components such as pop-up menus, dialogue boxes, forms and iconic buttons. An example screen from a prototype version of Dialogue Adviser is shown in Figure 37.4.

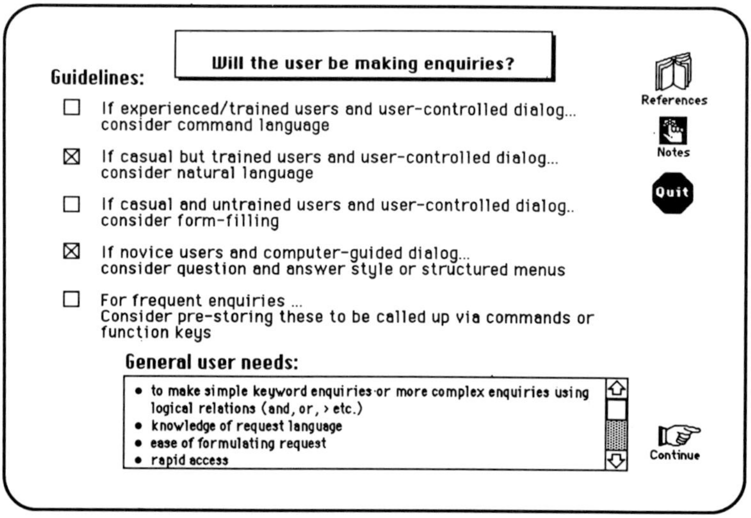

Figure 37.4 Example screen from Dialogue Adviser prototype.

The HUFIT tools in practice

The following brief case study shows how the tools are used in the design process. The example is based on a real design project – the Training Access Points (TAP) project. The example is now used successfully in training seminars to teach designers how to use the HUFIT tools.

The TAP project

The UK Government Training Agency wanted to extend the use of technology in providing information to the public about training courses. A series of local databases had been set up to give on-line access to information about all the training courses, vocational and leisure courses, available in the area. The Training Agency approached HUSAT to develop a suitable user interface to allow public use of these local TAP databases in public areas such as community centres, job centres, careers offices and shopping centres. It was also envisaged that the TAP system should be used in small businesses to help them find suitable training for their employees.

The user-centred design approach which has been built into the HUFIT tools was taken throughout. The interface was designed iteratively, with evaluations conducted at many stages. The information which the HUFIT tools elicited during this design exercise is too extensive to reproduce in full. Examples of the information recorded and conclusions reached by using each tool are given briefly to allow the reader to get the flavour of the Toolset in use.

The user requirements tools

The first stage of the User Requirements Program as shown in Figure 37.2 identifies the users and stakeholders. For the TAP interface there will be three main types of user:

1. The public: any member of the public over school age.
2. Professional intermediaries: careers advisers and personnel officers will be using the system on behalf of clients or employees.
3. TAP agents: these agents maintain the local data base and will look after the public interface units.

The Remote Users heading reminds the designer that the Training Agency will want data on system usage to monitor the results of their investment.

The User Description Section at the beginning of stage 2 ensures that the designer considers the characteristics of the direct user groups identified in Stage 1. For instance there is a wide range of language abilities in the British public and also that there may be users with some physical handicaps.

The remaining steps in stage 2 lead the designer to consider various user and task characteristics and the resulting product requirements. Some examples of these are:

(a) very supportive dialogues to help the many first-time users in the public,
(b) short-cuts for the frequent users, e.g. careers advisers,
(c) quick results, even for first-time users as they may just give up and go away if the system is difficult or time-consuming to learn and use,
(d) highly robust housing and software to withstand possible vandalization in unattended public locations,

(e) users have different constraints so that they will want to search for courses with a variety of keys, e.g. day of the week, type of qualification, as well as the subject of the course.

Summarizing these product requirements at the end of the User Requirements Program also has the effect of developing ideas for solutions. For example the need for an easy-to-use, straightforward interface which will also be usable by people with reduced manual dexterity or other partial handicaps may be translated into a proposal to develop a simplified keypad with large, well-spaced keys, this being one solution to the problem.

Usability Specification for Evaluation

Before the product plans are handed over to developers the Usability Specification for Evaluation table should be completed. This tool has not been described in detail since it is still a paper-based tool. Essentially, it leads the planners/designers to state specific product goals in a form which is testable. The goals, tests, measures and acceptance criteria are set together. This should be done in the planning stage so that designers and testers are working with the same goals.

For the TAP interface a number of goals can be identified, some of which are firm requirements which must be met, for example, (i) users must be able to learn to use the system and get results within a few minutes, and (ii) the user must not be able to crash out of the program into the system. The tests, measures and criteria are set to reflect the critical nature of these goals.

Other goals may simply be desirable, for example, novice users may learn to use the system without the tutorial, and users may remember how to use the system a month or more after learning. The tests and criteria for these would be less stringent. There might not be a minimum acceptance level for non-critical goals of this sort: the goals elicited and tests which will be used are useful for developers, the goals can be expanded to provide design suggestions, and the product testers would take the USE table and collate the tests and measures to create a high level testing program. This would be developed into more detailed plans later in the design process.

The Functionality Matrix

Having completed the User Requirements Specification procedure the Functionality Matrix was used to assess the degree to which the initial functional specification for TAP met the underlying requirements of the identified user and task characteristics.

Each user and task characteristic was summarized and cross-referenced using a simple rating scale against each of the functionality items; categorized, in this case, according to hardware, basic functions and software interface items. It is important to note that the system requirements derived from user analysis were also included in the functionality statements in order that their suitability and appropriateness could also be assessed.

The rating procedure itself tends to throw up a number of design considerations which result in greater definition of the functionality items themselves. In TAP, the printer type had not been specified but consideration of its use in libraries, for instance, demanded the provision of a thermal line printer (meeting the other demands of cheapness and reliability) rather than a daisywheel or dot matrix printer. Laser printers were, of course, too expensive. Similarly, the intended tutorials on operation and content were only assessed as meeting the user and task requirements if they were an optional rather than mandatory front end.

Frequent or expert users would wish to bypass the tutorial in subsequent sessions. Perhaps the most useful outcome of the rating procedure was to confirm the various options for function keys on the proposed custom keypad. An advantage of a spread-sheet based version of the matrix is that options can be assessed by opening additional columns and comparing the ratings made, closing down the columns for the discarded items when assessment is complete. By this method, the provision of a touch sensitive pad with the seven large-target area function keys was confirmed, together with the need for a smaller alpha-numeric pad for 'expert' entry.

The last item followed also from the identified need for two levels of access to the database. Whereas one set of users would, requiring infrequent access, be computer naive, have few keyboard skills, etc., an increasing number of users would become 'expert' and require the ability for fast shortcuts to access the specific known items of interest. The knock-on effect of dual navigation, different yet simultaneous dialogue provision (i.e. form-filling and command language, etc.) for instance, for coding and style, could quickly be assessed using the matrix.

As the individual functionality items are rated, so designers are forced to consider the design constraints which may affect the usability of a given item and, as a consequence, a significant output of the matrix is a separate set of proposals which relate to this design implementation. This was certainly true of the TAP interface design process itself where consideration of the high-level statements of required functionality in direct relation to the individual user and task requirements, led to a greater clarification of the ways in which the different design usability considerations could be developed.

The User–Computer Interface Design tool

The designer now works through the User–Computer Interface (UCI) Design tool, using the information from the User Requirements Program and Functionality Matrix. The User Requirements information provides the basis for answering the questions in the first stage, filling in the detail of which tasks fit in which category from task content analysis.

The first stage was to decide on the locus of control: computer or user. In this instance the prime user of the system was considered to be a casual user, so the computer was to guide the interaction. The major activities of the user were identified as making enquiries and reading information. The UCI tool suggests that if the casual user is to make enquiries on a computer-guided system then structured menus are an appropriate dialogue style. Structured menus are also recommended for browsing through a hierarchically organized information system, and so this was chosen as the predominant dialogue style.

The next step was to design the overall appearance of the application. Graphics were considered superfluous, clearly structured text located in consistent screen areas and colour coded, for example using green to denote help information, met most user needs. Following guidance in the tool, fixed window areas were felt preferable to full-function user-controlled windows, as windowing is a difficult concept for computer novices to grasp, and one of the major considerations was to keep the interface consistent and simple.

The dialogue structure was then created. This involved linking the functions identified using the Functionality Matrix using a flowchart-like notation and then analysing the ways in which shortcuts, navigation facilities and user support could be added. Figure 37.5 shows the UCI structure created including the additional features that were added (the inset shows an earlier outline structure). This proved particularly useful when considering how the system would intervene and display course details directly when the number of data base 'hits' was at a sensibly low level.

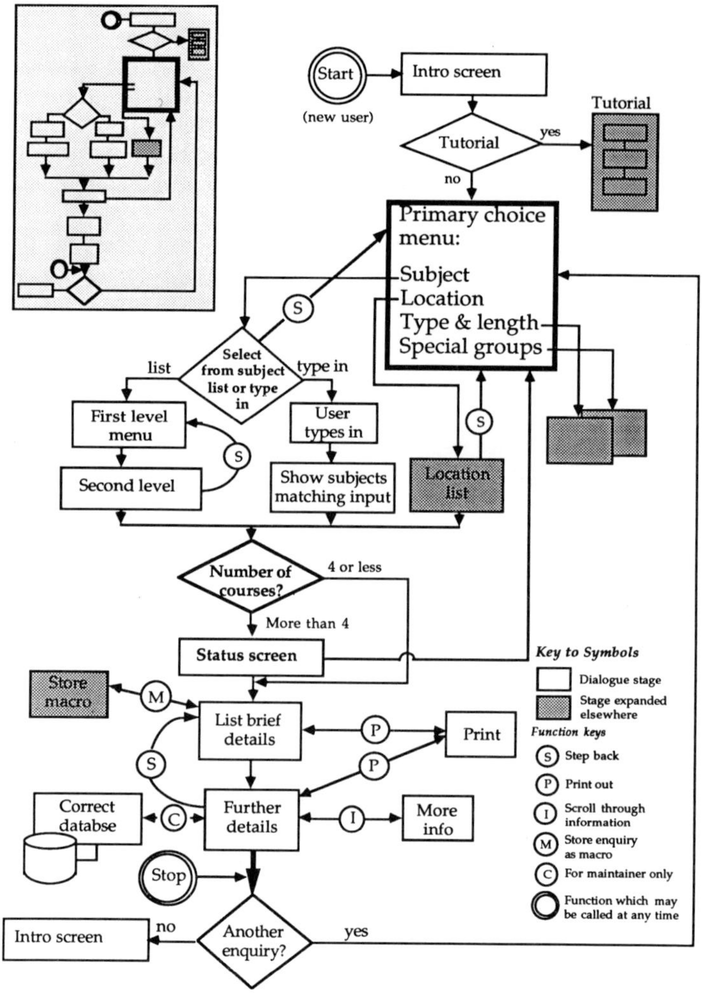

Figure 37.5 UCI structure. Inset shows early outline.

Then came the painstaking job of considering how each individual screen was to be constructed. The next section of the UCI tool contains details of dialogue styles and construction principles. Thus selection of menu options was always to be done by moving a highlighted bar over the option in order to give intermediate feedback to the user and reduce errors. As the user moved the selection bar (with arrow keys) a supplementary window was updated dynamically to display both the number of courses within each option and any following menu that would be displayed. The basic screen layout is shown in Figure 37.6

Finally, the user support was built-in. This included a status box that automatically appeared after the user finished selecting courses on any major criteria, i.e. subject, location, allowing them either to see the courses or develop their enquiry further. Help and error messages were displayed in the supplementary window and were designed to be neutral with no implication of blame or foolishness on the user's behalf.

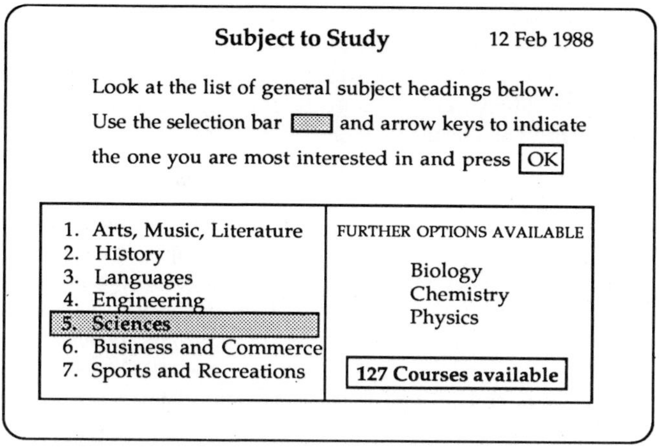

Figure 37.6 Dialogue style and screen layout.

Multi-method evaluation of TAP

Having developed the TAP system user interface, it was evaluated according to the usability goals described in the section above on the USE tool. Prototypes were made at an early stage and shown to careers officers, data base agents responsible for the upkeep of the system (users in their own right), and human factors experts for their appraisal. An improved prototype system was tested with the general public both in the laboratory and in a local field site.

This multi-method approach to product testing proved very successful, and considerable improvements were made to the interface. There is one note of caution, however. At every round of testing the data helped identify possible enhancements and it would have been possible to keep on changing the interface *ad infinitum*. This highlights the need for pre-determined usability criteria so that the developers can call a halt to the iterative design when the product meets these criteria.

The Documentation Design Guidelines and HyperGuidelines

The Documentation Design Guidelines and HyperGuidelines are a set of tools to aid the production of computer user documentation. Computer user documentation has long suffered from a very poor image. It is traditional that user documentation is afforded few resources by IT suppliers as it is seen as a high cost/low return feature of the product. The public opinion up until now of documentation appears to back up this view.

Today computer user documentation is increasingly recognized as being of vital importance to the success of a product. Recent changes in the information technology marketplace have meant that products are becoming more similar to one another. In addition, the public is becoming more sophisticated and the negative effect of poor documentation is beginning to have an impact on sales. Additionally, the increased awareness of the possibilities of on-line documentation coupled with the increase in the availability of mass storage systems means that most major companies are looking towards significant changes in their user documentation.

The problem of poor user documentation is seen as a lack of information, on the part of the writer or designer, of the human factors of documentation. There are well documented factors which affect the way different sorts of users react and learn from different aspects of documentation. These differences can relate to the physical aspect of the material, its structure or its content. The Documentation Design Guidelines have collated this information and present it in a form that is both useful and usable to the writers or designers of computer user documentation.

A user-centred approach was taken in the design of the Guidelines and documentation designers have been heavily involved in the process at all stages. As a starting point it was assumed that documentation designers wish to produce good documentation and that if problems existed these were directly related to a lack of information. Consequently 27 pieces of computer end-user documentation from a variety of companies and relating to a number of product types were tested to establish whether problems existed. These tests were carried out by interviews with users and by comparison. Analysis showed that a wide variety of problems existed in most documentation but that there were very few problems that were common to all documentation.

It was felt that these problems were due, at least in part, to the documentation designers' lack of awareness of the users' interaction with the documentation, i.e. the human factors aspects of good documentation design. Because of the disparate nature of the problems it was not feasible to produce structured training material which would be applicable to all documentation designers. For this reason a guideline format was proposed. This format involved the compilation of the human factors information documentation designers were lacking.

Development of the tools

Initially, information was collected from the literature dealing with documentation in general, documentation for instructions, computer documentation and computer users. A limitation on the information gathered was the consideration that it ought to be of a type that could be organized to offer the documentation designer useful information in a succinct form.

In line with the user-centred design approach, iterative refinement of the content of the guidelines was achieved by meetings with documentation designers in a number of IT supplier companies and other organizations. The designers gave their opinions of each guideline and made suggestions as to how it should be modified or expanded in order to facilitate its use or make the information it contained more valuable to them. This filtering of the information base resulted in the reduction of the initial information from 250 chunks to about 100. In addition a panel composing experts in the field of human factor aspects of documentation and instructional design was asked to comment on the validity of the recommendations made in the guidelines and their comments were taken into account in the ultimate choice of guidelines.

The format of the Guidelines

The delivery format of the tool was defined with reference to previously published guidelines and was modified following discussion with human factors experts and during the iterative development process with the documentation designers. It was concluded that if the tool was to be effective in guideline form the presentation would have to be developed to overcome the following problems.

Ensuring consistency between diagrams and text

Guideline

There must be no inconsistency between what is presented in the diagrams and the text.

Explanation

Inconsistencies often occur because the elevation of the diagram does not agree with the elevation dealt with in the text.

In addition, there are often cases where the diagram is wrong or has elements missing. Such inconsistencies cause confusion in users.

Example

Connect the parallel input to the socket. Press the on switch and wait for the red light, also on the rear of the machine, to light up.

Qualifications/Suggestions

Wherever possible, it would be best for the author to do at least the first sketches of diagrams.

Reference

Hammond, N., Hinton, G., Barnard, P., MacLean, A., Long, J. & Whitefield, A. (1985)
Evaluating the interface of a document processor: A comparison of expert judgement and user observation
in *Human-Computer Interaction INTERACT '84* (ed. B. Shackel) vol. pp.135
Elsevier Science Publishers B.V. (North-Holland) IFIP.

Figure 37.7 Example 'page' from the HyperGuidelines.

Guidelines as they had been presented elsewhere suffered from two major drawbacks. First, it was always difficult for the user to relate the information to the task in hand. There was seldom an example of the guideline in use and there was no explanation of the guideline in relation to the task of the writer. Second, it was unusual to have any formal qualifications for the guidelines given. This pre-supposed that guidelines were always applicable and valid. While this may be true of a very small proportion of guidelines material it is obviously much more beneficial to be able to make as generalized a guideline as possible, while anticipating and flagging cases where the guideline may be inappropriate or need modification.

In practice HUSAT produced the Advanced Guidelines Format. An example of the Advanced Guidelines Format can be seen in Figure 37.7. Essentially it consists of the following features:

1. A generally prescriptive guideline; limited to a paragraph in length and laying out in the clearest terms possible what ought and ought not to be done.
2. An explanation of each prescription; explaining wherever necessary why a guideline is needed and the effects it will have on the way the final documentation will be used.

3. Examples of good and bad practice (where applicable); these examples were drawn wherever possible from existing documentation in order that the writer may visualize and identify with the guideline.
4. Qualifications and suggestions (where applicable); in order that any cases where the guideline does not apply or may need modification may be flagged.
5. Bibliographical references so that the guidelines may avoid the spurious validity of purporting to come from research without referring to it. This allows the writer to refer back to the original work for more information.

The Documentation Design Guidelines were then widely distributed and presented to documentation designers in a variety of companies. Feedback received provided an indication of the refinements needed before formal experimentation and evaluation commenced.

A number of comments suggested that the delivery method would benefit from computerization. Documentation designers felt that the guidelines would benefit from being grouped according to task and subject. However, because of the overlap between different tasks it was not possible to do this in paper form without a great deal of duplication.

For this reason the HyperGuidelines software was developed. The HyperGuidelines contain the same material as the Documentation Design Guidelines and were initially produced using the hypercard application on the Macintosh™. The interface was produced in conjunction with expert advice and user testing. The testing of the interface has taken place with documentation designers and human factors experts at a number of locations across Europe.

Evaluating the Guidelines and the HyperGuidelines

At this stage there was little doubt that the Documentation Design Guidelines and the HyperGuidelines were sought after as useful tools, as they were already in use in a number of real-life documentation production projects while still in draft form. Following 9 months work in the development phase the Documentation Design Guidelines and the HyperGuidelines entered the evaluation phase. Testing took place at a number of sites in order to validate them with a range of potential users.

An example of testing procedures was that seen at Baddeley Associates, an information design consultancy company, where a 3-month testing programme was completed. The Documentation Design Guidelines were used by documentation designers at Baddeley Associates' Cambridge offices as part of their daily work practice. Their use of the Documentation Design Guidelines was monitored by a diary check system included in the tool itself. Concurrently the London branch of Baddeley Associates used the HyperGuidelines which had a self-monitoring system built into it.

At the end of the 3-month period the data were collected and analysis provided the use patterns of the guidelines in both media. These measures, in addition to interviews with the documentation designers taking part at fixed times throughout the course of the experiment gave as near a view of potential use of the tools as possible.

Further modifications were made to both tools at the end of the validation period in light of the data collected. One of the most crucial results was the need for increased content. The attitude of the users towards the two forms of the tools were most marked in this respect. The reaction to the paper-based tool was that it was nearing the limit of the information that could be reasonably accessed without producing sub-volumes on different subjects, thus defeating the object of a unified access tool. The general view of the HyperGuidelines, however, did not see the inclusion of more information as a problem of any sort, rather as a crucial step in the development of the tool.

The two tools have proved to be well received in the documentation design community. While a great difference exists between the perceived utility of the presentation formats, on-line and paper-based, it is felt that they both fulfil a valid rôle. The HyperGuidelines version of the tools has generated the most interest and suggestions have been made that it be enhanced to form the basis of a writer's workbench. Amongst the suggestions that are likely to be taken up are a facility to incorporate the user's house style and the ability to access this information while in a text editing package.

The information content of the two tools has generated a great deal of interest. However, it must be recognized that the information is not new, rather the concept of making the information available to the user in a useful and usable form is the basis of this work. This work has also generated interest in producing both BSI and ISO standards relating to computer user documentation.

Conclusion

The HUFIT Toolset provides an integrated human factors input to the design of IT products. The Toolset is oriented towards enabling designers to take a user-centred approach in the development of IT products. The full Toolset shown in Figure 37.8 provides tools for each stage of the product life cycle from product planning through specification, development, testing and user support. The tools are designed for use by the participants in the design process, within the time and resource constraints of the design process and by people who do not necessarily have human factors skills and expertise.

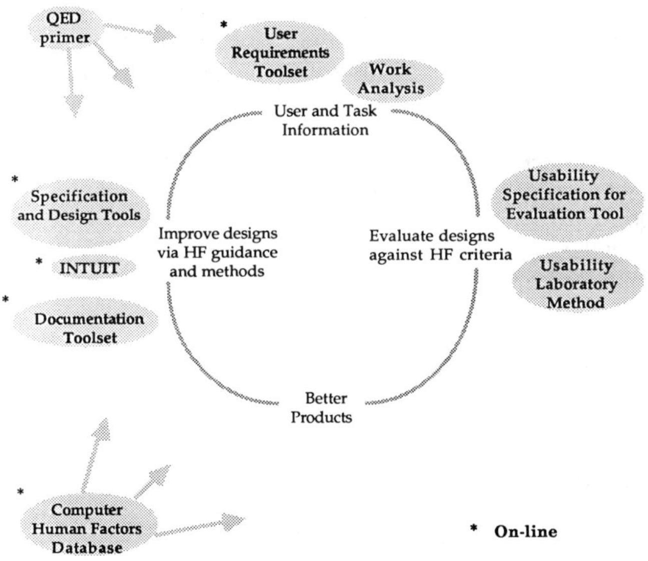

Figure 37.8 HUFIT Tools contribution to improving products.

The developers of the tools, themselves human factors experts, have taken a user-centred design approach and the users of the tools have been involved in their development and evaluation. The users include those participants in the design process including product planners, marketeers, systen developers, quality assurance, documentation design and so on.

The HUFIT Toolset was orginally developed as paper-based tools and, where appropriate to the requirements of the users and their tasks, on-line versions have been developed. The user can decide which formats are appropriate to their environment.

The HUFIT Toolset has been well received in Europe as a major contribution to the design of usable IT products.

References

Hutt, A., Donnelly, N., Macauley, L., Fowler, C. and Twigger, D., 1987, Describing a product opportunity: a methodology for understanding the users environment. In: *People and Computers III*, edited by Diaper and Winder (Cambridge: Cambridge University Press).

Smith, S. L. and Mosier, J. N., 1984, *Design Guidelines for User–System Interface Software*. Report ESTD-TR-84-190 MTR-9420. (Bedford, MAL: The MITRE Corporation).

Chapter 38
Human design and computer-aided design

N. R. KODALI
Engineering Design Department,
Tufts University,
Medford, MA 02155, U.S.A.

Abstract: A primer that gives the power of exploration when encountered with a CAD* system is described. Identifying the CAD process as a building rather than drawing operation, the author builds mental models in the reader's mind by formally exploring the conceptual model of a CAD system. The building process is performed within the computer's virtual space (3-D) by viewing it on the computer screen (2-D). The primer requires CAD users to ask and unravel the answers to the most rudimentary and often neglected questions: What do we build with? How do we build? How do we tell the computer to build for us? How do we manoeuvre and manipulate the objects in the computer's space?

A demonstration of the primer is VISUALCAD, an interactive technique with me playing the role of a computer. VISUALCAD guides the learners to explore and reinvent the concepts of CAD and build adequate mental models to use the system. Using a video camera and a TV to bridge the gap between 3-D construction and its 2-D representation, it reduces the learning time drastically. A paradigm to answer the conceptual questions is presented using Moran's Command Language Grammar and Veer's metacommunication principles.

Keywords: CAD training; Training design; Conceptual models; Mental models; VISUALCAD; CAD primer; Human factors in CAD.

Introduction

Were you ever excited about the new car you bought? Were you ever proud of driving it and showing it off? This is more or less true for anything new we buy or acquire. But not a new computer . . . definitely not a new CAD system! Regardless of it being from the sun or the moon we just wish to use it; this is not in reference to people who run canned demos.

However, 'CAD is not a tool that can be used freely without training – it takes between nine months to a year for a person to learn to use a CAD system as effectively as a drawing board and it can take a company anything from three to five years before it can expect any major benefits from a CAD system' (Burkill, 1986). It is assumed that the user is largely responsible for the efficiency of the system, and that productivity is a function of training, skills, aptitude and motivation of the users (Simon *et al.*, 1986). What is the training? This is the issue involved.

*We are concerned with mainly three-dimensional computer-aided design systems used in mechanical, industrial and architectural design.

Existing CAD training situation

Like any new technology, we face problems in CAD training (see Abbas, 1977; Arthur, 1982, 1984). The extent of the problem covers the CAD industry that requires over 100 000 operators (Burns, 1986) and a recent recorded market growth of 20% (Krouse, 1989).

It is not the intention to say that CAD training is non-existent. However, it is argued that the existing training methods are not adequate and appropriate. *Vocational* training provides familiarity with the equipment and *awareness* training provides the managers with the potentials of CAD technology. Both suffer from two defects; the breadth of subject and lack of realism – users asking 'what will it do for me'? (Watt, 1983).

Many companies using CAD fail to foresee the importance of training. Fitzgerald (1984) states 'it never ceases to amaze me the way companies will risk the success of a very expensive investment by skimping on the thing that positively controls that success – the training of the staff to use the equipment properly'. Consequently, with inadequate training it is the designer/user who is subjected to all the humiliation and constant pressure to achieve the salesman's productivity promises (Simon *et al.*, 1986). As Duus and Rader (1983) view the situation, for most CAD managers 'actually proving effectiveness is a kind of tight-rope act, on one side they are faced with exaggerated expectations from company leadership, on the other side, with resistance and reserve on the part of the users'.

Watt (1983) believes that only in the academic world is the essential re-training being initiated. The situation is not optimistic. Anderson (1986) explores this dilemma: 'how does the teacher keep up, on a limited time, budget and severely limited or no training?' There are 50 different courses offered for one CAD package by the vendor at an average cost of $1300–1400 per course (see Computervision, 1988).

At Tufts University we had the CADDS-4X system for six years. I never thought I would miss that system until it was recently replaced with I'DEAS, running on SUN workstations. I want to use it but not at the expense of a long and arduous retraining programme. Vendors might think we expect too much by asking to 'use' the system that we have purchased!

One presenter at the CHI-87 conference responding to my wish (use the CAD system) said '. . . there are more than one million lines of FORTRAN program running behind the screen with lots of functions, you can't expect to use it with no effort and no training'. I watched the training videotapes of the new CAD system and tried to do the tutorials/workshops. Conceptual paralysis, a phrase used by Carroll and Rosson (1987), best describes my experience which was similar to the one I had when I was learning the CADDS-4X system. With my four years of experience in learning/teaching three different CAD systems I was unable to infer the concepts of the new system from the manuals and the video training material.

The situation is complex and calls for a solution that is practical, realistic as well as longstanding. It is not to claim that there is any simple treatment of CAD learning to offer but I believe that CAD can be more comprehendible. The systems should be learnable and not require laborious formal training, in the form of 50 courses or 36 videotapes and 36 cookbook tutorials.

Let us take a fresh look at CAD as a 'tool' and study its intended purpose and find out how to use it. The following sections present CAD Primer that should give enough power to the reader to explore CAD systems and use them.

CAD Primer

CAD systems are *not* emulations of drawing boards on computers. Instead, they are simulations of an exact building process. We build geometric models on the computer the way we all build with Lego and Tinker toys. CAD systems of today can be best described as tools that assist a designer in physically realizing one's own design (Katz, 1985). The use of these computer models is never-ending; models can be put together to create more elaborate models. Further, these virtual models can be crushed for testing, put into animation, checked for mass computations in finding volume, surface areas and centroid. For users, constructing and manipulating these models is the rudimentary task . . . and then the fun starts.

CAD setup

The basic computer setup of today is a computer screeen, a keyboard and a digitizer or a mouse. In contrast to our task (building), being three-dimensional in nature, the system is two-dimensional – a flat screen, a mouse we chase on a table or a pen on a digitizer. With the new technology such as head-mounted displays and Myron Kruger's work on gestural interaction techniques (Kruger's work 1989) or DataGlove we will be getting into the next generation of 3-D CAD. The human factors issues of interacting with such new technologies is yet to be studied. For now, at least in the next 5–10 years we will be working with the screen, keyboard and mouse setup (see Figure 38.1).

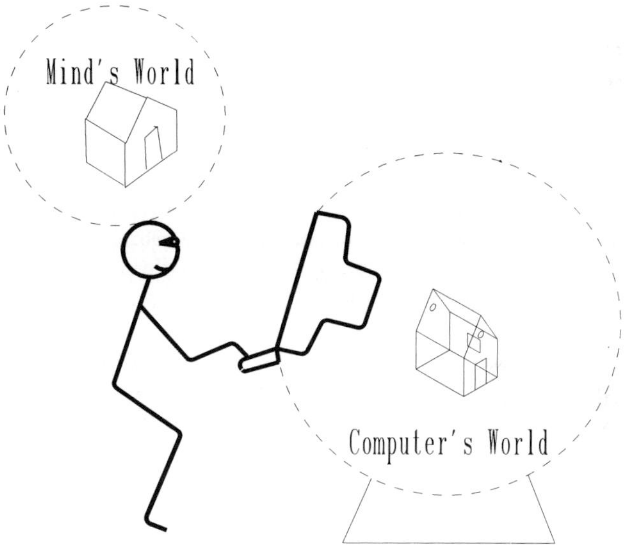

Figure 38.1 Computer-aided design. In addition to the physical set up; the computer screen, the user, and the input devices, there exist two virtual spaces, the mind's and the computer's. It shows a user building a house from his/her mind's virtual space into the computer's virtual space. The computer screen is only a window or a view into the computer's virtual space.

We need to have a control on the computer, instructing the computer to construct models and also to manipulate them in its virtual space. We should also be able to view the models from any desired orientation (top, front views, etc.) and distance (zoom in and out).

In Figure 38.1 CAD is pictured as an interaction between two virtual spaces, the mind's virtual space and the computer's virtual space. The communication is done via the real space we all live in, using the computer screen and keyboard, etc. A CAD operator is portrayed as building a house from the idea in his/her mind. Both the virtual spaces are shown in dotted lines implying no physical boundaries.

How and why CAD questions

Introducing CAD as just a building machine raises many questions about the semantics of a CAD system. What do we build with? Where do we build? How do we build? How do we tell the computer to build for us? How do we make any changes? How can we view on the computer screen? How can we manipulate the objects inside the computer's virtual space? Can we paint it . . . What else can we do . . . What are the tricks to build complex shapes?

The answers to these questions form a 'mental model' in the learner's mind that gives him/her the ability and skills to explore and use a CAD system. The developers and trainers should be equipped to answer these questions. The answers are there, remaining under the bulk of user manuals and tutorials for the learner to discover and interpret.

The answers to these conceptual questions are subtle and involve associated and abstract concepts; the entities we build with, the coordinate system to address space with, the screen we view on and the plane of the mouse inputs. The primary task that is three-dimensional in nature should be distinguished in the context of the inherent two-dimensional nature of the CAD setup. The following sections describe a formal approach for CAD training to equip the learners with adequate 'mental models'.

Paradigm for CAD training

Building adequate mental models in the learners' minds is the tenet of this training paradigm. Mental models allow one to understand and anticipate the behaviour of a system (Norman, 1983). The use of the term 'mental model' in this paper is distinguished from a 'conceptual model'. Mental models reside in the learner's mind while conceptual models are external to the learner's mind. The actual system, the documentation, the training programmes, the trainer's models are all external to the learner's mind and are referred to as conceptual models. There is a symbiotic relationship between the two models; good conceptual models cultivate adequate mental models. A bad conceptual model would have little functionality and would fail to communicate and explain important aspects of the system. Research findings provide ample support for the notion, that providing a user with a conceptual model of a system aids the novice user in building mental models (Hammond *et al.*, 1983; Kodali, 1987b; Sein *et al.*, 1987; Schmalhofer, 1987).

Moran's (1981) Command Language Grammar (CLG) aims at designing the user's model and provides a structured means of describing the system as the user sees it and understands it. It is made up of a sequence of description *levels*, each level being a refinement of previous levels. These levels are divided into two components; (i) the conceptual component and (ii) the communication component. The conceptual component contains the abstract concepts around which the system is organized, while the communication component contains the conversational dialogue. This model aims at the intended user's model and serves as a reference for the system developers, the document developers and trainers.

The paradigm of CAD training is shown in Figure 38.2. It requires one to describe the conceptual model and then to design 'metacommunication'. Metacommunication refers to all the activities of communicating the conceptual model of the system. Veer *et al.* (1985) give us an intuitive approach for designing metacommunications relating to different levels of

Moran's CLG. By selective metaphors and/or analogues and appropriately chosen medium (discussions, demonstrations, or doing) CAD concepts are communicated to the learner. An example of CAD training is presented in Table 38.1. A more elaborate description of CAD training is presented in Kodali and Coleman (1989).

Conceptual Model			Metacommunication	
Component		*Content*	*Analogue*	*Medium*
Conceptual	Task			
	Semantic			
communication	Syntactic			
	Interaction			

Figure 38.2 Paradigm of CAD training consisting of the conceptual model and the metacommunication of the CAD system. This requires the description of the conceptual model of the CAD system at all *the descriptive levels. Metacommunication constitutes appropriately chosen/designed analogues (Tinker toys, Lego, physical models) and the medium (discussions, demonstrations or learning by doing) for communicating the conceptual model.*

Table 38.1 Example of CAD training for CADDS–4X.

Task level	Realize design by constructing models. Constructing with Tinker Toys or Lego is analogous to constructing in CAD. These models can be assembled together, viewed from different directions, checked for physical properties, put in animation, checked for manufacturing
Semantic level	To construct models we have lines, circles and arcs. Just like we can have different coloured blocks in Lego, in CAD we can modify the attributes of entities, the colour and size. To locate and orientate entities we use the cartesian coordinate system. To do repetitive tasks, entities can be copied, rotated and translated. Example: for constructing a gear we create one tooth and rotate it, copying it around. To construct in an arbitrary plane in space the user should define a working plane called *construction plane*. To display multiple views, we define views on the computer screen corresponding to appropriate construction planes The physical models and demonstrations serve the purpose of communicating the above semantics. Just like we cannot verbally instruct someone to make a knot, we cannot document the semantics of CAD that deals with visual concepts
Syntactic	To instruct the computer we have a command structure. It is *Command Processor: Get Data*. Command Processor is the command given to the computer to act and the Get Data is the response of the computer, asking for any required data. For example, if *Draw a Line* is the given command processor, the computer will get back to you asking, *Where to where*?
Interaction	Commands can be selected from the menus or can be keyed in. There are three buttons on the light pen. The bottom button is used for selecting a command from the menu or for specifying a location. The other button can be user defined and the third is used to drag entities in dynamics.

Once having completed Figure 38.2, or 'paradigm for CAD training' the conceptual model of the system is complete and ready to be communicated to the learner. As mentioned earlier, the conceptual model assists the user to build a clear and accurate mental model.

Figure 38.3 is complementary to Table 38.1 in our building process of mental models. It shows someone actually communicating the conceptual, i.e. external model of the CAD system to the user who is building his/her mental model. Will the user benefit more by having smaller blocks of information carried to him/her through the four steps over several cycles, or will the learner gain more by one trip where a larger block of information is carried? This is a question yet to be answered. Through the description of the conceptual model we get some elementary idea on how much of it can be taught in any given time.

Carroll and Rosson (1987) raise issues about learners; they are not 'blank slates' for training designers to write upon and they get frustrated when training introduces them to concepts and does not allow them to use the system. They try to avoid reading. However, the visual concepts of CAD semantics such as viewing a 3-D scene on a 2-D screen, the relation between 2-D input devices and the virtual space behind the screen are beyond verbal description. VISUALCAD addresses such issues successfully demonstrating effective CAD training.

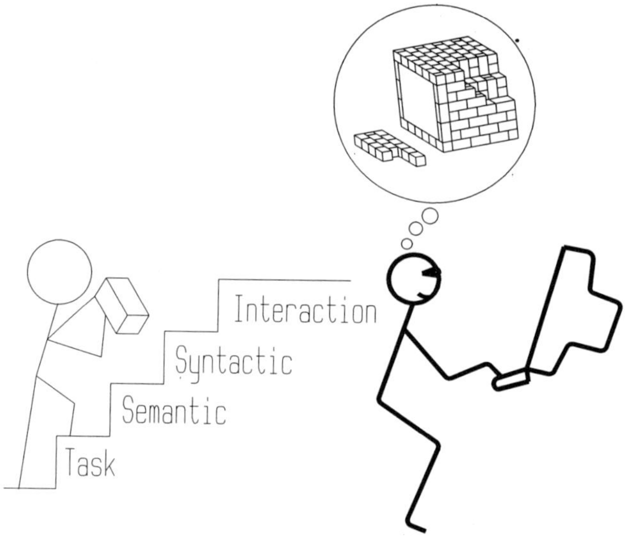

Figure 38.3 Figure depicting the building process of mental models. The trainer carries a block of the model through various levels. The conceptual component comes first and then the communication component. The block is initially communicated through discussions, and later by actually 'doing it' on the system.

VISUALCAD

VISUALCAD is an interactive technique developed by the author in 1986 to teach computer-aided design concepts to freshmen engineering students at Tufts University. The reader is asked to scrutinize the following material which claims a demonstration of 'almost ten times better performance'. The technique has been tested on six different groups for three years. A microanalysis is being done to investigate and deduce a more elaborate theory of CAD training.

Principle of VISUALCAD

Guiding the learner to discover the demand for certain concepts, and then by introducing the specific semantics of the CAD system, VISUALCAD helps him/her build adequate mental models. The following principles are instrumental in VISUALCAD presentation:

1. Using visuals to exploit the speed and accuracy of processing visual information and also its retention.
2. Using 3-D models, since humans learn 3-D skills more efficiently from 3-D information.
3. Using a video camera and a TV to help the learners to adapt to 3-D space on a 2-D computer screen.
4. Specific mental rotation exercise for fluency in manipulating objects on the 2-D screen.

Setup of VISUALCAD

The VISUALCAD setup (see Figure 38.4) constitutes a trainer and learners. The trainer, in this case the author, *discusses* and *demonstrates* CAD concepts. The trainer also acts in the role of the CAD computer following the commands of the learners in actually building with Lego

Figure 38.4 VISUALCAD setup showing the trainers and the learners. The trainer also plays the role of the CAD computer in building with Lego on the working table. The video camera connected a TV and directed at the working table. The coordinate system (X,Y,Z) identified on the table is seen on the TV that helps the learner to begin to identify the 3-D space on the 2-D screen.

and models. This also gives the learners an experience of actually *working* on CAD. A table is located on which construction with Lego and Tinker Toys is done. A video camera and a TV play a key communication role in showing a real 3-D construction process on the table and its corresponding 2-D representation on the TV screen. The camera is directed at the table to show the learners the view they see in reality. In the real CAD system the computer screen represents the TV while the computer and the keyboard replace the trainer's role.

VISUALCAD presentation

The presentation of VISUALCAD is shown below in the form of a dialogue, actually performed between the trainer and the learners.

Introductory dialogue

Trainer: Draw a line, write your name. How long did you take?
Learners: Almost no time.
T: Ask someone to draw the same line, ask someone to write your name, how long did you take?

Students tried to tell the person sitting next to them to draw the line they wanted and also to write their name. The objective was to guide them to appreciate that we have to give enough instructions and to accept that the process is slow.

Common reference system

T: Computers neither think and never design, we have to tell everything we want it to do for us. To tell a computer to draw a line, we have to tell it to hold a pen, grab a paper, start on the top left corner, draw a line towards the bottom right, of so much length . . . slow and needs more and all the information. What size paper? What colour pen???
T: Did you ever play with Lego?
L: Yes
T: CAD is just like playing with Lego! It is like asking the computer to construct with Lego for us (I threw some Lego on the table). Now you should tell me what I should do.
L: Build something.
T: The computer is not intelligent enough to build something.
L: Put that green block on the red one.
T: Which green one? (there were lots of Lego pieces on the table).
L: Hold the small square green block in your right hand.
T: OK.
L: No put it on the other side of the red one.
T: This way?
L: No . . . No other side,
T: This side?
L: No . . . other side . . .

Soon the students realized there has to be some common reference system. They were introduced to the right-hand rule cartesian coordinate system and even performed a short exercise on identifying the third axis given any two of X, Y, Z, -X, -Y, -Z. The directions X and Y were marked on the working table and the same could be seen on TV. The world was identified and the computer (myself) and the learners built some common reference.

Command structure

The command structure of the CAD system was introduced including the necessary syntax, such as colon and comma. VERB NOUN MODIFIER is the structure of the command we give the computer, for example; Insert (*verb*) Circle (*noun*) Radius (*modifier*). Any mistakes in the commands mean 'beeps' and error messages are explained.

T: From now you can give instructions only in the CADDS-4X way using verb noun modifier and with all the syntax. Your task is to build a cube with these sticks. Give me complete instructions to build the cube.
L: How does the computer know the working units . . . inches?
T: The computer does not know, we have to specify size and scale in the beginning.
L: INSert LINe: (0,0,0) to (6,0,0) to (6, 6, 0) to (0,6,0) to (0,0,0) . . .

Every participant was required to give commands and follow the construction process. As the cube grew bigger and went out of the TV screen, the concept of ZOOM was introduced.

Need for construction plane

T: Put a circle here (pointing to the table).
L: INSERT CIRCLE RADIUS 3: x3 y3 z3 . . . (I put a roll of masking tape on the table, i.e. XY plane).
T: Put a circle here (at a height of 6 units parallel to XY plane).
L: INSERT CIRCLE RADIUS 3: x3 y3 z6.
T: OK, now put a circle here (holding the tape role perpendicular to the table).
L: INSERT . . . CIRCLE . . . RADIUS . . . 3: x6 y3 z3 (gave the command hesitantly)

Meanwhile, some of the class started murmuring. They started asking questions. How does it know that we want to put the circle perpendicular to the plane? Can we put the circle and then rotate it? Can we tell the computer to work on XZ plane? Why can't we rotate 90° and then place it?

As soon as the learners realized they needed a new concept, I introduced the concepts of default planes and user-defined planes and gave the commands to that. The response was 'I thought about that, almost the same thing . . . '. This provided them with an incentive of discovery as they thought about alternate means of accomplishing the same task.

All the concepts were introduced only after they realized the *need* to perform a certain task. The VISUALCAD session is interactive in nature and the whole class participates and watches everyone's commands. After the session, the students were taken to the CAD laboratory. All the procedural issues (login, starting the part, starting a drawing, etc.,) were introduced.

The screen on the CAD system is similar to the TV screen in VISUALCAD showing the XY plane and Z identified by the right-hand rule. They worked in groups in building a box and writing their name on a plane, their section on one and the date on the other plane. This required them to be able to converse with CAD for basic construction and be able to define their working planes and changing from one to the other.

An assignment on 3-D construction was given for individuals. The same assignment was used as the seventh assignment in a graduate level CAD course at the end of the semester when I was taking the same course. VISUALCAD was tested on four groups (Kodali, 1987a) and found to be surprisingly effective. Apart from being able to finish the last assignment from a graduate level CAD course, freshmen engineering students found that CAD is fun and easy to use. Almost ten times better performance was observed. The content of this *two hour* VISUALCAD training can be compared to a *four and half day* basic mechanical design course offered by manufacturer of the same CAD system. (VISUALCAD presentation is available in the form of a videotape at Tufts University Library.)

After VISUALCAD

VISUALCAD demonstrates that CAD can be learnt by building the appropriate mental models in the users' minds. The successful demonstration of VISUALCAD (reported in 1987) raised questions on various elements that contributed to its success. Later this year the technique was experimented upon using one hour in-class presentation and not followed by the actual CAD lab session. It was found to be less effective in reinforcing the concepts they learned and practised in the classroom setup. Recently, a CAD training was designed and presented without using the video setup. Although the results were more inspiring than those of traditional CAD training sessions, it is claimed that video is a powerful communication tool in bridging a 3-D scene to a 2-D screen (Kodali, 1989). Therefore, I believe that CAD training is resolvable, and should be addressed by the developers, users, human factors researchers in providing effective 'minimal training'.

Future work – thoughts on DESIGN and computers

So far, CAD is seen as a building machine that supports the latter or even the last stages of the design process, i.e. building. However, design is an intellectual activity (Yoshikawa, 1983) that every human being does; one designs an aircraft, one designs furniture or one designs the decoration on a wedding cake, or a living room arrangement. It is the knowledge of these people that differs in designing various things. The design process itself is a cognitive process we do interactively. By interactively, I mean we interact with our own designs/ideas and refine them iteratively. We sketch and see, we scribble and see, think aloud and see, we make models and see . . . interacting with our externalized ideas. It is the externalization and the interaction that controls the characteristic of design. Sketching, scribbling, drawing, talking aloud, making models facilitates the design process. Of all the externalization methods which one is more natural than the other and what are the variables that play a role? Can we explore new means of externalization and computerize them?

CAD development would benefit by studying the natural ways we externalize our design thoughts. The research into the human imagery and the computer imagery would be mutually beneficial. The work on mental imagery is a promising approach at this juncture. Regarding computer graphics as a kind of 'externalized mental imagery' Weber and Kosslyn (1986) contrast both mental and computer imagery with a two-fold purpose: to explore ways in which graphics programs could be improved to externalize mental imagery better and to discover ways in which human imagery might be enhanced by utilizing ideas from computer graphics.

How exactly, can computer graphics serve our purpose of improving and overcoming the limitations of the human imagery? (i) Computer graphics excel in capacity for number and complexity of objects in a single display material indefinitely. (ii) Rotation is greatly complemented with real-time dynamic continuous rotation. The research behind the design process would be instrumental in formulating the CAD of the future.

Conclusions

The greatest misconception of all must be discarded: CAD is not an emulation of a drawing board. By building within the computer's 3-D virtual space and displaying it on the 2-D computer screen, the gap between 3-D and 2-D can be bridged. What we need is a minimal training and maximum emphasis on building the correct mental models.

The CAD primer relates to both the developer and the users by providing feedback to both corners. Developers of CAD systems must at least rise to the description of the conceptual model of the system. Only then can the users successfully build clear and accurate mental models. Without this communication there the 'conceptual paralysis' will persist and plague CAD development and usage. A structured way to implement the realization and communication of the conceptual model will make CAD learning/training an experience that is adaptable from system to system . . . making it a real design machine.

Acknowledgement

I acknowledge my partner Indrani who helped me 'externalize' my thoughts on CAD.

References

Abbas, S. A, Coultas, A. and Lee, B. S., 1977, *CAD ED – International Conference on Computer Aided Design Education*, (England: IPC Science and Technology Press).

Anderson, B., 1986, Resources for excellence in CAD Education – Computer-aided design instructors need extensive training for their field. *Industrial Education*, **75**, 19–22.

Arthur, P., 1982, *CADCAM in Education and Training – Proceedings of the CAD ED 82 Conference* (London: Kogan Page).

Arthur, P., 1984, *CADCAM: Training and Education Through the 80s – Proceedings of the CAD ED '84 Conference* (London: Kogan Page).

Burkill, A., 1986, Design creativity – Does CAD help or hinder? In: *New Horizons in Design Office Practice* (London: Mechanical Engineering Publications), pp. 4/1–5.

Burns, W. E., 1986, Computer-aided design (CAD) – Part of the computer-based technologies that are rapidly changing manufacturing. *Industrial Education*, **75**, 10–11.

Carroll, J. M. and Rosson, M. B., 1987, Paradox of the active user. In: *Interfacing Thought*, edited by J. M. Carroll (Cambridge: MIT Press), pp. 81–11.

Computervision, 1988, (Division of Prime Computer Inc.), *Course Descriptions* (Bedford, MA: Computervision).

Duus, W. and Rader, M., 1983, The tightrope act of proving effectiveness – some results on CAD. In: *Computer Applications in Production and Engineering*, edited by E. A. Warman (Amsterdam: North-Holland), pp. 1019–1029.

Fitzgerald, A., 1984, User training for CAE/CAD/CAM – a vendor view. In: *CADCAM Training and Education Through the '80s – Proceedings of the CAD ED '84*, edited by P. Arthur (London: Kogan Page), pp. 133–139.

Hammond, N., Morton, J., MacLean, A. and Barnard, P., 1983, Fragments and signposts: Users' models of the system. In: *Proceedings of the Tenth International Symposium on Human Factors in Telecommunications* (Boston: Information Gate Keepers), pp. 81–88.

Katz, R. H., 1985, *Information Management for Engineering Design* (Berlin: Springer).

Kodali, N. R. and Coleman, M. F., 1989, *Instruct to Construct: CAD Training, Proceedings of Interface 89* (Santa Monica: Human Factors Society), pp. 247–253.

Kodali, N. R., 1989, 3D Scene on 2D Screen: The VISUALCAD connection. *ACM/SIGCHI Workshop on Video as a Research and Design Tool*, 31 March – 1 April, MIT, Cambridge.

Kodali, V. N. R., 1987a, VISUALCAD: A visual aid for CAD. In: *Proceedings of Interface 87*, (Santa Monica: Human Factors Society), pp. 207–214.

Kodali, V. N. R., 1987b, Mental models and computer aided design (CAD). In: *Abridged Proceedings of the HCI International '87 – Second International Conference on Human-Computer Interaction*, Honolulu, Hawaii, 10–14 August 1987, pp. 98.

Krouse, J., 1989, Outlook for '89. *Computer-Aided Engineering*, **8**, 4 (Editorial).

Moran, T. P., 1981, The command language grammar: A representation for the user interface of interactive computer systems. *International Journal of Man–Machine Studies*, **15**, 3–50.

Norman, D. A., 1983, Some observations on mental models. In: *Mental Models*, edited by D. Gentner and A. L. Stevens (New Jersey: Lawrence Erlbaum), pp. 7–14.

Schmalhofer, F., 1987, Mental model and procedural elements approaches as guidelines for designing word processing instructions. In: *Human–Computer Interaction – INTERACT '87*, edited by H. J. Bullinger, B. Shackel and K. Kornwachs (Amsterdam: North-Holland), pp. 269–274.

Sein, M. K., Bostrom, R. P. and Olfman, L., 1987, Conceptual models in training novice users. In: *Human–Computer Interaction – INTERACT '87*, edited by H. J. Bullinger, B. Shackel and K. Kornwachs (Amsterdam: North-Holland), pp. 861–867.

Simon, M., Downs, J. E. and Leach, B., 1986, Human element in the design office. In: *New Horizons in Design Office Practice* (London: Mechanical Engineering Publications), 5/1–9.

Veer, G. C. van der., Tauber, M. J., Waern, Y. and Muylwijk, B. van, 1985, On the interaction between system and user characteristics. *Behaviour and Information Technology*, **4**, 289–308.

Watt, J. M., 1983, Re-training the designer. In: *Proceedings of the 6th Annual Conference on Design Engineering*, edited by B. J. Davies (England: IFS conferences), pp. 57–61.

Weber, R. J. and Kosslyn, S. M., 1986, Computer graphics and mental imagery. In: *Visual Languages*, edited by S. Chang, T. Ichikawa and P. A. Ligomenides (New York: Plenum Press), pp. 305–324.

Yoshikawa, H., 1983, CAD framework guided by general design theory. In: *CAD Systems Framework*, edited by K. Bo and F. M. Lillehagen (Amsterdam: North Holland), pp. 241–253.

Chapter 39
User-defined icon sets: a technique for the reduction of system complexity

R. C. MacGREGOR and R. J. CLARKE

Department of Information Systems,
The University of Wollongong,
P.O. Box 1144,
Wollongong, NSW 2500, Australia

Abstract: Psychological research has indicated that humans render meaning from complex situations by the evolution of intuitive practical notational systems. A system is described which allows a user arbitrarily to assign icons to system functions as part of an Object Oriented Interface (OOI). Such an approach enables the user to accommodate system complexity more easily. The user interface design is presented with a view to its incorporation into manufacturing and commercial organizations.

Keywords: User-defined; Object-oriented; Interface; Icons.

Introduction

In most commercial computing environments a person's work with a computer is a latent dialogue between the designer and the user, fixed in the machine programs. While a program may be able to transform previously acquired information to suit new eventualities, the results are ultimately presented within the framework of a universal rigidly-designed person–machine dialogue.

Recent studies (Turner and Karasek, 1984; Hirschheim, 1986; Johansson and Aronson cited in Turner and Karasek, 1984) have given rise to suggestions that person–machine dialogue style can have a detrimental effort upon systems performance both at the individual and the organizational level. At an individual level the complexity of dialogue style has been shown to affect operator task dimensions, specifically workload, uncertainty, interdependence and autonomy, which in turn affects operator learning effectiveness, performance reliability, fatigue and stress. These studies, then, suggest that the more difficult the computer system is to learn or operate, the more likely the operator becomes dependent upon the additional machine support or upon organizational mechanisms such as hierarchical supervision.

At an organizational level the complexities of the computer system manifest in the form of rigid task specialization (often the result of a 'one-off' approach to task allocation; Clarke and MacGregor, 1989). Since the person–machine interface lacks any 'intellectual resonance that comes as a result of synchronization between man and individual (sic)' (Venda, 1988, p. 321) and tasks tend to be allocated to people in an 'isolationist context' (Clarke and MacGregor, 1989), the informal structure of the organization is modified making it less able to cope with environmental change (Turner, 1980; MacGregor and Clarke, 1988).

Literature surrounding person–machine interfaces (Rouse, 1981; Schneidermann, 1980; Friend and Milojkovic, 1984; Weber, 1986) has provided alternatives by which potential person–machine dialogue might be developed. Some of the notable alternatives include:

1. Hierarchical menus: when the number of functional levels remains small, this is a viable interface method. With increased functionality however, user uncertainty and learning duration increases.
2. Pseudo WIMPS: these tend to be text-based and non-iconic, often requiring a degree of skill to use them adequately. This type of interface is further disadvantaged by software limits on the number of active and displayed windows which can be opened simultaneously.
3. WIMPS Interface: most optimal for graphics applications, but often used for text-based ones.

While the effectiveness of such approaches are in some cases inconclusive (Kling, 1978; Turner and Karasek, 1984), recent literature (Rogers, 1986; Reilly, 1987; Rogers *et al.*, 1988) suggests that commonly accepted interface standards fail to be based on the shared features of the target domain and tend to be perceived as difficult in the early learning of a new system.

Intuitive notational systems

Research (Kreitler and Kreitler, 1968; Gardner *et al.*, 1974; Harrison, 1974; Pribram, 1982; MacGregor and Clarke, 1988) suggests the existence of intuitive practical notational systems within the individual. Initial findings suggest that these exist as a 'vocabulary' of icons, able to be enhanced together with (or in spite of) language or cultural intervention.

Harrison (1974) suggests that these symbolic 'vocabularies' exist to enhance understanding by rendering meaning from complex situations since: (i) the amount of pre-learning is reduced, (ii) there is a stronger impact when compared with linguistic cues, (iii) the 'vocabulary' contains an inherent continuity, naturalness and similarity to referents, and (iv) the 'vocabulary' is less dense than standard symbols or language, reducing the possibility of early closure. Furthermore, these same studies indicate that these 'vocabularies' are utilized by the individual in both the learning and working environments.

Use of icons in person–machine interfaces

Attempts to include an iconic 'vocabulary' into person–machine interfaces have for the most part been a compromise between the 'average user', the organization and technical requirements and have resulted in a variety of interface metaphors (see Williams, 1983; Smith *et al.*, 1982; Hijne, 1986; Hammond and Allison, 1987; Card and Henderson, 1987). Rogers *et al.* (1988) suggest that interface metaphors can be categorized into a 'similarity/dissimilarity' continuum, where similarity of an icon to a system function allows specific tasks to be carried out, while dissimilarity between icons and their functions is used in more abstract goal requirements. Given the categorization schema derived by Rogers *et al.* (1988), they rightly suggest that the placement of an interface within the continuum is, at best, a concensus placement and in accord with the findings of Reilly (1987) they have shown that the placement within the continuum varied with the individual. Rogers *et al.* (1988) further showed that for the most part the use of interface metaphors promoted *closed loop learning* (feedback: action-> error-> reporting-> correction) while adding little towards the more desirable *open loop learning* strategy (feedforward: assessment-> prediction of error-> prevention).

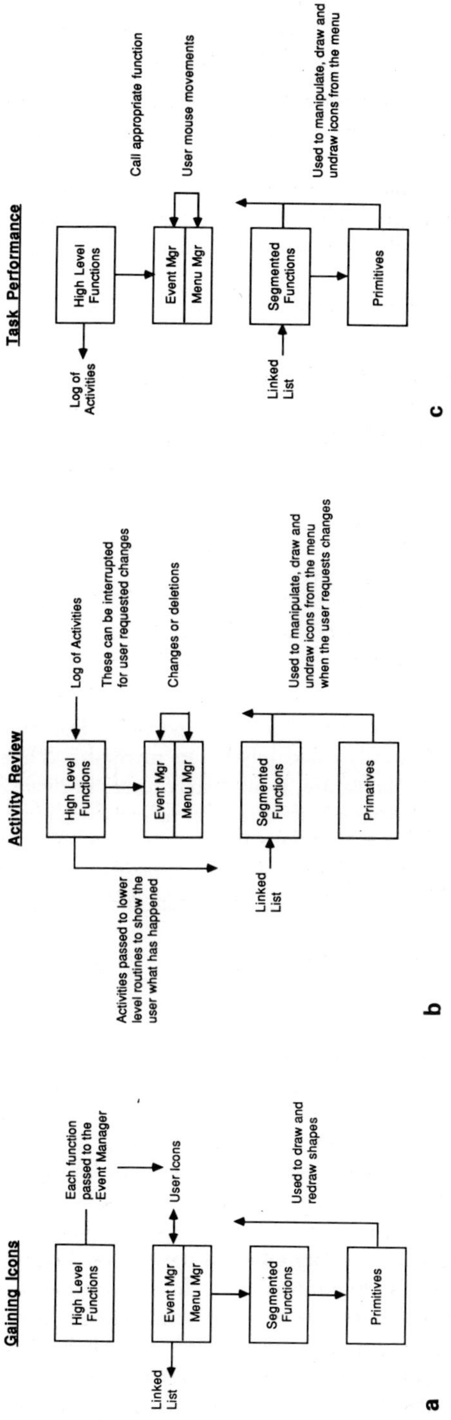

Figure 39.1. *(a) Gaining icons; (b) activity review; (c) task performance. (From MacGregor* et al., *1988.)*

Alternatives to metaphors

Recent attempts in software design have been aimed at providing maximum icon-function similarity by the use of individual notational systems. These design strategies (see MacGregor *et al.*, 1988) have been carried out in the field of music composition. The designs use a segmented graphics approach (see Sobelman and Krekelberg, 1985). User-defined icons are held as linked lists of coordinates and the user interface is handled by event managers and menu managers. The design (see Figure 39.1) incorporates three sections deemed necessary to promote user autonomy, these were:

1. User icon set acquisition – which involves a detailed analysis of the requirements of the system and an evaluation of the user's perception of the tasks to be carried out.
2. The operation of the software – where the user icon was provided in the form of a menu together with functions considered necessary for the reduction of complexity (UNDO, CLEAR, STOP, etc.).
3. A review function – where part or all of a completed activity was available for user analysis and modification.

Early evaluation suggests that the use of an individual icon vocabulary promotes open loop learning by forcing the user to view tasks in a more holistic fashion while allowing the user to view the operations from their own perspective.

Suggested enhancements

An evaluation of the interface metaphor and the individual icon enhancement shows both interface styles to be functionally-based job-design methods. While at an early phase of learning this may be a desirable feature, it ultimately forces software design into a total dissection of task requirement, paying little attention to social structure or the work environment (MacGregor and Clarke, 1988). An enhancement to the individual icon interface style would be the adoption of an 'Object Oriented Interface (OOI)' where both data and function can be provided with user-defined icons such that data drives function within the operation of the system.

Conclusion

Venda (1988) suggests that as a first step to the development of individualized person–machine dialogue, the structure of the work group, its adaptation to functions and its means of interaction need to be understood and maintained. This would allow the informal networks within a group to adapt more easily to technological intervention and would remove the 'isolationist' approach currently the practice with conventional dialogue design. The design of a user-defined icon-based OOI satisfies these criteria and therefore removes the total dissection of work which is a biproduct of conventional design methodologies and a typical organizational response to information technology at the work-unit level.

When designing tasks which involve person–machine interfaces there must be a synthesis and adaptation of the need of the group and the needs of the individual. Such designs must be pre-empted on an analysis and incorporation of the formal and informal work unit structure and function. For the software engineer, the user adaptable interface changes the emphasis away from the design phase to the analysis and implementation phases. No longer can the target population be described simply in terms of a heterogeneous mean, but design strategies must assume a population capable of learning and decision making.

References

Card, S. K. and Henderson, D. A., 1987, Catalogues: a metaphor for computer application delivery. In: *INTERACT 87* edited by H. J. Bullinger and B. Schakel (Elsevier: North Holland), pp. 959–964.

Clarke, R. J. and MacGregor, R. C., 1989, Task allocation in system design. *Professional Conference SOST '89*, Terrigal, edited by R. Clarke and J. Cameron (Australian Computer Society, Inc.), pp. 81–88.

Friend, J. and Milojkovic, J. D., 1984, Designing interactions between students and computers. In: *Instructional Software*, edited by D. F. Walker and R. D. Hess (California: Wadsworth).

Gardner, H., Howard, V. and Perkins, D., 1974, Symbol systems: a philosophical, psychological and educational investigation. In: *Media and Symbols: The Forms of Expression, Communication and Education*, 73rd Yearbook of the National Society for the Study of Education Part 1, edited by D. Olson (Chicago: University of Chicago Press), pp. 27–55.

Hammond, N. and Allison, A., 1987, The travel metaphor as design principle and training aid for navigating around complex systems. In: *Computers and People III*, edited by D. Diaper and R. Winder (Cambridge: Cambridge University Press).

Harrison, R. P., 1974, *Beyond Words*, (Englewood Cliffs, NJ: Prentice-Hall).

Hijne, C. H., 1986, *An Information System Metaphor for an Information System*. Eurohelp Document No COE/EUROHELP/019.

Hirschheim, R. A., 1986, Participative systems design: user experiences, evaluations and conclusions. *The Australian Computing Journal*, **18**(4), 166–173.

Kling, R., 1978, *The Impacts of Computing on the Works of Managers*. Data Analysts and Clerks working paper, Public Policy Research Organization, University of California at Irvine, California.

Kreitler, H. and Kreitler, S., 1968, The validation of psychodramatic behaviour against behaviour in life. *British Journal of Medical Psychology*, **41**, 185–192.

MacGregor, R. C., 1987, An information systems approach to the design of music compositional software for the young learner. Unpublished Honours Masters of Education, University of Wollongong.

MacGregor, R. C. and Clarke, R. J., 1988, The loss of the informal in systems design. *EURINFO 88*, Athens, pp. 242–246.

MacGregor, R. C., King, K. F. and Clarke, R. J., 1988, Individualising the man–machine interface. *Conference on Ergonomics and Hybrid Automation Systems*, Kentucky.

Pribram, K. H., 1982, Brain mechanisms in music. In: *Music, Mind and Brain, The Neuropsychology of Music*, edited by M. Clynes (New York: Plenum Press).

Reilly, R. G. (Ed.), 1987, *Communication Failure in Dialogue and Discourse* (Amsterdam: North-Holland).

Rogers, Y., 1986, The Potential of the Visual Image at the Interface. *Current Psychological Research and Reviews*, **5**, 105–119.

Rogers, Y., Leiser, R. and Carr, D., 1988, Evaluating metaphors for use at the user–system interface. *EURINFO 88*, Athens, pp. 158–163.

Rouse, R., 1981, Human computer interaction in the control of dynamic systems. *ACM Computer Surveys*, **13**, 71–99.

Schneiderman, B., 1980, *Software Psychology* (Cambridge MA: Winthrop).

Smith, D. C., Irby, C., Kimball, R. and Verplank, B., 1982, Designing the star user interface. *BYTE*, **7**, 242–282.

Sobelman, G. E. and Krekelberg, D. E., 1985, *Advanced C Techniques and Applications* (Indianapolis: Que Corp).

Turner, J. A., 1980, Computers in bank clerical functions: Implications for Productivity and the Quality of Life. Ph.D. Dissertation, University of Columbia.

Turner, J. A. and Karasek, R. A., Jr, 1984, Software ergonomics: the effects of computer application design parameters on operator task performance and health. *Ergonomics*, **27**(6), 663–690.

Venda, V. F., 1988, Design of individually adaptive man–computer dialogues in the hybrid intelligence systems. In: *Ergonomics of Hybrid Automation Systems 1*, edited by W., Karwowski, H. R. Parsaei, and M. R. Wilhelm (Amsterdam: Elsevier), pp. 319–327.

Weber, R., 1986, Planning and control in end user computing. *The Australian Computing Journal*, **18**(4), 159–165.

Williams, G., 1983, The Lisa computer system. *BYTE*, **8**, 33–50.

Index